Instructor's Edition

Statistical Reasoning

FOR EVERYDAY LIFE

FOURTH EDITION

Statistical Reasoning
FOR EVERYDAY LIFE

Jeffrey Bennett
University of Colorado at Boulder

William L. Briggs
University of Colorado at Denver

Mario F. Triola
Dutchess Community College

PEARSON

Boston • Columbus • Indianapolis • New York • San Francisco • Upper Saddle River
Amsterdam • Cape Town • Dubai • London • Madrid • Milan • Munich • Paris • Montréal • Toronto
Delhi • Mexico City • São Paulo • Sydney • Hong Kong • Seoul • Singapore • Taipei • Tokyo

Editor in Chief: Deirdre Lynch
Acquisitions Editor: Christopher Cummings
Content Editor: Elizabeth Bernardi
Editorial Assistant: Sonia Ashraf
Senior Managing Editor: Karen Wernholm
Senior Production Project Manager: Tracy Patruno
Associate Director of Design: Andrea Nix
Senior Designer and Cover Designer: Barbara T. Atkinson
Digital Assets Manager: Marianne Groth
Production Coordinator: Katherine Roz
Media Producer: Aimee Thorne
Software Developers: Mary Durnwald and Robert Carroll
Marketing Manager: Erin Lane
Marketing Assistant: Kathleen DeChavez
Senior Author Support/Technology Specialist: Joe Vetere
Image Manager: Rachel Youdelman
Procurement Manager: Evelyn Beaton
Procurement Specialist: Linda Cox
Production Coordination, Composition, Illustrations: Integra
Text Design: Jenny Willingham/Infiniti
Cover Image: Corey Holms/Getty Images

Credits appear on page 383, which constitutes a continuation of the copyright page.

Library of Congress Cataloging-in-Publication Data

Bennett, Jeffrey O.
  Statistical reasoning for everyday life / Jeffrey O. Bennett, William L. Briggs,
Mario F. Triola.—4th ed.
    p. cm.
  Includes index.
  ISBN 978-0-321-81762-4 [student edition]
  1. Statistics.   I. Briggs, William L.   II. Triola, Mario F.   III. Title.
  QA276.12.B45 2013
  519.5—dc23

                                                    2012013041

1 2 3 4 5 6 7 8 9 10—DOJC—16 15 14 13 12

www.pearsonhighered.com

ISBN 10:   0-321-81768-0
ISBN 13: 978-0-321-81768-6

This book is dedicated to everyone
who will try to make the world
a better place. We hope that your
study of statistics will be useful
to your efforts.

● ● ●

And it is dedicated to those who
make our own lives brighter,
especially Lisa, Grant, Brooke, Julie,
Katie, Ginny, Marc, and Scott.

CONTENTS

10.3 Analysis of Variance (One-Way Anova), 363

USING TECHNOLOGY: ANALYSIS OF VARIANCE, 366

FOCUS ON CRIMINOLOGY: CAN YOU TELL A FRAUD WHEN YOU SEE ONE?, 372

FOCUS ON EDUCATION: WHAT CAN A FOURTH-GRADER DO WITH STATISTICS?, 375

*"Statistical thinking will one day be as necessary for efficient citizenship
as the ability to read and write."*

—H. G. Wells

Why Study Statistics?

The future imagined by fiction writer H. G. Wells in the above quote is here. Statistics is now an important part of everyday life, unavoidable whether you are starting a new business, deciding how to plan for your financial future, or simply watching the news on television. Statistics comes up in everything from opinion polls to economic reports to the latest research on cancer prevention. Understanding the core ideas behind statistics is therefore crucial to your success in the modern world.

What Kind of Statistics Will You Learn in This Book?

Statistics is a rich field of study—so rich that it is possible to study it for a lifetime and still feel as if there's much left to learn. Nevertheless, you can understand the core ideas of statistics with just a quarter or semester of academic study. This book is designed to help you learn these core ideas. The ideas you'll study in this book represent the statistics that you'll *need* in your everyday life—and that you can reasonably learn in one course of study. In particular, we've designed this book with three specific purposes:

1. To provide you with the understanding of statistics you'll need for **college** courses, particularly in social sciences such as economics, psychology, sociology, and political science.

2. To help you develop the ability to reason using statistical information—an ability that is crucial to almost any **career** in the modern world.

3. To provide you with the power to evaluate the many news reports of statistical studies that you encounter in your daily **life**, thereby helping you to form opinions about their conclusions and to decide whether the conclusions should influence the way you live.

Who Should Read This Book?

We hope this book will be useful to everyone, but it is designed primarily for students who are *not* planning to pursue advanced course work in statistics. In particular, this book should provide a suitable introduction to statistics for students majoring in a broad range of fields that require statistical literacy, including most disciplines in the humanities and social sciences. The level of this text should be appropriate to anyone who has completed two years of high school mathematics.

Approach

This book takes an approach designed to help you understand important statistical ideas qualitatively, using quantitative techniques only when they clarify those ideas. Here are a few of the key pedagogical strategies that guided the creation of this book.

Start with the Big Picture. Most people entering a statistics course have little prior knowledge of the subject, so it is important to keep sight of the overall purpose of statistics while learning individual ideas or methods. We therefore begin this book with a broad overview of statistics in Chapter 1, in which we explain the relationship between samples and populations, discuss sampling methods and the various types of statistical study, and show numerous examples designed to help you decide whether to believe a statistical study. This "big picture" overview of statistics provides a solid foundation for the more in-depth study of statistical ideas in the rest of the book.

Build Ideas Step by Step. The goal of any course in statistics is to help students understand real statistical issues. However, it is often easier to begin by investigating simple examples in order to build step-by-step understanding that can then be applied to more complex studies. We apply this strategy within every section and every chapter, gradually building toward real examples and case studies.

Use Computations to Enhance Understanding. The primary goal of this book is to help students understand statistical concepts and ideas, but we firmly believe that this goal is best achieved by doing at least some computation. We therefore include computational techniques wherever they will enhance understanding of the underlying ideas.

Connect Probability to Statistics. Many statistics courses include coverage of probability, but to students the concept of probability often seems disconnected from the rest of the subject matter. This is a shame, since probability plays such an integral role in the science of statistics. We discuss this point beginning in Chapter 1, with the basic structure of statistical studies, and then revisit it throughout the book—particularly in Chapter 6, where we present many ideas of probability. For those courses in which coverage of probability is not emphasized, Chapter 6 is designed to be optional.

Stay on Goal: Applying Statistical Reasoning to Everyday Life. Because statistics is such a rich subject, it can be difficult to decide how far to go with any particular statistical topic. In making such decisions for this book, we always turned back to the goal reflected in the title: This book is supposed to help you with the statistical reasoning needed in everyday life. If we felt that a topic was not often encountered in everyday life, we left it out. In the same spirit, we included a few topics—such as a discussion of percentages in Chapter 2 and an in-depth study of graphics in Chapter 3—that are not often covered in statistics courses but are a major part of the statistics encountered in daily life.

Modular Structure

Although we have written this book so that it can be read as a narrative from beginning to end, we recognize that many instructors might wish to teach material in a different order than we have chosen or to cover only selected portions of the text as time allows for classes of different length or with students at different levels. We have therefore organized the book with a modular structure that allows instructors to create a customized course. The 10 chapters are organized broadly by conceptual areas. Each chapter, in turn, is divided into a set of self-contained sections, each devoted to a particular topic or application. In most cases, you may cover the sections or chapters in any order or skip sections that do not fit well into your course. Please note the following specific structure within each chapter:

Learning Goals. Each chapter begins with a one-page overview of its subject matter, including a list mapping each section to learning goals.

Numbered Sections. Each chapter is subdivided into a set of numbered sections (e.g., Sections 1.1, 1.2, ...). To facilitate use of these sections in any order, each section ends with its own set of exercises specific only to that section. Exercises are divided into subgroups with headings that should be self-explanatory, including "Statistical Literacy and Critical Thinking," "Concepts and Applications," "Projects for the Internet and Beyond," and "In the News." Answers to most odd-numbered exercises appear in the back of the book. The numbered sections also include the following pedagogical features:

- **Examples and Case Studies.** Numbered examples within each section are designed to build understanding and to offer practice with the types of questions that appear in the exercises. Case studies, which always focus on real issues, go into more depth than the numbered examples.

- **Time Out to Think.** The "Time Out to Think" features pose short conceptual questions designed to help students reflect on important new ideas. They also serve as excellent starting points for classroom discussions and, in some cases, can be used as a basis for clicker questions.

Chapter Review Exercises. A set of review exercises is included near the end of each chapter. These exercises are designed primarily for self-study, with answers to all of them appearing in the back of the book.

Chapter Quiz. The main part of each chapter ends with chapter quiz questions that are designed to require relatively short answers. These questions address topics found throughout the chapter, and all answers are included in the back of the book.

Focus Topics. Each chapter concludes with two sections entitled "Focus on..." that go into depth on important statistical issues of our time. The topics of these sections were chosen to demonstrate the great variety of fields in which statistics plays a role, including history, environmental studies, agriculture, and economics. Each of these Focus sections includes a set of questions for assignment or discussion.

Technology. The STATDISK statistics software program can be downloaded at www. STATDISK.org. The XLSTAT Excel add-in can be downloaded at www.XLSTAT.com. An access code is required to install XLSTAT.

About the Fourth Edition

We've developed this Fourth Edition of *Statistical Reasoning for Everyday Life* with the help of many users and reviewers. In addition to editing and redesigning the entire book to make it even more student-friendly, we have made the following major changes for this edition:

- Because this book is intended to show the relevance of statistics to everyday life, it is critical that discussions and examples be up to date. We have therefore revised or replaced many dozens of in-text and numbered examples and case studies to be sure they reflect the latest data and topics of interest.

- We have replaced or rewritten six of the 20 in-depth Focus topics and substantially rewritten several others to bring them up to date.

- We have thoroughly reworked the Exercise sets, completely replacing nearly half of the exercises and revising or updating data in most of the others.

- We have added numerous new Using Technology sections, which offer help in using Excel, STATDISK, and the TI-83/84 Plus family of calculators. While the previous edition of this text had Using Technology sections at the end of relevant chapters, this edition has more Using Technology sets of instructions located in relevant sections.

Acknowledgments

Writing a textbook requires the efforts of many people besides the authors. This book would not have been possible without the help of many people. We'd particularly like to thank our editors at Pearson, Greg Tobin and Chris Cummings, whose faith allowed us to create this book. We'd also like to thank the rest of the team at Pearson who helped produce this book, including Elizabeth Bernardi, Sonia Ashraf, Tracy Patruno, Aimee Thorne, Erin Lane, Kathleen DeChavez, and Barbara Atkinson.

For helping to ensure the accuracy of this text, we thank Jamis Perrett and Laura Shick. For reviewing this or earlier editions of this text and providing invaluable advice, we thank the following individuals:

Jennifer Beineke
Western New England College

Matthew Bognar
University of Iowa

Dale Bowman
University of Mississippi

Pat Buchanan
Pennsylvania State University

Robert Buck
Western Michigan University

Antonius H. N. Cillessen
University of Connecticut

Olga Cordero-Brana
Arizona State University

Terry Dalton
University of Denver

Jim Daly
California Polytechnic State University

Robert Dobrow
Carleton College

Mickle Duggan
East Central University

Juan Estrada
Metropolitan State University, Minneapolis–St. Paul

Beverly J. Ferrucci
Keene State College

Jack R. Fraenkel
San Francisco State University

Frank Grosshans
West Chester University

Silas Halperin
Syracuse University

Golde Holtzman
Virginia Polytechnic Institute and State University

Susan Janssen
University of Minnesota–Duluth

Colleen Kelly
San Diego State University

Jim Koehler

Becky Ladd
Arizona State University

Christopher Leary
SUNY Geneseo

Stephen Lee
University of Idaho

Kung-Jong Lui
San Diego State University

Carrie M. Margolin
The Evergreen State College

Judy Marwick
Prairie State College

Craig McCarthy
Ohio University

Richard McGrath
Pennsylvania State University

Abdelelah Mostafa
University of South Florida

Todd Ogden
University of South Carolina

Thomas Petee
Auburn University

Nancy Pfenning
University of Pittsburgh

William S. Rayens
University of Kentucky

Steve Rein
California Polytechnic State University

Lawrence D. Ries
University of Missouri–Columbia

Larry Ringer
Texas A&M University

Pali Sen
University of North Florida

Donald Hugh Smith
Old Dominion University

John Spurrier
University of South Carolina

Gwen Terwilliger
University of Toledo

David Wallace
Ohio University

Elizabeth Walters
Loyola College of Maryland

Larry Wasserman
Carnegie Mellon University

Sheila O'Leary Weaver
University of Vermont

Robert Wolf
University of San Francisco

Fancher Wolfe
Metropolitan State University, Minneapolis–St. Paul

Ke Wu
University of Mississippi

SUPPLEMENTS

Student Supplements

Student's Solutions Manual. This manual provides detailed, worked-out solutions to all odd-numbered text exercises and chapter quiz problems. ISBN-13: 978-0-321-81763-1; ISBN-10: 0-321-81763-X.

Companion Website. The companion website contains additional resources for students, including data sets and web links from the text. The URL is www.pearsonhighered.com/mathstatsresources.

Instructor Supplements

Instructor's Edition. This version of the text includes the answers to all exercises and quiz problems. (The Student Edition contains answers to only the odd-numbered ones.)

Instructor's Solutions Manual. This comprehensive manual contains solutions to all text exercises and chapter quizzes.

Online Test Bank. The Test Bank, available in Pearson Education's online catalog contains four tests to accompany every chapter of the text.

TestGen®. TestGen (www.pearsonhighered.com/testgen) enables instructors to build, edit, print, and administer tests using a computerized bank of questions developed to cover all the objectives of the text. TestGen is algorithmically based, allowing instructors to create multiple but equivalent versions of the same question or test with the click of a button. Instructors can also modify test bank questions or add new questions. Tests can be printed or administered online. The software and test bank are available for download from Pearson Education's online catalog.

Active Learning Questions. Formatted as PowerPoint® slides, these questions can be used with classroom response systems. Several multiple-choice questions are available for each section of the book, allowing instructors to quickly assess mastery of material in class. Slides are available to download from within MyStatLab and from Pearson Education's online catalog.

PowerPoint® Lecture Slides. These slides present key concepts and definitions from the text. Slides are available to download from within MyStatLab and from Pearson Education's online catalog.

Technology Resources

MyStatLab™ Online Course (access code required)
MyStatLab is a course management system that delivers **proven results** in helping individual students succeed.

- MyStatLab can be successfully implemented in any environment—lab-based, hybrid, fully online, traditional—and demonstrates the quantifiable difference that integrated usage has on student retention, subsequent success, and overall achievement.

- MyStatLab's comprehensive online gradebook automatically tracks students' results on tests, quizzes, and homework and in the study plan. Instructors can use the gradebook to provide positive feedback or intervene if students have trouble. Gradebook data can be easily exported to a variety of spreadsheet programs, such as Microsoft Excel. You can determine which points of data you want to export and then analyze the results to determine success.

MyStatLab provides **engaging experiences** that personalize, stimulate, and measure learning for each student. In addition to the resources below, each course includes a full interactive online version of the accompanying textbook.

- **Tutorial Exercises with Multimedia Learning Aids.** The homework and practice exercises in MyStatLab align with the exercises in the textbook, and they regenerate algorithmically to give students unlimited opportunity for practice and mastery. Exercises offer immediate helpful feedback, guided solutions, sample problems, animations, videos, and eText clips for extra help at point of use.

- **StatTalk Videos: *24 Conceptual Videos to Help You Actually Understand Statistics*.** Fun-loving statistician Andrew Vickers takes to the streets of Brooklyn, NY to demonstrate important statistical concepts through interesting stories and real-life events. These fun and engaging videos will help students actually understand statistical concepts. Available with an instructor's user guide and assessment questions.

- **Getting Ready for Statistics.** A library of questions now appears within each MyStatLab course to offer the developmental math topics students need for the course. These can be assigned as a prerequisite to other assignments, if desired.

- **Conceptual Question Library.** In addition to algorithmically regenerated questions that are aligned with your textbook, there is a library of 1,000 Conceptual Questions available in the assessment manager that require students to apply their statistical understanding.

- **StatCrunch™.** MyStatLab integrates the web-based statistical software, StatCrunch, within the online assessment platform so that students can easily analyze data sets from exercises and the text. In addition, MyStatLab includes access to **www.StatCrunch.com,** a website where users can access more than 14,500 shared data sets, conduct online surveys, perform complex analyses using the powerful statistical software, and generate compelling reports.

- **Statistical Software Support.** Knowing that students often use external statistical software, we make it easy to copy our data sets, both from the ebook and the MyStatLab questions, into software such as StatCrunch, Minitab, Excel, and more. Students have access to a variety of support tools—Technology Tutorial Videos, Technology Study Cards, and Technology Manuals for select titles—to learn how to effectively use statistical software.

- **Expert Tutoring.** Although many students describe the whole of MyStatLab as "like having your own personal tutor," students also have access to live tutoring from Pearson. Qualified statistics instructors provide tutoring sessions for students via MyStatLab.

And MyStatLab comes from a **trusted partner** with educational expertise and an eye on the future. Knowing that you are using a Pearson product means knowing that you are using quality content. That means that our eTexts are accurate and our assessment tools work. Whether you are just getting started with MyStatLab or have a question along the way, we're here to help you learn about our technologies and how to incorporate them into your course.

To learn more about how MyStatLab combines proven learning applications with powerful assessment, visit **www.mystatlab.com** or contact your Pearson representative.

MyStatLab™ Ready to Go Course (access code required)

These new Ready to Go courses provide students with all the same great MyStatLab features that you're used to, but make it easier for instructors to get started. Each course includes pre-assigned homework and quizzes to make creating your course even simpler. Ask your Pearson representative about the details for this particular course or to see a copy of this course.

StatCrunch™

StatCrunch is powerful web-based statistical software that allows users to perform complex analyses, share data sets, and generate compelling reports of their data. The vibrant online community offers more than 15,500 data sets for students to analyze.

- **Collect.** Users can upload their own data to StatCrunch or search a large library of publicly shared data sets, spanning almost any topic of interest. Also, an online survey tool allows users to quickly collect data via web-based surveys.

- **Crunch.** A full range of numerical and graphical methods allow users to analyze and gain insights from any data set. Interactive graphics help users understand statistical concepts and are available for export to enrich reports with visual representations of data.

- **Communicate.** Reporting options help users create a wide variety of visually appealing representations of their data.

Full access to StatCrunch is available with a MyStatLab kit, and StatCrunch is available by itself to qualified adopters. For more information, visit our website at **www.StatCrunch.com** or contact your Pearson representative.

How to Succeed in Your Statistics Course

If you are reading this book, you probably are enrolled in a statistics course of some type. The keys to success in your course include approaching the material with an open and optimistic frame of mind, paying close attention to how useful and enjoyable statistics can be in your life, and studying effectively and efficiently. The following sections offer a few specific hints that may be of use as you study.

Using This Book

Before we get into more general strategies for studying, here are a few guidelines that will help you use *this* book most effectively.

- Before doing any assigned exercises, read assigned material *twice*.
  — On the first pass, read quickly to gain a "feel" for the material and concepts presented.
  — On the second pass, read the material in more depth and work through the examples carefully.

- During the second reading, take notes that will help you when you go back to study later. In particular:
  — *Use the margins!* The wide margins in this textbook are designed to give you plenty of room for making notes as you study.
  — Don't highlight—underline! Using a pen or pencil to underline material requires greater care than highlighting and therefore helps keep you alert as you study.

- You'll learn best by *doing*, so after you complete the reading be sure to do plenty of the end-of-section exercises and the end-of-chapter review exercises. In particular, try some of the exercises that have answers in the back of the book, in addition to any exercises assigned by your instructor.

- If you have access to MyStatLab with this book, be sure to take advantage of the many additional study resources available on this website.

Budgeting Your Time

A general rule of thumb for college classes is that you should expect to study about 2 to 3 hours per week *outside* class for each unit of credit. Based on this rule of thumb, a student taking 15 credit hours should expect to spend 30 to 45 hours each week studying outside of class. Combined with time in class, this works out to a total of 45 to 60 hours spent on academic work—not much more than the time required of a typical job, and you get to choose your own hours. Of course, if you are working while you attend school, you will need to budget your time carefully. Here are some rough guidelines for how you might divide your studying time.

If your course is	Time for reading the assigned text (per week)	Time for homework assignments (per week)	Time for review and test preparation (average per week)	Total study time (per week)
3 credits	1 to 2 hours	3 to 5 hours	2 hours	6 to 9 hours
4 credits	2 to 3 hours	3 to 6 hours	3 hours	8 to 12 hours
5 credits	2 to 4 hours	4 to 7 hours	4 hours	10 to 15 hours

If you find that you are spending fewer hours than these guidelines suggest, you could probably improve your grade by studying more. If you are spending more hours than these guidelines suggest, you may be studying inefficiently; in that case, you should talk to your instructor about how to study more effectively.

General Strategies for Studying

- Budget your time effectively. One or two hours each day is more effective, and far less painful, than studying all night before homework is due or before exams.

- Engage your brain. Learning is an active process, not a passive experience. Whether you are reading, listening to a lecture, or working on assignments, always make sure that your mind is actively engaged. If you find your mind drifting or falling asleep, make a conscious effort to revive yourself, or take a break if necessary.

- Don't miss class. Listening to lectures and participating in class activities and discussions are much more effective than reading someone else's notes. Active participation will help you retain what you are learning.

- Be sure to complete any assigned reading *before* the class in which it will be discussed. This is crucial because class lectures and discussions are designed to help reinforce key ideas from the reading.

- Start your homework early. The more time you allow yourself, the easier it is to get help if you need it. If a concept gives you trouble, first try additional reading or studying beyond what has been assigned. If you still have trouble, ask for help: You surely can find friends, peers, or teachers who will be glad to help you learn.

- Working together with friends can be valuable in helping you understand difficult concepts. However, be sure that you learn *with* your friends and do not become dependent on them.

- Don't try to multitask. A large body of research shows that human beings simply are not good at multitasking: When we attempt it, we do more poorly at all of the individual tasks. And in case you think you are an exception, the same research found that those people who believed they were best at multitasking were actually the worst! So when it is time to study, turn off your electronic devices, find a quiet spot, and give your work a focused effort of concentration.

Preparing for Exams

- Rework problems and other assignments; try additional questions to be sure you understand the concepts. Study your performance on assignments, quizzes, or exams from earlier in the term.

- Study your notes from lectures and discussions, and reread relevant sections in your textbook. Pay attention to what your instructor expects you to know for an exam.

- Study individually *before* joining a study group with friends. Study groups are effective only if every individual comes prepared to contribute.

- Don't stay up too late before an exam. Don't eat a big meal within an hour of the exam (thinking is more difficult when blood is being diverted to the digestive system).

- Try to relax before and during the exam. If you have studied effectively, you are capable of doing well. Staying relaxed will help you think clearly.

Presenting Homework and Writing Assignments

All work that you turn in should be of *collegiate quality:* neat and easy to read, well organized, and demonstrating mastery of the subject matter. Future employers and teachers will expect this quality of work. Moreover, although submitting homework of collegiate quality requires "extra" effort, it serves two important purposes directly related to learning:

1. The effort you expend in clearly explaining your work solidifies your learning. In particular, research has shown that writing and speaking trigger different areas of your brain. By writing something down—even when you think you already understand it—your learning is reinforced by involving other areas of your brain.

2. By making your work clear and self-contained (that is, making it a document that you can read without referring to the questions in the text), you will have a much more useful study guide when you review for a quiz or exam.

The following guidelines will help ensure that your assignments meet the standards of collegiate quality:

- Always use proper grammar, proper sentence and paragraph structure, and proper spelling. Do not use texting shorthand.

- All answers and other writing should be fully self-contained. A good test is to imagine that a friend is reading your work and to ask yourself whether the friend would understand exactly what you are trying to say. It is also helpful to read your work out loud to yourself, making sure that it sounds clear and coherent.

- In problems that require calculation:

  - Be sure to *show your work* clearly. By doing so, both you and your instructor can follow the process you used to obtain an answer. Also, please use standard mathematical symbols, rather than "calculator-ese." For example, show multiplication with the $\times$ symbol (not with an asterisk), and write 10^5, not 10^5 or 10E5.

  - *Word problems should have word answers*. That is, after you have completed any necessary calculations, any problem stated in words should be answered with one or more *complete sentences* that describe the point of the problem and the meaning of your solution.

  - Express your word answers in a way that would be *meaningful* to most people. For example, most people would find it more meaningful if you express a result of 720 hours as 1 month. Similarly, if a precise calculation yields an answer of 9,745,600 years, it may be more meaningful in words as "nearly 10 million years."

- Include illustrations whenever they help explain your answer, and make sure your illustrations are neat and clear. For example, if you graph by hand, use a ruler to make straight lines. If you use software to make illustrations, be careful not to make them overly cluttered with unnecessary features.

- If you study with friends, be sure that you turn in your own work stated in your own words—you should avoid anything that might even give the *appearance* of possible academic dishonesty.

APPLICATIONS INDEX

(CS = CASE STUDY, E = EXAMPLE, F = FOCUS, IE = IN-TEXT EXAMPLE, P = PROBLEM, PR = PROJECT)

BUSINESS AND ECONOMICS

SURVEYS AND OPINION POLLS

MISCELLANEOUS (INCLUDING SPORTS)

Speaking of Statistics

Is your drinking water safe? How many people approve of the President's budget plan? Are we getting good value for our health care dollars? Questions like these can be addressed only through statistical studies. In this first chapter, we will discuss basic principles of statistical research and lay a foundation for the more detailed study of statistics that follows in the rest of this book. Along the way, we will consider a variety of examples that show how well-designed statistical studies can provide guidance for social policy and personal decisions, as well as a few cases in which statistics can be misleading.

Statistical thinking will one
day be as necessary for
efficient citizenship as the
ability to read and write.

—H. G. Wells

LEARNING GOALS

1.1 What Is/Are Statistics?

Understand the two meanings of the term *statistics* and the basic ideas behind any statistical study, including the relationships among the study's population, sample, sample statistics, and population parameters.

1.2 Sampling

Understand the importance of choosing a representative sample and become familiar with several common methods of sampling.

1.3 Types of Statistical Studies

Understand the differences between observational studies and experiments; recognize key issues in experiments, including the selection of treatment and control groups, the placebo effect, and blinding.

1.4 Should You Believe a Statistical Study?

Be able to evaluate statistical studies that you find in the media, so that you can decide whether the results are meaningful.

FOCUS TOPICS

1.1 WHAT IS/ARE STATISTICS?

The subject of statistics is often stereotyped as dry or technical, but it touches on almost everything in modern society. Statistics can tell us whether a new drug is effective in treating cancer, it can help agricultural inspectors ensure that our food is safe, and it is essential for conducting and interpreting opinion polls. Businesses use statistics in market research and advertising. We even use statistics in sports, often as a way of ranking teams and athletes. Indeed, you'll be hard-pressed to think of any topic that is not linked with statistics in some important way.

The primary goal of this text is to help you learn the core ideas behind statistical methods. These basic ideas are not difficult to understand, although mastery of the details and theory behind them can require years of study. One of the great things about statistics is that even the small amount of theory covered in this text will give you the power to understand the statistics you encounter in the news, in your classes or workplace, and in your everyday life.

A good place to start is with the term *statistics* itself, which can be either singular or plural and has different meanings in the two cases. When it is singular, *statistics* is the *science* that helps us understand how to collect, organize, and interpret numbers or other information about some topic; we refer to the numbers or other pieces of information as *data*. When it is plural, *statistics* are the actual data that describe some characteristic. For example, if there are 30 students in your class and they range in age from 17 to 64, the numbers "30 students," "17 years," and "64 years" are all statistics that describe your class in some way.

BY THE WAY

Although you'll sometimes see the word *data* used as a singular synonym for *information*, technically it is plural: One piece of information is called a *datum*, and two or more pieces are called *data*.

Two Definitions of Statistics

- Statistics is the *science* of collecting, organizing, and interpreting data.

- Statistics are the *data* (numbers or other pieces of information) that describe or summarize something.

How Statistics Works

According to news reports, 111.3 million Americans watched the New York Giants win Super Bowl XLVI, which explains why the networks can now ask advertisers to pay more than $3 million for a 30-second commercial. But you may wonder: Who counted all these million people?

The answer is *no one*. The claim that 111.3 million people watched the Super Bowl came from statistical studies conducted by a company called Nielsen Media Research. This company compiles its famous *Nielsen ratings* by monitoring the television viewing habits of people in only about 5,000 homes.

If you are new to the study of statistics, Nielsen's conclusion may seem like a stretch. How can anyone draw a conclusion about millions of people by studying just a few thousand? However, statistical science shows that this conclusion can be quite accurate, as long as the statistical study is conducted properly. Let's take the Nielsen ratings of the Super Bowl as an example and ask a few key questions that will illustrate how statistics works in general.

BY THE WAY

Statistics originated with the collection of census and tax data, which are affairs of state. That is why the word *state* is at the root of the word *statistics*.

What Is the Goal of the Research?

Nielsen's goal is to determine the total number of Americans who watched the Super Bowl. In the language of statistics, we say that Nielsen is interested in the **population** of all Americans. The number that Nielsen hopes to determine—the number of people who watched the Super Bowl—is a particular characteristic of the population. In statistics, characteristics of the population are called **population parameters.**

Although we usually think of a population as a group of people, a statistical population can be any kind of group—people, animals, or things. For example, in a study of automobile safety, the population might be *all cars on the road*. Similarly, the term *population parameter* can refer to any characteristic of a population. In the case of automobile safety, the population parameters might include the total number of cars on the road during a certain time period, the accident rate among cars on the road, or the range of weights of cars on the road.

> **Definitions**
>
> The **population** in a statistical study is the *complete* set of people or things being studied.
>
> **Population parameters** are specific numbers describing characteristics of the population.

EXAMPLE **1** Populations and Population Parameters

For each of the following situations, describe the population being studied and identify some of the population parameters that would be of interest.

a. You work for Farmers Insurance and you've been asked to determine the average amount paid to accident victims in cars without side-impact air bags.

b. You've been hired by McDonald's to determine the weights of the potatoes delivered each week for French fries.

c. You are a business reporter covering Genentech Corporation and you are investigating whether its new treatment is effective against childhood leukemia.

SOLUTION

a. The population consists of people who have received insurance payments for accidents in cars that lacked side-impact air bags. The relevant population parameter is the average amount paid to these people. (Later, the term "average" will be replaced by the more correct term "mean.")

b. The population consists of all the potatoes delivered each week for French fries. Relevant population parameters include the average weight of the potatoes and the variation of the weights (for example, are most of them close to or far from the average?).

c. The population consists of all children with leukemia. Important population parameters are the percentage of children who recover *without* the new treatment and the percentage of children who recover with the new treatment. $\cdots\bullet$

What Actually Gets Studied?

If researchers at Nielsen were all-powerful, they might determine the number of people watching the Super Bowl by surveying every individual American. But no one can do that, so instead they try to estimate the number of Americans watching by studying a relatively small group of people. Nielsen attempts to learn about the population of all Americans by carefully monitoring the viewing habits of a much smaller **sample** of Americans. More specifically, Nielsen uses recording devices in about 5,000 homes, so the people who live in these homes are the sample of Americans that Nielsen studies.

The individual measurements that Nielsen collects from the people in the 5,000 homes constitute the **raw data.** Nielsen collects much raw data—for example, when and how long each TV in the household is on, what show it is tuned to, and who in the household is watching. Nielsen then consolidates these raw data into a set of numbers that characterize the sample, such as the percentage of viewers in the sample who watched each individual television show or the total number of people in the sample who watched the Super Bowl. These numbers are called **sample statistics.**

BY THE WAY

Arthur C. Nielsen founded his company and invented market research in 1923. He introduced the Nielsen Radio Index to rate radio programs in 1942 and extended his methods to television programming in the 1960s. The company now also tracks other media (Internet, smart phones, etc.) and must constantly adapt its methodology to new media technologies.

> **Definitions**
>
> A **sample** is a subset of the population from which data are actually obtained.
>
> The actual measurements or observations collected from the sample constitute the **raw data**.
>
> **Sample statistics** are numbers describing characteristics of the sample found by consolidating or summarizing the raw data.

EXAMPLE 2 Unemployment Survey

The U.S. Labor Department defines the *civilian labor force* as all those people who are either employed or actively seeking employment. Each month, the Labor Department reports the unemployment rate, which is the percentage of people actively seeking employment within the entire civilian labor force. To determine the unemployment rate, the Labor Department surveys 60,000 households. For the unemployment reports, describe each of the following.

a. population **b.** sample **c.** raw data **d.** sample statistics **e.** population parameters

SOLUTION

a. The *population* is the group that the Labor Department wants to learn about, which is all the people who make up the civilian labor force.

b. The *sample* consists of all the people among the 60,000 households surveyed.

c. The *raw data* consist of all the information collected in the survey.

d. The *sample statistics* summarize the raw data for the sample. In this case, the relevant sample statistic is the percentage of people in the sample who are actively seeking employment. (The Labor Department also calculates similar sample statistics for subgroups in the population, such as the percentages of teenagers, men, women, and veterans who are unemployed.)

e. The *population parameters* are the characteristics of the entire population that correspond to the sample statistics. In this case, the relevant population parameter is the actual unemployment rate. Note that the Labor Department does *not* actually measure this population parameter, because data are collected only for the sample and then are used to estimate the population parameter.   · · ●

How Do Sample Statistics Relate to Population Parameters?

Suppose Nielsen finds that 31% of the people in the 5,000 homes in its sample watched the Super Bowl. This "31%" is a sample statistic, because it characterizes the sample. But what Nielsen really wants to know is the corresponding population parameter, which is the percentage of all Americans who watched the Super Bowl.

There is no way for Nielsen researchers to know the exact value of the population parameter, because they've studied only a sample. However, Nielsen researchers hope that they've done their work correctly so that the sample statistic is a good estimate of the population parameter. In other words, they would like to conclude that because 31% of the sample watched the Super Bowl, approximately 31% of the population also watched the Super Bowl. One of the primary purposes of statistics is to help researchers assess the validity of this type of conclusion.

> **TIME (◐)UT TO THINK**
>
> Suppose Nielsen concludes that 30% of Americans watched the Super Bowl. How many people does this represent? (The population of the United States is approximately 310 million.)

Statistical science provides methods that enable researchers to determine how well a sample statistic estimates a population parameter. For example, results from surveys or opinion polls are usually quoted along with a value called the **margin of error.** By adding and subtracting the margin of error from the sample statistic, we find a range of values, or **confidence interval,** that is *likely* to contain the population parameter. In most cases, the margin of error is defined so that we can have 95% confidence that this range contains the population parameter. We'll discuss the precise meaning of "likely" and "95% confidence" in Chapter 8, but for now you might find an explanation given by the *New York Times* useful (Figure 1.1). In the case of the Nielsen ratings, the margin of error is about 1 percentage point. Therefore, if 31% of the sample was watching the Super Bowl, then we can be 95% confident that the range from 30% to 32% contains the actual percentage of the population watching the Super Bowl.

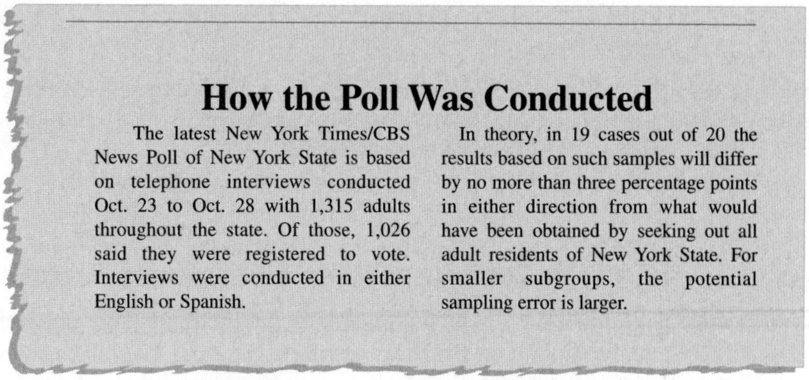

How the Poll Was Conducted

The latest New York Times/CBS News Poll of New York State is based on telephone interviews conducted Oct. 23 to Oct. 28 with 1,315 adults throughout the state. Of those, 1,026 said they were registered to vote. Interviews were conducted in either English or Spanish.

In theory, in 19 cases out of 20 the results based on such samples will differ by no more than three percentage points in either direction from what would have been obtained by seeking out all adult residents of New York State. For smaller subgroups, the potential sampling error is larger.

Figure 1.1 The margin of error in a survey or opinion poll usually describes a range that is likely (with 95% confidence, meaning in 19 out of 20 cases) to contain the population parameter. This excerpt from the *New York Times* explains a margin of error of 3 percentage points.

One of the most remarkable findings of statistical science is that it is possible to get meaningful results from surprisingly small samples. Nevertheless, larger sample sizes are better (when they are feasible), because the margin of error is generally smaller for larger samples. For example, the margin of error for a 95% confidence interval in a well-conducted poll is typically about 5 percentage points for a sample size of 400, but drops to 3 percentage points for a sample size of 1,000 and to 1 percentage point for a sample of 10,000. (See Chapter 8 to understand how margins of error are calculated.)

Definition

The **margin of error** in a statistical study is used to describe the range of values, or **confidence interval,** likely to contain the population parameter. We find this confidence interval by adding and subtracting the margin of error from the sample statistic obtained in the study. That is, the range of values likely to contain the population parameter is

$$\text{from} \quad (\text{sample statistic} - \text{margin of error})$$
$$\text{to} \quad (\text{sample statistic} + \text{margin of error})$$

The margin of error is usually defined to give a 95% confidence interval, meaning that 95% of samples of the size used in the study would contain the actual population parameter (and 5% would not).

EXAMPLE 3 Sex and Politics

The Pew Research Center for People and the Press interviewed 1,002 adult Americans and asked about the reason for a recent increase in sex scandals among elected officials. Fifty-seven percent of the respondents claimed that the increase is due to greater scrutiny by the media, while 19% felt that the increase is due to declining moral standards. The margin of error for the poll was 3 percentage points. Describe the population and the sample for this survey, and explain the meaning of the sample statistic of 57%. What can we conclude about the percentage of the population that believes the increase in political sex scandals is due to greater media scrutiny?

SOLUTION   The population is all adult Americans and the sample consists of the 1,002 people who were interviewed. The sample statistic of 57% is the *actual* percentage of people in the sample who answered that greater media scrutiny is responsible for the increase in political sex scandals. The 57% sample statistic and the margin of error of 3 percentage points tell us that the range of values

$$\text{from} \quad 57\% - 3\% = 54\%$$
$$\text{to} \quad 57\% + 3\% = 60\%$$

is likely (with 95% confidence) to contain the population parameter, which in this case is the true percentage of all adult Americans who believe that greater media scrutiny is responsible for the increase in political sex scandals.   • • ●

TIME ⏲ OUT TO THINK

In the poll described in Example 3, the respondents were given the two possible explanations, greater media scrutiny and lower moral standards. Do you think the results might have been different if respondents were asked to provide their own explanations? Explain.

Putting It All Together: The Process of a Statistical Study

The process used by Nielsen Media Research is similar to that used in many statistical studies. Figure 1.2 and the box below summarize the basic steps in a statistical study. Keep in mind that these steps are somewhat idealized, and the actual steps may differ from one study to another. Moreover, the details hidden in the basic steps are critically important. For example, a poorly chosen sample in Step 2 can render the entire study meaningless, and great care must be taken in inferring conclusions about a population from results found for the much smaller sample of that population.

BY THE WAY

Statisticians often divide their subject into two major branches: **descriptive statistics,** which deals with *describing* raw data in the form of graphics and sample statistics, and **inferential statistics,** which deals with *inferring* (or estimating) population parameters from sample data. In this book, Chapters 2 through 5 primarily cover descriptive statistics, while Chapters 6 through 10 focus on inferential statistics.

Basic Steps in a Statistical Study

Step 1.   State the goal of your study precisely; that is, determine the population you want to study and exactly what you'd like to learn about it.

Step 2.   Choose a representative sample from the population.

Step 3.   Collect raw data from the sample, and summarize these data by finding sample statistics of interest.

Step 4.   Use the sample statistics to make inferences about the population.

Step 5.   Draw conclusions; determine what you learned and whether you achieved your goal.

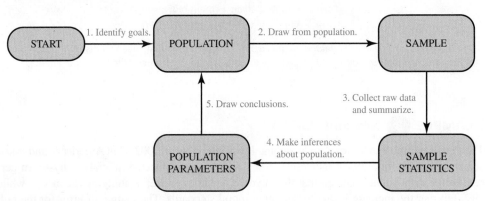

Figure 1.2   The process of a statistical study.

EXAMPLE 4   Identifying the Steps

Identify how researchers applied the five basic steps in the survey from Example 3.

SOLUTION   The steps apply as follows.

1.   The researchers had a goal of learning what Americans think about the causes of recent political scandals. They chose adult Americans as the population, deliberately leaving out children.

2.   They chose 1,002 adult Americans for their sample. Although we are not told how the sample was drawn, we will assume that it was drawn so that the 1,002 adult Americans are typical of the entire adult American population.

3. They collected the raw data by asking a simple question of the people in the sample. The raw data are the individual responses to the question. They summarized these data with sample statistics, such as the overall percentages of people in the sample who chose each answer.

4. Techniques of statistical science allowed the researchers to infer population characteristics. In this case, the inference consisted of estimating the relevant population parameter and calculating the margin of error.

5. By making sure that the study was conducted properly and interpreting the estimates of the population parameters, the researchers drew overall conclusions about Americans' attitudes concerning recent scandals.

Statistics: Decisions for an Uncertain World

Most of the examples we've discussed so far involve surveys or polls, but the subject of statistics encompasses much more, including experiments designed to test new medical treatments, analyses of the dangers of global warming, and even assessments of the value of a college education. Indeed, it is fair to say that the primary purpose of statistics is to help us make good decisions whenever we are confronted with a variety of possible options.

> **The Purpose of Statistics**
> Statistics has many uses, but perhaps its most important purpose is to help us make good decisions about issues that involve uncertainty.

This purpose will be clear in most of the case studies and examples we consider in this text, but occasionally we'll have to discuss a bit of theory that may seem somewhat abstract at first. If you keep the overall purpose of statistics in mind, you'll be rewarded in the end when you see how the theory helps us understand our world. The following case study will give you a taste of what lies ahead. It involves several important theoretical ideas that led to one of the 20th century's greatest accomplishments in public health.

CASE STUDY The Salk Polio Vaccine

If you had been a parent in the 1940s or 1950s, one of your greatest fears would have been the disease known as polio. Each year during this long polio epidemic, thousands of young children were paralyzed by the disease. In 1954, a large experiment was conducted to test the effectiveness of a new vaccine created by Dr. Jonas Salk (1914–1995). The experiment involved a sample of 400,000 children chosen from the population of all children in the United States. Half of these 400,000 children received an injection of the Salk vaccine. The other half received an injection that contained only salt water. (The salt water injection was a *placebo*; see Section 1.3.) Among the children receiving the Salk vaccine, only 33 contracted polio. In contrast, there were 115 cases of polio among the children who did not get the Salk vaccine. Using techniques of statistical science that we'll study later, the researchers concluded that the vaccine was effective at preventing polio. They therefore decided to launch a major effort to improve the Salk vaccine and distribute it to the population of *all* children. Thanks to this vaccine (and improved ones developed later), the horror of polio is now largely a memory of the past.

BY THE WAY

Polio quickly became rare in the United States after the development of the Salk vaccine, but it remained common in less-developed countries. A global effort to vaccinate children against polio began in 1998 and has achieved great success, though it has not yet reached its goal of completely eradicating the disease.

The greatest reward for doing is the opportunity to do more.

—Jonas Salk

Section 1.1 Exercises

Statistical Literacy and Critical Thinking

1. **Population and Sample.** What is a *population*, what is a *sample*, and what is the difference between them?

2. **Statistic and Statistics.** Suppose that, in a discussion, one person refers to baseball *statistics* and another refers to the use of *statistics* in showing that a particular drug is an effective treatment. Do both uses of the term *statistics* have the same meaning? If not, how do they differ?

3. **Statistics and Parameters.** What is a sample statistic, what is a population parameter, and what is the difference between them?

4. **Margin of Error.** What is the margin of error in a statistical study, and why is it important?

Does It Make Sense? For Exercises 5–10, decide whether the statement makes sense (or is clearly true) or does not make sense (or is clearly false). Explain clearly; not all of these have definitive answers, so your explanation is more important than your chosen answer.

5. **Statistics and Parameters.** My professor conducted a statistical study in which he was unable to measure any sample statistics, but he succeeded in determining the population parameters with a very small margin of error.

6. **Poor Poll.** A poll conducted two weeks before the election found that Smith would get 70% of the vote, with a margin of error of 3%, but he ended up losing the election anyway.

7. **Poll Certainty.** There is no doubt that Johnson won the election, because an exit poll showed that she received 54% of the vote and the margin of error is only 3 percentage points.

8. **Beating Nielsen.** A new startup company intends to compete with Nielsen Media Research by providing data with a larger margin of error for the same price.

9. **Depression Sample.** The goal of my study is to learn about depression among people who have suffered through a family tragedy, so I plan to choose a sample from the population of patients in support groups for loss of a spouse.

10. **New Product.** Our market research department surveyed 1,000 consumers on their attitude toward our new product. Because the people in this sample were so enthusiastic in their desire to purchase the product, we have decided to roll out a nationwide advertising campaign.

Concepts and Applications

Population, Sample, Statistic, and Parameter. Exercises 11–14 each describe a statistical study. In each case, identify the sample, the population, the sample statistic, and the population parameter.

11. **Smoking Poll.** In a Gallup poll of 1,018 adults in the United States, it was found that 22% smoked cigarettes in the past week.

12. **Birth Weights.** For 186 randomly selected babies, the average (mean) of their birth weights is 3,103 grams (based on data from "Cognitive Outcomes of Preschool Children with Prenatal Cocaine Exposure," by Singer et al., *Journal of the American Medical Association,* Vol. 291, No. 20).

13. **Garlic and Cholesterol.** In a test of the effectiveness of garlic for lowering cholesterol, 47 adult subjects were treated with Garlicin, which is garlic in a processed tablet form. Cholesterol levels were measured before and after the treatment. The changes in their levels of LDL cholesterol (in mg/dL) have an average (mean) of 3.2 (based on data from "Effect of Raw Garlic vs Commercial Garlic Supplements on Plasma Lipid Concentrations in Adults With Moderate Hypercholesterolemia," by Gardner et al, *Archives of Internal Medicine*, Vol. 167).

14. **Job Interview Mistakes.** In an Accountemps survey of 150 senior executives, 47% said that the most common job interview mistake is to have little or no knowledge of the company.

Identifying the Range of Values. In Exercises 15–18, use the given statistics and margin of error to identify the range of values (confidence interval) likely to contain the true value of the population parameter.

15. **Wake Up.** A Braun Research poll asked 1,000 office workers how they wake up in time for work; 60% of them said that they use an alarm clock. The margin of error of was 3 percentage points.

16. **Wash Up.** *USA Today* reported that among 6,028 adults observed in restrooms, 85% washed their hands. The margin of error was 1 percentage point.

17. **Claim to Wash Up.** In a Harris Interactive survey of 1,006 adults, 96% say that they wash their hands when in a public restroom. The margin of error was 3 percentage points.

18. **Body Temperatures.** One hundred and six adults are randomly selected and tested for their body temperatures. Based on that sample, researchers estimated that the average (mean) body temperature is 98.2° F with a margin of error of 0.1° F.

19. **Global Warming Poll.** A Pew Research Center poll asked 1,708 randomly selected adults whether "global warming is a problem that requires immediate government action." Results

showed that 55% of those surveyed said yes. The margin of error was 2 percentage points. Can a news reporter safely write that the majority (more than 50%) of people believe that immediate government action is required?

20. **Nielsen Survey.** In a survey of 25,047 Super Bowl viewers, 51% of them said that they enjoyed commercials more than the game. The margin of error is 0.6%. Can we conclude that the majority of Super Bowl viewers prefer commercials to the game? Why or why not?

21. **Do People Lie About Voting?** In a survey of 1,002 people, 701 (or 70%) said that they voted in the last presidential election (based on data from ICR Research Group). The margin of error was 3 percentage points. However, actual voting records show that only 61% of all eligible voters actually voted. Does this imply that people lied when they responded in the survey? Explain.

22. **Why the Discrepancy?** An Eagleton Institute poll asked men if they agreed with this statement: "Abortion is a private matter that should be left to women to decide without government intervention." Among the men who were interviewed by women, 77% agreed with the statement. Among the men who were interviewed by men, 70% agreed with the statement. Assuming that the discrepancy is significant, how might that discrepancy be explained?

Interpreting Real Studies. For each of Exercises 23–26, do the following:

a. Based on the given information, state what you think was the goal of the study. Identify a possible population and the population parameter of interest.

b. Briefly describe the sample, raw data, and sample statistic for the study.

c. Based on the sample statistic and the margin of error, identify the range of values (confidence interval) likely to contain the population parameter of interest.

23. **Death Penalty.** A Gallup poll asked 511 randomly selected adults if they favor the death penalty for a person convicted of murder, with 64% saying yes. The margin of error is 4 percentage points.

24. **Prescription Drugs.** A study of 3,005 adults ages 57 to 85 showed that 82% of them use at least one prescription drug. The margin of error is 2 percentage points (based on data from "Use of Prescription and Over-the-Counter Medications and Dietary Supplements Among Older Adults in the United States," by Qato, et al., *Journal of the American Medical Association*, Vol. 300, No. 24).

25. **Super Bowl.** In a recent Super Bowl, a Nielsen report found that 45% of the 9,000 surveyed U.S. households had TV sets tuned to the game. The margin of error is 1 percentage point.

26. **Piercings and Tattoos.** A Harris Interactive survey of 514 human resources professionals showed that 46% of them say that piercings or tattoos make job applicants less likely to be hired. The margin of error for the survey was 4 percentage points.

Five Steps in a Study. Describe how you would apply the five basic steps in a statistical study (as listed in the box on p. 6) to the issues in Exercises 27–30.

27. **Cell Phones and Driving.** You want to determine the percentage of drivers who use cell phones while they are driving.

28. **Credit Scores.** FICO (Fair Isaac Corporation) scores are routinely used to rate the quality of consumer credit. You want to determine the average (mean) FICO score of an adult in the United States.

29. **Passenger Weight.** Recognizing that overloading commercial aircraft would lead to unsafe flights, you want to determine the average (mean) weight of airline passengers.

30. **Pacemaker Batteries.** Because the batteries used in heart pacemakers are so critically important, you want to determine the average (mean) length of time that such batteries last before failure.

PROJECTS FOR THE INTERNET & BEYOND

31. **Current Nielsen Ratings.** Find the Nielsen ratings for the past week. What were the three most popular television shows? Explain the meaning of the "rating" and the "share" for each show.

32. **Nielsen Methods.** Nielsen Media Research supplies statistical data that is very important to advertisers and marketers. Visit the Nielsen Web site and read about one or more of their data products. Write a brief report on how they collect the data and how the data are used.

33. **Comparing Airlines.** The U.S. Department of Transportation routinely publishes on-time performance, lost baggage rates, and other statistics for different airline companies. Find a recent example of such statistics. Based on what you find, is it fair to say that any particular airline stands out as better or worse than others? Explain.

34. **Labor Statistics.** Use the Bureau of Labor Statistics Web site to find monthly unemployment rates over the past 12 months. If you assume that the monthly survey has a margin of error of about 0.2 percentage point, has there been a noticeable change in the unemployment rate over the past year? Explain.

35. **Statistics and Safety.** Identify a study that has been done (or should be done) to improve the safety of car drivers and passengers. Briefly describe the importance of statistics to the study.

36. **Pew Research Center.** The Pew Research Center for the People and the Press studies public attitudes toward the press, politics, and public policy issues. Go to its Web site and find the latest survey about attitudes. Select a particular recent survey, and write a summary of what was surveyed, how the survey was conducted, and what was found.

IN THE NEWS

37. Statistics in the News. Identify three stories from the past week that involve statistics in some way. In each case, write a brief statement describing the role of statistics in the story.

38. Statistics in Your Major. Write a brief description of some ways in which you think that the science of statistics can be used in your major field of study. (If you have not yet selected a major, answer the question for a major that you are considering.)

39. Statistics and Entertainment. The Nielsen ratings are well known for their role in gauging television viewing. Identify another way that statistics are used in the entertainment industry. Briefly describe the role of statistics in this application.

40. Statistics in Sports. Choose a sport and describe at least three different statistics commonly tracked by participants in or spectators of the sport. In each case, briefly describe the importance of the statistic to the sport.

41. Economic Statistics. The government regularly publishes many different economic statistics, such as the unemployment rate, the inflation rate, and the surplus or deficit in the federal budget. Study recent newspapers and identify five important economic statistics. Briefly explain the purpose of each of these five statistics.

1.2   SAMPLING

The only way to know the true value of a population parameter is to observe *every* member of the population. For example, to learn the exact mean height of all students at your school, you'd need to measure the height of every student. A collection of data from every member of a population is called a **census.** Unfortunately, conducting a census is often impractical. In some cases, the population is so large that it would be too expensive or time-consuming to collect data from every member. In other cases, a census may be ruled out because it would interfere with a study's overall goals. For example, a study designed to test the quality of candy bars before shipping could not involve a census because that would mean testing a piece of every candy bar, leaving none intact to sell.

Not everything that can be counted counts, and not everything that counts can be counted.

—Albert Einstein

> **Definition**
>
> A **census** is the collection of data from *every* member of a population.

Fortunately, most statistical studies can be done without going to the trouble of conducting a census. Instead of collecting data from every member of the population, we collect data from a sample and use the sample statistics to make inferences about the population. Of course, the inferences will be reasonable only if the members of the sample represent the population fairly, at least in terms of the characteristics under study. That is, we seek a **representative sample** of the population.

> **Definition**
>
> A **representative sample** is a sample in which the relevant characteristics of the sample members are generally the same as the characteristics of the population.

EXAMPLE ❶ A Representative Sample for Heights

Suppose you want to determine the mean height of all students at your school. Which is more likely to be a representative sample for this study: the men's basketball team or the students in your statistics class?

SOLUTION   The men's basketball team is not a representative sample for a study of height, both because it consists only of men and because basketball players tend to be taller than average. The mean height of the students in your statistics class is much more likely to be close to the mean height of all students, so the members of your class make a more representative sample than the members of the men's basketball team.   ••●

Bias

Imagine that, for the 5,000 homes in its sample, Nielsen chose only homes in which the primary wage earners worked a late-night shift. Because late-night workers aren't home to watch late-night television, Nielsen would find late-night shows to be unpopular among the homes in this sample. Clearly, this sample would *not* be representative of all American homes, and it would be wrong to conclude that late-night shows were unpopular among all Americans. We say that such a sample is *biased* because the homes in the sample differed in a specific way from "typical" American homes. (In reality, Nielsen takes great care to avoid such obvious bias in the sample selection.) More generally, the term **bias** refers to any problem in the design or conduct of a statistical study that tends to favor certain results. We cannot trust the conclusions of a biased study.

> **Definition**
>
> A statistical study suffers from **bias** if its design or conduct tends to favor certain results.

Bias can arise in many ways. For example:

- A sample is biased if the members of the sample differ in some specific way from the members of the general population. In that case, the results of the study will reflect the unusual characteristics of the sample rather than the actual characteristics of the population.
- A researcher is biased if he or she has a personal stake in a particular outcome. In that case, the researcher might intentionally or unintentionally distort the true meaning of the data.
- The data set itself is biased if its values were collected intentionally or unintentionally in a way that makes the data unrepresentative of the population.
- Even if a study is done well, it may be reported in a biased fashion. For example, a graph representing the data may tell only part of the story or depict the data in a misleading way (see Section 3.4).

Preventing bias is one of the greatest challenges in statistical research. Looking for bias is therefore one of the most important steps in evaluating a statistical study or media reports about a statistical study.

EXAMPLE 2 Why Use Nielsen?

Nielsen Media Research earns money by charging television stations and networks for its services. For example, NBC pays Nielsen to provide ratings for its television shows. Why doesn't NBC simply do its own ratings, instead of paying a company like Nielsen to do them?

SOLUTION   The cost of advertising on a television show depends on the show's ratings. The higher the ratings, the more the network can charge for advertising—which means NBC would have a clear bias if it conducted its own ratings. Advertisers therefore would not trust ratings that NBC produced on its own. By hiring an independent source, such as Nielsen, NBC can provide information that advertisers are more likely to believe.   ••●

BY THE WAY

Many medical studies are experiments designed to test whether a new drug is effective. In an article published in the *Journal of the American Medical Association*, the authors found that studies with positive results (the drug is effective) are more likely to be published than studies with negative results (the drug is not effective). This "publication bias" tends to make new drugs, as a group, seem more effective than they really are.

Sampling Methods

A good statistical study *must* have a representative sample. Otherwise the sample is biased and conclusions from the study are not trustworthy. Let's examine a few common sampling methods that, at least in principle, can provide a representative sample.

Simple Random Samples

In most cases, the best way to obtain a representative sample is by choosing *randomly* from the population. A **random sample** is one in which every member of the population has an equal chance of being selected to be part of the sample. For example, you could obtain a random sample by having everyone in a population roll a die and choosing those people who roll a 6. In contrast, the sample would not be random if you chose everyone taller than 6 feet, because not everyone would have an equal chance of being selected.

In statistics, we usually decide in advance the sample size that is needed. With **simple random sampling,** every possible sample of a particular size has an equal chance of being selected. For example, to choose a simple random sample of 100 students from all the students in your school, you could assign a number to each student in your school and choose the sample by drawing 100 of these numbers from a hat. As long as each student's number is in the hat only once, every sample of 100 students has an equal chance of being selected. As a faster alternative to using a hat, you might choose the student numbers with the aid of a computer or calculator that has a built-in *random number generator.*

Because simple random sampling gives every sample of a particular size the same chance of being chosen, it is likely to provide a representative sample, as long as the sample size is large enough.

EXAMPLE 3 Local Resident Sampling

You want to conduct an opinion poll in which the population is all the residents in a town. Could you choose a simple random sample by randomly selecting names from local property tax records?

SOLUTION   A sample drawn from property tax records is not a simple random sample of the town population because these records would only list people who own property in the town. These records are therefore missing many town residents, and might also include people who live elsewhere but own property in the town.   · · ●

Systematic Sampling

Simple random sampling is effective, but in many cases we can get equally good results with a simpler technique. Suppose you are testing the quality of microchips produced by Intel. As the chips roll off the assembly line, you might decide to test every 50th chip. This ought to give a representative sample because there's no reason to believe that every 50th chip has any special characteristics compared with other chips. This type of sampling, in which we use a system such as choosing every 50th member of a population, is called **systematic sampling.**

EXAMPLE 4 Museum Assessment

When the National Air and Space Museum wanted to test possible ideas for a new solar system exhibit, a staff member interviewed a sample of visitors selected by systematic sampling. She interviewed a visitor exactly every 15 minutes, choosing whoever happened to enter the current solar system exhibit at that time. Why do you think she chose systematic sampling rather than simple random sampling? Was systematic sampling likely to produce a representative sample in this case?

SOLUTION Simple random sampling might occasionally have selected two visitors so soon after each other that the staff member would not have had time to interview each of them. The systematic process of choosing a visitor every 15 minutes prevented this problem from arising. Because there's no reason to think that the people entering at a particular moment are any different from those who enter a few minutes earlier or later, this process is likely to give a representative sample of the population of visitors during the time of the sampling. ⋅ ⋅ ●

EXAMPLE 5 When Systematic Sampling Fails

You are conducting a survey of students in a co-ed dormitory in which males are assigned to odd-numbered rooms and females are assigned to even-numbered rooms. Can you obtain a representative sample when you choose every 10th room?

SOLUTION No. If you start with an odd-numbered room, every 10th room will also be odd-numbered (such as room numbers 3, 13, 23, …). Similarly, if you start with an even-numbered room, every 10th room will also be even-numbered. You will therefore obtain a sample consisting of either all males or all females, neither of which is representative of the co-ed population. ⋅ ⋅ ●

BY THE WAY

The sampling described in Example 4 was undertaken prior to the construction of the Voyage Scale Model Solar System, a permanent scale model exhibit that stretches along the National Mall from the National Air and Space Museum to the Smithsonian Castle. The photo below shows the son of one of the exhibit creators (also an author of this text) touching the scale model Sun.

TIME ⏲ UT TO THINK

Suppose you chose every fifth room, rather than every 10th room, in Example 5. Would the sample then be representative?

Convenience Samples

Systematic sampling is easier than simple random sampling but might also be impractical in many cases. For example, suppose you want to know the proportion of left-handed students at your school. It would take great effort to select a simple random sample or a systematic sample, because both require drawing from all the students in the school. In contrast, it would be easy to use the students in your statistics class as your sample—you could just ask the left-handed students to raise their hands. This type of sample is called a **convenience sample** because it is chosen for convenience rather than by a more sophisticated procedure. For trying to find the proportion of left-handed people, the convenience sample of your statistics class is probably fine; there is no reason to think that there would be a different proportion of left-handed students in a statistics class than anywhere else. But if you were trying to determine the proportions of students with different majors, this sample would be biased because some majors require a statistics course and others do not. In general, convenience sampling tends to be more prone to bias than most other forms of sampling.

EXAMPLE 6 Salsa Taste Test

A supermarket wants to decide whether to carry a new brand of salsa, so it offers free tastes at a stand in the store and asks people what they think. What type of sampling is being used? Is the sample likely to be representative of the population of all shoppers?

SOLUTION The sample of shoppers stopping for a taste of the salsa is a convenience sample because these people happen to be in the store and are willing to try the new product. (This type of convenience sample, in which people choose whether or not to be part of the sample, is also called a *self-selected sample*. We will study self-selected samples further in Section 1.4.) This sample is unlikely to be representative of the population of all shoppers, because different types of people may shop at different times (for example, stay-at-home parents are more likely to shop at midday than are working parents) and only people who like salsa are likely to participate. The data might be still be useful, however, because the opinions of people who like salsa are probably the most important ones in this case.   ·· ●

Cluster Samples

Cluster sampling involves the selection of *all* members in randomly selected groups, or *clusters*. Imagine that you work for the Department of Agriculture and wish to determine the percentage of farmers who use organic farming techniques. It would be difficult and costly to collect a simple random sample or a systematic sample because either would require visiting many individual farms that are located far from one another. A convenience sample of farmers in a single county would be biased because farming practices vary from region to region. You might therefore decide to select a few dozen counties at random from across the United States and survey *every* farmer in each of those counties. We say that each county contains a *cluster* of farmers, and the sample consists of *every* farmer within the randomly selected clusters.

EXAMPLE 7 Gasoline Prices

You want to know the mean price of gasoline at gas stations located within a mile of rental car locations at airports. Explain how you might use cluster sampling in this case.

SOLUTION You could randomly select a few airports around the country. For these airports, you would check the gasoline price at *every* gas station within a mile of the rental car location.   ·· ●

Stratified Samples

Suppose you are conducting a poll to predict the outcome of the next U.S. presidential election. The population under study is all likely voters, so you might choose a simple random sample from this population. However, because presidential elections are decided by electoral votes cast on a state-by-state basis, you'll get a better prediction if you determine voter preferences within each state. Your overall sample should therefore consist of separate random samples from each of the 50 states. In statistical terminology, the populations of the 50 states represent subgroups, or **strata,** of the total population. Because your overall sample consists of randomly selected members from each stratum, you've used **stratified sampling.**

EXAMPLE 8 Unemployment Data

The U.S. Labor Department surveys 60,000 households each month to compile its unemployment report (see Example 2 in Section 1.1). To select these households, the department first groups cities and counties into about 2,000 geographic areas. It then randomly selects households to survey within these geographic areas. How is this an example of stratified sampling? What are the strata? Why is stratified sampling important in this case?

SOLUTION The unemployment survey is an example of stratified sampling because it first breaks the population into subgroups. The subgroups, or strata, are the people in the 2,000 geographic regions. Stratified sampling is important in this case because unemployment rates are likely to differ in different geographic regions. For example, unemployment rates in rural Kansas may be very different from those in Silicon Valley. By using stratified sampling, the Labor Department ensures that its sample fairly represents all geographic regions.   ·· ●

BY THE WAY

As mandated by the U.S. Constitution, voting for the President is actually done by a small group of people called *electors*. Each state may select as many electors as it has members of Congress (counting both senators and representatives). When you cast a ballot for President, you actually cast a vote for your state's electors, each of whom has promised to vote for a particular presidential candidate. The electors cast their votes a few weeks after the general election.

Summary of Sampling Methods

The following box and Figure 1.3 summarize the five sampling methods we have discussed. No single method is "best," as each one has its uses. (Some studies even combine two or more types of sampling.) But regardless of how a sample is chosen, keep in mind the following three key ideas:

- A study can be successful only if the sample is representative of the population.
- A biased sample is unlikely to be a representative sample.
- Even a well-chosen sample may still turn out to be unrepresentative just because of bad luck in the actual drawing of the sample.

Common Sampling Methods

- **Simple random sampling:** We choose a sample of items in such a way that every sample of the same size has an equal chance of being selected.

- **Systematic sampling:** We use a simple system to choose the sample, such as selecting every 10th or every 50th member of the population.

- **Convenience sampling:** We use a sample that happens to be convenient to select.

- **Cluster sampling:** We first divide the population into groups, or clusters, and select some of these clusters at random. We then obtain the sample by choosing *all* the members within each of the selected clusters.

- **Stratified sampling:** We use this method when we are concerned about differences among sub-groups, or *strata*, within a population. We first identify the strata and then draw a random sample within each stratum. The total sample consists of all the samples from the individual strata.

Simple Random Sampling:
Every sample of the same size has an equal chance of being selected. Computers are often used to generate random numbers.

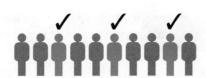

Systematic Sampling:
Select every *k*th member.

Convenience Sampling:
Use results that are readily available.

Election precincts in Carson County

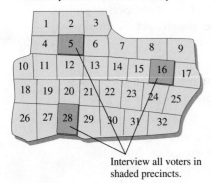

Interview all voters in shaded precincts.

Cluster Sampling:
Divide the population into clusters, randomly select some of those clusters, then choose all members of the selected clusters.

Stratified Sampling:
Partition the population into at least two strata, then draw a sample from each.

Figure 1.3 Common sampling methods.

EXAMPLE 9 Sampling Methods

Identify the type of sampling used in each of the following cases.

a. The apple harvest from an orchard is collected in 1,200 baskets. An agricultural inspector randomly selects 25 baskets and then checks every apple in each of these baskets for worms.

b. An educational researcher wants to know whether, at a particular college, men or women tend to ask more questions in class. Of the 10,000 students at the college, she interviews 50 randomly selected men and 50 randomly selected women.

c. In trying to learn about planetary systems, astronomers conduct a survey by looking for planets among 100 nearby stars.

d. To determine who will win autographed footballs, a computer program randomly selects the ticket numbers of 11 people in a stadium filled with people.

SOLUTION

a. The apple inspection is an example of cluster sampling because the inspector begins with a randomly selected set of clusters (baskets) and then checks every apple in the selected clusters.

b. The groups of men and women represent two different strata for this study, so this is an example of stratified sampling.

c. The astronomers presumably focus on nearby stars because they are easier to study in this case, so this is an example of convenience sampling.

d. Because the computer selects the 11 ticket numbers at random, every ticket number has an equal chance of being chosen. This is an example of simple random sampling.   ·· ●

Section 1.2 Exercises

Statistical Literacy and Critical Thinking

1. **Census and Sample.** What is a *census,* what is a sample, and what is the difference between them?

2. **Biased Sample.** In a survey before a general election (as opposed to a primary), a sample is drawn randomly from a list of registered Democrats. Is there anything wrong with this sampling method?

3. **Cluster and Stratified Sampling.** Cluster sampling and stratified sampling both involve selecting subjects in subgroups of the population. What is the difference between those two types of sampling?

4. **Sample of Students.** A college statistics teacher conducted a study by recording whether each student in his class was right-handed. The objective was to form a conclusion about the proportion of college students that is right-handed. What type of sample was obtained? Is this sample likely to be biased? Why or why not?

Does It Make Sense? For Exercises 5–8, decide whether the statement makes sense (or is clearly true) or does not make sense (or is clearly false). Explain clearly; not all of these have definitive answers, so your explanation is more important than your chosen answer.

5. **Graduation Age.** For a statistics class project, I conducted a census to determine the mean age of students when they earn their bachelor's degrees.

6. **Convenience Sample.** For a statistics class project, I used a convenience sample, but the results may still be meaningful.

7. **Biased Sample.** The study must have been biased, because it concluded that 75% of Americans are more than 6 feet tall.

8. **Death Row.** There are currently 3,242 convicts on death row (based on 2010 data from the Bureau of Justice Statistics). We obtained a simple random sample of those convicts by compiling a numbered list, then using a computer to randomly generate 20 numbers between 1 and 3,242, then selecting the convicts that correspond to the generated numbers.

Concepts and Applications

Census. In Exercises 9–12, determine whether a census is practical in the situations described. Explain your reasoning.

9. Laker Heights. You want to determine the mean height of all basketball players on the LA Lakers team.

10. High School Heights. You want to determine the mean height of all high school basketball players in the United States.

11. IQ Scores. You want to determine the mean IQ score of all statistics instructors in the United States.

12. Instructor Ages. You want to determine the mean age of all statistics instructors at the University of Colorado.

Representative Samples? In Exercises 13–16, identify the sample, population, and sampling method. Then comment on whether you think it is likely that the sample is representative of the population.

13. Senate Terms. A political scientist randomly selects 4 of the 100 senators currently serving in Congress, then finds the lengths of time that they have served.

14. Super Bowl. During the Super Bowl game, Nielsen Media Research conducts a survey of 5,108 randomly selected households and finds that 44% of them have television sets tuned to the Super Bowl.

15. Gun Ownership. In a Gallup poll of 1,059 randomly selected adults, 39% answered "yes" when asked "Do you have a gun in your home?"

16. Mail Survey. A graduate student conducts a research project about how adult Americans send thank-you notes. She uses the U.S. Postal Service to mail a survey to 500 adults that she knows and asks them to mail back a response to this question: "Do you prefer to send thank you notes by e-mail or snail mail (the U.S. Postal Service)?" She gets back 65 responses, with 42 of them indicating a preference for snail mail.

Evaluate the Sample Choices. Exercises 17–18 each describe the goal of a study, then offer you four options for obtaining a sample. In each case, decide which sample is most likely to be a representative sample, and explain why. Then explain why each of the other choices is *not* likely to make a representative sample for the study.

17. Credit Card Debt. You want to determine the mean amount of credit card debt owed by adult consumers in Florida.

- Sample 1: The Florida drivers who own and have registered Land Rover vehicles

- Sample 2: The first 1,000 Florida residents listed in the Fort Lauderdale phone book

- Sample 3: The first 1,000 Florida residents in a complete list of all Florida telephone numbers

- Sample 4: The Florida residents who mail back a survey printed in the *Miami Herald*

18. California Voters. You want to conduct a survey to determine the proportion of eligible voters in California likely to vote for the Democratic presidential candidate in the next election.

- Sample 1: All eligible voters in San Diego County

- Sample 2: All eligible voters in the city of Sonoma

- Sample 3: All eligible voters who respond to a CNN Internet survey

- Sample 4: Every 1,000th person on a complete list of all eligible voters in California

Bias. Are there sources of bias in the situations described in Exercises 19–22? Explain.

19. Movie Critic. A film critic for ABC News gives her opinion of the latest movie from Disney, which also happens to own ABC.

20. Car Reviews. *Consumer Reports* magazine prints a review of new cars and does not accept free products or advertising from anyone.

21. GMO Soybeans. Monsanto hires independent university scientists to determine whether its new, genetically engineered soybean poses any threat to the environment.

22. Drug Study Funding. The *Journal of the American Medical Association* prints an article evaluating a drug, and some of the physicians who wrote the article received funding from the pharmaceutical company that produces the drug.

Sampling Methods. In Exercises 23–38, identify which of the following applies: simple random sample, systematic sample, convenience sample, stratified sample, or cluster sample. In each case, state whether you think the procedure is likely to yield a representative sample or a biased sample, and explain why.

23. Clinical Trial. In phase II testing of a new drug designed to increase the red blood cell count, a researcher obtains envelopes with the names and addresses of all treated subjects. She wants to increase the dosage in a sub-sample of 12 subjects, so she thoroughly mixes all of the envelopes in a bin, then pulls 12 of those envelopes to identify the subjects to be given the increased dosage.

24. Sobriety Checkpoint. Police set up a sobriety checkpoint at which every fifth driver is stopped and interviewed.

25. Exit Polls. On days of presidential elections, the news media organize an exit poll in which specific polling stations are randomly selected and all voters are surveyed as they leave the premises.

26. Education and Sports. A researcher for the Spaulding athletic equipment company is studying the relationship between level of education and participation sports. She conducts a survey of 40 randomly selected golfers, 40 randomly selected tennis players, and 40 randomly selected swimmers.

27. Ergonomics. An engineering student measures the strength of fingers used to push buttons by testing family members.

28. Tax Cheating. An Internal Revenue Service researcher investigates false reporting of tip income by waiters and waitresses by surveying all waiters and waitresses at 20 randomly selected restaurants.

29. MTV Survey. A marketing expert for MTV is planning a survey in which 500 people will be randomly selected from each age group of 10–19, 20–29, and so on.

30. Credit Card Data. A professor surveyed students in her class to obtain sample data consisting of the number of credit cards students possess.

31. Fundraising. Fundraisers for the College of Newport test a new telemarketing campaign by obtaining an alphabetical list of all alumni and selecting every 100th name on that list.

32. Telephone Poll. In a Gallup poll of 1,059 adults, the interview subjects were selected by using a computer to randomly generate telephone numbers that were then called.

33. Market Research. A market researcher has partitioned all California residents into categories of unemployed, employed full time, and employed part time. She is surveying 50 people from each category.

34. Student Drinking. Motivated by a student who died from binge drinking, the College of Newport conducts a study of student drinking by randomly selecting 10 different classes and interviewing all of the students in each of those classes.

35. Magazine Survey. *People* magazine chooses its "best-dressed celebrities" by compiling responses from readers who mail in a survey printed in the magazine.

36. Heart Transplants. A medical researcher at Johns Hopkins University obtains a numbered list of all patients waiting for a heart transplant, then uses a computer to select the patients corresponding to the 50 numbers randomly generated by computer.

37. Quality Control. A sample of manufactured CDs is obtained by using a computer to randomly generate a number between 1 and 1,000 for each CD, and the CD is selected if the generated number is 1,000.

38. Seat Belts. Every 500th seat belt is tested by stressing it until it fails.

Choose a Sampling Method. For each of Exercises 39–42, suggest a sampling method that is likely to produce a representative sample. Explain why you chose this method over other methods.

39. Student Election. You want to predict the winner of an upcoming election for student body president.

40. Blood Type. You want to determine the percentage of people in this country in each of the four major blood groups (A, B, AB, and O).

41. Heart Deaths. You want to determine the percentage of deaths due to heart disease each year.

42. Mercury in Tuna. You want to determine the average mercury content of the tuna fish consumed by U.S. residents.

PROJECTS FOR THE INTERNET & BEYOND

43. Public Opinion Poll. Use information available on the Web site of a polling organization, such as Gallup, Harris, Pew, or Yankelovich, to answer the following questions.

   a. How exactly is a sample of subjects selected?

   b. Based on what you have learned, do you think the poll results are reliable? If so, why? If not, why not?

44. Unemployment Sample. Use the Bureau of Labor Statistics Web page to find details on how the bureau chooses the sample of households in its monthly survey. Write a short summary of the procedure and why it is likely to yield a representative sample.

45. Selective Voting. The Academy Awards, the Heisman Trophy, and the *New York Times* "Bestseller List" are just three examples of selections that are determined by the votes of specially selected individuals. Pick one of these selection processes, and describe who votes and how those people are chosen. Discuss sources of bias in the process.

IN THE NEWS

46. Sampling in the News. Find a recent news report about a statistical study that you find interesting. Write a short summary of how the sample for the study was chosen, and briefly discuss whether you think the sample was representative of the population under study.

47. Opinion Poll Sample. Find a recent news report about an opinion poll carried out by a news organization (such as Gallup, Harris, *USA Today*, *New York Times*, or CNN). Briefly describe the sample and how it was chosen. Was the sample chosen in a way that was likely to introduce any bias? Explain.

48. Political Polls. Find results from a recent poll conducted by a political organization (such as the Republican or Democratic party or an organization that seeks to influence Congress on some particular issue). Briefly describe the sample and how it was chosen. Was the sample chosen in a way that was likely to introduce any bias? Should you be more concerned about bias in such a poll than you would be in a poll conducted by a news organization? Explain.

1.3 TYPES OF STATISTICAL STUDIES

Statistical studies are conducted in many different ways. In all cases, the people, animals (or other living things), or objects chosen for the sample are called the **subjects** of the study. If the subjects are people, it is common to refer to them as **participants** in the study.

> **Definition**
>
> The **subjects** of a study are the people, animals (or other living things), or objects chosen for the sample; if the subjects are people, they may be called the **participants** in the study.

You can observe a lot by just watching.

—Yogi Berra

There are two basic types of statistical study: observational studies and experiments. In an **observational study,** we observe or measure specific characteristics while trying to be careful to avoid influencing or modifying the characteristics we are observing. The Nielsen ratings are an example of an observational study, because Nielsen uses devices to *observe* what the subjects are watching on TV, but does not try to influence what they watch.

Note that an observational study may involve activities that go beyond the usual definition of *observing*. Measuring people's weights requires interacting with them, as in asking them to stand on a scale. But in statistics, we consider these measurements to be observations because the interactions do not change people's weights. Similarly, an opinion poll in which researchers conduct in-depth interviews is considered observational as long as the researchers attempt only to learn people's opinions, not to change them.

In contrast, consider a medical study designed to test whether large daily doses of vitamin C help prevent colds. To conduct this study, the researchers must ask some people in the sample to take large doses of vitamin C every day. This type of statistical study is called an **experiment.** The purpose of an experiment is to study the effects of some **treatment**—in this case, large daily doses of vitamin C.

> **Two Basic Types of Statistical Study**
>
> - In an **observational study,** researchers observe or measure characteristics of the subjects, but do not attempt to influence or modify these characteristics.
>
> - In an **experiment,** researchers apply some **treatment** and observe its effects on the subjects of the experiment.

EXAMPLE 1 Type of Study

Identify the study as an observational study or an experiment.

a. The Salk polio vaccine study (see the Case Study on page 7)

b. A poll in which college students are asked if they commute or live on campus

SOLUTION

a. The Salk vaccine study was an *experiment* because researchers tested a treatment—in this case, the vaccine—to see whether it reduced the incidence of polio.

b. The poll is an *observational study* because it attempts to determine how college students go to their classes, and it does not try to influence how they get there.   ••●

Identifying the Variables

Statistical studies—whether observations or experiments—generally are attempts to measure what we call **variables of interest.** The term *variable* refers to an item or quantity that can vary or take on different values, and variables of interest are those we seek to learn about. For example, variables of interest in the Nielsen studies of television viewing habits include *show being watched* and *number of viewers*. The variable *show being watched* can take on different values such as "Super Bowl" or "*60 Minutes*." The variable *number of viewers* depends on the popularity of a particular show. In essence, the raw data in any statistical study are the different values of the variables of interest.

In cases where we think cause and effect may be involved, we sometimes subdivide the variables of interest into two categories. For example, each person in the study of vitamin C and colds may take a different daily dose of vitamin C and may end up with a different number of colds over some period of time. Because we are trying to learn if vitamin C causes a lower number of colds, we say that *daily dose of vitamin C* is an **explanatory variable**—it may explain or cause a change in the number of colds. Similarly, we say that *number of colds* is a **response variable,** because we expect it to respond to changes in the explanatory variable (the dose of vitamin C).

> **Definitions**
>
> A variable is any item or quantity that can vary or take on different values.
>
> The **variables of interest** in a statistical study are the items or quantities that the study seeks to measure.
>
> When cause and effect may be involved, an **explanatory variable** is a variable that may explain or cause the effect, while a **response variable** is a variable that responds to changes in the explanatory variable.

EXAMPLE ② Identify the Variables

Identify the variables of interest for each study.

a. The Salk polio vaccine study

b. A poll in which college students are asked if they commute or live on campus

SOLUTION

a. The two variables of interest in the Salk vaccine study are *vaccine* and *polio*. They are variables because they can take on two different values: A child either did or did not get the vaccine and either did or did not contract polio. In this case, because the study seeks to determine whether the vaccine prevents polio, we say that *vaccine* is the explanatory variable (it may explain a change in the incidence of polio) and *polio* the response variable (it is supposed to change in response to the vaccine).

b. The variables of interest are the responses to the question asking how students get to their classes, and the proportions of responses to the possible different choices (such as commute, live on campus, refuse to answer, don't know). (There is no cause and effect involved in this study, so we do not need to decide whether the variable is explanatory or response.)   · · ●

Observational Studies

The observational studies we have discussed up to this point, such as Nielsen ratings, opinion polls, and determining the mean height of students, are studies in which the data are all generally collected around the same time. Sometimes, however, observational studies look at past data or are designed to look at future data over a long period of time.

A **retrospective study** (also called a *case-control* study) is an observational study that uses data from the past—such as official records or past interviews—to learn about some issue of concern. Retrospective studies are especially valuable in cases where it may be impractical or unethical to perform an experiment. For example, suppose we want to learn how alcohol consumed during pregnancy affects newborn babies. Because it is already known that consuming alcohol during pregnancy can be harmful, it would be highly unethical to ask pregnant mothers to test the "treatment" of consuming alcohol. However, because many mothers consumed alcohol in past pregnancies (either before the dangers were known or choosing to ignore the dangers), we can do a retrospective study in which we compare children born to those mothers to children born to mothers who did not consume alcohol.

Sometimes, the data we need to reach clear conclusions are not available in past records. In those cases, researchers may set up a **prospective study** (sometimes called a *longitudinal* study) designed to collect observations in the future from groups that share common factors. A classic example of a prospective study is the Harvard Nurses' Health Study, which was started in 1976 in order to collect data about how different lifestyles affect women's health (see the "Focus on Public Health" section at the end of this chapter). The study, still ongoing today, has followed thousands of nurses over more than three decades, collecting data about their lifestyles and health.

Variations on Observational Studies

The most familiar observational studies are those in which data are collected all at once (or as close to that as possible). Two variations on observational studies are also common:

- A **retrospective** (or **case-control**) **study** uses data from the past, such as official records or past interviews.

- A **prospective** (or **longitudinal**) **study** is set up to collect data in the future from groups that share common factors.

EXAMPLE 3 Observational Study

You want to know whether children born prematurely do as well in elementary school as children born at full term. What type of study should you do?

SOLUTION An observational, retrospective study is the only real option in this case. You would collect data on past births and compare the elementary school performance of those born prematurely to that of those born at full term.
··●

Experiments

Because experiments require active intervention, such as applying a treatment, we must take special care to ensure that they are designed in ways that will provide the information we seek. Let's examine a few of the issues that arise in the design of experiments.

The Need for Controls

Consider an experiment that gives some participants in a study vitamin C to determine its effect on colds. Suppose the people taking vitamin C daily get an average of 1.5 colds in a three-month period. How can the researchers know whether the subjects would have gotten more colds without the vitamin C? To answer this type of question, the researchers must conduct their experiment with two (or more) groups of subjects: One group takes large doses of vitamin C daily and another group does not. As we'll discuss shortly, in most cases it is important that participants be *randomly* assigned to the two groups.

The group of people who are randomly assigned to take vitamin C is called the **treatment group** because its members receive the treatment being tested (vitamin C). The group of people who do *not* take vitamin C is called the **control group.** The researchers can be confident that vitamin C is an effective treatment only if the people in the treatment group get significantly fewer colds than the people in the control group.

BY THE WAY

The control group gets its name from the fact that it helps control the way we interpret experimental results.

Treatment and Control Groups

The **treatment group** in an experiment is the group of subjects who receive the treatment being tested.

The **control group** in an experiment is the group of subjects who do *not* receive the treatment being tested.

EXAMPLE ④ Treatment and Control

Look again at the Salk polio vaccine Case Study (page 7). What was the treatment? Which group of children constituted the treatment group? Which constituted the control group?

SOLUTION   The treatment was the Salk vaccine. The treatment group consisted of the children who received the Salk vaccine. The control group consisted of the children who did not get the Salk vaccine and instead got an injection of salt water.   ·· •

EXAMPLE ⑤ Mozart Treatment

A study divided college students into two groups. One group listened to Mozart or other classical music before being assigned a specific task and the other group simply was assigned the task without listening to the music. Researchers found that those listening to the classical music performed the task slightly better, but only if they did the task within a few minutes of listening to the music. (The two groups performed equally on tasks given later). Identify the treatment and the control and treatment groups.

SOLUTION   The treatment was the classical music. The treatment group consisted of the students who listened to the music. The control group consisted of the students who did not listen to the music.   ·· •

Confounding Variables

Using control groups helps to ensure that we account for known variables that could affect a study's results. However, researchers may be unaware of or be unable to account for other important variables. Consider an experiment in which a statistics teacher seeks to determine whether students who study collaboratively (in study groups with other students) earn higher grades than students who study independently. The teacher chooses five students who will study collaboratively (the treatment group) and five others who will study independently (the control group). To ensure that the students all have similar abilities and will study diligently, the teacher chooses only students with high grade-point averages. At the end of the semester, the teacher finds that the students who studied collaboratively earned higher grades.

The variables of interest for this study are *collaborative study* (whether they do so or not) and *final grade*. But suppose that, unbeknownst to the teacher, the collaborative students all lived in a dormitory where a curfew ensured that they got plenty of sleep. This fact introduces a new variable—which we might call *amount of sleep*—that might partially explain the results. In other words, the experiment's conclusion may *seem* to support the benefits of collaborative study, but this conclusion is not justified because the teacher did not account for how much students slept.

In statistical terminology, this study suffers from **confounding.** The higher grades may be due either to the variable of interest (*collaborative study*) or to the differing amounts of sleep or to a combination of both. Because the teacher did not account for differences in the amount of sleep, we say that *amount of sleep* is a **confounding variable** for this study. You can probably think of other potentially confounding variables that could affect a study like this one.

> **Definition**
> A study suffers from **confounding** if the effects of different variables are mixed so we cannot determine the specific effects of the variables of interest. The variables that lead to the confusion are called **confounding variables.**

CASE STUDY   Confounding Drug Results

An advisory panel of the Federal Drug Administration (FDA) recently recommended revoking the approval of the advanced breast cancer drug Avastin, which at the time was the world's best-selling cancer drug. Early trials had found that, when used in combination with a certain chemotherapy drug, Avastin delayed the growth of tumors by five months compared to using the chemotherapy drug alone. However, additional trials found less significant delays and no improvement in the length or quality of the lives of those taking Avastin. The differing results in different trials suggest that confounding variables have not all been accounted for. One possible confounding variable may be the choice of chemotherapy drug. Indeed, studies in Europe found Avastin to be effective, but were based on using it in combination with a different chemotherapy drug than the one approved in the United States. As a result, European regulators were approving Avastin for expanded use at the same time that regulators in the United States were recommending against it. The lesson should be clear: Even when thousands of lives and millions of dollars are on the line, confounding variables may be very difficult to eliminate, making decisions far more uncertain than doctors or patients (or government regulators) would like.

Assigning Treatment and Control Groups

As the collaborative study experiment illustrates, results are almost sure to suffer from confounding if the treatment and control groups differ in important ways (other than receiving or not receiving the treatment). Researchers generally employ two strategies to prevent such differences and thereby ensure that the treatment and control groups can be compared fairly. First, they assign participants to the treatment and control groups *at random*, meaning that they use a technique designed to ensure that each participant has an equal chance of being assigned to either group. When the participants are randomly assigned, it is less likely that the people in the treatment and control groups will differ in some way that will affect the study results.

Second, researchers try to ensure that the treatment and control groups are sufficiently large. For example, in the collaborative study experiment, including 50 students in each group rather than five would have made it much less likely that all the students in one group would live in a special dormitory.

> ### Strategies for Selecting Treatment and Control Groups
>
> - **Select groups at random.** Make sure that the subjects of the experiment are assigned to the treatment or control group at random, meaning that each subject has an equal chance of being assigned to either group.
>
> - **Use sufficiently large groups.** Make sure that the treatment and control groups are both sufficiently large that they are unlikely to differ in a significant way (aside from the fact that one group gets the treatment and the other does not).

EXAMPLE 6   Salk Study Groups

Briefly explain how the two strategies for selecting treatment and control groups were used in the Salk polio vaccine study.

SOLUTION   A total of about 400,000 children participated in the study, with half receiving an injection of the Salk vaccine (the treatment group) and the other half receiving an injection of salt water (the control group). The first strategy was implemented by choosing children for the two groups randomly from among all the children. The second strategy was implemented by using a large number of participants (200,000 in each group) so that the two groups were unlikely to differ by chance.   • • •

The Placebo Effect

When an experiment involves people, effects can occur simply because people know they are part of the experiment. For example, suppose you are testing the effectiveness of a new anti-depression drug. You find 500 people who suffer from depression and randomly divide them into a treatment group that receives the new drug and a control group that does not. A few weeks later, interviews with the patients show that people in the treatment group tend to be feeling much better than people in the control group. Can you conclude that the new drug works?

Unfortunately, it's quite possible that the mood of people receiving the drug improved simply because they were happy to be getting some kind of treatment, which means you cannot be sure that the drug really helped. This type of effect, in which people improve because they believe that they are receiving a useful treatment, is called the **placebo effect.** (The word *placebo* comes from the Latin "to please.")

To distinguish between results caused by a placebo effect and results that are truly due to the treatment, researchers try to make sure that the participants do not know whether they are part of the treatment or control group. To accomplish this, the researchers give the people in the control group a **placebo**: something that looks or feels just like the treatment being tested, but lacks its active ingredients. For example, in a test of a drug that comes in pill form, the placebo might be a pill of the same shape and size that contains sugar instead of the real drug. In a test of an injected vaccine, the placebo might be an injection that contains only a saline solution (salt water) instead of the real vaccine. In a recent test of the effectiveness of acupuncture, the placebo consisted of treatment with needles as in real acupuncture, except the needles were not put in the special places that acupuncturists claim to be important.

As long as the participants do not know whether they received the real treatment or a placebo, the placebo effect ought to affect the treatment and control groups equally. If the results for the two groups are significantly different, it is reasonable to believe that the differences can be attributed to the treatment. For example, in the study of the anti-depression drug, we would conclude that the drug was effective only if the control group received a placebo and members of the treatment group improved much more than members of the control group. For even better control, some experiments use three groups: a treatment group, a placebo group, and a control group. The placebo group is given a placebo while the control group is given nothing.

> **Definitions**
>
> A placebo lacks the active ingredients of a treatment being tested in a study, but looks or feels like the treatment so that participants cannot distinguish whether they are receiving the placebo or the real treatment.
>
> The **placebo effect** refers to the situation in which patients improve simply because they believe they are receiving a useful treatment.
>
> Note: Although participants should not know whether they belong to the treatment or control group, for ethical reasons it is very important that participants be told that some of them will be given a placebo, rather than the real treatment.

EXAMPLE **7** **Vaccine Placebo**

What was the placebo in the Salk polio vaccine study? Why did researchers use a placebo in this experiment?

SOLUTION   The placebo was the salt water injection given to the children in the control group. To understand why the researchers used a placebo for the control group, suppose that a placebo had *not* been used. When improvements were observed in the treatment group, it would have been impossible to know whether the improvements were due to the vaccine or to the placebo effect. In order to remove this confounding, all participants had to believe that they were being treated in the same way. This ensured that any placebo effect would occur in both groups equally, so that researchers could attribute any remaining differences to the vaccine.   · · ●

TIME (◯)UT TO THINK

Although participants should not know whether they belong to the treatment or control (placebo) group, for ethical reasons they should be told that some of them will receive a placebo rather than the real treatment. This was not always the case in decades past, when participants were sometimes told that they all received a treatment when in fact some received a placebo. Should researchers be allowed to use results of past studies that do not meet today's ethical criteria? Defend your opinion.

Experimenter Effects

Even if the study subjects don't know whether they received the real treatment or a placebo, the *experimenters* may still have an effect. In testing an anti-depression drug, for example, experimenters will probably interview patients to find out whether they are feeling better. But if the experimenters know who received the real drug and who received the placebo, they may inadvertently smile more at the people in the treatment group. Their smiles might improve those participants' moods, making it seem as if the treatment worked when in fact the improvement was caused by the experimenter. This type of confounding, in which the experimenter somehow influences the results, is called an **experimenter effect** (or a Rosenthal effect). The only way to avoid experimenter effects is to make sure that the experimenters don't know which subjects are in which group.

Definition

An **experimenter effect** occurs when a researcher or experimenter somehow influences subjects through such factors such as facial expression, tone of voice, or attitude.

EXAMPLE 8 Child Abuse?

In a famous case, two couples from Bakersfield, California, were convicted of molesting dozens of preschool-age children at their daycare center. The evidence for the abuse came primarily from interviews with the children. However, the conviction was overturned—after one man had served 14 years in prison—when a judge re-examined the interviews and concluded that the children had given answers that they thought the interviewers wanted to hear. If we think of the interviewers as experimenters, this is an example of an experimenter effect because the interviewers influenced the children's answers through the tone and style of their questioning.   •‥●

Blinding

In statistical terminology, the practice of keeping people in the dark about who is in the treatment group and who is in the control group is called **blinding.** A **single-blind** experiment is one in which the participants don't know which group they belong to, but the experimenters do know. If neither the participants nor the experimenters know who belongs to each group, the study is said to be **double-blind.** Of course, *someone* has to keep track of the two groups in order to evaluate the results at the end. In a double-blind experiment, the researchers conducting the study typically hire experimenters to make any necessary contact with the participants. The researchers thereby avoid any contact with the participants, ensuring that they cannot influence them in any way. The Salk polio vaccine study was double-blind because neither the participants (the children) nor the experimenters (the doctors and nurses giving the injections and diagnosing polio) knew who got the real vaccine and who got the placebo.

Blinding in Experiments

An experiment is **single-blind** if the participants do not know whether they are members of the treatment group or members of the control group, but the experimenters do know.

An experiment is **double-blind** if neither the participants nor any experimenters know who belongs to the treatment group and who belongs to the control group.

BY THE WAY

Many similar cases of supposedly widespread child abuse at daycare centers and preschools are being re-examined to see if experimenter effects (by those who interviewed the children) may have led to wrongful convictions. Similar claims of experimenter effects have been made in cases involving repressed memory, in which counseling supposedly helped people retrieve lost memories of traumatic events.

EXAMPLE 9   What's Wrong with This Experiment?

For each of the experiments described below, identify any problems and explain how the problems could have been avoided.

a. A new drug for attention deficit disorder (ADD) is supposed to make affected children more polite. Randomly selected children suffering from ADD are divided into treatment and control groups. The experiment is single-blind. Experimenters evaluate how polite the children are during one-on-one interviews.

b. Researchers wonder whether drinking coffee before an exam improves performance. Fifty coffee-drinkers are told not to drink coffee in the four hours before an exam, and 50 non-coffee-drinkers are asked to drink at least two cups of coffee in the four hours before an exam. Both groups had the same average performance on the exam.

c. Researchers wonder if the effects of a rare degenerative disease can be slowed by exercise. They identify six people suffering from the disease and randomly assign three to a treatment group that exercises every day and three to a control group that avoids exercise. After six months, they compare the amounts of degeneration in each group.

d. A chiropractor performs adjustments on 25 patients with back pain. Afterward, 18 of the patients say they feel better. He concludes that the adjustments are an effective treatment.

SOLUTION

a. The experimenters assess politeness in interviews, but because they know which children received the real drug, they may inadvertently speak differently to these children during the interviews. Or, they might interpret the children's behavior differently because they know which subjects received the real drug. These are experimenter effects that can confound the study results. The experiment should have been double-blind.

b. Both groups took the exam under conditions that were unnatural for them: Coffee-drinkers were deprived of coffee and non-coffee-drinkers were forced to drink it. This fact almost certainly introduced confounding variables that were not taken into account in interpreting the results. In fact, this is an experiment that is difficult to design flawlessly.

c. The results of this study will be difficult to interpret because the sample sizes are not sufficiently large. In addition, if those who exercise are asked not to (or vice versa), it may create other health problems that could confound the results.

d. The 25 patients who receive adjustments represent a treatment group, but this study lacks a control group. The patients may be feeling better because of a placebo effect rather than any real effect of the adjustments. The chiropractor might have improved his study by hiring an actor to do a fake adjustment (one that feels similar, but doesn't actually conform to chiropractic guidelines) on a control group. Then he could have compared the results in the two groups to see whether a placebo effect was involved.   ··•

EXAMPLE 10   Identifying the Study Type

For each of the following questions, what type of statistical study is most likely to lead to an answer? Be as specific as possible.

a. What is the average (mean) income of stock brokers?

b. Do seat belts save lives?

c. Can lifting weights improve runners' times in a 10-kilometer (10K) race?

d. Does skin contact with a particular glue cause a rash?

e. Can a new herbal remedy reduce the severity of colds?

f. Do supplements of resveratrol (an extract from red grapes) increase life span?

With proper treatment, a cold can be cured in a week. Left to itself, it may linger for seven days.

—Medical Folk Saying

SOLUTION

a. An *observational study* can tell us the mean income of stock brokers. We need only survey the brokers, and the survey itself will not change their incomes.

b. It would be unethical to do an experiment in which some people were told to wear seat belts and others were told *not* to wear them. A study to determine whether seat belts save lives must therefore be *observational*. Because some people choose to wear seat belts and others choose not to, we can conduct a *retrospective study*. By comparing the death rates in accidents between those who do and do not wear seat belts, we can learn whether seat belts save lives. (They do.)

c. We need an *experiment* to determine whether lifting weights can improve runners' 10K times. We select randomly from a group of runners to create a treatment group of runners who are put on a weight-lifting program and a control group that is asked to stay away from weights. We must try to ensure that all other aspects of their training are similar. Then we can see whether the runners in the lifting group improve their times more than those in the control group. We cannot use blinding in this experiment because there is no way to prevent participants from knowing whether they are lifting weights.

d. An *experiment* can help us determine whether skin contact with the glue causes a rash. In this case, it's best to use a *single-blind experiment* in which we apply the real glue to participants in one group and apply a placebo that looks the same, but lacks the active ingredient, to members of the control group. There is no need for a double-blind experiment because it seems unlikely that the experimenters could influence whether a person gets a rash. (However, if the question of whether the subject *has* a rash is subject to interpretation, the experimenter's knowledge of who got the real treatment could affect this interpretation.)

e. We should use a *double-blind experiment* to determine whether a new herbal remedy can reduce the severity of colds. Some participants get the actual remedy, while others get a placebo. We need the double-blind conditions because the severity of a cold may be affected by mood or other factors that researchers might inadvertently influence. In the double-blind experiment, the researchers do not know which participants belong to which group and thus cannot treat the two groups differently.

f. Resveratrol has been identified and made available in supplement form only recently; we will need many years of data to determine whether it has an effect on life span. We therefore should use a prospective study designed to monitor participants over many years. The participants could keep written records regarding whether and how much of the supplement they take, and eventually researchers could analyze the data to see if resveratrol has an effect on life span.   ⋯●

BY THE WAY

Experiments have found that mice live longer and have greater endurance when taking large doses of resveratrol, an extract from the skin of red grapes. However, it is not yet known whether high doses of resveratrol would have the same effects for humans, nor is it known whether such doses are safe for humans. Moreover, some scientists question whether the results from the mice studies have been properly interpreted.

Meta-Analysis

All individual statistical studies are either observational studies or experiments. In recent years, however, statisticians have found it useful to "mine" groups of past studies to see if we can learn something that we were unable to learn from the individual studies. For example, hundreds of studies have considered the possible effects of vitamin C on colds, so researchers might decide to review the data from many of these studies as a group. This type of study, in which researchers review many past studies as a group, is called a **meta-analysis.**

Definition

In a **meta-analysis,** researchers review many past studies. The meta-analysis considers these studies as a combined group, with the aim of finding trends that were not evident in the individual studies.

CASE STUDY   Music and Autism: A Meta-Analysis

Numerous individual studies with different approaches have suggested that music is beneficial to children and adolescents with autism, but the results have varying levels of significance. Rather than conducting a separate new study, a researcher at Florida State University decided to conduct a meta-analysis that combined data from all the existing studies. This meta-analysis provided a stronger correlation between music therapy and behavioral improvement than the individual studies provided by themselves, and it helped lead to much greater use of music therapy for autistic children. Of course, a meta-analysis can only be as good as the data from the studies that go into it, and in this case some psychologists remain unconvinced of the benefits of music therapy. They argue that all the studies have ignored confounding variables, especially the possibility that the children are responding more to the attention and enthusiasm of the therapists than to the music itself. While meta-analysis can be a very useful technique, it is still subject to the "garbage in, garbage out" phenomenon, so we must be just as careful in interpreting the conclusions of a meta-analysis as we are with any other type of statistical study.

Section 1.3 Exercises

Statistical Literacy and Critical Thinking

1. **Placebo.** What is a placebo, and why is it important in an experiment to test the effectiveness of a drug?

2. **Blinding.** What is blinding, and why is it important in an experiment to test the effectiveness of a drug?

3. **Confounding.** In testing the effectiveness of a new vaccine, suppose that researchers used males for the treatment group and females for the placebo group. What is confounding, and how would it affect such an experiment?

4. **Ethics.** A clinical trial of a new drug designed to treat hypertension (high blood pressure) is designed to last for three years, but after the first year it becomes clear that the drug is highly successful. Is it ethical to continue the trial with the result that some hypertensive subjects continue to receive a placebo instead of the effective treatment?

5. **Clothing Color.** A researcher plans to investigate the belief that people are more comfortable in the summer sun when they wear clothing with light colors instead of clothing with dark colors. Does it make sense to use a double-blind experiment in this case? Is it easy to implement blinding in this case? Explain.

6. **Lawn Treatment.** A researcher plans to test the effectiveness of a new fertilizer on grass growth. Does it make sense to use a double-blind experiment in this case?

7. **Treating Depression.** A psychologist has developed a procedure for modifying behavior so that subjects suffering from depression can greatly improve that condition. In formal tests of the effectiveness of the treatment, what is an experimenter effect, and how might it be avoided?

8. **Improving IQ Scores.** A psychologist develops a procedure for improving IQ scores by training subjects to become better at taking tests. A standard IQ test is used for evaluating the effectiveness of the procedure. In this case, is it necessary to take precautions against an experimenter effect? Why or why not?

Concepts and Applications

Type of Study. For Exercises 9–20, state the type of study. Be as specific as possible.

9. **Quality Control.** Apple selects a simple random sample of iPhone batteries. The voltage of each battery is measured.

10. **Quality Control.** Apple selects a simple random sample of iPhone batteries. The voltage of each battery is measured after being heated to 43°C.

11. **Magnetic Bracelets.** Some cruise ship passengers are given magnetic bracelets, which they agree to wear in an attempt to eliminate or diminish the effects of motion sickness. Others are given similar bracelets that have no magnetism.

12. **Touch Therapy.** Nine-year-old Emily Rosa became an author of an article in the *Journal of the American Medical Association* after she tested professional touch therapists (see the "Focus on Education" section in Chapter 10).

Using a cardboard partition, she held her hand above the therapist's hand, and the therapist was asked to identify the hand that Emily chose.

13. **Twins.** In a study of hundreds of Swedish twins, it was determined that the level of mental skills was more similar in identical twins (twins coming from a single egg) than in fraternal twins (twins coming from two separate eggs) (*Science*).

14. **Texting and Driving.** A study of 2,500 fatal car crashes identified those that involved drivers who were texting and those who were not.

15. **Gender Selection.** In a study of the YSORT gender selection method developed by the Genetics & IVF Institute, 152 couples using the method had 127 baby boys and 25 baby girls.

16. **AOL Poll.** An America OnLine (AOL) poll resulted in 1,651 responses to the question that asked which of four organizations has the most unethical people in charge. Among the respondents, 36% chose government.

17. **GMO Corn.** Researchers at New York University found that the genetically modified corn known as Bt corn releases an insecticide through its roots into the soil, while the insecticide was not released by corn that was not genetically modified (*Nature*).

18. **Hygiene Survey.** In a Harris Interactive survey of 1,006 randomly selected subjects, 96% of adults said that they wash their hands in public restrooms.

19. **Magnet Treatment.** In a study of the effects of magnets on back pain, some subjects were treated with magnets while others were given non-magnetic devices with a similar appearance. The magnets did not appear to be effective in treating back pain (*Journal of the American Medical Association*).

20. **Power Lines and Cancer.** Hundreds of separate and individual scientific and statistical studies have been done to determine whether high-voltage overhead power lines increase the incidence of cancer among those living nearby. A summary study based on many previous studies concluded that there is no significant link between power lines and cancer (*Journal of the American Medical Association*).

What's Wrong with This Experiment? For each of the (hypothetical) studies described in Exercises 21–28, identify any problems that are likely to cause confounding and explain how the problems could be avoided. Discuss any other problems that might affect the results.

21. **Poplar Tree Growth.** An experiment is designed to evaluate the effectiveness of irrigation and fertilizers on poplar tree growth. Fertilizer is used with one group of poplar trees in a moist region, and irrigation is used with poplar trees in a dry region.

22. **Internet Shopping.** Two hundred volunteers are recruited for a study of how Internet shopping affects purchases. Each person is allowed to choose whether to be in the Internet user group or the group that agrees not to use the Internet for shopping. After one month, the purchases of the two groups are compared.

23. **Octane Rating.** In a comparison of gasoline with different octane ratings, 24 vans are driven with 87 octane gasoline, while 28 sport utility vehicles are driven with 91 octane gasoline. After being driven for 250 miles, the amount of gasoline consumed is measured for each vehicle.

24. **Aspirin Trial.** In Phase I of a clinical trial designed to test the effectiveness of aspirin in preventing heart attacks, aspirin is given to three people and a placebo to seven other people.

25. **Treating Back Pain.** A physician conducts a clinical trial of the effectiveness of running as a treatment for back pain. One group undergoes the running treatment while a control group does not.

26. **Athlete's Foot.** In a clinical trial of the effectiveness of a lotion used to treat tinea pedis (athlete's foot), the physicians who evaluate the results know which subjects were given the treatment and which were given a placebo.

27. **Weight Lifting.** In a test of the effects of lifting heavy weights on blood pressure, one group undergoes a treatment consisting of a weight-lifting program while another group lifts tennis balls.

28. **Paint Mixtures.** In durability tests of Benjamin Moore paint and Sherwin Williams paint, the researchers who evaluate the results know which samples are from each of the two different brands.

Analyzing Experiments. Exercises 29–32 present questions that might be addressed in an experiment. If you were to design the experiment, how would you choose the treatment and control groups? Should the experiment be single-blind, double-blind, or neither? Explain your reasoning.

29. **Beethoven and Intelligence.** Does listening to Beethoven make infants more intelligent?

30. **Lipitor and Cholesterol.** Does the drug Lipitor result in lower cholesterol levels?

31. **Ethanol and Mileage.** Does an ethanol additive in gasoline cause reduced mileage?

32. **Home Siding.** Does aluminum siding on a home last longer than wood siding?

PROJECTS FOR THE INTERNET & BEYOND

33. **Experimenter Effects in Repressed Memory Cases.** Search the Internet for articles and information about the controversy regarding recovering repressed memories. Briefly summarize one or two of the most interesting cases and, based on what you read, express your own opinion as to whether the allegedly recovered memories are being influenced by experimenter effects.

34. **Ethics in Experiments.** In an infamous study conducted in Tuskegee, Alabama, from 1932 to 1972, African American males were told that they were receiving treatment for

syphilis, but in fact they were not. The researchers' hidden goal was to study the long-term effects of the disease. Use the Internet to learn about the history of the Tuskegee syphilis study. Have a class discussion about the ethical issues involved in this case, or write a short essay summarizing the case and its ethical lessons.

35. **Debate: Should We Use Data from Unethical Experiments?** Past research often did not conform to today's ethical standards. In extreme cases, such as research conducted by doctors in Nazi Germany, the researchers sometimes killed the subjects of their experiments. While this past unethical research clearly violated the human rights of the experimental subjects, in some cases it led to insights that could help people today. Is it ethical to use the results of unethical research?

36. **Study Stopped Early.** It sometimes happens that study is stopped early before its completion. Use the Internet to find an example of such a study. Why was the study stopped? Should it have been stopped, or would it be better to complete the study?

IN THE NEWS

37. **Observational Studies.** Search through recent newspapers or journals and find an example of a statistical study that was observational. Briefly describe the study and summarize its conclusions.

38. **Experimental Studies.** Search through recent newspapers or journals and find an example of a statistical study that involved an experiment. Briefly describe the study and summarize its conclusions.

39. **Retrospective Studies.** Search through recent newspapers or journals and find an example of an observational, retrospective study. Briefly describe the study and summarize its conclusions.

40. **Meta-Analysis.** Search through recent newspapers or journals and find an example of a meta-analysis. Briefly describe the study and summarize its conclusions.

1.4   SHOULD YOU BELIEVE A STATISTICAL STUDY?

Much of the rest of this book is devoted to helping you build a deeper understanding of the concepts and definitions we've studied up to this point. But already you know enough to achieve one of the major goals of this text: being able to answer the question "Should you believe a statistical study?"

Most researchers conduct their statistical studies with honesty and integrity, and most statistical research is carried out with diligence and care. Nevertheless, statistical research is sufficiently complex that bias can arise in many different ways, making it very important that we always examine reports of statistical research carefully. There is no definitive way to answer the question "Should I believe a statistical study?" However, in this section we'll look at eight guidelines that can be helpful. Along the way, we'll also introduce a few more definitions and concepts that will prepare you for discussions to come later.

If I have ever made any valuable discoveries, it has been owing more to patient attention, than to any other talent.

—Isaac Newton

Eight Guidelines for Critically Evaluating a Statistical Study

1. *Get a Big Picture View of the Study.* For example, you should understand the goal of the study, the population that was under study, and whether the study was observational or an experiment.

2. *Consider the Source.* In particular, look for any potential biases on the part of the researchers.

3. *Look for Bias in the Sample.* That is, decide whether the sampling method was likely to produce a representative sample.

4. *Look for Problems Defining or Measuring the Variables of Interest.* Ambiguity in the variables can make it difficult to interpret reported results.

5. *Beware of Confounding Variables.* If the study neglected potential confounding variables, its results may not be valid.

6. *Consider the Setting and Wording in Surveys.* In particular, look for anything that might tend to produce inaccurate or dishonest responses.

7. *Check That Results Are Presented Fairly.* For example, check whether the study really supports the conclusions that are presented in the media.

8. *Stand Back and Consider the Conclusions.* For example, evaluate whether study achieved its goals. If so, do the conclusions make sense and have practical significance?

News reports do not always provide enough information for you to apply all eight guidelines, but you can usually find additional information on the Web. Look for clues such as "reported by NASA" or "published in the *New England Journal of Medicine*" to help you track down original sources or other relevant information.

Guideline 1: Get a Big Picture View of the Study

The first thing you should do when you hear about a statistical study is figure out what it was all about; that is, get a big picture view of the study that will allow you to consider its results in an appropriate context. A good starting point for gaining a big picture view comes in trying to answer these basic questions:

- What was the goal of the study?
- What was the population under study? Was the population clearly and appropriately defined?
- Was the study an observational study, an experiment, or a meta-analysis? If it was an observational study, was it retrospective? If it was an experiment, was it single- or double-blind, and were the treatment and control groups properly randomized? Given the goal, was the type of study appropriate?

EXAMPLE 1 Appropriate Type of Study?

Imagine the following (hypothetical) newspaper report: "Researchers gave 100 participants their individual astrological horoscopes and asked whether the horoscopes were accurate. 85% of the participants said their horoscopes were accurate. The researchers concluded that horoscopes are valid most of the time." Analyze this study according to Guideline 1.

SOLUTION   The goal of the study was to determine the validity of horoscopes. Based on the news report, it appears that the study was *observational*: The researchers simply asked the participants about the accuracy of the horoscopes. However, because the accuracy of a horoscope is somewhat subjective, the study goal would have been better served by a controlled experiment in which some people were given their actual horoscope and others were given a fake horoscope (which would serve as a placebo). Then the researchers could have looked for differences between the two groups. Moreover, because researchers could easily influence the results by how they questioned the participants, the experiment should have been double-blind. In summary, the type of study was inappropriate to the goal and its results are meaningless. $\cdots$•

> ### TIME ◯UT TO THINK
> Try your own test of horoscopes. Find yesterday's horoscope for each of the 12 signs and put each one on a separate piece of paper, without anything identifying the sign. Shuffle the pieces of paper randomly, and ask a few people to guess which one was supposed to be their personal horoscope. How many people choose the right one? Discuss your results.

EXAMPLE 2 Does Aspirin Prevent Heart Attacks?

A study reported in the *New England Journal of Medicine* (Vol. 318, No. 4) sought to determine whether aspirin is effective in preventing heart attacks. It involved 22,000 male physicians considered to be at risk for heart attacks. The men were divided into a treatment group that took aspirin and a control group that did not. The results were so convincing in favor of the benefits of aspirin that the experiment was stopped for ethical reasons before it was completed, and the subjects were informed of the results. Many news reports led with the headline that taking aspirin can help prevent heart attacks. Analyze this headline according to Guideline 1.

SOLUTION   The study was an experiment, which is appropriate, and its results appear convincing. However, the fact that the sample consisted only of men means that the results should be considered to apply only to the population of men. Because results of medical tests on men do not necessarily apply to women, the headlines misstated the results when they did not qualify the population. $\cdots$•

BY THE WAY

Surveys show that nearly half of Americans believe their horoscopes. However, in controlled experiments, the predictions of horoscopes come true no more often than would be expected by chance.

BY THE WAY

Many recent studies show substantial differences in the ways men and women respond to the same medical treatments. For example, aspirin is more effective at thinning blood in men than in women (thinning is thought to help prevent heart attacks in some people). Morphine controls pain better in women, but ibuprofen is more effective for men. And women reject heart transplants more often than men. These differences may stem from interactions with hormones that differ in men and women or from differences in the rates at which men and women metabolize different drugs.

Guideline 2: Consider the Source

Statistical studies are supposed to be objective, but the people who carry them out and fund them may be biased. It is therefore important to consider the source of a study and evaluate the potential for biases that might invalidate the study's conclusions.

Bias may be obvious in cases where a statistical study is carried out for marketing, promotional, or other commercial purposes. For example, a toothpaste advertisement that claims "4 out of 5 dentists prefer our brand" appears to be statistically based, but we are given no details about how the survey was conducted. Because the advertisers obviously want to say good things about their brand, it's difficult to take the statistical claim seriously without much more information about how the result was obtained.

Other cases of bias may be more subtle. For example, suppose that a carefully conducted study concludes that a new drug helps cure cancer. On the surface, the study might seem quite believable. But what if the study was funded by a drug company that stands to gain billions of dollars in sales if the drug is proven effective? The researchers may well have carried out their work with great integrity despite the source of funding, but it might be worth a bit of extra investigation to be sure.

Major statistical studies are usually evaluated by experts who are supposed to be unbiased. The most common process by which scientists examine each other's research is called **peer review** (because the scientists who do the evaluation are *peers* of those who conducted the research). Reputable scientific journals require all research reports to be peer reviewed before the research is accepted for publication. Peer review does not guarantee that a study is valid, but it lends credibility because it implies that other experts agree that the study was carried out properly.

> **Definition**
>
> **Peer review** is a process in which several experts in a field evaluate a research report before the report is published.

BY THE WAY

After decades of arguing to the contrary, in 1999 the Philip Morris Company—the world's largest seller of tobacco products—publicly acknowledged that smoking causes lung cancer, heart disease, emphysema, and other serious diseases. Shortly thereafter, Philip Morris changed its name to Altria.

EXAMPLE 3 Is Smoking Healthy?

By 1963, research had so clearly shown the health dangers of smoking that the Surgeon General of the United States publicly announced that smoking is bad for health. Research done since that time built further support for this claim. However, while the vast majority of studies showed that smoking is unhealthy, a few studies found no dangers from smoking and perhaps even health *benefits*. These studies generally were carried out by the Tobacco Research Institute, funded by the tobacco companies. Analyze these studies according to Guideline 2.

SOLUTION   Even in a case like this, it can be difficult to decide whom to believe. However, the studies showing smoking to be unhealthy came primarily from peer-reviewed research. In contrast, the studies carried out at the Tobacco Research Institute had a clear potential for bias. The *potential* for bias does not mean the research was biased, but the fact that it contradicts virtually all other research on the subject should be cause for concern.   · · ●

EXAMPLE 4 Press Conference Science

Suppose the nightly TV news shows scientists at a press conference announcing that they've discovered evidence that a newly developed chemical can stop the aging process. The work has not yet gone through the peer review process. Analyze this study according to Guideline 2.

SOLUTION   Scientists often announce the results of their research at a press conference so that the public may hear about their work as soon as possible. However, a great deal of expertise may be required to evaluate their study for possible biases or other errors—which is the goal of the peer review process. Until the work is peer reviewed and published in a reputable journal, any findings should be considered preliminary—especially about an astonishing claim such as being able to stop the aging process.   · · ●

Guideline 3: Look for Bias in the Sample

A statistical study cannot be valid unless the sample is representative of the population under study. Poor sampling methods almost guarantee a biased sample that makes the study results useless.

Biased samples can arise in many ways, but two closely related problems are particularly common. The first problem, called **selection bias** (or a **selection effect**), occurs whenever researchers *select* their sample in a way that tends to make it unrepresentative of the population. For example, a pre-election poll that surveys only registered Republicans has selection bias because it is unlikely to reflect the opinions of non-Republican voters.

The second problem, called **participation bias,** can arise when people *choose* to be part of a study—that is, when the participants are volunteers. The most common form of participation bias occurs in **self-selected surveys** (or **voluntary response surveys**)—surveys or polls in which people decide for themselves whether to participate. In such cases, people who feel strongly about an issue are more likely to participate, and their opinions may not represent the opinions of the larger population that has less emotional attachment to the issue.

> **Definitions**
>
> **Selection bias** (or a selection effect) occurs whenever researchers *select* their sample in a biased way.
>
> **Participation bias** occurs any time participation in a study is voluntary.
>
> A **self-selected survey** (or **voluntary response survey**) is one in which people decide for themselves whether to be included in the survey.

CASE STUDY   The 1936 Literary Digest Poll

The *Literary Digest*, a popular magazine of the 1930s, successfully predicted the outcomes of several elections using large polls. In 1936, editors of the *Literary Digest* conducted a particularly large poll in advance of the presidential election. They randomly chose a sample of 10 million people from various lists, including names in telephone books and rosters of country clubs. They mailed a postcard "ballot" to each of these 10 million people. About 2.4 million people returned the postcard ballots. Based on the returned ballots, the editors of the *Literary Digest* predicted that Alf Landon would win the presidency by a margin of 57% to 43% over Franklin Roosevelt. Instead, Roosevelt won with 62% of the popular vote. How did such a large survey go so wrong?

The sample suffered from both selection bias and participation bias. The selection bias arose because the *Literary Digest* chose its 10 million names in ways that favored affluent people. For example, selecting names from telephone books meant choosing only from those who could afford telephones back in 1936. Similarly, country club members are usually quite wealthy. The selection bias favored the Republican Landon because affluent voters of the 1930s tended to vote for Republican candidates.

The participation bias arose because return of the postcard ballots was voluntary, so people who felt strongly about the election were more likely to be among those who returned their ballots. This bias also tended to favor Landon because he was the challenger—people who did not like President Roosevelt could express their desire for change by returning the postcards. Together, the two forms of bias made the sample results useless, despite the large number of people surveyed.

EXAMPLE 5   Self-Selected Poll

The television show *Nightline* conducted a poll in which viewers were asked whether the United Nations headquarters should be kept in the United States. Viewers could respond to the poll by paying 50 cents to call a special phone number with their opinions. The poll drew

186,000 responses, of which 67% favored moving the United Nations out of the United States. Around the same time, a poll using simple random sampling of 500 people found that 72% wanted the United Nations to *stay* in the United States. Which poll is more likely to be representative of the general opinions of Americans?

SOLUTION   The *Nightline* sample was severely biased. It had selection bias because its sample was drawn only from the show's viewers, rather than from all Americans. The poll itself was a self-selected survey in which viewers not only chose whether to respond, but also had to *pay* 50 cents to participate. This cost made it even more likely that respondents would be those who felt a need for change. Despite its large number of respondents, the *Nightline* survey was therefore unlikely to give meaningful results. In contrast, the simple random sample of 500 people is quite likely to be representative, so the finding of this small survey has a better chance of representing the true opinions of all Americans. $\cdots$●

BY THE WAY

NASA's *Kepler* mission, launched in 2009, is an orbiting telescope that has made the first detections of planets as small as Earth around other stars. *Kepler* looks for slight dimming of a star's light each time an orbiting planet passes in front of it (a "transit"), which means it can detect planets only for the small fraction of stars that happen to have their planetary systems aligned with our line of sight.

EXAMPLE 6   Planets Around Other Stars

Until the mid-1990s, astronomers had never found conclusive evidence for planets outside our own solar system. But improving technology made it possible to begin finding such planets, and more than 2,000 had been discovered by mid-2012. The existing technology makes it easier to find large planets than small ones and easier to find planets that orbit close to their stars than planets that orbit far from their stars. According to the leading theory of solar system formation, large planets should form far from their stars, not close by. But large planets in close orbits are quite common among the planets discovered to date. Does this mean there is something wrong with the leading theory of solar system formation?

SOLUTION   Even if large planets in close orbits are relatively rare, the fact that current technology makes these rare cases the easiest ones to find introduces a *selection effect* that biases the sample (of discovered planets) toward that type. In fact, most astronomers think the existing solar system formation theory is still correct and that large planets in close-in orbits are not as common as they appear. Those that do exist can be explained by physical interactions that can cause large planets to migrate inward as a planetary system forms. $\cdots$●

Guideline 4: Look for Problems Defining or Measuring the Variables of Interest

Results of a statistical study may be difficult to interpret if the variables under study are difficult to define or measure. For example, imagine trying to conduct a study of how exercise affects resting heart rates. The variables of interest would be *amount of exercise* and *resting heart rate*. Both variables are difficult to define and measure. In the case of *amount of exercise*, it's not clear what the definition covers—does it include walking to class? Even if we specify the definition, how can we measure *amount of exercise* given that some forms of exercise are more vigorous than others?

> TIME ◖◗UT TO THINK
> What are the challenges of defining and measuring resting heart rate?

EXAMPLE 7   Can Money Buy Love?

A Roper poll reported in *USA Today* involved a survey of the wealthiest 1% of Americans. The survey found that these people would pay an average of $487,000 for *true love*, $407,000 for *great intellect*, $285,000 for *talent*, and $259,000 for *eternal youth*. Analyze this result according to Guideline 4.

SOLUTION   The variables in this study are very difficult to define. How, for example, do you define *true love*? And does it mean true love for a day, a lifetime, or something else?

Similarly, does the ability to balance a spoon on your nose constitute *talent*? Because the variables are so poorly defined, it's likely that different people interpreted them differently, making the results very difficult to interpret.   · · ●

EXAMPLE 8 Illegal Drug Supply

A commonly quoted statistic is that law enforcement authorities succeed in stopping only about 10% to 20% of the illegal drugs entering the United States. Should you believe this statistic?

SOLUTION   There are essentially two variables in a study of illegal drug interception: *quantity of illegal drugs intercepted* and *quantity of illegal drugs NOT intercepted*. It should be relatively easy to measure the quantity of illegal drugs that law enforcement officials intercept. However, because the drugs are illegal, it's unlikely that anyone is reporting the quantity of drugs that are *not* intercepted. How, then, can anyone know that the intercepted drugs are 10% to 20% of the total? In a *New York Times* analysis, a police officer was quoted as saying that his colleagues refer to this type of statistic as "PFA" for "pulled from the air."   · · ●

Guideline 5: Beware of Confounding Variables

Variables that are *not intended* to be part of a study can make it difficult to interpret results properly. As discussed in Section 1.3, it's not always easy to discover these *confounding variables*. Sometimes they are discovered only years after a study is completed, and other times they are not discovered at all, in which case a study's conclusion may be accepted even though it's not correct. Fortunately, confounding variables are sometimes more obvious and can be discovered simply by thinking hard about factors that may have influenced a study's results.

EXAMPLE 9 Radon and Lung Cancer

Radon is a radioactive gas produced by natural processes (the decay of uranium) in the ground. The gas can leach into buildings through the foundation and can accumulate to relatively high concentrations if doors and windows are closed. Imagine a (hypothetical) study that seeks to determine whether radon gas causes lung cancer by comparing the lung cancer rate in Colorado, where radon gas is fairly common, with the lung cancer rate in Hong Kong, where radon gas is less common. Suppose the study finds that the lung cancer rates are nearly the same. Would it be reasonable to conclude that radon is *not* a significant cause of lung cancer?

SOLUTION   The variables of interest are *amount of radon* (an explanatory variable in this case) and *lung cancer rate* (a response variable) However, radon gas is not the only possible cause of lung cancer. For example, smoking can cause lung cancer, so *smoking rate* may be a confounding variable in this study—especially because the smoking rate in Hong Kong is much higher than the smoking rate in Colorado. As a result, we cannot draw any conclusions about radon and lung cancer without taking the smoking rate into account (and perhaps other variables as well). In fact, careful studies have shown that radon gas *can* cause lung cancer, and the U.S. Environmental Protection Agency (EPA) recommends taking steps to prevent radon from building up indoors.   · · ●

Guideline 6: Consider the Setting and Wording in Surveys

Even when a survey is conducted with proper sampling and with clearly defined terms and questions, you should watch for problems in the setting or wording that might produce inaccurate or dishonest responses. Dishonest responses are particularly likely when the survey concerns sensitive subjects, such as personal habits or income. For example, the question "Do you cheat on your income taxes?" is unlikely to elicit honest answers from those who cheat, unless they are assured of complete confidentiality (and perhaps not even then).

In other cases, even honest answers may not really be accurate if the wording of questions invites bias. Sometimes just the order of words in a question can affect the outcome. A poll conducted in Germany asked the following two questions.

- Would you say that traffic contributes more or less to air pollution than industry?
- Would you say that industry contributes more or less to air pollution than traffic?

BY THE WAY

Many hardware stores sell simple kits that you can use to test whether radon gas is accumulating in your home. If it is, the problem can be eliminated by installing an appropriate "radon mitigation" system, which usually consists of a fan that blows the radon out from under the house before it can get into the house.

BY THE WAY

People are more likely to choose the item that comes first in a survey because of what psychologists call the *availability error*—the tendency to make judgments based on what is *available* in the mind. Professional polling organizations take great care to avoid this problem; for example, they may pose a question with two choices in one order to half the people in the sample and in the opposite order to the other half.

The only difference is the order of the words *traffic* and *industry*, but this difference dramatically changed the results: With the first question, 45% answered traffic and 32% answered industry. With the second question, only 24% answered traffic while 57% answered industry.

EXAMPLE 10 Do You Want a Tax Cut?

The Republican National Committee commissioned a poll to find out whether Americans supported their proposed tax cuts. Asked "Do you favor a tax cut?," a large majority of respondents answered *yes*. Should we conclude that Americans supported the proposal?

SOLUTION   A question like "Do you favor a tax cut?" is biased because it does not give other options or discuss any consequences. In fact, other polls conducted at the same time showed a similarly large majority expressing great concern about federal deficits. Indeed, support for the tax cuts was far lower when the question was asked by independent organizations in the form "Would you favor a tax cut even if it increased the federal deficit?

··●

EXAMPLE 11 Sensitive Survey

Two surveys asked Catholics in the Boston area whether contraceptives should be made available to unmarried women. The first survey involved in-person interviews, and 44% of the respondents answered *yes*. The second survey was conducted by mail and telephone, and 75% of the respondents answered *yes*. Which survey was more likely to be accurate?

SOLUTION   Contraceptives are a sensitive topic, particularly among Catholics (because the Catholic Church officially opposes contraceptives). The first survey, with in-person interviews, may have encouraged dishonest responses. The second survey made responses seem more private and therefore was more likely to reflect the respondents' true opinions.

··●

Guideline 7: Check That Results Are Presented Fairly

Even when a statistical study is done well, it may be misrepresented in graphics or concluding statements. Researchers occasionally misinterpret the results of their own studies or jump to conclusions that are not supported by the results, particularly when they have personal biases. News reporters may misinterpret a survey or jump to unwarranted conclusions that make a story seem more spectacular. Misleading graphics are especially common (we will devote much of Chapter 3 to this topic). You should always look for inconsistencies between the interpretation of a study (in pictures and in words) and any actual data given along with it.

EXAMPLE 12 Does the School Board Need a Statistics Lesson?

The school board in Boulder, Colorado, created a hubbub when it announced that 28% of Boulder school children were reading "below grade level" and hence concluded that methods of teaching reading needed to be changed. The announcement was based on reading tests on which 28% of Boulder school children scored below the national average for their grade. Do these data support the board's conclusion?

SOLUTION   The fact that 28% of Boulder children scored below the national average for their grade implies that 72% scored at or above the national average. Thus, the school board's ominous statement about students reading "below grade level" makes sense only if "grade level" means the national average score for a particular grade. This interpretation of "grade level" is curious because it would imply that half the students in the nation are always below grade level—no matter how high the scores. It may still be the case that teaching methods needed to be improved, but these data did not justify that conclusion.

··●

Guideline 8: Stand Back and Consider the Conclusions

Finally, even if a study seems reasonable according to all the previous guidelines, you should stand back and consider the conclusions. Ask yourself questions such at these:

- Did the study achieve its goals?
- Do the conclusions make sense?
- Can you rule out alternative explanations for the results?
- If the conclusions make sense, do they have any practical significance?

EXAMPLE 13 Extraordinary Claims

A recent study by a respected psychologist, Daryl J. Bem (Cornell University) claimed to find evidence for the existence of extrasensory perception, or ESP. Dr. Bem's results showed relatively small effects; for example, in one experiment in which subjects were asked to identify a picture hidden behind a screen, 53% gave the right answer versus the 50% expected by pure chance. Nevertheless, Dr. Bem claimed that the results were statistically significant, and his results were peer reviewed and published in a respected scholarly journal. Should you conclude that ESP really exists?

Extraordinary claims require extraordinary evidence.

—Carl Sagan

SOLUTION Although the study may well have been done carefully, the claim of ESP should be considered what scientists often call an "extraordinary claim," because decades of study have never previously turned up any indisputable evidence for ESP and its existence would seem to violate some well-established laws of physics. Moreover, Dr. Bem's claim that the results were statistically significant has been disputed by other scientists and statisticians, who argue that he used the wrong type of analysis and that a proper analysis shows the results were consistent with pure chance. Clearly, for a claim as extraordinary as one of ESP, we would need far stronger evidence before concluding that it actually exists.   ··•

EXAMPLE 14 Practical Significance

An experiment is conducted in which the weight losses of people who try a new "Fast Diet Supplement" are compared to the weight losses of a control group of people who try to lose weight in other ways. After eight weeks, the results show that the treatment group lost an average of one-half pound more than the control group. Assuming that it has no dangerous side effects, does this study suggest that the Fast Diet Supplement is a good treatment for people wanting to lose weight?

SOLUTION Compared to the average person's body weight, a weight loss of one-half pound hardly matters at all. So while loss results may be interesting, they don't seem to have much practical significance.   ··•

Section 1.4 Exercises

Statistical Literacy and Critical Thinking

1. **Peer Review,** What is peer review? How is it useful?

2. **Selection Bias and Participation Bias,** Describe and contrast selection bias and participation bias in sampling.

3. **Self-Selected Surveys.** Why are self-selected surveys almost always prone to participation bias?

4. **Confounding Variables.** What are confounding variables, and what problems can they cause?

Does It Make Sense? For Exercises 5–8, decide whether the statement makes sense (or is clearly true) or does not make sense (or is clearly false). Explain clearly; not all of these have definitive answers, so your explanation is more important than your chosen answer.

5. **Large Survey.** A survey involving a larger sample of subjects is always better than one involving a smaller sample.

6. **Survey Location.** The survey of the use of credit among adult Americans suffered from selection bias because the questionnaires were handed out only on college campuses.

7. **Vitamin C and Colds.** My experiment proved that vitamin C can reduce the

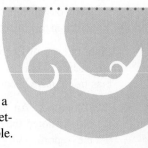

severity of colds, because I controlled the experiment carefully for every possible confounding variable.

8. **Diet Effectiveness.** The Simon diet is effective because it was used by a sample of 1,000 subjects and there was a mean weight loss of 1.7 pounds during a six-month study.

Concepts and Applications

Applying Guidelines. In Exercises 9–16, determine which of the eight guidelines appears to be most relevant. Explain your reasoning.

9. **Hygiene.** The Winslow Supply Company manufactures deodorants and sponsored a survey showing that good personal hygiene is critically important for success in a job interview.

10. **Smoking.** A clinical trial involved the use of a nicotine gum as an aid to help smokers stop smoking. The clinical trial involved 1,000 college students who were paid for their participation, and the results showed that the nicotine gum treatment was highly successful. The researchers concluded that the use of their nicotine gum would be a successful treatment for the population of all smokers.

11. **Goodness.** In a study of 1,200 college students, each was asked whether he or she was a good person.

12. **Agriculture.** Researchers conclude that an irrigation system used to grow tomatoes in California is more effective than a competing system used in Arizona.

13. **New York City Subway Survey.** The New York City Transit Authority routinely conducts a survey of satisfaction by distributing surveys on subways. The passengers take the surveys home and return them by mail.

14. **Election Poll.** Under the headline "Turner predicted to win in a landslide," it was reported that 54% of voters in a pre-election poll prefer Turner, compared with 46% for her opponent.

15. **Nuclear Energy Poll.** Randomly selected adults were asked: "Do you agree or disagree with increasing the production of nuclear energy that could potentially kill thousands of innocent people?"

16. **Counterfeit Goods.** A consortium of manufacturers plans a study designed to compare the value of counterfeit goods produced in the United States in the year 2000 to the current year.

Bias. In each of Exercises 17–20, identify and explain at least one source of bias in the study described. Then suggest how the bias might have been avoided.

17. **Chocolate.** An article in *Journal of Nutrition* (Vol. 130, No. 8) noted that chocolate is rich in flavonoids. The article reports that "regular consumption of foods rich in flavonoids may reduce the risk of coronary heart disease." The study received funding from Mars, Inc., the candy company, and the Chocolate Manufacturers Association.

18. **Famous Book.** When author Shere Hite wrote *Woman and Love: A Cultural Revolution in Progress*, she based conclusions about the general population of all women on 4,500 replies that she received after mailing 100,000 questionnaires to various women's groups.

19. **Political Polling.** You receive a call in which the caller claims to be conducting a national opinion research poll. You are asked if your opinion about congressional candidate John Sweeney would change if you knew that Sweeney once had a car crash while driving under the influence of alcohol.

20. **Survey Method.** You conduct a survey to find the percentage of people in your state who can name the lieutenant governor, who plans to run for the United States Senate. You obtain addresses from a list of property owners in the state and you mail a survey to 850 randomly selected people from the list.

21. **It's All in the Wording.** Princeton Survey Research Associates did a study for *Newsweek* magazine illustrating the effects of wording in a survey. Two questions were asked:

• Do you personally believe that abortion is wrong?

• Whatever your own personal view of abortion, do you favor or oppose a woman in this country having the choice to have an abortion with the advice of her doctor?

To the first question, 57% of the respondents replied yes, while 36% responded no. In response to the second question, 69% of the respondents favored the choice, while 24% opposed the choice. Discuss why the two questions produced seemingly contradictory results. How could the results of the questions be used selectively by various groups?

22. **Tax or Spend?** A Gallup poll asked the following two questions:

• Do you favor a tax cut or "increased spending on other government programs"? *Result:* 75% for the tax cut.

• Do you favor a tax cut or "spending to fund new retirement savings accounts, as well as increased spending on education, defense, Medicare and other programs"? *Result:* 60% for the spending.

Discuss why the two questions produced seemingly contradictory results. How could the results of the questions be used selectively by various groups?

What Do You Want to Know? Exercises 23–26 pose two related questions that might form the basis of a statistical study. Briefly discuss how the two questions differ and how these differences would affect the goal of a study and the design of the study.

23. **Internet Dating**

First question: What percentage of Internet dates lead to marriage?

Second question: What percentage of marriages begin with Internet dates?

24. **Full-Time Faculty**

First question: What percentage of introductory classes on campus are taught by full-time faculty members?

Second question: What percentage of full-time faculty members teach introductory classes?

25. Binge Drinking

First question: How often do college students do binge drinking?

Second question: How often is binge drinking done by college students?

26. Statistics Courses

First question: What is the proportion of college graduates who have taken a statistics course?

Second question: What is the proportion of statistics courses taken by college students?

Accurate Headlines? Exercises 27 and 28 give a headline and a brief description of the statistical news story that accompanied the headline. In each case, discuss whether the headline accurately represents the story.

27. Headline: "Drugs shown in 98 percent of movies"

Story summary: A "government study" claims that drug use, drinking, or smoking was depicted in 98% of the top movie rentals (Associated Press).

28. Headline: "Sex more important than jobs"

Story summary: A survey found that 82% of 500 people interviewed by phone ranked a satisfying sex life as important or very important, while 79% ranked job satisfaction as important or very important (Associated Press).

Stat-Bytes. Politicians commonly believe that they must make their political statements (often called sound-bytes) very short because the attention span of listeners is so short. A similar effect occurs in reporting statistical news. Major statistical studies are often reduced to one or two sentences. The summaries of statistical reports in Exercises 29–32 are taken from various news sources. Describe what crucial information is missing in the given statement and what more you would want to know before you acted on the report.

29. Confidence in Military. *USA Today* reports on a Harris poll claiming that the percentage of adults with a "great deal of confidence" in military leaders stands at 54%.

30. Top Restaurants. CNN reports on a Zagat Survey of America's Top Restaurants that found that "only nine restaurants achieved a rare 29 out of a possible 30 rating and none of those restaurants are in the Big Apple."

31. Forecasting Weather. A *USA Today* headline reported that "More companies try to bet on forecasting weather." The article gave examples of companies believing that long-range forecasts are reliable, and four companies were cited.

32. Births in China. A *USA Today* headline reported that "China thrown off balance as boys outnumber girls," and an accompanying graph showed that for every 100 girls born in China, 116.9 boys are born.

PROJECTS FOR THE INTERNET & BEYOND

33. Analyzing a Statistical Study. Find a detailed report on some recent statistical study of interest to you. Write a short report applying each of the eight guidelines given in this section. (Some of the guidelines may not apply to the particular study you are analyzing; in that case, explain why the guideline is not applicable.)

34. Twin Studies. Researchers doing statistical studies in biology, psychology, and sociology are grateful for the existence of twins. Twins can be used to study whether certain traits are inherited from parents (nature) or acquired from one's surroundings during upbringing (nurture). Identical twins are formed from the same egg in the mother and have the same genetic material. Fraternal twins are formed from two separate eggs and share roughly half of the same genetic material. Find a published report of a twin study. Discuss how identical and fraternal twins are used to form case and control groups. Apply Guidelines 1–8 to the study and comment on whether you find the conclusions of the report convincing.

35. Professional Journals. Consult an issue of a professional journal. Select one specific article and use the ideas of this section to summarize and evaluate the study.

IN THE NEWS

36. Applying the Guidelines. Find a recent newspaper article or television report about a statistical study on a topic that you find interesting. Write a short report applying each of the eight guidelines given in this section. (Some of the guidelines may not apply to the particular study you are analyzing; in that case, explain why the guideline is not applicable.)

37. Believable Results. Find a recent news report about a statistical study whose results you believe are meaningful and important. In one page or less, summarize the study and explain why you find it believable.

38. Unbelievable Results. Find a recent news report about a statistical study whose results you *don't* believe are meaningful and important. In one page or less, summarize the study and explain why you don't believe its claims.

CHAPTER REVIEW EXERCISES

1. **Tats.** A Harris poll surveyed 2,320 adults in the United States, among which 14% said that they have at least one tattoo. The margin of error is 2 percentage points.

   a. Interpret the margin of error by identifying the range of values likely to contain the percentage of adults with tattoos.

   b. Identify the population.

   c. Is this study an experiment or an observational study? Explain. Identify the variable of interest.

   d. Is the reported value of 14% a population parameter or a sample statistic? Why?

   e. If you learned that survey subjects responded to a magazine article asking readers to phone in their responses, would you consider the survey results to be valid? Why or why not?

   f. Describe a procedure for selecting the survey subjects using a simple random sample.

   g. Describe a procedure for selecting similar survey subjects using stratified sampling.

   h. Describe a procedure for selecting similar survey subjects using cluster sampling.

   i. Describe a procedure for selecting similar survey subjects using systematic sampling.

   j. Describe a procedure for selecting similar survey subjects using a convenience sample.

2. **Simple Random Sample.** An important element of this chapter is the concept of a simple random sample.

   a. What is a simple random sample?

   b. When the Bureau of Labor Statistics conducts a survey, it begins by partitioning the United States adult population into 2,007 groups called *primary sampling units*. Assume that these primary sampling units all contain the same number of adults. If you randomly select one adult from each primary sampling unit, is the result a simple random sample? Why or why not?

   c. Refer to the primary sampling units described in part b and describe a sampling plan that results in a simple random sample.

3. **Clinical Trial of Bystolic.** In clinical trials of the drug Bystolic used to treat hypertension (high blood pressure), 677 Bystolic users were observed for adverse reactions. It was found that among those treated with Bystolic, 7% experienced headaches.

   a. Based on the given information, can you conclude that in some cases, Bystolic causes headaches? Why or why not?

   b. While 677 subjects were treated with Bystolic, another 205 subjects were given a placebo, and 6% of the placebo group experienced headaches. What does this additional information suggest about headaches as an adverse reaction to the use of Bystolic?

   c. Is this clinical trial an observational study or an experiment? Explain.

   d. In this clinical trial, what is blinding and why is it important in testing the effects of Bystolic?

   e. What is an experimenter effect, and how might this effect be minimized?

4. **Wording of a Survey Question.** In *The Superpollsters*, David W. Moore describes an experiment in which different subjects were asked if they agree with the following statements:

   i. Too little money is being spent on welfare.

   ii. Too little money is being spent on assistance to the poor.

   Even though it is the poor who receive welfare, only 19% agreed when the word "welfare" was used, but 63% agreed with "assistance to the poor."

   a. Which of the two questions should be used in a survey? Why?

   b. If you are working on a campaign for a conservative candidate for Congress, and you want to emphasize opposition to the use of federal funds for assistance to the poor, which of the two questions would you use? Why?

   c. Is it ethical to deliberately word a survey question so that it influences responses? Why or why not?

CHAPTER QUIZ

Choose the best answer to each of the following questions. Explain your reasoning with one or more complete sentences.

1. You conduct a poll in which you randomly select 1,200 college students in California and ask if they have taken an online course. The *population* for this study is: (a) All students who have taken an online course; (b) the 1200 college students that you interview; (c) all college students in California.

2. For the poll described in Exercise 1, which sampling plan would likely yield results that are *most* biased: (a) Mail the survey to college students in California and use the returned responses; (b) randomly select 20 colleges in California, then randomly select 60 students at each college; (c) obtain a numbered list of all college students in California and select every 100th name until a sample size of 1,200 is obtained.

3. When we say that a sample is *representative* of the population, we mean that: (a) the results found for the sample are similar to those we would find for the entire population; (b) the sample is very large; (c) the sample was chosen in the best possible way.

4. Consider an experiment designed to see whether cash incentives can improve school attendance. The researcher chooses two groups of 100 high school students: She offers one group $10 for every week of perfect attendance. She tells the other group that they are part of an experiment but does not give them any incentive. The students who do not receive an incentive represent: (a) the treatment group; (b) the control group; (c) the observation group.

5. The experiment described in Exercise 4 is: (a) single-blind; (b) double-blind; (c) not blind.

6. The purpose of a *placebo* is: (a) to prevent participants from knowing whether they belong to the treatment group or the control group; (b) to distinguish between the cases and the controls in a case-control study; (c) to determine whether diseases can be cured without any treatment.

7. If we see a *placebo effect* in an experiment to test a new treatment designed to cure warts, it means: (a) the experiment was not properly double-blind; (b) the experimental groups were too small; (c) warts were cured among members of the control group.

8. An experiment is single-blind if: (a) it lacks a treatment group; (b) it lacks a control group; (c) the participants do not know whether they belong to the treatment or control group.

9. Poll X predicts that Powell will receive 49% of the vote, while Poll Y predicts that she will receive 53% of the vote. Both polls have a margin of error of 3 percentage points. What can you conclude? (a) one of the two polls must have been conducted poorly; (b) the two polls are consistent with one another; (c) Powell will receive 51% of the vote.

10. A survey reveals that 24% of adults believe that the most fun way to flirt is through instant messages. The margin of error is 3 percentage points. The confidence interval for this poll is: (a) from 18% to 30%; (b) from 24% to 27%; (c) from 21% to 27%.

11. A study conducted by the oil company Exxon Mobil shows that there was no lasting damage from a large oil spill in Alaska. This conclusion: (a) is definitely invalid, because the study was biased; (b) may be correct, but the potential for bias means you should look very closely at how the conclusion was reached; (c) could be correct if it falls within the confidence interval of the study.

12. The television show *American Idol* selects winners from votes cast by anyone who wants to vote. This means the winner: (a) is the person most Americans want to win; (b) may or may not be the person most American want to win, because the voting is subject to participation bias; (c) may or may not be the person most Americans want to win, because the voting should have been double-blind.

13. Consider an experiment in which you measure the weights of randomly selected cars. The variable of interest in this study is: (a) the size of the sample; (b) the weights of the cars; (c) the average (mean) weight of all cars.

14. Imagine a survey of randomly selected people in which it is found that people who use sunscreen were *more* likely to have been sunburned in the past year. Which explanation for this result seems most likely? (a) sunscreen is useless; (b) the people in this study all used sunscreen that had passed its expiration date; (c) people who use sunscreen are more likely to spend time in the sun.

15. If a statistical study is carefully conducted in every possible way, then: (a) its results must be correct; (b) we can have confidence in its results, but it is still possible that the results are not correct; (c) we say that the study is perfectly biased.

F CUS ON

PSYCHOLOGY

Use this and other "Focus" sections found at the end of each chapter to focus in on particular topics of interest in statistics.

Are You Driving "Drunk" on Your Cell Phone?

One of the hottest topics in statistics deals with how cell phones and other types of distraction affect the abilities and reactions of drivers, and whether these distractions lead to more accidents and fatalities than would occur otherwise. These questions are studied by scientists in a variety of disciplines but are especially important to psychologists, because the answers depend on how the human brain reacts to different types of stimuli.

Many different types of statistical study have looked at the issue. Some researchers study accident reports to see what fraction of accidents involve talking on a cell phone, texting, or other distractions. Some have looked at cell phone records of people who have been in crashes, to see if the phone was in use at the time of the crash. These studies clearly indicate higher crash and fatality rates associated with distracted driving. Perhaps most notably, the University of Utah's Applied Cognition Lab has conduced a series of studies in which subjects use simulators to drive cars under a variety of distraction conditions, including use of a hand-held cell phone, use of a hands-free cell phone, and while texting. They compare these results with those in which the same subjects are not distracted and are either sober or intoxicated. Their astonishing conclusion: Even with hands-free devices, talking on a cell phone makes drivers as dangerous as drunk drivers, and texting and other distractions can make drivers even more likely to cause a crash.

Should we believe this claim that using your cell phone essentially makes you a drunk driver? As always, there are many ways to evaluate the claim, but for practice, let's use the eight guidelines given in Section 1.4 (p. 30).

Guideline 1: *Get a Big Picture View of the Study.* The goal of the Utah studies is to learn about the relative danger of different types of distraction, especially of cell phones, and the population under study is all drivers. The study is an experiment, because it has the same subjects drive simulators under different conditions of distraction, with the control being the results when they are undistracted and sober.

Guideline 2: *Consider the Source.* We can generally assume that a university lab operates independently, and we have no reason to suspect that any pressure is put on researchers to come up with a particular result. Indeed, if there were pressure, it would likely come from cell phone companies that would have a bias toward *not* finding danger in their products,

which is the opposite of what the researchers concluded. We similarly have no reason to suspect bias on the part of the researchers themselves. It's almost impossible to be sure that no bias is involved, but in this case the source seems likely to be trustworthy.

Guideline 3: *Look for Bias in the Sample.* For the results to be valid, the group of subjects must be a representative sample of all drivers. Media reports of the Utah studies rarely talk about the sample selection method, but in this case it does not seem like it should be difficult to get a representative sample, since it's unlikely that any particular group of people would be especially prone or immune to distractions. So unless we see evidence to the contrary, it seems safe to assume that the sample was well chosen.

Guideline 4: *Look for Problems Defining or Measuring the Variables of Interest.* The Utah researchers study variables such as the reaction times of drivers in surprising situations and how often they get into crashes in their simulated drives. These variables are straightforward to define and measure.

Guideline 5: *Beware of Confounding Variables.* It's almost impossible to eliminate all possible confounding variables, and we can easily think of some variables that might affect these studies. For example, the driver responses might depend on the order and time at which different tests are conducted, as a driver may be fatigued as tests progress or might not be feeling well on a particular day. However, the researchers can in principle avoid most of these problems by changing the order of the tests for different drivers on different days and by having a large enough test group so that things like minor illnesses should not affect the overall results.

Guideline 6: *Consider the Setting and Wording in Surveys.* The Utah studies are not survey based, so this guideline does not apply. However, studies such as those that look at real crash rates are essentially surveys based on crash reports compiled by police or insurance companies, so it's worth considering potential biases in them. For example, could it be that the role of cell phones in crashes might be either overreported or underreported? A little thought shows the former is unlikely, because no one would be expected to tell police they were on a cell phone if they weren't. But the opposite seems quite likely, because denying cell phone use might be a way of deflecting potential blame for a crash. We conclude that if the surveys are biased at all, it is most likely toward *under*estimating the danger of distracted driving.

Guideline 7: *Check That Results Are Presented Fairly.* The Utah studies have found clear evidence of the danger of cell phone use, with or without hand-free devices. The fact that these results are consistent with actual data from accidents gives further reason to believe that the results have been presented fairly.

Guideline 8: *Stand Back and Consider the Conclusions.* This last step is essentially our summary. We've found that the Utah studies have achieved their goals and lead to clear conclusions about the danger of using a cell phone while driving. In fact, the studies also show that similar or greater danger arises from texting while driving, programming your GPS while driving, finding music on your iPod while driving, and a variety of other distractions. Do these results have practical significance? They should, as they make it very clear that if you talk on your cell phone or are otherwise distracted while driving, you are putting your own life and the lives of others at risk.

QUESTIONS FOR DISCUSSION

1. Do *you* ever drive while talking on your cell phone, texting, or programming your GPS? If so, have you ever noticed evidence of your distraction, such as missing a turn or an especially close call for a collision? Have you seen such evidence of distraction when being driven by friends or family or taxi drivers?

2. Public safety advocates and many insurance companies are using the data about distracted driving to argue in favor of laws banning the use of cell phones while driving. Do you support or oppose such laws? Defend your opinion.

3. The fact that the studies found essentially no difference in danger between hand-held and hands-free devices comes as a surprise to most people. Research studies on brain activity that have provided an explanation for this surprising fact, often called "inattention blindness." Discuss how and why talking on a hands-free cell phone turns out to increase the risk of crashes, while talking to a passenger does not.

4. Find the latest statistics from the U.S. Department of Transportation concerning the number of crashes and fatalities estimated to be caused by distracted driving each year. Do you think the estimates are likely to be accurate? Why or why not?

F★CUS ON

PUBLIC HEALTH

Is Your Lifestyle Healthy?

Consider the following findings from statistical studies:

- Smoking increases the risk of heart disease.
- Eating margarine can increase the risk of heart disease.
- One glass of wine per day can protect against heart disease but increases the risk of breast cancer.
- Potato chips and sugary sodas are the foods most strongly associated with weight gain.

You are probably familiar with some of these findings, and perhaps you've even altered your lifestyle as a result of them. But where do they come from? Remarkably, these and hundreds of other important findings on public health come from huge prospective studies that have provided data for hundreds of smaller statistical studies. The longest-running of these is the Harvard Nurses' Health Study, which began in 1976 when Dr. Frank E. Speizer decided to study the long-term effects of oral contraceptives. He mailed questionnaires to approximately 370,000 registered nurses and received more than 120,000 responses. He chose to survey nurses because he believed that their medical training would make their responses more reliable than those of the general public.

As Dr. Speizer and his colleagues sifted through the data in the returned questionnaires, they realized that the study could be expanded to include more than just the effects of contraceptives. Today, this research team continues to follow many of the original 120,000 respondents.

Annual questionnaires are still a vital part of the study, allowing researchers to gather data about what the nurses eat; what medicines and vitamins they take; whether and how much they exercise, drink, and smoke; and what illnesses they have contracted. Some of the nurses also provide blood samples, which are used to measure such things as cholesterol level, hormone levels, genetic variations, and residues from pesticides and environmental pollutants. Dr. Speizer's faith in nurses has proven justified, as they reliably complete surveys and almost always provide properly drawn and labeled blood samples upon request.

After more than three decades of correspondence, both the researchers and the nurses say they feel a sense of closeness. Many of the nurses look forward to hearing from the researchers and say that the study has helped them to pay more attention to how they live their lives. Today, as the original nurses become elderly, the study is beginning to turn out results that should shed light on factors that influence longevity and health in old age.

The success of the Harvard Nurses Study has spurred its expansion and many similar studies of large groups. When you see statistical reports based on these studies, remember the hundreds of thousands of people whose willingness to participate in these studies is making life better for everyone.

QUESTIONS FOR DISCUSSION

1. Consider some of the results that are likely to come from the Harvard Nurses' Health Study over the next 10 to 20 years. What types of results do you think will be most important? Do you think the findings will alter the way you live your life?

2. Explain why the Harvard Nurses' Health Study is an observational study. Critics sometimes say that the results would be more valid if obtained by experiments rather than observations. Discuss whether it would be possible to gather similar data by carrying out experiments in a practical and ethical way.

3. In principle, the Harvard Nurses' Health Study is subject to participation bias because only 120,000 of the original 370,000 questionnaires were returned. Should the researchers be concerned about this bias? Why or why not?

4. Another potential pitfall comes from the fact that the questionnaires often deal with sensitive issues of personal health, and researchers have no way to confirm that the nurses answer honestly. Do you think that dishonesty could be leading researchers to incorrect conclusions? Defend your opinion.

5. All of the participants in the Harvard Nurses' Health Study were women. Do you think that the results also are of use to men? Why or why not?

6. Do a Web search for news articles that discuss results from the Harvard Nurses' Health Study or other similar studies. Choose one recent result that interests you, and discuss what it means and how it may affect public health or your own health in the future.

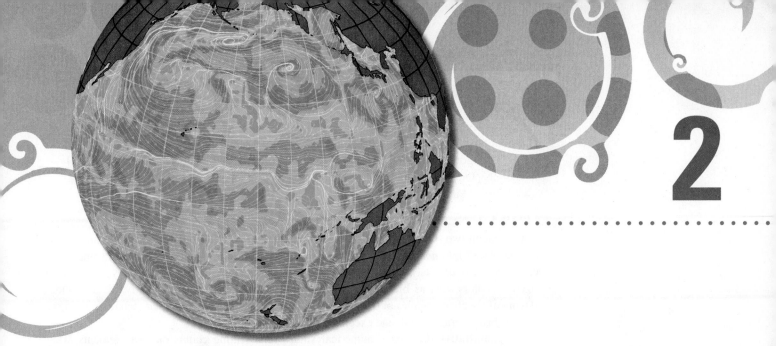

Measurement in Statistics

We all know how to measure quantities such as height, weight, and temperature. However, in statistical studies there are many other kinds of measurements, and we must be sure that they are defined, obtained, and reported carefully. In this chapter, we will discuss a few important concepts associated with measurements and statistics. As you'll see, these concepts are very useful to understanding the statistical reports you encounter in your daily life.

Practically no one knows what they're talking about when it comes to numbers in the newspapers. And that's because we're always quoting other people who don't know what they're talking about, like politicians and stock-market analysts.

—Molly Ivins (1944–2007), newspaper columnist

LEARNING GOALS

2.1 Data Types and Levels of Measurement

Identify data as qualitative or quantitative, discrete or continuous, and by level of measurement (nominal, ordinal, interval, or ratio).

2.2 Dealing with Errors

Understand the difference between random and systematic errors, describe errors by their absolute and relative sizes, and know the difference between accuracy and precision in measurements.

2.3 Uses of Percentages in Statistics

Understand how percentages are used to report statistical results and recognize ways in which they are sometimes misused.

2.4 Index Numbers

Understand the concept of an index number and the use of the Consumer Price Index (CPI).

FOCUS TOPICS

2.1 DATA TYPES AND LEVELS OF MEASUREMENT

One of the challenges in statistics is deciding how best to summarize and display data. Different types of data call for different types of summaries. In this section, we'll discuss how data are categorized, an idea that will help us when we consider data summaries and displays in later chapters.

Data Types

Data come in two basic types: qualitative and quantitative. **Qualitative data** have values that can be placed into *nonnumerical categories.* (For this reason, qualitative data are sometimes called *categorical* data.) For example, eye color data are qualitative because they are categorized by colors such as blue, brown, and hazel. Other examples of qualitative data include flavors of ice cream, names of employers, genders of animals, and movie or restaurant ratings, such as bad, average, good, and excellent.

    Quantitative data have numerical values representing counts or measurements. The times of runners in a race, the incomes of college graduates, and the numbers of students in different classes are all examples of quantitative data.

Data Types

Qualitative (or **categorical**) **data** consist of values that can be placed into nonnumerical categories.

Quantitative data consist of values representing counts or measurements.

EXAMPLE ① Data Types

Classify each of the following sets of data as qualitative or quantitative.

a. Brand names of shoes in a consumer survey

b. Scores on a multiple-choice exam

c. Letter grades on an essay assignment

d. Numbers on uniforms that identify players on a basketball team

SOLUTION

a. Brand names are categories and therefore represent qualitative data.

b. Scores on a multiple-choice exam are quantitative because they are counts of the number of correct answers.

c. Letter grades on an essay assignment are qualitative because they represent different categories of performance (failing through excellent).

d. The players' uniform numbers are qualitative because they do not represent a count or measurement; they are used solely for identification. You can tell that these numbers are qualitative rather than quantitative because you could not use them for computations. For example, it would make no sense to add or subtract the uniform numbers of different players.
        ·· ●

Discrete versus Continuous Data

Quantitative data can be further classified as continuous or discrete. Data are **continuous** if they can take on *any* value in a given interval. For example, a person's weight can be anything between 0 and a few hundred pounds, so data that consist of weights are continuous. Data are **discrete** if they can take on only particular values and not other values in between. For example, the number of students in your class is discrete because it must be a whole number, and shoe sizes are discrete because they take on only integer and half-integer values such as 7, $7\frac{1}{2}$, 8, and $8\frac{1}{2}$. (Actual foot lengths are continuous, but such shoe sizes are discrete.)

Discrete versus Continuous Data

Continuous data can take on *any* value in a given interval.

Discrete data can take on only particular, distinct values and not other values in between.

EXAMPLE 2   Discrete or Continuous?

For each data set, indicate whether the data are discrete or continuous.

a. Measurements of the time it takes to walk a mile

b. The numbers of calendar years (such as 2013, 2014, 2015)

c. The numbers of dairy cows on different farms

d. The amounts of milk produced by dairy cows on a farm

SOLUTION

a. Time can take on any value, so measurements of time are continuous.

b. The numbers of calendar years are discrete because they cannot have fractional values. For example, on New Year's Eve of 2016, the year will change from 2016 to 2017; we'll never say the year is $2016\frac{1}{2}$.

c. Each farm has a whole number of cows that we can count, so these data are discrete. (You cannot have fractional cows, for example.)

d. The amount of milk that a cow produces can take on any value in some range, so the milk production data are continuous.   ·· ●

Levels of Measurement

Another way to classify data is by their *level of measurement*. The simplest level of measurement applies to variables such as eye color, ice cream flavors, or gender of animals. These variables can be described solely by names, labels, or categories. We say that such data are at a **nominal level of measurement**. (The word *nominal* refers to *names* for categories.) The nominal level of measurement does not involve any ranking or ordering of the data. For example, we could not say that blue eyes come before brown eyes or that vanilla ranks higher than chocolate.

When we describe data with a ranking or ordering scheme, such as star ratings of movies or restaurants, we are using an **ordinal level of measurement**. (The word *ordinal* refers to *order*.) Such data generally cannot be used in any meaningful way for computations. For example, it doesn't make sense to add star ratings—watching three one-star movies is not equivalent to watching one three-star movie.

TIME OUT TO THINK

Consider a survey that asks "What's your favorite flavor of ice cream?" We've said that ice cream flavors represent data at the nominal level of measurement. But suppose that, for convenience, the researchers enter the survey data into a computer by assigning numbers to the different flavors. For example, they assign 1 = vanilla, 2 = chocolate, 3 = cookies and cream, 4 = cherry garcia, and so on. Does this change the ice cream flavor data from nominal to ordinal? Why or why not?

The ordinal level of measurement provides a ranking system, but it does not allow us to determine precise differences between measurements. For example, there is no way to determine the exact difference between a three-star movie and a two-star movie. In contrast, a temperature of 81°F is hotter than 80°F by the same amount that 28°F is hotter than 27°F. Temperature data are at a higher level of measurement, because the *intervals* (differences) between units on a temperature scale always mean the same definite amount. However, while intervals (which involve subtraction) between Fahrenheit temperatures are meaningful, *ratios* (which involve division) are not. For example, it is *not true* that 20°F is twice as hot as 10°F or that −40°F is twice as cold as −20°F. The reason ratios are meaningless on the Fahrenheit scale is that its *zero point is arbitrary* and does not represent a state of "no heat." If intervals are meaningful but ratios are not, as is the case with Fahrenheit temperatures, we say that the data are at the **interval level of measurement**.

When both intervals and ratios are meaningful, we say that data are at the **ratio level of measurement**. For example, data consisting of distances are at the ratio level of measurement because a distance of 10 kilometers really is twice as far as a distance of 5 kilometers. In general, the ratio level of measurement applies to any scale with a *true zero*, which is a value that means *none* of whatever is being measured. In the case of distances, a distance of zero means "no distance." Other examples of data at the ratio level of measurement include weights, speeds, and incomes.

Note that data at the nominal or ordinal level of measurement are always qualitative, while data at the interval or ratio level are always quantitative (and can therefore be either continuous or discrete). Figure 2.1 summarizes the possible data types and levels of measurement.

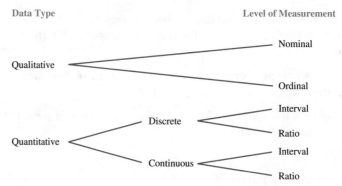

Figure 2.1 Data types and levels of measurement.

Levels of Measurement

The **nominal level of measurement** is characterized by data that consist of names, labels, or categories only. The data are qualitative and cannot be ranked or ordered.

The **ordinal level of measurement** applies to qualitative data that can be arranged in some order (such as low to high). It generally does not make sense to do computations with data at the ordinal level of measurement.

The **interval level of measurement** applies to quantitative data in which intervals are meaningful, but ratios are not. Data at this level have an arbitrary zero point.

The **ratio level of measurement** applies to quantitative data in which both intervals and ratios are meaningful. Data at this level have a true zero point.

EXAMPLE ③ Levels of Measurement

Identify the level of measurement (nominal, ordinal, interval, ratio) for each of the following sets of data.

a. Numbers on uniforms that identify players on a basketball team

b. Student rankings of cafeteria food as excellent, good, fair, or poor

c. Calendar years of historic events, such as 1776, 1945, or 2001

d. Temperatures on the Celsius scale

e. Runners' times in the Boston Marathon

SOLUTION

a. As discussed in Example 1, numbers on uniforms don't count or measure anything. They are at the nominal level of measurement because they are labels and do not imply any kind of ordering.

b. A set of rankings represents data at the ordinal level of measurement because the categories (excellent, good, fair, or poor) have a definite order.

c. An interval of one calendar year always has the same meaning. But ratios of calendar years do not make sense because the choice of the year 0 is arbitrary and does not mean "the beginning of time." Calendar years are therefore at the interval level of measurement.

d. Like Fahrenheit temperatures, Celsius temperatures are at the interval level of measurement. An interval of $1°$ C always has the same meaning, but the zero point ($1°$ C = freezing point of water) is arbitrary and does not mean "no heat."

e. Marathon times have meaningful ratios—for example, a time of 6 hours really is twice as long as a time of 3 hours—because they have a true zero point at a time of 0 hours.   ··•

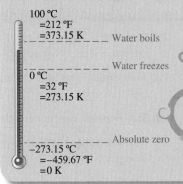

BY THE WAY

Scientists often measure temperatures on the Kelvin scale. Data on the Kelvin scale are at the ratio level of measurement, because the Kelvin scale has a true zero. A temperature of 0 Kelvin really is the coldest possible temperature. Called *absolute zero*, 0 K is equivalent to about −273.15°C or −459.67°F. (The degree symbol is not used for Kelvin temperatures.)

Section 2.1 Exercises

Statistical Literacy and Critical Thinking

1. **Qualitative/Quantitative.** What is the difference between qualitative data and quantitative data?

2. **Quantitative/Qualitative.** A football player is taking a statistics course and states that the names of the players on his team are qualitative, but they can be made quantitative by using the numbers on the jerseys of their uniforms. Is he correct? Why or why not?

3. **Qualitative/Quantitative.** Is a researcher correct when she argues that all data are either qualitative or quantitative? Explain.

4. **ZIP Codes.** A researcher argues that ZIP codes are quantitative data because they measure location, with low numbers in the east and high numbers in the west. Is she correct? Why or why not?

Concepts and Applications

Qualitative vs. Quantitative Data. In Exercises 5–16, determine whether the data described are qualitative or quantitative and explain why.

5. **Blood Groups.** The blood groups of A, B, AB, and O

6. **White Blood Cells.** The white blood cell counts of different people, consisting of the numbers of white blood cells per microliter of blood

7. **Reaction Times.** Braking reaction times (in seconds) are measured as part of a driver education program.

8. **Physicians.** The specialties of physicians (cardiac surgeon, pediatrician, etc.)

9. **Multiple Choice Test Questions.** The answers (a, b, c, d, e) to multiple choice test questions

10. **Survey Responses.** The responses (yes, no, refuse to answer) from survey subjects when asked a question

11. **Nielsen Survey.** The television shows being watched by households surveyed by Nielsen Media Research

12. **Nielsen Ratings.** The number of households with a television in use when surveyed by Nielsen Media Research

13. **Head Circumferences.** In studying different societies, an archeologist measures head circumferences of skulls

14. **Shoe Sizes.** The shoe sizes (such as 8 or $10\frac{1}{2}$) of test subjects

15. **GPA.** The grade point averages of randomly selected college students

16. Area Codes. The area codes (such as 617) of the telephones of survey subjects

Discrete or Continuous. In Exercises 17–28, state whether the data described are discrete or continuous and explain why.

17. Aircraft Baggage. The numbers of checked bags on flights between San Francisco and Atlanta

18. Aircraft Baggage. The weights of checked bags on flights between San Francisco and Atlanta

19. Flights. The total numbers of flights by different airlines between San Francisco and Atlanta in the past month

20. Flights. The lengths in minutes of each of the flights between San Francisco and Atlanta in the past month

21. Chemistry. An experiment in chemistry is repeated, and the times it takes for a reaction to occur are recorded.

22. Test Times. The times required by students to complete a statistics test

23. Test Scores. The numerical scores on a statistics test

24. Traffic Count. Number of cars crossing the Golden Gate Bridge each hour

25. Car Speeds. The speeds of cars as they pass the center of the Golden Gate Bridge

26. Movie Ratings. The movie ratings by a critic, with 0 stars, 1/2 star, 1 star, and so on

27. Stars. Number of stars in each galaxy in the universe

28. Cola. The exact amounts of cola in different cans

Levels of Measurement. For the data described in Exercises 29–40, identify the level of measurement (nominal, ordinal, interval, or ratio).

29. Weights of Textbooks. Weights of college textbooks

30. Movie Ratings. A critic's movie recommendations of "must see," "good," "fair," "poor," or "avoid"

31. Movie Types. Types of movies (drama, comedy, etc.)

32. Temperatures. Body temperatures in Fahrenheit of all students in a statistics class

33. Cars. Classifications of cars by size as subcompact, compact, intermediate, full-size

34. Clinical Trial. Results from a clinical trial consisting of "true positive," "false positive," "true negative," or "false negative"

35. Grades. Final course grades of A, B, C, D, F

36. Distances. Distances traveled by college students as they drive from their homes to their colleges

37. SSN. Social Security numbers

38. Weights. Weights of the cola in cans of Diet Coke

39. Word Counts. Numbers of words spoken in a day by a sample of males

40. Car Safety Ratings. *Consumer Reports* safety ratings of cars: 0 = unsafe up to 3 = safest

Meaningful Ratios? In Exercises 41–48, determine whether the given statement represents a meaningful ratio, so that the ratio level of measurement applies. Explain.

41. Movie Rating. A movie with a 4-star rating is twice as good as one with a 2-star rating.

42. Wind Speed. Wind with a speed of 40 mi/h moves four times as fast as wind with a speed of 10 mi/h.

43. IQ Score as a Measure of Intelligence. One subject has an IQ score of 140 while another subject has an IQ score of 70, so the first subject is twice as intelligent as the second subject.

44. Temperatures. On August 6, it was $80°$ F in New York City, so it was twice as hot as on December 7, when it was $40°$ F.

45. Art Dating. Using carbon dating, one sculpture is found to be 1,000 years old while a second sculpture is found to be 500 years old, so the first sculpture is twice as old as the second.

46. Carbon Dating. Using carbon dating, one sample of wood is found to be twice as old as another, because the first sample is found to be 200 years old while the other sample is 100 years old.

47. Salary. An employee with a salary of $150,000 earns twice as much as one with a $75,000 salary.

48. SAT Scores. A person with an SAT score of 2200 is twice as qualified for college as a person with a score of 1100.

Complete Classification. In Exercises 49–56, determine whether the data described are qualitative or quantitative and give their level of measurement. If the data are quantitative, state whether they are continuous or discrete. Give a brief explanation.

49. Marathon Times. Finish times of the New York City Marathon

50. Marathon Runners. Home nations (such as U. S., France, Kenya) of runners in a marathon

51. Employee ID Numbers. The employees of the Telektronics Corporation have six-digit identification numbers that are randomly generated.

52. Employee Service Times. Seniority of each employee at the Telektronics Corporation is based on the length of time that has passed since the employee was first hired.

53. Employee Hiring Years. The years in which employees were hired (such as 2000, 1995, 2012) are used to determine their pension plan.

54. Political Survey. In a survey of voter preferences, the political parties of respondents are recorded as coded numbers 1, 2, 3, 4, or 5 (where 1 = Democrat, 2 = Republican, 3 = Liberal, 4 = Conservative, 5 = other).

55. Product Ratings. *Consumer Reports* magazine lists ratings of "best buy," "recommended," or "not recommended" for each of several different computers.

56. Quality Control. Apple tests each of its manufactured iPhones and labels each as acceptable or defective.

2.2 DEALING WITH ERRORS

We turn now to the issue of how to deal with errors in measurement. First we will consider the various types of errors that can occur, and then we'll discuss how to account for possible errors when we state results. Note that while we will phrase most of this discussion in terms of measurements, it applies equally well to estimates or projections, such as population estimates or projected revenues for a corporation.

Mistakes are the portals of discovery.

—James Joyce

Types of Error: Random and Systematic

Broadly speaking, measurement errors fall into two categories: random errors and systematic errors. An example will illustrate the difference.

Suppose you work in a pediatric office and use a digital scale to weigh babies. If you've ever worked with babies, you know that they usually aren't very happy about being put on a scale. Their thrashing and crying tends to shake the scale, making the readout jump around. For the case shown in Figure 2.2a, you could conceivably record the baby's weight as anything between about 14.5 and 15.0 pounds. We say that the shaking of the scale introduces a **random error** because any particular measurement may be either too high or too low.

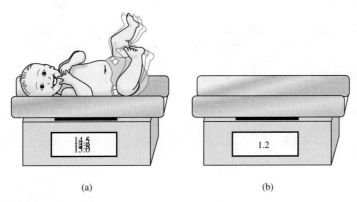

(a)   (b)

Figure 2.2   (a) The baby's motion introduces random errors. (b) The scale reads 1.2 pounds when empty, introducing a systematic error that makes all measurements 1.2 pounds too high.

Now suppose you have been measuring weights of babies all day long with the scale shown in Figure 2.2b. At the end of the day, you notice that the scale reads 1.2 pounds even when there is nothing on it. If you assume that this problem had been present all day, then every measurement you made was high by 1.2 pounds. This type of error is called a **systematic error** because it is caused by an error in the measurement *system*—an error that consistently (systematically) affects all measurements.

BY THE WAY

A systematic error in which a scale's measurements differ consistently from the true values is called a *calibration error.* You can test the calibration of a scale by putting known weights on it, such as weights that are 2, 5, 10, and 20 pounds, and making sure that the scale gives the expected readings.

> **Two Types of Measurement Error**
>
> **Random errors** occur because of random and inherently unpredictable events in the measurement process.
>
> **Systematic errors** occur when there is a problem in the measurement system that affects all measurements in the same way.

A systematic error affects all measurements in the same way, such as making them all too high or all too low. If you discover a systematic error, you can go back and adjust the affected measurements. In contrast, the unpredictable nature of random errors makes it impossible to correct for them. However, you can minimize the effects of random errors by making many measurements and averaging them. For example, if you measure the baby's weight 10 times, your measurements will probably be too high in some cases and too low in others. You can therefore get a better value by averaging the 10 individual measurements.

BY THE WAY

The fact that urban areas tend to be warmer than they would be in the absence of human activity is often called the *urban heat island effect*. Major causes of this effect include heat released by burning fuel in automobiles, homes, and industry and the fact that pavement and large masonry buildings tend to retain heat from sunlight.

EXAMPLE 1 Errors in Global Warming Data

Scientists studying global warming need to know how the average temperature of the entire Earth, or the *global average temperature*, has changed with time. Consider two difficulties in trying to interpret historical temperature data from the early 20th century: (1) Temperatures were measured with simple thermometers and the data were recorded by hand, and (2) most temperature measurements were recorded in or near urban areas, which tend to be warmer than surrounding rural areas because of heat released by human activity. Discuss whether each of these two difficulties produces random or systematic errors, and consider the implications of these errors.

SOLUTION   The first difficulty involves *random errors* because people undoubtedly made occasional errors in reading the thermometers, in calibrating the thermometers, and in recording temperature readings. There is no way to predict whether any individual reading is correct, too high, or too low. However, if there are several readings for the same region on the same day, averaging these readings can minimize the effects of the random errors.

The second difficulty involves a *systematic error* because the excess heat in urban areas always causes the temperature reading to be higher than it would be otherwise. If the researchers can estimate how much this systematic error affected the temperature readings, they can correct the data for this error.

⋅ ⋅ ●

◯ASE STUDY The Census

The Constitution of the United States mandates a census of the population every 10 years. The U.S. Census Bureau conducts the census (and also does many other demographic studies).

In attempting to count the population, the Census Bureau relies largely on a survey that is supposed to include every household in the United States. However, many *random errors* occur in this survey process. For example, some people fill out survey forms incorrectly and some responses are recorded incorrectly by Census Bureau employees.

The census also is subject to several types of *systematic error.* For example, the difficulty of counting the homeless and the fact that undocumented aliens may try to hide their presence both tend to lead to undercounts in the population. Other systematic errors tend to cause overcounts—such as college students being counted both at home and at their school residences, or children of divorced parents being counted in both households.

The 2010 census, for example, found a United States population of about 308.7 million people, but this reflects only the actual survey results. In principle, the Census Bureau could use statistical studies to correct for the systematic errors and make the count more accurate, but that is not allowed by current law.

A little inaccuracy sometimes saves a ton of explanation.

—H. H. Munro (Saki)

TIME (Ⓢ)UT TO THINK

The question of whether the Census Bureau should be allowed to adjust its "official" count on the basis of statistical surveys is very controversial. The Constitution calls for an "actual enumeration" of the population (Article 1, Section 2, Subsection 2). Do you believe that this wording precludes or allows the use of statistical surveys in the official count? Defend your opinion. Also discuss reasons why Democrats tend to favor the use of sampling methods while Republicans tend to oppose it.

Size of Errors: Absolute versus Relative

Besides wanting to know whether an error is random or systematic, we often want to know whether an error is big enough to be of concern or small enough to be unimportant. An example should clarify the idea.

Suppose you go to the store and buy what you think is 6 pounds of hamburger, but because the store's scale is poorly calibrated, you actually get only 4 pounds. You'd probably be upset by this 2-pound error. Now suppose you are buying hamburger for a huge town barbeque and you order 3,000 pounds of hamburger, but you actually receive only 2,998 pounds. You are short by the same 2 pounds as before, but in this case the error probably doesn't seem very important.

In more technical language, the 2-pound error in both cases is an **absolute error**—it describes how far the claimed or measured value lies from the true value. A **relative error** compares the size of the absolute error to the true value. The relative error for the first case is fairly large because the absolute error of 2 pounds is half the true weight of 4 pounds; we say that the relative error is 2/4, or 50%. In contrast, the relative error for the second case is the absolute error of 2 pounds divided by the true hamburger weight of 2,998 pounds, which is only $2/2998 \approx 0.00067$, or 0.067%.

Absolute and Relative Error

The **absolute error** describes how far a claimed or measured value lies from the true value:

$$\text{absolute error} = \text{claimed or measured value} - \text{true value}$$

The **relative error** compares the size of the absolute error to the true value. It is often expressed as a percentage:

$$\text{relative error} = \frac{\text{absolute error}}{\text{true value}} \times 100\%$$

Note that the absolute and relative errors are *positive* when the claimed or measured value is greater than the true value, and *negative* when the claimed or measured value is less than the true value.

EXAMPLE ② Absolute and Relative Error

Find the absolute and relative error in each case.

a. Your true weight is 100 pounds, but a scale says you weigh 105 pounds.

b. The government claims that a program costs $99.0 billion, but an audit shows that the true cost is $100.0 billion.

SOLUTION

a. The measured value is the scale reading of 105 pounds and the true value is 100 pounds:

$$\text{absolute error} = \text{measured value} - \text{true value}$$
$$= 105 \text{ lb} - 100 \text{ lb}$$
$$= 5 \text{ lb}$$

$$\text{relative error} = \frac{\text{absolute error}}{\text{true value}} \times 100\%$$
$$= \frac{5 \text{ lb}}{100 \text{ lb}} \times 100\%$$
$$= 5\%$$

The measured weight is too high by 5 pounds, or by 5%.

b. The claimed cost is $99.0 billion and the true cost is $100.0 billion:

$$\text{absolute error} = \text{claimed value} - \text{true value}$$
$$= \$99.0 \text{ billion} - \$100.0 \text{ billion}$$
$$= -\$1.0 \text{ billion}$$

$$\text{relative error} = \frac{\text{absolute error}}{\text{true value}} \times 100\%$$
$$= \frac{-\$1.0 \text{ billion}}{\$100.0 \text{ billion}} \times 100\%$$
$$= -1\%$$

The claimed cost is too low by $1.0 billion, or by 1%.

$\cdots \bullet$

> *A billion here, a billion there; soon you're talking real money.*
>
> —attributed to Senator Everett Dirksen

Describing Results: Accuracy and Precision

Two key ideas about any reported value are its *accuracy* and its *precision*. Although these terms are often used interchangeably in English, they are not quite the same thing.

The goal of any measurement is to obtain a value that is as close as possible to the *true value*. **Accuracy** describes how close the measured value lies to the true value. **Precision** describes the amount of detail in the measurement. For example, suppose a census says that the population of your hometown is 72,453 but the true population is 96,000. The census value of 72,453 is quite precise because it seems to tell us the exact count, but it is not very accurate because it is nearly 25% smaller than the actual population of 96,000. Note that accuracy is usually defined by relative error rather than absolute error. For example, if a company projects sales of $7.30 billion and true sales turn out to be $7.32 billion, we say the projection was quite accurate because it was off by less than 1%, even though the error of $0.02 billion represents $20 million.

> **Definitions**
>
> **Accuracy** describes how closely a measurement approximates a true value. An accurate measurement is close to the true value. (*Close* is generally defined as a small *relative* error, rather than a small absolute error.)
>
> **Precision** describes the amount of detail in a measurement.

We generally assume that the precision with which a number is reported reflects what was actually measured. If you say that you weigh 132 pounds, we assume that you measured your weight only to the nearest pound. In that case, a more precise measurement might find that you weighed, say, 132.3 pounds or 131.6 pounds (both of these would round to 132). In contrast, if you say that you weigh 132.0 pounds, we assume that you measured your weight to the nearest tenth of a pound. This assumption means that you should never report a measurement with more precision than is justified. For example, if you actually measured your weight only to the nearest pound, it would be wrong to say that you weighed 132.0 pounds, because that would imply you made a measurement to the nearest tenth of a pound.

EXAMPLE ③ Accuracy and Precision in Your Weight

Suppose that your true weight is 102.4 pounds. The scale at the doctor's office, which can be read only to the nearest quarter pound, says that you weigh $102\frac{1}{4}$ pounds. The scale at the gym, which gives a digital readout to the nearest 0.1 pound, says that you weigh 100.7 pounds. Which scale is more *precise*? Which is more *accurate*?

SOLUTION   The scale at the gym is more *precise* because it gives your weight to the nearest tenth of a pound, whereas the doctor's scale gives your weight only to the nearest quarter pound. However, the scale at the doctor's office is more *accurate* because its value is closer to your true weight.   ⋯●

TIME ⏱ UT TO THINK

In Example 3, we need to know your true weight to determine which scale is more accurate. But how would you know your true weight? Can you ever be sure that you know it? Explain.

CASE STUDY Understanding Census Results

Upon completing the 2010 census, the U.S. Census Bureau reported a population of 308,745,538 (on April 1, 2010), thereby implying an exact count of everyone living in the United States. Unfortunately, such a precise count could not possibly be as accurate as it seems to imply.

Even in principle, the only way to get an exact count of the number of people living in the United States would be to count everyone *instantaneously*. Otherwise, the count would be off because, for example, an average of about eight births and four deaths occur every minute in the United States.

In fact, the Census is conducted over a period of many months, so the actual population on a given date could not really be known. Moreover, the Census results are affected by both random and systematic errors (see the Case Study on p. 52). A more honest report of the Census result would use much less precision—for example, stating the population as "about 310 million." In fairness to the Census Bureau, their detailed reports explained the uncertainties in the population count, but these uncertainties were rarely mentioned by the press.

BY THE WAY

The digits in a number that were actually measured are called *significant digits*. All the digits in a number are significant except for zeros that are included so that the decimal point can be properly located. For example, 0.001234 has four significant digits; the zeros are required for proper placement of the decimal point. The number 1,234,000,000 has four significant digits for the same reason. The number 132.0 has four significant digits; the zero is significant because it is not required for the proper placement of the decimal point and therefore implies an actual measurement of zero tenths.

CASE STUDY Trillions Disappear in Unjustified Budget Precision

In early 2001, politicians heralded a projected U.S. federal budget *surplus* of $5.6 trillion over the next 10 years. When 2011 actually arrived, rather than having gained trillions in surplus money, the government had instead gone another $9 trillion into debt. In other words, the projection proved to be in error by nearly $15 trillion (going from *positive* $5.6 trillion to *negative* $9 trillion).

You've probably heard or read news reports about this multi-trillion-dollar turn in government fortunes, and people of different political persuasions may feel quite differently about its cause. For our purposes here, the most remarkable thing about it was the unjustified precision of the original projection. After all, the projection of $5.6 trillion was quoted precise to the nearest *tenth* of a trillion dollars (which means to the nearest hundred billion dollars), implying that the actual value would be between about $5.5 and $5.7 trillion. Now that's what you might call an "Ooops!"

In fairness to the economists responsible for the projection, they were well aware of the associated uncertainties. Moreover, changes that occurred after the projection was made—including the terrorist attacks of 9/11, wars in Afghanistan and Iraq, substantial cuts in taxes, and a major recession—make it unsurprising that the projection turned out so far off base. But the lesson is clear: Next time you hear politicians or the media talking about future federal budgets with great precision, remember it is quite likely that their numbers will prove to be abysmally inaccurate.

Summary: Dealing with Errors

The ideas we've covered in this section are a bit technical, but very important to understanding measurements and errors. Let's briefly summarize how the ideas relate to one another.

- Errors can occur in many ways, but generally can be classified into one of two basic types: random errors or systematic errors.
- Whatever the source of an error, its size can be described in two different ways: as an absolute error or as a relative error.
- Once a measurement is reported, we can evaluate it in terms of its accuracy and its precision.

Section 2.2 Exercises

Statistical Literacy and Critical Thinking

1. **Error Type.** When recording the weight of a watermelon, a supermarket clerk writes the wrong number. Is this type of error a random error or a systematic error? Explain.

2. **Standard Weight.** A standard weight defined to represent exactly 1 kg is kept by the National Institute of Standards and Technology. If you put this 1-kilogram true weight on a scale and the scale says it is 1.002 kg, what is the absolute error and the relative error of the measurement?

3. **Standard Weight.** Using the same standard weight from Exercise 2, assume that you put the weight on a scale and record the weight as 1.2034278 kg. Describe the accuracy and precision of the recorded weight. Explain.

4. **World Population.** At a particular moment, the U.S. Census Bureau population clock shows that the global population is 7,118,233,027 people. Describe the accuracy and precision of that population number.

Does It Make Sense? For Exercises 5–8, decide whether the statement makes sense (or is clearly true) or does not make sense (or is clearly false). Explain clearly; not all of these have definitive answers, so your explanation is more important than your chosen answer.

5. **Species of Fish.** There are 24,627 species of fish on Earth.

6. **Relative Error.** The relative error that a microbiologist makes in measuring a cell must be less than the relative error that an astronomer makes in measuring a galaxy, because cells are smaller than galaxies.

7. **Scanner Error.** The Jenkins supermarket manager claims that the scanning errors on purchased items are random, and about half of the errors are in favor of the supermarket.

8. **Aircraft Baggage.** An American Airlines agent tells you that you must pay a surcharge because your checked bag weighs 23.018 kg, which exceeds the limit of 23 kg, and that there's no doubt that the scale is correct because it measures to the thousandth of a kilogram.

Concepts and Applications

9. **Tax Audit.** A tax auditor reviewing a tax return looks for several kinds of problems, including these two: (1) mistakes made in entering or calculating numbers on the tax return

and (2) places where the taxpayer reported income dishonestly. Discuss whether each problem involves random or systematic errors.

10. **Safe Air Travel.** Before taking off, a pilot is supposed to set the aircraft altimeter to the elevation of the airport. A pilot leaves from Denver (altitude 5,280 feet) with her altimeter set to 2,500 feet. Explain how this affects the altimeter readings throughout the flight. What kind of error is this?

11. **Technical Specifications.** An iPod battery is supposed to provide 3.7 volts. An aftermarket supplier manufactures 5,000 replacement batteries and finds that they have a mean of 3.7 V, but about half of the batteries have less than 3.7 volts and half have more than 3.7 volts. Does the error appear to be random or systematic? Explain.

12. **Drunk Driving Data.** For data collected on car driving fatalities, a researcher claims that while many fatalities are recorded as involving alcohol, many others are missed because the deceased are not tested for alcohol consumption. If this is true, what kind of error is introduced and how does it affect the values of the data?

Sources of Errors. For each measurement described in Exercises 13–20, identify at least one likely source of random errors and also identify at least one likely source of systematic errors.

13. **Contributions.** A survey asks people for the amount of money they donated to charity in the past year.

14. **Tax Returns.** The annual incomes of 200 people are obtained from their tax returns.

15. **Passenger Weights.** For a flight on a small plane, the pilot asks passengers what they weigh.

16. **M&Ms.** The weights of individual M&M plain candies were obtained by placing each candy in a paper cup, then obtaining the weight without accounting for the weight of the cup.

17. **Radar Speeds.** Speeds of cars are recorded by a police officer who uses a radar gun.

18. **Counterfeit Products.** The police commissioner in New York City estimates the annual value of counterfeit goods sold in the city.

19. **Cigarette Sales.** The health commissioner of Los Angeles estimates the number of cigarettes smoked in her city from data for taxes collected on sales of cigarettes.

20. **Measuring Length.** A groundskeeper measures the length and width of a school's athletic field using a ruler that is 1 foot long.

Absolute and Relative Errors. In Exercises 21–24, find the values of the absolute and relative errors.

21. **Credit Card Bill.** You receive a Visa credit card bill for $2,995, but it includes a charge of $1,750 that was not valid. (That is, the true value is $1,750 less than the bill claims.)

22. **Steak Weight.** A steak at a restaurant actually weighs 18 ounces (the true value), but the menu claims that it is a 20-ounce steak.

23. **Wrong Change.** When purchasing lunch in a cafeteria, the actual (true value) of change due is $2.75, but the incorrect amount of $1.75 is given instead.

24. **Baker's Dozen.** The bakery menu claims that there are 12 doughnuts in a bag, but the baker always puts 13 doughnuts (the true value) in each bag.

25. **Minimizing Errors.** Twenty-five people, including yourself, are to measure the length of a room to the nearest tenth of a millimeter. Assume that everyone uses the same well-calibrated measuring device, such as a tape measure.

   a. All 25 measurements are not likely to be exactly the same; thus, the measurements will contain some sources of error. Are these errors systematic or random? Explain.

   b. If you want to minimize the effect of random errors in determining the length of the room, which is the better choice: to report your own personal measurement as the length of the room or to report the average of all 25 measurements? Explain.

   c. Describe any possible sources of systematic errors in the measurement of the room length.

   d. Can the process of averaging all 25 measurements help reduce any systematic errors? Why or why not?

26. **Minimizing Errors.** When weighing a model 22F car battery, the measuring instrument is very precise, and the weight is obtained 10 consecutive times.

   a. All 10 measurements are not likely to be exactly the same; thus, the measurements will contain some sources of error. Are these errors systematic or random? Explain.

   b. If you want to minimize the effect of random errors in determining the true weight of the battery, which is the better choice: to choose one of the 10 measurements at random or to report the average of all 10 measurements? Explain.

   c. Describe any possible sources of systematic errors in the 10 measurements.

   d. Can the process of averaging all 10 weights help reduce any systematic errors? Why or why not?

27. **Accuracy and Precision in Corvette Weight.** A new Corvette weighs 3,273 lb. A manufacturer's scale that is accurate to the nearest 10 lb gives the weight as 3,250 lb, while the U.S. Department of Transportation uses a scale that is accurate to the nearest 0.1 lb and obtains a weight of 3,298.2 lb. Which measurement is more *precise*? Which is more *accurate*? Explain.

28. **Accuracy and Precision in Height.** Assume that your statistics professor has a height of exactly 175.2 cm. Assume that this height is measured with a tape measure that can be read to the nearest mm (or 1/10 cm) and results from two different measurements are reported as 175 cm and 175.5 cm. Which measurement is more *precise*? Which is more *accurate*? Explain.

29. Accuracy and Precision in Weight. Suppose your weight is 52.55 kilograms. A scale at a health clinic that gives weight measurements to the nearest half kilogram gives your weight as 53 kilograms. A digital scale at the gym that gives readings to the nearest 0.01 kilogram gives your weight as 52.88 kilograms. Which measurement is more *precise*? Which is more *accurate*? Explain.

30. Accuracy and Precision in Weight. Suppose your weight is 52.55 kilograms. A scale at a health clinic that gives weight measurements to the nearest half kilogram gives your weight as $52\frac{1}{2}$ kilograms. A digital scale at the gym that gives readings to the nearest 0.01 kilogram gives your weight as 51.48 kilograms. Which measurement is more *precise*? Which is more *accurate*? Explain.

Believable Facts? Exercises 31–38 give statements of "fact" coming from statistical measurements. For each statement, briefly discuss possible sources of error in the measurement. Then, considering the precision with which the measurement is given, discuss whether you think the fact is believable.

31. Population. The population of the United States in 1860 was 31,443,321.

32. Motor Vehicle Deaths. Last year there were 38,929 deaths in the United States due to motor vehicle crashes.

33. Population of China. Last year, the population of China was 1,339,414,205 people.

34. Tallest Building. The Burj Khalifa in Dubai is 2,717 feet tall, making it the world's tallest building.

35. Gateway Arch. The St. Louis Gateway Arch is 630.2377599694 feet tall.

36. Cell Phones. The *Newport Chronicle* reported that there are now 5 billion cell phones in use.

37. College Students. Wikipedia reports that there are currently 14,261,778 college students in the United States.

38. Threatened Species. The U.S. government now lists 1,879 endangered or threatened species of animals and plants.

PROJECTS FOR THE INTERNET & BEYOND

39. The Census. Go to the Web site for the U.S. Census Bureau and learn about the census conducted every 10 years. How and when will data be collected for the next census? Are any significant changes in the collection process planned?

40. Census Controversies. Use the Library of Congress's "Thomas" Web site to find out about any pending legislation concerning the collection or use of census data. If you find more than one legislative bill pending, choose one to study in depth. Summarize the proposed legislation, and briefly discuss arguments both for and against it.

41. Wristwatch Errors. Use a Web site that gives you the local time (such as www.time.gov) to set a watch to the nearest second. Then compare the time on your watch with the times on friends' watches. Record the errors with positive signs for watches that are ahead of the true time and negative signs for those watches that are behind the true time. Use the concepts of this section to describe the accuracy of the wristwatches in your sample.

IN THE NEWS

42. Random and Systematic Errors. Find a recent news report that gives a quantity that was measured statistically (for example, a report of population, average income, or the number of homeless people). Write a short description of how the quantity was measured, and briefly describe any likely sources of either random or systematic errors. Overall, do you think that the reported measurement was accurate? Why or why not?

43. Absolute and Relative Errors. Find a recent news report that describes some mistake in a measured, estimated, or projected number (for example, a budget projection that turned out to be incorrect). In words, describe the size of the error in terms of both absolute error and relative error.

44. Accuracy and Precision. Find a recent news article that causes you to question accuracy or precision. For example, the article might report a figure with more precision than you think is justified, or it might cite a figure that you know is inaccurate. Write a summary of the report, and explain why you question its accuracy or precision (or both).

2.3 USES OF PERCENTAGES IN STATISTICS ························

Statistical results are often stated with percentages. A percentage is simply a way of expressing a fraction; the words *per cent* literally mean "divided by 100." However, percentages are often used in subtle ways. Consider a statement that appeared in a front-page article in the *New York Times*:

> *The percentage of smokers among 8th graders is up 44 percent, to 10.4 percent.*

Although the statement uses percentages in a valid way, it can be difficult to understand what it means by "up 44%, to 10.4%." In this section, we will investigate some of the subtle uses and abuses of percentages. Before we begin, you should review the following basic rules regarding conversions between fractions and percentages.

Conversions Between Fractions and Percentages

To convert a percentage to a common fraction: Replace the % symbol with division by 100; simplify the fraction if necessary.

$$\textit{Example:}\quad 25\% = \frac{25}{100} = \frac{1}{4}$$

To convert a percentage to a decimal: Drop the % symbol and move the decimal point two places to the left (that is, divide by 100).

$$\textit{Example:}\quad 25\% = 0.25$$

To convert a decimal to a percentage: Move the decimal point two places to the right (that is, multiply by 100) and add the % symbol.

$$\textit{Example:}\quad 0.43 = 43\%$$

To convert a common fraction to a percentage: First convert the common fraction to a decimal; then convert the decimal to a percentage.

$$\textit{Example:}\quad \frac{1}{5} = 0.2 = 20\%$$

EXAMPLE ① Newspaper Survey

A newspaper reports that 54% of 1,069 people surveyed said that the President is doing a good job. How many people said that the President is doing a good job?

SOLUTION   The 54% represents the fraction of respondents who said the President is doing a good job. Because "of" usually indicates multiplication, we multiply:

$$54\% \times 1{,}069 = 0.54 \times 1{,}069 = 577.26 \approx 577$$

About 577 out of the 1,069 people said the President is doing a good job. Note that we round the answer to 577 to obtain a whole number of people, because the number of people must be a discrete (integer) value. (The symbol $\approx$ means "approximately equal to.")   ···●

Using Percentages to Describe Change

Percentages are often used to describe how data change with time. For example, suppose the population of a town was 10,000 in 1970 and 15,000 in 2000. We can express this change in two basic ways:

- Because the population rose by 5,000 people (from 10,000 to 15,000), we say that the **absolute change** in the population was 5,000 people.
- Because the increase of 5,000 people was 50% of the starting population of 10,000, we say that the **relative change** in the population was 50%.

In general, calculating an absolute or relative change always involves two numbers: a starting number, or **reference value**, and a **new value**. Once we identify these two values, we can calculate the absolute and relative change with the following formulas. A change is positive if the new value is greater than the reference value and negative if the new value is less than the reference value.

Absolute and Relative Change

The **absolute change** is the actual increase or decrease from a reference value to a new value:

$$\text{absolute change} = \text{new value} - \text{reference value}$$

The **relative change** is the size of the absolute change in comparison to the reference value and can be expressed as a percentage:

$$\text{relative change} = \frac{\text{new value} - \text{reference value}}{\text{reference value}} \times 100\%$$

TIME (🕐)UT TO THINK

Compare the formulas for absolute and relative change to the formulas for absolute and relative error, given in Section 2.2. Why are they so similar?

EXAMPLE 2   World Population Growth

Estimated world population in 1950 was 2.6 billion. By the end of 2010, it had reached 6.9 billion. Describe the absolute and relative change in world population from 1950 to 2010.

SOLUTION   The reference value is the 1950 population of 2.6 billion and the new value is the 2010 population of 6.9 billion.

$$\text{absolute change} = \text{new value} - \text{reference value}$$

$$= 6.9 \text{ billion} - 2.6 \text{ billion}$$

$$= 4.3 \text{ billion}$$

$$\text{relative change} = \frac{\text{new value} - \text{reference value}}{\text{reference value}} \times 100\%$$

$$= \frac{6.9 \text{ billion} - 2.6 \text{ billion}}{2.6 \text{ billion}} \times 100\%$$

$$= 165.4\%$$

World population increased by 4.3 billion people, or by about 165%, during the 60-year period from 1950 to 2010.   ··●

Using Percentages for Comparisons

Percentages are also commonly used to compare two numbers:

- The **reference value** is the number that we are using as the basis for a comparison.
- The **compared value** is the other number, which we compare to the reference value.

BY THE WAY

According to United Nations and U.S. Census Bureau estimates, world population passed 7 billion in early 2012—only 13 years after passing the 6 billion mark. The population is still growing by more than 75 million people each year, which means it takes only about four years for the world to add a population equivalent to that of the entire United States.

We can then express the absolute or relative difference between these two values with formulas very similar to those for absolute and relative change. The difference is positive if the compared value is greater than the reference value and negative if the compared value is less than the reference value.

Absolute and Relative Difference

The **absolute difference** is the difference between the compared value and the reference value:

$$\text{absolute difference} = \text{compared value} - \text{reference value}$$

The **relative difference** describes the size of the absolute difference in comparison to the reference value and can be expressed as a percentage:

$$\text{relative difference} = \frac{\text{compared value} - \text{reference value}}{\text{reference value}} \times 100\%$$

EXAMPLE ③ Russian and American Life Expectancy

According to United Nations data, life expectancy for American men is about 76 years, while life expectancy for Russian men is about 63 years. Using the life expectancy of Russian men as the reference value, compare the life expectancy of American men with that of Russian men in absolute and relative terms. (See Section 6.4 for a discussion of the meaning of life expectancy.)

SOLUTION   We want to compare the American male life expectancy with the Russian male life expectancy, so the Russian male life expectancy is the reference value and the American male life expectancy is the compared value:

$$\text{absolute difference} = \text{compared value} - \text{reference value}$$

$$= 76 \text{ years} - 63 \text{ years}$$

$$= 13 \text{ years}$$

$$\text{relative difference} = \frac{\text{compared value} - \text{reference value}}{\text{reference value}} \times 100\%$$

$$= \frac{76 \text{ years} - 63 \text{ years}}{63 \text{ years}} \times 100\%$$

$$\approx 21\%$$

The life expectancy of American men is 13 years greater in absolute terms and 21% greater in relative terms than the life expectancy of Russian men.   ·· •

BY THE WAY

No one knows all the reasons for the low life expectancy of Russian men, but one contributing factor is alcoholism, which is much more common in Russia than in America.

Of versus *More Than*

Consider a population that *triples* in size from 200 to 600. There are two equivalent ways to state this change with percentages:

• Using *more than*: The new population is 200% *more than* the original population. Here, we are looking at the relative change in the population:

$$\text{relative change} = \frac{\text{new value} - \text{reference value}}{\text{reference value}} \times 100\%$$

$$= \frac{600 - 200}{200} \times 100\%$$

$$= 200\%$$

- Using *of*: The new population is 300% *of* the original population, which means it is three times the original population. Here, we are looking at the *ratio* of the new population to the original population:

$$\frac{\text{new population}}{\text{original population}} = \frac{600}{200} = 3.00 = 300\%$$

Notice that the percentages in the "more than" and "of" statements are related by $300\% = 100\% + 200\%$. This leads to the following general relationship.

> **Of versus More Than (or Less Than)**
>
> - If the new or compared value is *P% more than* the **reference value**, then it is $(100 + P)\%$ *of* the reference value.
>
> - If the new or compared value is *P% less than* the reference value, then it is $(100 - P)\%$ *of* the reference value.

For example, 40% *more than* the reference value is 140% *of* the reference value, and 40% *less than* the reference value is 60% *of* the reference value. When you hear statistics quoted with percentages, it is very important to listen carefully for the key words *of* and *more than* (or *less than*)—and hope that the speaker knows the difference.

EXAMPLE 4 World Population

In Example 2, we found that world population in 2010 was about 165% more than world population in 1950. Express this change with an "of " statement.

SOLUTION   World population in 2010 was 165% more than world population in 1950. Because $(100 + 165)\% = 265\%$, the 2010 population was 265% *of* the 1950 population. This means that the 2010 population was 2.65 times the 1950 population. $\cdots\bullet$

EXAMPLE 5 Sale!

A store is having a "25% off " sale. How does a sale price compare to an original price?

SOLUTION   The "25% off " means that a sale price is 25% *less than* the original price, which means it is $(100 - 25)\% = 75\%$ *of* the original price. For example, if an item's original price was $100, its sale price is $75. $\cdots\bullet$

> **TIME OUT TO THINK**
>
> One store advertises "1/3 off everything!" Another store advertises "Sale prices just 1/3 of original prices!" Which store is having the bigger sale? Explain.

Percentages of Percentages

Percentage changes and percentage differences can be particularly confusing when the values *themselves* are percentages. Suppose your bank increases the interest rate on your savings account from 3% to 4%. It's tempting to say that the interest rate increases by 1%, but that statement is ambiguous at best. The interest rate increases by 1 *percentage point*, but the relative change in the interest rate is 33%:

$$\frac{4\% - 3\%}{3\%} \times 100\% = 0.33 \times 100\% = 33\%$$

You can therefore say that the bank raised your interest rate by 33%, even though the actual rate increased by only 1 percentage point (from 3% to 4%).

> **Percentage Points versus %**
>
> When you see a change or difference expressed in *percentage points*, you can assume it is an *absolute* change or difference. If it is expressed as a percentage, it probably is a *relative* change or difference.

EXAMPLE 6 Care in Wording

Assume that 40% of the registered voters in Carson City are Republicans. Read the following questions carefully and give the most appropriate answers.

a. The percentage of voters registered as Republicans is 25% higher in Freetown than in Carson City. What percentage of the registered voters in Freetown are Republicans?

b. The percentage of voters registered as Republicans is 25 percentage points higher in Freetown than in Carson City. What percentage of the registered voters in Freetown are Republicans?

If you can't convince them, confuse them.

—Harry S. Truman

SOLUTION

a. We interpret the "25%" as a relative difference, and 25% of 40% is 10% (because $0.25 \times 0.40 = 0.10$). Therefore, the percentage of registered Republicans in Freetown is $40\% + 10\% = 50\%$.

b. In this case, we interpret the "25 percentage points" as an absolute difference, so we simply add this value to the percentage of Republicans in Carson City. Therefore, the percentage of registered Republicans in Freetown is $40\% + 25\% = 65\%$.   ·· ●

Section 2.3

Statistical Literacy and Critical Thinking

1. Percentages. Last year's budget for the legislative branch of the U.S. government was $4919 million, and this year it is $5333 million. Consider last year's budget of $4919 million to be the reference value.

   a. What is the absolute change in the budget from last year to this year?

   b. What is the relative change in the budget from last year to this year?

   c. Next year's budget is estimated to be $5185 million. What is the percentage decrease from this year's budget of $5333?

   d. If next year's budget is changed so that it is 5% less than this year's budget of $5333 million, what is the amount of next year's budget?

2. Percentage A *New York Times* editorial criticized a chart caption that described a dental rinse as one that "reduces plaque on teeth by over 300%." If the dental rinse removes all of the plaque, what percentage is removed? Is it possible to reduce plaque by over 300%?

3. Percentage Points. A Ridgid survey of 1,023 high school students showed that 25% of them plan to enter the field of information technology, and the margin of error is 3 percentage points. Why is it misleading to state that the margin of error is 3% instead of 3 percentage points?

4. *Of* and *More Than*. In an Opinion Research poll, 1,072 adults were asked what they would

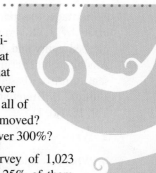

do with their old cell phones, and 44.0% of them said that they would donate them to charity. What is the actual number of respondents who plan to donate their old cell phones to charity? If another poll is to be conducted with a sample size that is 5% greater than the sample size of 1,072 adults, how many subjects will be included in this new poll?

Does It Make Sense? For Exercises 5–8, decide whether the statement makes sense (or is clearly true) or does not make sense (or is clearly false). Explain clearly; not all of these have definitive answers, so your explanation is more important than your chosen answer.

5. **Cell Phones.** The percentage of people with cell phones increased by 1.2 million people.

6. **Salary Percentages.** The CEO of the Brandon Marketing Group announces that all employees must take a 5% cut in pay this year, but they will all get a 5% raise next year, so the salaries will then be the same as they are now.

7. **Interest Rate.** The Jefferson Valley Bank increased its new-car loan rate by 100%.

8. **Interest Rate.** The Jefferson Valley Bank increased its new-car loan rate (annual) by 100 percentage points.

Concepts and Applications

9. **Fractions, Decimals, Percentages.** Express each of the following numbers in the three forms of a fraction, decimal, and percentage.

   a. 75%   **b.** 3/8   **c.** 0.4   **d.** 80%

10. **Fractions, Decimals, Percentages.** Express the following numbers in the three forms of fraction, decimal, and percentage.

    a. 350%   **b.** 2.5   **c.** −0.44   **d.** −200%

11. **Percentage Practice.** A study was conducted of pleas made by 1,348 criminals. Among those criminals, 956 pleaded guilty and 392 of them were sentenced to prison. Among 72 other criminals who pleaded not guilty, 58 were sent to prison (based on data from "Does It Pay to Plead Guilty?" by Brereton and Casper, *Law and Society Review*, Vol. 16, No. 1).

    a. What percentage of the criminals pleaded guilty?

    b. What percentage of the criminals were sent to prison?

    c. Among those who pleaded guilty, what is the percentage who were sent to prison?

    d. Among those who pleaded not guilty, what is the percentage who were sent to prison?

12. **Percentage Practice.** A study was conducted to determine whether flipping a penny or spinning a penny has an effect on the proportion of heads. Among 49,437 trials, 29,015 involved flipping pennies, and 14,709 of those pennies turned up heads. The other 20,422 trials involved spinning pennies, and 9,197 of those pennies turned up heads (based on data from Robin Lock as given in *Chance News*).

    a. What percentage of the trials involved flipping pennies?

    b. What percentage of the trials involved spinning pennies?

    c. Among the pennies that were flipped, what is the percentage that turned up heads?

    d. Among the pennies that were spun, what is the percentage that turned up heads?

Relative Change. Exercises 13–20 each provide two values. For each pair of values, use a percentage to express their relative change or difference. Use the second given value as the reference value, and express results to the nearest percentage point. Also, write a statement describing the result.

13. **Newspapers.** The number of daily newspapers in the United States is now 1,387, and it was 2,226 in 1900.

14. **Cars.** There are now 143,781,202 registered passenger cars, and in 1980 there were 121,601,000.

15. **Airline Flights.** This January there were 751,183 scheduled passenger flights in the United States, and in January of 1996 there were 634,343.

16. **Bankruptcies.** There were 1,531,997 bankruptcy cases filed last year, and in the year 2000 there were 1,276,900 bankruptcy cases filed.

17. **Newspapers.** The daily circulation of the *Wall Street Journal* is currently 2.09 million (the largest in the country). The daily circulation of *USA Today* is currently 1.83 million (the second largest in the country).

18. **Car Sales.** In the current month, 18,830 Toyota Camry cars were sold, and there were 18,341 Honda Civic cars sold.

19. **Airports.** Chicago's O'Hare Airport handled 67 million passengers last year. As the busiest airport in the world, Atlanta's Hartsfield Airport handled 89 million passengers last year.

20. **Tourists.** Last year, France ranked as the number one tourist destination with 78 million international arrivals. The United States ranked second with 58 million international arrivals.

Surveys. Some important analyses of survey results require that you know the actual number of subjects whose responses fall into a particular category. In Exercises 21–24, find the actual number of respondents corresponding to the given percentage.

21. **Personal Calls.** In an At-A-Glance survey of 1,385 office workers, 4.8% said that they do not make personal phone calls.

22. **Interview Mistakes.** In an Accountemps survey of 150 executives, 47% said that the most common interview mistake is to have little or no knowledge of the company.

23. **Televisions.** In a Frank N. Magid Associates survey of 1,005 adults, 83% reported having more than one television at home.

24. **Cell Phones.** In a Harris Interactive survey of 9,132 adults, 89% reported being cell phone users.

Of vs. *More Than.* Fill in the blanks in Exercises 25–28. Briefly explain your reasoning in each case.

25. **Weights.** If a truck weighs 40% more than a car, then the truck's weight is ____% of the car's weight.

26. **Areas.** If the area of Norway is 24% more than the area of Colorado, then Norway's area is ____% of Colorado's area.

27. **Population.** If the population of Montana is 20% less than the population of New Hampshire, then Montana's population is ____% of New Hampshire's population.

28. **Salary.** The Vice President's salary is currently 42% less than the President's salary, so the Vice President's salary is ____% of the President's salary.

29. **Margin of Error.** A Gallup poll of 1,012 adults showed that 89% of Americans say that human cloning should not be allowed. The margin of error was 3 percentage points. Would it matter if a newspaper reported the margin of error as "3%"? Explain.

30. **Margin of Error.** A Pew Research Center survey of 3,002 adults showed that the percentage who listen to National Public Radio is probably between 14% and 18%. How should a newspaper report the margin of error? Explain.

Percentages of Percentages. Exercises 31–34 describe changes in which the measurements themselves are percentages. Express each change in two ways: (1) as an absolute difference in terms of percentage points and (2) as a relative difference in terms of percent.

31. The percentage of high school seniors using alcohol decreased from 68.2% in 1975 to 52.7% now.

32. The percentage of the world's population living in developed countries decreased from 27.1% in 1970 to 19.5% now.

33. The five-year survival rate for Caucasians for all forms of cancer increased from 39% in the 1960s to 61% now.

34. The five-year survival rate for Blacks for all forms of cancer increased from 27% in the 1960s to 48% now.

PROJECTS FOR THE INTERNET & BEYOND

35. **World Population.** Find the current estimate of world population on the U.S. Census Bureau's world population clock. Describe the percentage change in population since the 6 billion mark was passed during 1999. Also find how the population clock estimates are made, and discuss the uncertainties in estimating world population.

36. **Drug Use Statistics.** Go to the Web site for the National Center on Addiction and Substance Abuse (CASA) and find a recent report giving statistics on substance abuse. Write a summary of the new research, giving at least some of the conclusions in terms of percentages.

IN THE NEWS

37. **Percentages.** Find three recent news reports in which percentages are used to describe statistical results. In each case, describe the meaning of the percentage.

38. **Percentage Change.** Find a recent news report in which percentages are used to express the change in a statistical result from one time to another (such as an increase in population or in the number of children who smoke). Describe the meaning of the change. Be sure to watch for key words such as *of* or *more than*.

39. **Quote Interpretation.** Consider this quote: "The rate [of smoking] among 10th graders jumped 45 percent, to 18.3 percent, and the rate for 8th graders is up 44 percent, to 10.4 percent." Briefly explain the meaning of each of the percentages in this statement.

2.4

INDEX NUMBERS

You've probably heard about **index numbers**, such as the Consumer Price Index, the Producer Price Index, or the Consumer Confidence Index. Index numbers are very common in statistics because they provide a simple way to compare measurements made at different times or in different places. In this section, we'll investigate the meaning and use of index numbers, focusing on the Consumer Price Index (CPI). Let's start with an example using gasoline prices.

Table 2.1 shows the average price of gasoline in the United States for selected years from 1960 to 2010. (These are real prices from those years; that is, they have *not* been adjusted for inflation.) Suppose that, instead of the prices themselves, we want to know how the price of gasoline in different years compares with the 1980 price. One way to compare the prices would

BY THE WAY

By showing only 10-year increments, Table 2.1 hides substantial price variations that occurred within each decade. For example, while the 1990 average gasoline price was nearly identical to the 1980 price, actual prices during the 1980s ranged from below $0.90 to almost $1.50 per gallon. Similarly, during the 2000s the price ranged from a low of just over $1 (in late 2001) to a high of over $4 (in mid-2008) per gallon.

TABLE 2.1	Average Gasoline Prices (per gallon)		
Year	Price	Price as a percentage of 1980 price	Price index (1980 = 100)
1960	$0.31	25.4%	25.4
1970	$0.36	29.5%	29.5
1980	$1.22	100.0%	100.0
1990	$1.23	100.8%	100.8
2000	$1.56	127.9%	127.9
2010	$2.84	232.8%	232.8

Source: U.S. Department of Energy; prices are the year-long average for all grades of gasoline.

be to express each year's price as a percentage of the 1980 price. For example, by dividing the 1970 price by the 1980 price, we find that the 1970 price was 29.5% of the 1980 price:

$$\frac{1970 \text{ price}}{1980 \text{ price}} = \frac{\$0.36}{\$1.22} = 0.295 = 29.5\%$$

Proceeding similarly for each of the other years, we can calculate all the prices as percentages of the 1980 price. The third column of Table 2.1 shows the results. Note that the percentage for 1980 is 100%, because we chose the 1980 price as the reference value.

Now look at the last column of Table 2.1. It is identical to the third column, except we dropped the % signs. This simple change converts the numbers from percentages to a *price index*, which is one type of index number. The statement "1980 = 100" in the column heading shows that the reference value is the 1980 price. In this case, there's really no difference between stating the comparisons as percentages and as index numbers—it's a matter of choice and convenience. However, as we'll see shortly, it's traditional to use index numbers rather than percentages in cases where many factors are being considered simultaneously.

BY THE WAY

The term *index* is commonly used for almost any kind of number that provides a useful comparison, even when the numbers are not standard index numbers. For example, body mass index (BMI) provides a way of comparing people by height and weight, but is defined without any reference value. Specifically, body mass index is defined as weight (in kilograms) divided by height (in meters) squared.

Index Numbers

An **index number** provides a simple way to compare measurements made at different times or in different places. The value at one particular time (or place) must be chosen as the *reference value* (or *base value*). The index number for any other time (or place) is

$$\text{index number} = \frac{\text{value}}{\text{reference value}} \times 100$$

EXAMPLE ❶ Finding an Index Number

Suppose the cost of gasoline today is $3.60 per gallon. Using the 1980 price as the reference value, find the price index number for gasoline today.

SOLUTION We use the 1980 price of $1.22 per gallon (see Table 2.1) as the reference value to find the index number for the $3.60 gasoline price today:

$$\text{index number} = \frac{\text{current price}}{\text{1980 price}} \times 100 = \frac{\$3.60}{\$1.22} \times 100 = 295.1$$

This index number for the current price is 295.1, which means the current gasoline price is 295.1% of the 1980 price. $\cdots\bullet$

TIME OUT TO THINK

Find the actual price of gasoline today at a nearby gas station. What is the gasoline price index for today's price, with the 1980 price as the reference value?

Making Comparisons with Index Numbers

The primary purpose of index numbers is to facilitate comparisons. For example, suppose we want to know how much more expensive gas was in 2000 than in 1980. We can get the answer easily from Table 2.1, which uses the 1980 price as the reference value. This table shows that the price index for 2000 was 127.9, which means that the price of gasoline in 2000 was 127.9% of the 1980 price. Equivalently, we can say that the 2000 price was 1.279 times the 1980 price.

We can also do comparisons when neither value is the reference value. For example, suppose we want to know how much more expensive gas was in 1990 than in 1960. We find the answer by dividing the index numbers for the two years:

$$\frac{\text{index number for 1990}}{\text{index number for 1960}} = \frac{100.8}{25.4} = 3.97$$

A study of economics usually reveals that the best time to buy anything is last year.

—Marty Allen

The 1990 price was 3.97 times the 1960 price, or 397% of the 1960 price. In other words, the same amount of gas that cost $1.00 in 1960 would have cost $3.97 in 1990.

EXAMPLE 2 Using the Gas Price Index

Use Table 2.1 to answer the following questions.

a. Suppose that it cost $16.00 to fill your gas tank in 1980. How much did it cost to buy the same amount of gas in 2010?

b. Suppose that it cost $20.00 to fill your gas tank in 2000. How much did it cost to buy the same amount of gas in 1960?

SOLUTION

a. Table 2.1 shows that the price index (1980 = 100) for 2010 was 232.8, which means that the price of gasoline in 2010 was 232.8% of the 1980 price. So the 2010 price of gas that cost $16.00 in 1980 was

$$232.8\% \times \$16.00 = 2.328 \times \$16.00 = \$37.25$$

b. Table 2.1 shows that the price index (1980 = 100) for 2000 was 127.9 and the index for 1960 was 25.4. Dividing the index numbers, we find that the price of gasoline in 1960 was a fraction

$$\frac{\text{index number for 1960}}{\text{index number for 2000}} = \frac{25.4}{127.9} = 0.1986$$

of the price in 2000. Therefore, gas that cost $20.00 in 2000 cost $0.1986 \times \$20.00 = \$3.97$ in 1960. $\cdots\bullet$

Next, we compare the average baseball salaries for those two years:

$$\frac{\text{average baseball salary for 2011}}{\text{average baseball salary for 1987}} = \frac{\$3,319,000}{\$412,000} = 8.06$$

During the same period of time that average prices (as measured by the CPI) approximately doubled, the mean baseball salary rose by more than 800%. In other words, the mean salaries of major league baseball players rose more than four times as much as the overall rate of inflation.

$\cdots\bullet$

Other Index Numbers

The Consumer Price Index is only one of many index numbers that you'll see in news reports. Some are also price indices, such as the Producer Price Index (PPI), which measures the prices that producers (manufacturers) pay for the goods they purchase (rather than the prices that consumers pay). Other indices attempt to measure more qualitative variables. For example, the Consumer Confidence Index is based on a survey designed to measure consumer attitudes so that businesses can gauge whether people are likely to be spending or saving. New indices are created frequently by groups attempting to provide simple comparisons.

BY THE WAY

Thinking of becoming a comedian? Then you'll probably want to check the Cost of Laughing Index (there really is one!), which tracks costs of items such as rubber chickens, Groucho Marx glasses, and admission to comedy clubs.

Section 2.4

Statistical Literacy and Critical Thinking

1. **Index Number.** A newspaper reports that the Gas Price Index in 2011 was $3.92 per gallon. What is wrong with that statement?

2. **Index Number.** If computer costs in the year 2000 are set equal to 100 so they can be used as the basis for determining index numbers, and the index number for the year 2012 is 15, what do we know about computer costs in 2012 compared with computer costs in 2000?

3. **CPI.** If the prices of goods, services, and housing increase, must the Consumer Price Index increase? Explain.

4. **CPI.** If the Consumer Price Index increases, must wages also increase? Explain.

Concepts and Applications

Gasoline Price Index. In Exercises 5–8, use the gasoline price index from Table 2.1. Briefly explain your reasoning in each case.

5. **Current Data.** Suppose the cost of gasoline today is $5.00 per gallon. What is the price index number for gasoline today, with the 1980 price as the reference value? (See the data in Table 2.1.)

6. **2006 Index.** The average price of a gallon of gas was $2.62 in 2006. What is the price index for gasoline in 2006, with the 1980 price as the reference value?

7. **1998 Price.** Using the 1980 price as the reference value, the gasoline price index for 1998 is 0.9. What was the cost of a gallon of gasoline in 1998?

8. **2005 Price.** Using the 1980 price as the reference value, the gasoline price index for 2005 is 189.3. What was the cost of a gallon of gasoline in 2005?

9. **Reconstructing the Gasoline Price Index.** Identify the six price indexes in Table 2.1 that result from using the price from 2000 as the reference value. (*Hint*: Create a column for price as a percentage of 2000 price and another column giving the price index with 2000 = 100.)

10. **Reconstructing the Gasoline Price Index.** Identify the six price indexes in Table 2.1 that result from using the price from 1970 as the reference value. (*Hint:* Create a column for price as a percentage of 1970 price and another column giving the price index with 1970 = 100.)

11. **Using the Gas Price Index.** If it cost $19.52 to fill your gas tank in 1980, how much would it have cost to fill the same tank in 2010?

12. **Using the Gas Price Index.** If it cost $23.40 to fill your gas tank in 2000, how much would it have cost to fill the same tank in 2010?

13. **Private College Costs.** The average annual cost (tuition, fees, and room and board) at four-year private universities rose from

$5,600 in 1980 to $37,000 in 2010. Calculate the percentage rise in cost from 1980 to 2010, and compare it with the overall rate of inflation as measured by the Consumer Price Index.

14. Public College Costs. The average annual cost (tuition, fees, and room and board) at four-year public universities rose from $2,550 in 1980 to $16,100 in 2010. Calculate the percentage rise in cost from 1980 to 2010, and compare it with the overall rate of inflation as measured by the Consumer Price Index.

15. Home Prices—South. The typical (median) price of a new single-family home in the South (United States) rose from $75,300 in 1990 to $122,000 in 2010. Calculate the percentage rise in cost of a home from 1990 to 2010, and compare it with the overall rate of inflation as measured by the Consumer Price Index.

16. Home Prices—West. The typical (median) price of a new single-family home in the West (United States) rose from $129,600 in 1990 to $235,600 in 2010. Calculate the percentage rise in cost of a home from 1990 to 2010, and compare it with the overall rate of inflation as measured by the Consumer Price Index.

Housing Price Index. Realtors use an index to compare housing prices in major cities throughout the country. The index numbers for several cities are given in the table below. If you know the price of a particular home in your town, you can use the index to find the price of a comparable house in another town:

$$\begin{array}{c} \text{price} \\ \text{(other town)} \end{array} = \begin{array}{c} \text{price} \\ \text{(your town)} \end{array} \times \frac{\text{index in other town}}{\text{index in your town}}$$

Use the housing price index in Exercises 17–20.

City	Index	City	Index
Denver	100	Boston	358
Miami	194	Las Vegas	101
Phoenix	86	Dallas	81
Atlanta	90	Cheyenne	60
Baltimore	150	San Francisco	382

17. Housing Prices. If you see a house valued at $300,000 in Denver, find the price of a comparable house in Miami and Cheyenne.

18. Housing Prices. If you see a house valued at $500,000 in Boston, find the price of a comparable house in Baltimore and Phoenix.

19. Housing Prices If you see a house valued at $250,000 in Cheyenne, find the price of a comparable house in San Francisco and Boston.

20. Housing Prices. If you see a house valued at $1,000,000 in Boston, find the price of a comparable house in San Francisco and Cheyenne.

PROJECTS FOR THE INTERNET & BEYOND

21. Consumer Price Index. Go to the Consumer Price Index home page and find the latest news release with updated figures for the CPI. Summarize the news release and any important trends in the CPI.

22. Producer Price Index. Go to the Producer Price Index (PPI) home page. Read the overview and recent news releases. Write a short summary describing the purpose of the PPI and how it is different from the CPI. Also summarize any important recent trends in the PPI.

23. Consumer Confidence Index. Use a search engine to find recent news about the Consumer Confidence Index. After studying the news, write a short summary of what the Consumer Confidence Index is trying to measure and describe any recent trends in the Consumer Confidence Index.

24. Human Development Index. The United Nations Development Programme regularly releases its Human Development Report. A closely watched finding of this report is the Human Development Index (HDI), which measures the overall achievements in a country in three basic dimensions of human development: life expectancy, educational attainment, and adjusted income. Find the most recent copy of this report and investigate exactly how the HDI is defined and computed.

25. Convenience Store Index. Go to a local supermarket and find the prices of a few staples, such as bread, milk, juice, and coffee. Compute the total cost of those items. Then go to a few smaller convenience stores and find the prices of the same items. Using the supermarket total as the reference value, compute the index numbers for the convenience stores.

IN THE NEWS

26. Consumer Price Index. Find a recent news report that includes a reference to the Consumer Price Index. Briefly describe how the Consumer Price Index is important in the story.

27. Index Numbers. Find a recent news report that includes an index number other than the Consumer Price Index. Describe the index number and its meaning, and discuss how the index is important in the story.

1. **Astrology.** In a Harris Interactive poll of 2,303 adults, 26% said that they believe in astrology.

   a. Find the true number of the poll respondents who believe in astrology.

   b. Is the number of respondents who believe in astrology from a discrete data set or a continuous data set? Explain.

   c. Of the 2,303 adults who were polled, 1,382 believe in the devil. What percentage of respondents believe in the devil?

   d. Among the respondents, 1,382 believe in the devil, and the number who believe in UFOs is 53% of that number. How many believe in UFOs?

   e. If we compile the ages of all respondents, what is the level of measurement (nominal, ordinal, interval, ratio) of those ages?

   f. If we classify the respondents according to gender, what is the level of measurement (nominal, ordinal, interval, ratio) of the gender data?

2. **AOL Poll.** In an America OnLine poll of 671 people, 36% answered yes when asked if today's NFL games are too long.

   a. What is the number of respondents who answered yes?

   b. Among the respondents, 40 answered by saying that they were not sure. What is the percentage of "not sure" responses?

   c. Given that the possible responses are "yes, no, not sure," what is the level of measurement of those responses? (nominal, ordinal, interval, ratio)

   d. Given that the poll was conducted by asking America OnLine users to respond to a question that was posted on the Web site, what do you conclude about the poll results? Is it likely that the results reflect the opinion of the general population?

3. **Health Care Spending.** Total spending on health care in the United States rose from $80 billion in 1973 to $2.5 trillion in 2010. The Consumer Price Index was 44.4 in 1973, and it was 218.1 in 2010 (with 1982–1984= 100). Compare the change in health care spending from 1973 to 2010 to the overall rate of inflation as measured by the Consumer Price Index.

4. **Minimum Wage.** The accompanying table lists the federal hourly minimum in both actual dollars at the time and 1996 dollars. The table entries correspond to years in which the minimum wage changed (based on data from the Department of Labor).

Year	Actual Dollars	1996 Dollars
1938	0.25	2.78
1939	0.30	3.39
1945	0.40	3.49
1950	0.75	4.88
1956	1.00	5.77
1961	1.25	6.41
1967	1.40	6.58
1968	1.60	7.21
1974	2.00	6.37
1976	2.30	6.34
1978	2.65	6.38
1979	2.90	6.27
1981	3.35	5.78
1990	3.50	4.56
1991	4.25	4.90
1996	4.75	4.75
1997	5.15	5.03
2009	7.25	5.22

   a. According to the table, how much is $0.25 in 1938 dollars worth in 1996 dollars?

   b. According to the table, how much is $1.00 in 1956 dollars worth in 1996 dollars?

   c. Note that the minimum wage for 2009 in actual dollars is greater than the minimum wage for 2009 in 1996 dollars. What does that tell us about the minimum wage in 2009?

1. Sitting heights of 750 randomly selected licensed drivers are measured. Are those values continuous or discrete?

2. What is the level of measurement of the values described in the preceding exercise? (nominal, ordinal, interval, ratio)

3. A researcher measures the sitting height of a subject and records a value of 91.4 cm, but the subject's actual sitting height is 89.0 cm. What is the absolute error?

4. A researcher measures the sitting height of a subject and records a value of 91.4 cm, but the subject's actual sitting height is 89.0 cm. What is the relative error?

5. Recorded survey results include the states in which respondents reside. What is the level of measurement of those data? (nominal, ordinal, interval, ratio)

6. Sitting heights of 750 randomly selected licensed drivers are measured and 38 of those heights are greater than 97.2 cm. What is the percentage of sitting heights greater than 97.2 cm?

7. In a Gallup poll of 1,038 adults, 5% of the respondents said that second-hand smoke is not at all harmful. How many respondents said that second-hand smoke is not at all harmful?

8. Two different students measure the height of an instructor who is actually 178.44 cm tall. The first student obtains a measurement of 178 cm and the second student obtains a measurement of 179.18 cm. Which measurement is more accurate? Which measurement is more precise?

9. The Telektronics Company has been in business for five years, and the table lists the net profits in each of those years. Using the first year as a reference, find the index number for the net profit in the second year.

Year	Net Profit
1	$12,335
2	$15,257
3	$23,444
4	$31,898
5	$47,296

10. Refer to the same table used for Exercise 9. If the net profit in the sixth year is projected to be 12% more than in the fifth year, what is that projected net profit for the sixth year?

F   CUS ON
POLITICS

Who Benefited Most from Lower Tax Rates?

Politicians have a remarkable capacity to cast numbers in whatever light best supports their beliefs. There are many ways in which numbers can be made to support a particular position, but one of the most common occurs through selective use of relative (percentages) and absolute numbers.

Consider the two charts shown in Figure 2.4. Both purport to show effects of the tax rate cuts originally enacted under President Bush in 2001 and renewed under President Obama in 2010. (Under the 2001 law, the tax cuts would have expired at the end of 2010. The 2010 law extended them through 2012.) The chart in Figure 2.4a, created by supporters of the tax cuts, indicates that the rich ended up paying more under the tax cuts than they would have otherwise. Figure 2.4b, created by opponents of the tax cuts, shows that the rich received far more benefit from the tax cuts than lower-income taxpayers. The two charts therefore seem contradictory, because the first seems to indicate that the rich paid more while the second seems to indicate that they paid less.

Which story is right? In fact, both of the graphs are accurate and show data from reputable sources. (The Department of the Treasury, cited as the source of the data for Figure 2.4a, is an agency of the federal government; the Joint Committee on Taxation, cited in Figure 2.4b, is a nonpartisan committee of the U.S. Congress.) The seemingly opposing claims arise from the way in which each group chose its data.

The tax cut supporters show the *percentage of total taxes* that the rich paid with the cuts and what calculations suggest they would have paid without them. The title stating that the "rich pay more" therefore means that the tax cuts led them to pay a higher percentage of total taxes. However, if total tax revenue also was lower than it would have been without the cuts (as it was), a higher percentage of total taxes could still mean lower absolute dollars. It is these lower absolute dollars that are shown by the opponents of the tax cut.

Politicians and government officials usually abuse numbers and logic in the most elementary ways. They simply cook figures to suit their purpose, use obscure measures of economic performance, and indulge in horrendous examples of chart abuse, all in the name of disguising unpalatable truths.

—A. K. Dewdney, *200% of Nothing*

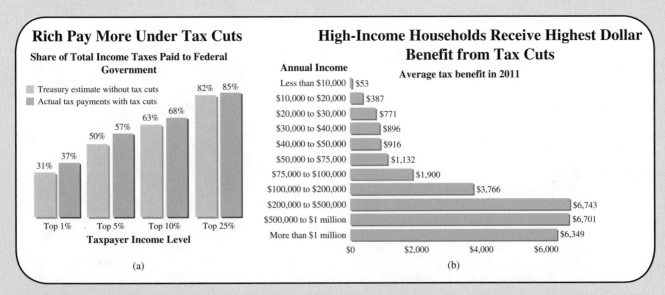

Rich Pay More Under Tax Cuts

Share of Total Income Taxes Paid to Federal Government

- Treasury estimate without tax cuts
- Actual tax payments with tax cuts

Top 1%: 31%, 37%
Top 5%: 50%, 57%
Top 10%: 63%, 68%
Top 25%: 82%, 85%

Taxpayer Income Level

(a)

High-Income Households Receive Highest Dollar Benefit from Tax Cuts

Annual Income — Average tax benefit in 2011

Less than $10,000	$53
$10,000 to $20,000	$387
$20,000 to $30,000	$771
$30,000 to $40,000	$896
$40,000 to $50,000	$916
$50,000 to $75,000	$1,132
$75,000 to $100,000	$1,900
$100,000 to $200,000	$3,766
$200,000 to $500,000	$6,743
$500,000 to $1 million	$6,701
More than $1 million	$6,349

$0 $2,000 $4,000 $6,000

(b)

Figure 2.4   Both charts purport to show effects on higher-income households of tax cuts enacted in 2001 and renewed in 2010, but they are designed to support opposing conclusions. (a) Adapted from a graph published in *The American—The Journal of the American Enterprise Institute.* (b) Adapted from a graph published by the Center on Budget and Policy Priorities (cbpp.org).

Which side was being more fair? Neither, really. The supporters have deliberately focused on a percentage in order to mask the absolute change, which would be less favorable to their position. The opponents focused on the absolute change, but neglected to mention the fact that the wealthy pay most of the taxes. Unfortunately, this type of "selective truth" is very common when it comes to numbers, especially those tied up in politics.

QUESTIONS FOR DISCUSSION

1. Discuss the chart in Figure 2.4a. Why does it only show taxpayer income levels from the top 1% to the top 25%? What does it tell us about the relative tax burden on each group shown? Do you think its claim that the "rich pay more" is an honest depiction or a distortion of the facts? Defend your opinion.

2. Discuss the chart in Figure 2.4b. Why does it show actual incomes rather than incomes as a percentage of all taxpayers? What does it tell us about the effects of the tax cuts on each income level? Do you think it supports the claim in its title? Defend your opinion.

3. Do you think that either of the charts in Figure 2.4 accurately portrays the overall "fairness" of the tax cuts? If so, which one and why? If not, how do you think the data could have been portrayed more fairly?

5. Do a Web search on "benefit of tax cuts" and find an article arguing for or against a particular tax cut proposal or policy. Briefly summarize the article and discuss whether you think the article made its case well.

6. Have any tax law changes been proposed by the U.S. Congress or the President this year? Discuss the fairness of current proposals.

• • • • • • • • • • • • • • • • • • • •

FOCUS ON

ECONOMICS

Can a Redefined CPI Help Solve the Budget Crisis?

We're all aware that the U.S. government has a sky-high debt of more than $15 trillion, which amounts to approximately $50,000 for every man, woman, and child in the nation. The only way to pay off the debt is through some combination of increased government revenues or decreased government spending. Unfortunately, a quick look at the news will show that neither option is politically popular, because some people are adamantly opposed to tax increases while others are adamantly opposed to cuts in government benefits. But what if part of the problem exists only because of a misunderstanding and misapplication of the Consumer Price Index (CPI)?

Benefits are popular. Paying for benefits is extremely unpopular.

—John Danforth, former U.S. Senator
(Republican–Missouri)

This question arises because most economists believe that the CPI overstates the true effects of inflation. The thrust of the economic argument is that the standard calculation of the CPI contains at least two systematic errors that overstate how inflation affects the cost of living. First, from one month to the next, the CPI is based on changes in the prices of particular items at particular stores. In reality, however, if the price of an item rises at one store, consumers often buy it more cheaply at another store, and if the price rises at all stores, consumers may substitute a similar but lower-priced item (such as a different brand of the same product). This "price substitution" effect means that consumers don't find their actual costs rising as much as the CPI indicates. Second, the CPI tracks changes in the price of "typical" items purchased by consumers at any given time, but it does not account for the effects of changes or improvements in these items with time. For example, the data used in computing the CPI may show that a typical cell phone is 10% more expensive than a cell phone from a few years ago, but these data do not account for the fact that today's cell phones have many more capabilities. The data therefore overstate the effect of the cell phone price rise, because you are getting so much more for your money today.

The accuracy of the CPI as a measure of the increase in the cost of living may seem like an academic debate, but it has real consequences for the government budget. The reason is that many budget items are tied to inflation, including one that is particularly important to government revenue and another that is particularly important to government spending. On the revenue side, the government raises the income thresholds for different tax rates each year, with the goal of making sure that people's tax rates go up only if they actually increase their standard of living, as opposed to going up just because inflation has made living more costly. For decades, the increases in the thresholds have been tied to the CPI. But if the CPI overstates the effects of inflation on living standards, then instead of holding rates steady, the threshold changes have effectively lowered tax rates. On the spending side, the government annually increases the checks provided to recipients of Social Security and other benefits, again with the goal of making sure that the benefits reflect changes in the cost of living. (The annual increase is called a "cost of living adjustment," or COLA for short.) If the CPI is overstating the effects of inflation, then these cost of living adjustments have been larger than is really needed to maintain the same standard of living.

Clearly, the federal government's budget picture could be dramatically improved by linking changes in tax rates and cost of living adjustments to a value that more accurately reflects inflation than the current CPI. In fact, the government already computes something called a "chained CPI" that was designed specifically to address this problem. Figure 2.5 shows how both revenue and spending would be different if the government tied changes to the chained CPI rather than the standard CPI. If you add all the revenue increases and spending savings that are shown in the chart, you'll find that this simple change would reduce the government's deficit by more than $200 billion over the next decade. The savings would be even greater in the following decade, because each year's savings builds on the previous year's.

QUESTIONS FOR DISCUSSION

1. Find examples in your own spending or the spending of friends and family of substitution effects and the purchase of products that were unavailable or lower quality

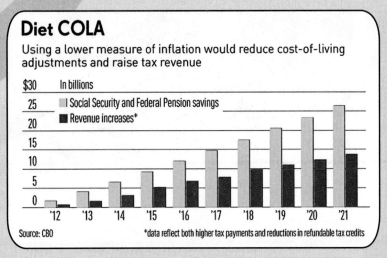

Diet COLA

Using a lower measure of inflation would reduce cost-of-living adjustments and raise tax revenue

Source: CBO   *data reflect both higher tax payments and reductions in refundable tax credits

Figure 2.5 Revenue increases and spending savings that would arise from trying to account for the ways in which the CPI tends to overstate inflation. Adapted from "Obama Mulls COLA Switch, Debt Talks Will Continue," by Jed Graham, from *Investor's Business Daily*, July 7, 2011. Copyright © 2011 by *Investor's Business Daily*. Reprinted with permission.

a few years ago. Overall, do you think these examples support the claim that the CPI overstates the effects of inflation? Defend your opinion.

2. Because adjusting the CPI would lead to a slower rise in the income thresholds for various tax rates, conservatives often claim that this change would mean "tax increases." Similarly, because the change would reduce the annual increases in cost of living adjustments, liberals often claim that it would mean a "benefit cut" to Social Security recipients. Do you think either claim is accurate? Defend your opinion.

3. As this text goes to press, the government is considering making a change to the chained CPI. Has it happened yet? Investigate the current status of the debate over whether the CPI should be adjusted for the purposes of setting tax rates and cost of living adjustments.

4. In Figure 2.5, notice that the revenue increases and spending savings get larger each year. Discuss why this occurs, focusing on how a change in the CPI compounds over time, much like compound interest in a bank account.

5. While most economists believe that the CPI overstates the effects of inflation, a few believe hat it actually understates the effects, especially on lower-income people. Research why some economists come to this conclusion, then draw your own conclusion: Does the CPI sometimes understate the effects of inflation?

3

Visual Displays of Data

LEARNING GOALS

3.1 Frequency Tables

Be able to create and interpret frequency tables.

3.2 Picturing Distributions of Data

Be able to create and interpret basic bar graphs, dotplots, pie charts, histograms, stemplots, line charts, and time-series graphs.

3.3 Graphics in the Media

Understand how to interpret the many types of more complex graphics that are commonly found in news media.

3.4 A Few Cautions About Graphics

Critically evaluate graphics and identify common ways in which graphics can be misleading.

FOCUS TOPICS

Whether you look at a newspaper, a corporate annual report, or a government study, you are almost sure to see tables and graphs of statistical data. Some of these tables and graphs are very simple; others can be quite complex. Some make it easy to understand the data; others may be confusing or misleading. In this chapter, we'll study some of the many ways in which statistical data are commonly displayed in tables and graphs. Because the ability to convey concepts through graphs is so valuable in today's data-driven society, the skills developed in this chapter are crucial for success in nearly every profession.

> The greatest value of a picture is when it forces us to notice what we never expected to see.
>
> —John Tukey

3.1 FREQUENCY TABLES ·······················

A teacher records the following list of the grades she gave to her 25 students on a set of essays:

A C C B C D C C F D C C C
B B A B D B A A B F C B

This list contains all of the grades, but it isn't easy to read. A better way to display these data is with a **frequency table** (Table 3.1)—a table showing the number of times, or **frequency**, that each grade appears. The five possible grades (A, B, C, D, F) are called the **categories** (or classes) for the table.

TABLE 3.1	Frequency Table for a Set of Essay Grades
Grade	**Frequency**
A	4
B	7
C	9
D	3
F	2
Total	**25**

> **Definitions**
>
> A basic **frequency table** has two columns:
>
> - The first column lists all the **categories** of data.
>
> - The second column lists the **frequency** of each category, which is the number of data values in the category.

EXAMPLE 1 Taste Test

The Rocky Mountain Beverage Company wants feedback on its new product, Coral Cola, and sets up a taste test with 20 people. Each individual is asked to rate the taste of the cola on a 5-point scale:

(bad taste) 1 2 3 4 5 (excellent taste)

The 20 ratings are as follows:

1 3 3 2 3 3 4 3 2 4
2 3 5 3 4 5 3 4 3 1

Construct a frequency table for these data.

SOLUTION The variable of interest is *taste*, and this variable can take on five values: the taste categories 1 through 5. (Note that the data are qualitative and at the ordinal level of measurement.) We construct a table with these five categories in the left column and their frequencies in the right column, as shown in Table 3.2. ··•

TABLE 3.2	Taste Test Ratings
Taste scale	**Frequency**
1	2
2	3
3	9
4	4
5	2
Total	**20**

Relative and Cumulative Frequency

Look again at the essay grades listed in Table 3.1. In some cases, we might be more interested in the fraction (or percentage) of students, rather than the actual number of students, who received each grade. Such a fraction is called a **relative frequency**. For example, because 4 of the 25 students received A grades, the relative frequency of A grades is 4/25, or 0.16, or 16%.

In other cases, we might want to know the answer to questions like How many students got a grade of C or better? The answer to this question is the sum of the frequencies of A, B, and C grades, which we call the **cumulative frequency**. Table 3.3 repeats the data from Table 3.1, but this time with added columns showing the calculations for the relative and cumulative frequencies.

TABLE 3.3	Essay Grade Table with Relative and Cumulative Frequency		
Grade	**Frequency**	**Relative frequency**	**Cumulative frequency**
A	4	4/25 = 0.16	4
B	7	7/25 = 0.28	7 + 4 = 11
C	9	9/25 = 0.36	9 + 7 + 4 = 20
D	3	3/25 = 0.12	3 + 9 + 7 + 4 = 23
F	2	2/25 = 0.08	2 + 3 + 9 + 7 + 4 = 25
Total	**25**	**1**	**25**

> **TECHNICAL NOTE**
>
> Most frequency tables start with the lowest category, but tables of grades commonly start with the highest category (A). Note that category can affect cumulative frequency. For example, if we reversed the order in Table 3.3, the cumulative frequency for C would be the number of grades of C or lower, rather than C or higher.

Note that the sum of the relative frequencies must equal 1 (or 100%), because each individual relative frequency is a fraction of the total frequency. (Rounding sometimes causes the total to be slightly different from 1.) The cumulative frequency for the last category must always equal the total number of data values, because it represents the total number of data values in that category *and* all the preceding categories.

> **Definitions**
>
> The **relative frequency** of any category is the fraction (or percentage) of the data values that fall in that category:
>
> $$\text{relative frequency} = \frac{\text{frequency in category}}{\text{total frequency}}$$
>
> The **cumulative frequency** of any category is the number of data values in that category *and all preceding* categories.

Keep in mind that cumulative frequencies make sense only for data categories that have a clear order. That is, we can use cumulative frequencies for data at the ordinal, interval, and ratio levels of measurement, but not for data at the nominal level of measurement.

EXAMPLE 2   Taste Test: Relative and Cumulative Frequency

Using the taste test data from Example 1, create a frequency table with columns for the relative and cumulative frequencies. What percentage of the respondents gave the cola the highest rating? What percentage gave the cola one of the three lowest ratings?

SOLUTION   We find the relative frequencies by dividing the frequency in each category by the total frequency of 20. We find the cumulative frequencies by adding the frequency in each category to the sum of the frequencies in all preceding categories. Table 3.4 shows the results. The relative frequency column shows that 0.10, or 10%, of the respondents gave the cola the highest rating. The cumulative frequency column shows that 14 out of 20 people, or 70%, gave the cola a rating of 3 or lower.

> **TECHNICAL NOTE**
>
> A cumulative frequency divided by the total frequency, such as the 70% of respondents in the taste test giving a rating of 3 or higher, is called a *relative cumulative frequency*.

TABLE 3.4	Relative and Cumulative Frequencies		
Taste scale	**Frequency**	**Relative frequency**	**Cumulative frequency**
1	2	2/20 = 0.10	2
2	3	3/20 = 0.15	3 + 2 = 5
3	9	9/20 = 0.45	9 + 3 + 2 = 14
4	4	4/20 = 0.20	4 + 9 + 3 + 2 = 18
5	2	2/20 = 0.10	2 + 4 + 9 + 3 + 2 = 20
Total	20	1	20

Binning Data

When we deal with quantitative data categories, it's often useful to group, or **bin**, the data into categories that cover a range of possible values. For example, in a table of income levels, it might be useful to create bins of $0 to $20,000, $20,001 to $40,000, and so on. In this case, the frequency of each bin is simply the number of people with incomes in that bin.

USING TECHNOLOGY—FREQUENCY TABLES

Excel Excel is easy to use for statistical tables and calculations. The following steps work with the essay grade data from Table 3.1. The screenshot on the left shows the Excel table with the formulas, and the one on the right shows the results of the formulas.

1. Create columns for the grade and frequency data, which you must type in manually (columns B and C below). At the bottom of column C, use the SUM function to compute the total frequency, which is entered in cell C8.

2. Compute the relative frequency (Column D) by dividing each frequency in Column C by the total frequency from cell C8. You can enter the formula for the first row (= C3/$C$8) and then use the "fill down" editing option to put the correct formulas in the remaining rows. Note: When using the "fill down" option, you must include the dollar signs in front of C and 8 to make the reference to cell C8 an "absolute cell reference." Without these dollar signs, the "fill down" option would make the cell reference shift down (becoming C9, C10, etc.) in each row, which would be incorrect in this case.

3. Cumulative frequency (Column E) is the total of all the frequencies up to a given category. The first row shows "= C3" because cell C3 contains the frequency for A grades. The next row (= E3 + C4) starts with the value in the prior row (cell E3) and add the frequency for B grades (cell C4). The pattern continues for the remaining rows, which you can fill with the "fill down" option.

	A	B	C	D	E
1					
2		Grade	Frequency	Relative Frequency	Cumulative Frequency
3		A	4	=C3/$C$8	=C3
4		B	7	=C4/$C$8	=E3+C4
5		C	9	=C5/$C$8	=E4+C5
6		D	3	=C6/$C$8	=E5+C6
7		F	2	=C7/$C$8	=E6+C7
8		Total	=SUM(C3:C7)	=SUM(D3:D7)	=C8

	A	B	C	D	E
1					
2		Grade	Frequency	Relative Frequency	Cumulative Frequency
3		A	4	16%	4
4		B	7	28%	11
5		C	9	36%	20
6		D	3	12%	23
7		F	2	8%	25
8		Total	25	100%	25

EXAMPLE 3   Binned Exam Scores

Consider the following set of 20 scores from a 100-point exam:

76 80 78 76 94 75 98 77 84 88 81 72 91 72 74 86 79 88 72 75

Determine appropriate bins and make a frequency table. Include columns for relative and cumulative frequency, and interpret the cumulative frequency for this case.

SOLUTION   The scores range from 72 to 98. One way to group the data is with 5-point bins. The first bin represents scores from 95 to 99, the second bin represents scores from 90 to 94, and so on. Note that there is no overlap between bins. We then count the frequency (the number of scores) in each bin. For example, only 1 score is in bin 95 to 99 (the high score of 98) and 2 scores are in bin 90 to 94 (the scores of 91 and 94). Table 3.5 shows the complete frequency table. In this case, we interpret the cumulative frequency of any bin to be the total number of scores in *or above* that bin. For example, the cumulative frequency of 6 for the bin 85 to 89 means that there were 6 scores of 85 or higher.

TABLE 3.5	Frequency Table for Binned Exam Scores		
Scores	**Frequency**	**Relative frequency**	**Cumulative frequency**
95 to 99	1	1/20 = 0.05	1
90 to 94	2	2/20 = 0.10	2 + 1 = 3
85 to 89	3	3/20 = 0.15	3 + 2 + 1 = 6
80 to 84	3	3/20 = 0.15	3 + 3 + 2 + 1 = 9
75 to 79	7	7/20 = 0.35	7 + 3 + 3 + 2 + 1 = 16
70 to 74	4	4/20 = 0.20	4 + 7 + 3 + 3 + 2 + 1 = 20
Total	**20**	**1**	**20**

"Data! Data! Data!" he cried impatiently. "I can't make bricks without clay."

— Sherlock Holmes in Sir Arthur Conan Doyle's *The Adventure of the Copper Beeches*

Statistical Literacy and Critical Thinking

1. **Frequency Table.** What is a frequency table? Explain what we mean by the categories (or classes) and frequencies.

2. **Relative Frequency.** A frequency table of grades has five classes (A, B, C, D, F) with frequencies of 4, 12, 16, 6, and 2. What are the relative frequencies of the five classes?

3. **Cumulative Frequency.** A frequency table of grades has five classes with frequencies of 4, 12, 16, 6, and 2. What are the corresponding cumulative frequencies of the five classes?

4. **Frequency Table.** The first class in a frequency table of incomes shows a frequency of 24 corresponding to the range of values from $0 to $999. Using only this information about the frequency table, is it possible to identify the original 24 sample values that are summarized by this class? Explain.

Does It Make Sense? For Exercises 5–8, decide whether the statement makes sense (or is clearly true) or does not make sense (or is clearly false). Explain clearly. Not all of these statements have definitive answers, so your explanation is more important than your chosen answer.

5. **Frequency Table.** A friend tells you that her frequency table has two columns labeled *State* and *Median Income*.

6. **Relative Frequency.** The relative frequency of category A in a table is 0.27 or 27%.

7. **Cumulative Frequency.** The cumulative frequency of a category in a table is 25.5.

8. **Bins.** For a given data set of IQ scores, if you increase the width of the bins, the number of bins decreases.

Concepts and Applications

9. **Frequency Table Practice.** Professor Diaz records the following final grades in one of her courses:

    A A A B B B B B B B B B B C C
    C C C C C C C C C D D F F F F

    Construct a frequency table for these grades. Include columns for relative frequency and cumulative frequency.

10. **Frequency Table Practice.** A guidebook for San Francisco lists 2 one-star restaurants, 12 two-star restaurants, 18 three-star restaurants, 6 four-star restaurants, and 2 five-star restaurants. Make a frequency table for these ratings. Include columns for relative frequency and cumulative frequency.

11. **Weights of Coke.** Construct a frequency table for the weights (in pounds) given below of 36 cans of regular Coke.

Start the first bin at 0.7900 pound and use a bin width of 0.0050 pound. Discuss your findings.

0.8192	0.8194	0.8211	0.8176
0.8062	0.8143	0.8110	0.8152
0.7901	0.8152	0.8079	0.8161
0.8161	0.8163	0.8194	0.8247
0.8165	0.8172	0.8150	0.8264
0.8207	0.8073	0.8295	0.8170
0.8150	0.8189	0.8181	0.8284
0.8128	0.8229	0.8251	0.8244
0.8244	0.8126	0.8044	0.8192

12. **Weights of Diet Coke.** Construct a frequency table for the weights (in pounds) given below of 36 cans of Diet Coke. Start the first bin at 0.7750 pound and use a bin width of 0.0050 pound. Discuss your findings.

0.7773	0.7874	0.7868	0.7802
0.7806	0.7907	0.7879	0.7833
0.7923	0.7910	0.7813	0.7859
0.7822	0.7896	0.7839	0.7861
0.7874	0.7852	0.7870	0.7826
0.7837	0.7872	0.7923	0.7760
0.7758	0.7822	0.7844	0.7892
0.7830	0.7771	0.7881	0.7822
0.7852	0.7879	0.7885	0.7811

13. **Oscar-Winning Actors.** The following data show the ages of all Academy Award–winning male actors at the time when they won their award, through 2012. Make a frequency table for the data, using bins of 20–29, 30–39, and so on. Discuss your findings.

    44 41 62 52 41 34 34 52 41 37 38 34 32 40 43
    56 41 39 49 57 41 38 42 52 51 35 30 39 41 44
    49 35 47 31 47 37 57 42 45 42 44 62 43 42 48
    49 56 38 60 30 40 42 36 76 39 53 45 36 62 43
    51 32 42 54 52 37 38 32 45 60 46 40 36 47 29
    43 37 38 45 50 48 60 50 39

14. **Body Temperatures.** The following data show the body temperatures (°F) of randomly selected subjects. Construct a frequency table with seven classes: 96.9–97.2, 97.3–97.6, 97.7–98.0, and so on.

98.6	98.6	98.0	98.0	99.0	98.4	98.4	98.4
98.4	98.6	98.6	98.8	98.6	97.0	97.0	98.8
97.6	97.7	98.8	98.0	98.0	98.3	98.5	97.3
98.7	97.4	98.9	98.6	99.5	97.5	97.3	97.6
98.2	99.6	98.7	99.4	98.2	98.0	98.6	98.6
97.2	98.4	98.6	98.2	98.0	97.8	98.0	98.4
98.6	98.6						

15. **Missing Information.** The following table shows grades for a term paper in an English class. The table is incomplete. Use the information given to fill in the missing entries and complete the table.

Category	Frequency	Relative frequency
A	?	?
B	?	18%
C	?	24%
D	11	?
F	6	?
Total	50	?

16. **Missing Information.** The following table shows grades for performances in a drama class. The table is incomplete. Use the information given to fill in the missing entries and complete the table.

Category	Frequency	Cumulative frequency
A	?	1
B	6	?
C	7	?
D	?	23
F	?	25
Total	?	?

17. **Loaded Die.** One of the authors drilled a hole in a die, filled it with a lead weight, and then proceeded to roll it. The results are given in the following frequency table.

    a. According to the data, how many times was the die rolled?

    b. How many times was the outcome greater than 2?

    c. What percentage of outcomes were 6?

    d. List the relative frequencies that correspond to the given frequencies.

    e. List the cumulative frequencies that correspond to the given frequencies.

Outcome	Frequency
1	27
2	31
3	42
4	40
5	28
6	32

18. **Interpreting Family Data.** Consider the following frequency table for the number of children in American families.

    a. According to the data, how many families are there in America?

    b. How many families have two or fewer children?

    c. What percentage of American families have no children?

    d. What percentage of American families have three or more children?

Number of children	Number of families (millions)
0	35.54
1	14.32
2	13.28
3	5.13
4 or more	1.97

19. **Computer Keyboards.** The traditional keyboard configuration is called a *Qwerty* keyboard because of the positioning of the letters QWERTY on the top row of letters. Developed in 1872, the Qwerty configuration supposedly forced people to type slower so that the early typewriters would not jam. Developed in 1936, the Dvorak keyboard supposedly provides a more efficient arrangement by positioning the most used keys on the middle row (or "home" row), where they are more accessible.

A *Discover* magazine article suggested that you can measure the ease of typing by using this point rating system: Count each letter on the home row as 0, count each letter on the top row as 1, and count each letter on the bottom row as 2. For example, the word *statistics* would result in a rating of 7 on the Qwerty keyboard and 1 on the Dvorak keyboard, as shown below.

	S	T	A	T	I	S	T	I	C	S	
Qwerty keyboard	0	1	0	1	1	0	1	1	2	0	(sum − 7)
Dvorak keyboard	0	0	0	0	0	0	0	0	1	0	(sum − 1)

Using this rating system with each of the 52 words in the Preamble to the Constitution, we get the rating values below.

Qwerty Keyboard Word Ratings:

2	2	5	1	2	6	3	3	4	2	4	0	5
7	7	5	6	6	8	10	7	2	2	10	5	8
2	5	4	2	6	2	6	1	7	2	7	2	3
8	1	5	2	5	2	14	2	2	6	3	1	7

Dvorak Keyboard Word Ratings:

2	0	3	1	0	0	0	0	2	0	4	0	3
4	0	3	3	1	3	5	4	2	0	5	1	4
0	3	5	0	2	0	4	1	5	0	4	0	1
3	0	1	0	3	0	1	2	0	0	0	1	4

    a. Create a frequency table for the Qwerty word ratings data. Use bins of 0–2, 3–5, 6–8, 9–11, and 12–14 Include a column for relative frequency.

    b. Create a frequency table for the Dvorak word ratings data, using the same bins as in part a. Include a column for relative frequency.

    c. Based on your results from parts a and b, which keyboard arrangement is easier for typing? Explain.

20. Double Binning. The students in a statistics class conduct a transportation survey of students in their high school. Among other data, they record the age and mode of transportation between home and school for each student. The following table gives some of the data that were collected. For age: 1 = 14 years, 2 = 15 years, 3 = 16 years, 4 = 17 years, 5 = 18 years. For transportation: 1 = walk, 2 = school bus, 3 = public bus, 4 = drive, 5 = other.

Student	Age	Transportation	Student	Age	Transportation
1	1	1	11	3	5
2	5	1	12	5	5
3	2	2	13	1	2
4	3	5	14	5	5
5	4	3	15	5	5
6	1	1	16	4	4
7	5	2	17	2	2
8	2	1	18	3	1
9	3	4	19	3	3
10	1	3	20	1	4

a. Classify the two variables, *age* and *transportation*, as qualitative or quantitative, and give the level of measurement for each.

b. In order to be analyzed or displayed, the data must be binned with respect to both variables. Count the number of students in each of the 25 age/transportation categories and fill in the blank cells in the following table.

		Transportation				
		1	2	3	4	5
Age	1					
	2					
	3					
	4					
	5					

PROJECTS FOR THE INTERNET & BEYOND

21. Energy Table. The U.S. Energy Information Administration (EIA) Web site offers dozens of tables relating to energy use, energy prices, and pollution. Explore the selection of tables. Find a table of raw data that is of interest to you and convert it to an appropriate frequency table. Briefly discuss what you can learn from the frequency table that is less obvious in the raw data table.

22. Endangered Species. The Web site for the World Conservation Monitoring Centre in Great Britain provides data on extinct, endangered, and threatened animal species. Explore these data and summarize some of your more interesting findings with frequency tables.

23. Navel Data. The *navel ratio* is defined to be a person's height divided by the height (from the floor) of his or her navel. An old theory says that, on average, the navel ratio of humans is the golden ratio: $(1+\sqrt{5})/2$. Measure the navel ratio of each person in your class. What percentage of students have a navel ratio within 5% of the golden ratio? What percentage of students have a navel ratio within 10% of the golden ratio? Does the old theory seem reliable?

24. Your Own Frequency Table (Unbinned). Collect your own frequency data for some set of categories that will *not* require binning. (For example, you might collect data by asking friends to do a taste test on some brand of cookie.) State how you collected your data, and make a list of all your raw data. Then summarize the data in a frequency table. Include a column for relative frequency, and also include a column for cumulative frequency if it is meaningful.

25. Your Own Frequency Table (Binned). Collect your own frequency data for some set of categories that *will* require binning (for example, weights of your friends or scores on a recent exam). State how you collected your data, and make a list of all your raw data. Then summarize the data in a frequency table. Include columns for relative frequency and cumulative frequency.

IN THE NEWS

26. Frequency Tables. Find a recent news article that includes some type of frequency table. Briefly describe the table and how it is useful to the news report. Do you think the table was constructed in the best possible way for the article? If so, why? If not, what would you have done differently?

27. Relative Frequencies. Find a recent news article that gives at least some data in the form of relative frequencies. Briefly describe the data, and discuss why relative frequencies were useful in this case.

28. Cumulative Frequencies. Find a recent news article that gives at least some data in the form of cumulative frequencies. Briefly describe the data, and discuss why cumulative frequencies were useful in this case.

29. Temperature Data. Look for a weather report that lists yesterday's high temperatures in many American cities. Choosing appropriate bins, make a frequency table for the high temperature data. Include columns for relative frequency and cumulative frequency. Briefly describe how and why you chose your bins.

PICTURING DISTRIBUTIONS OF DATA ·······················

A frequency table shows how a variable is distributed over chosen categories, so we say that it summarizes the **distribution** of data. We can often gain deeper insight into a distribution with a picture or graph. In this section, we'll study a few of the most common ways of visualizing data distributions.

> **Definition**
>
> The **distribution** of a variable refers to the way its values are spread over all possible values. We can summarize a distribution with a table or a graph.

Bar Graphs, Dotplots, and Pareto Charts

A **bar graph** uses a set of bars to represent the frequency (or relative frequency) of each category: the higher the frequency, the longer the bar. The bars can be either vertical or horizontal. Figure 3.1 shows a vertical bar graph based on the essay grade data in Table 3.1; there are five bars because there are five data categories (the grades A, B, C, D, F). Several key features help make the graph look good:

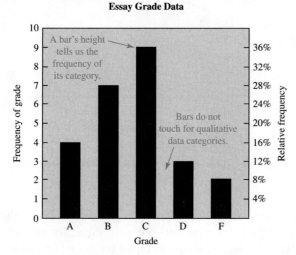

Figure 3.1 Bar graph for the essay grade data in Table 3.1.

- Because the highest frequency is 9 (for C grades), running the vertical scale from 0 to 10 allows all the bars to fit well.
- The height of each bar is proportional to its frequency. The graph is easy to read because we chose a total height of 5 centimeters, so that each centimeter of height corresponds to a frequency of 2.
- Because the data are qualitative, the bars do not need to touch. The widths of the bars have no special meaning. We therefore draw them with uniform widths.
- The graph is clearly labeled, in accord with the rules in the following summary.

> **Important Labels for Graphs**
>
> **Title/caption:** The graph should have a title or caption (or both) that explains what is being shown and, if applicable, lists the source of the data.
>
> **Vertical scale and label:** Numbers along the vertical axis should clearly indicate the scale and line up with the *tick marks*—marks along the axis that precisely locate the numerical values. Include an axis title that describes the variable that the numbers represent.
>
> **Horizontal scale and label:** The categories should be clearly indicated along the horizontal axis; tick marks are not necessary for qualitative data but should be used with quantitative data. Include an axis title that describes the variable that the categories represent.
>
> **Legend:** If multiple data sets are displayed on a single graph, include a legend or key to identify the individual data sets.

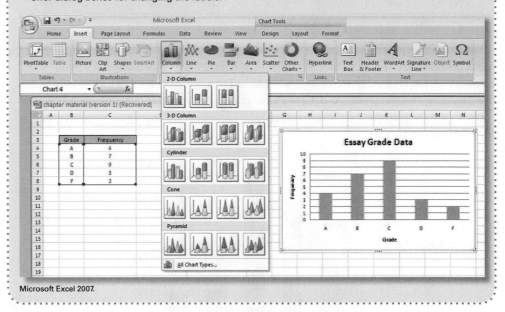

A **dotplot** is a variation on a bar graph in which we use dots rather than bars to represent the frequencies. Each dot represents one data value; for example, a stack of 4 dots means a frequency of 4. Figure 3.2 shows a dotplot for the essay data set. Dotplots are convenient when making graphs of raw data by hand, because you can tally the data by making a dot for each data value. You may then convert the graph to a bar chart for a formal report.

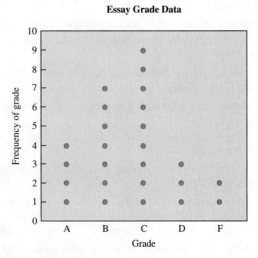

Figure 3.2 Dotplot for the essay grade data in Table 3.1.

A **Pareto chart** is a bar graph in which the bars are arranged in frequency order. This can be useful for data at the nominal level of measurement. For example, Figure 3.3 shows both a bar chart (with the cities in alphabetical order) and a Pareto chart for the populations of the five

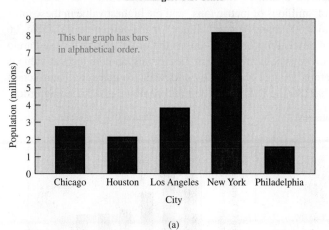

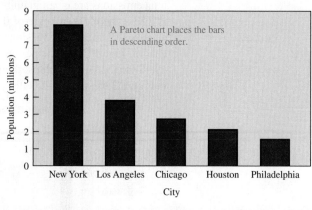

Figure 3.3 (a) Bar graph showing populations for the five largest cities in the United States (2010 census). (b) Pareto chart for the same data. *Source:* U.S. Census Bureau.

largest cities in the United States. Although the only difference is in the order of the bars, most people find the Pareto chart easier to study. Note that Pareto charts do not work for data at the ordinal level of measurement or higher; for example, it would not make sense to make a Pareto chart for the essay grade data, because the grades (A, B, C, D, F) already have a natural order.

Definitions

A **bar graph** consists of bars representing frequencies (or relative frequencies) for particular categories. The bar lengths are proportional to the frequencies.

A **dotplot** is similar to a bar graph, except each individual data value is represented with a dot.

A **Pareto chart** is a bar graph with the bars arranged in frequency order. Pareto charts make sense only for data at the nominal level of measurement.

TIME ◔UT TO THINK

Would it be practical to make a dotplot for the population data in Figure 3.3? Would it make sense to make a Pareto chart for data concerning SAT scores? Explain.

BY THE WAY

Pareto charts were invented by Italian economist Vilfredo Pareto (1848–1923). Pareto is best known for developing methods of analyzing income distributions, but his most important contributions probably were in developing new ways of applying mathematics and statistics to economic analysis.

EXAMPLE ❶ Carbon Dioxide Emissions

Carbon dioxide (CO_2) is released into the atmosphere by the combustion of fossil fuels (oil, coal, natural gas). Table 3.6 lists the eight countries that emit the most carbon dioxide each year. Construct Pareto charts for the total emissions and for the average emissions per person. Why do the two charts appear to be so different?

TABLE 3.6	The World's Eight Leading Emitters of Carbon Dioxide	
Country/region	Total CO_2 emissions (millions of metric tons)	Per person CO_2 emissions (metric tons)
United States	5,833	19.18
China	6,534	4.91
Russia	1,729	12.29
Japan	1,495	1.31
India	1,214	9.54
Germany	829	10.06
Canada	574	17.27
United Kingdom	572	9.38

Source: U.S. Department of Energy, *Annual Energy Outlook 2011* (provides data for 2008 emissions).

SOLUTION The categories are the countries and the frequencies are the data values. The total emissions are given in units of "millions of metric tons" and the highest value in these units is 6,534; therefore, a range of 0 to 7,000 makes a good choice for the vertical scale. Table 3.6 already lists the total emissions in descending order, so this is the order we use for the Pareto chart in Figure 3.4a. The per person emissions are given in metric tons, and the highest value is 19.18 for the United States; therefore, a range of 0 to 20 works well. To make the Pareto chart in Figure 3.4b, we put the tallest bar (for the U.S.) at the left and continue in descending order.

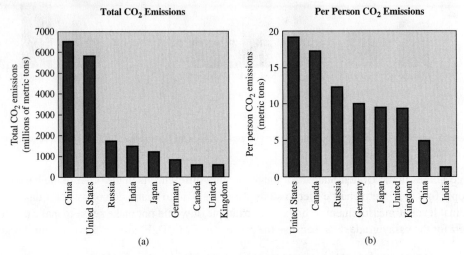

Figure 3.4 Pareto charts for (a) carbon dioxide emissions by country and (b) per person carbon dioxide emissions.

The fact that the two Pareto charts have the countries in different orders tells us that the biggest total emitters of carbon dioxide are not necessarily the biggest per person emitters. In particular, although China has the largest total emissions, the emissions per person are far higher in the United States and Canada.   ·· ●

> TIME ◔UT TO THINK
>
> Most people around the world aspire to a standard of living like that in the United States. Suppose that to achieve this, the rest of the world's per person carbon dioxide emissions rose to the same level as that in the United States. What consequences might this have for the world? Defend your opinions.

Pie Charts

Pie charts are commonly used to show relative frequency distributions. The entire pie represents the total relative frequency of 100%, so the sizes of the individual slices, or wedges, represent the relative frequencies of the various categories. As a simple example, Figure 3.5 shows a pie chart for the essay grade data. Note that the size of each wedge reflects the relative frequencies that we found in Table 3.3. To make comparisons easier, the relative frequencies (as percentages) are written on the wedges.

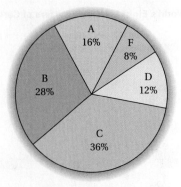

Figure 3.5 Pie chart for the essay grade data in Tables 3.1 and 3.3.

⏻ **USING TECHNOLOGY—PIE CHARTS**

Excel Making a pie chart in Excel is very similar to making a bar graph (p. 86), except:

- For a pie chart, you will probably want to select the relative frequencies rather than the frequencies (though both will work); it may be helpful to cut and paste these data so they are next to the grade letters.

- Choose a pie chart rather than a column chart from the Insert menu.

- The labeling process is different from that for bar graphs, as are the options for colors and other decorative features. The accompanying screen display shows one set of options used in Excel for Windows; you should experiment with other options to learn about the possibilities.

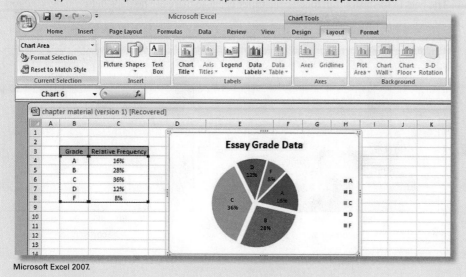

Microsoft Excel 2007.

Definition

A **pie chart** is a circle divided so that each wedge represents the *relative frequency* of a particular category. The wedge size is proportional to the relative frequency. The entire pie represents the total relative frequency of 100%.

EXAMPLE 2 **Simple Pie Chart**

The registered voters of Rochester County are 25% Democrats, 25% Republicans, and 50% independents. Construct a pie chart to represent the party affiliations.

**Registered Voters
in Rochester County**

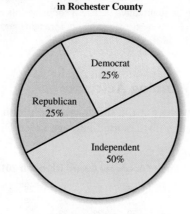

Figure 3.6 Pie chart for Example 2.

SOLUTION Because Democrats and Republicans each represent 25% of the voters, the wedges for Republicans and Democrats each occupy 25%, or one-fourth, of the pie. Independents represent half of the voters, so their wedge occupies the remaining half of the pie. Figure 3.6 shows the result. As always, note the importance of clear labeling.   • • •

TECHNICAL NOTE

Different books define the terms *histogram* and *bar graph* differently, and there are no universally accepted definitions. In this text, a bar graph is any graph that uses bars, and histograms are the types of bar graphs used for quantitative data categories.

Histograms and Line Charts

As we've seen, bar graphs and pie charts are used primarily for data in which the categories are qualitative, such as letter grades or countries. For quantitative data categories, the two most common types of graphics are *histograms* and *line charts*.

A **histogram** is essentially a bar graph in which the data categories are quantitative. The bars in a histogram must follow the natural order of the numerical categories. The widths of the bars must be equal, and they must have a specific meaning. For example, Figure 3.7a shows a histogram for the binned exam data of Table 3.5. Notice that the width of each bar represents 5 points on the exam. The bars in the histogram touch each other because there are no gaps between the categories.

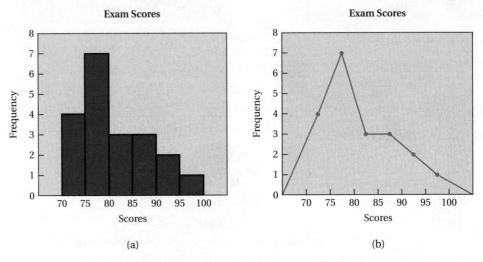

Figure 3.7   The numerical grade data from Table 3.5 shown as (a) a histogram and (b) a line chart.

TECHNICAL NOTE

A line chart created from a set of frequency data is often called a *frequency polygon* because it consists of many straight line segments that take the shape of a many-sided figure.

Figure 3.7b shows a **line chart** for the same data. To make the line chart, we use a dot (instead of a bar) to represent the frequency of each data category; that is, the dots go in the places where the tops of the bars go on the histogram. Because the data are binned into 5-point bins, we place the dot at the center of each bin. For example, the dot for the data category 70–75 goes at 72.5 along the horizontal axis. After the dots are placed, we connect them with straight lines. To make the graph look complete, we connect the points at the far left and far right back down to a frequency of zero.

> **Definitions**
>
> A **histogram** is a bar graph showing a distribution for quantitative data (at the interval or ratio level of measurement); the bars have a natural order and the bar widths have specific meaning.
>
> A **line chart** shows a distribution of quantitative data as a series of dots connected by lines. For each dot, the horizontal position is the *center* of the bin it represents and the vertical position is the frequency value for the bin.

EXAMPLE 3   Oscar-Winning Actresses

The following data show the ages (at the time when they won the award) of all Academy Award–winning actresses through 2012, sorted into age order. Display the data using 10-year bins. Discuss the results.

Ages of Actresses at Time of Academy Award (through 2012)

21	22	24	24	25	25	25	25	26	26	26
26	26	27	27	27	27	28	28	28	28	29
29	29	29	29	29	29	29	30	30	30	31
31	31	32	32	32	32	33	33	33	33	33
33	34	34	34	35	35	35	35	35	36	36
37	37	38	38	38	38	39	39	40	41	41
41	41	41	42	42	45	45	48	49	49	54
60	61	61	61	62	63	74	80			

Source: Academy of Motion Picture Arts and Sciences.

SOLUTION The first step is to bin the data. If you make a tally using 10-year bins of ages 20–29, 30–39, and so on, you'll find the frequencies shown in Table 3.7. Because the data are quantitative, either a histogram or line chart can be made to show the frequencies. Figure 3.8 shows both, with the line chart overlaying the histogram. Note that the vertical axis spans the range 0 to 40, which easily fits the largest frequency of 34 (for ages 30–39), and the dots for the line chart are placed at the center of each bin.

The data show that most actresses win the award at a fairly young age, which stands in contrast to the older ages of most male winners of Best Actor (see Exercise 13 in Section 3.1). Many actresses believe this difference arises because Hollywood producers rarely make movies that feature older women in strong character roles.

TABLE 3.7	Frequency table for Ages of Actresses at Time of Academy Award (through 2012)
Age	**Number of actresses**
20–29	29
30–39	34
40–49	13
50–59	1
60–69	6
70–79	1
80–89	1

Ages of Academy Award–Winning Actresses

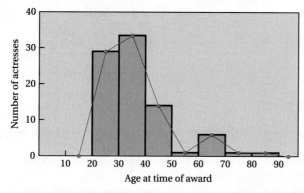

Figure 3.8 Histogram and line chart (overlaying the histogram) for the ages of Academy Award–winning actresses.

Variations on the Theme

There are many variations of histograms and line charts, but we'll discuss just two here. A histogram or line chart for which the data categories on the horizontal axis are time intervals is called a **time-series graph**. You've already seen some time-series diagrams in this book, such as the graph in Figure 2.3 showing how gasoline prices have changed with time.

Another common variation, at least among statisticians if not the media, is called a **stemplot** (or stem-and-leaf plot). A stemplot looks somewhat like a histogram turned sideways, except in place of bars we see a listing of the raw data values. Stemplots can be made in several ways, but one common way is to separate the data values into two parts: the stem (such as the leftmost digit) and the leaf (such as the rightmost digit). The following example shows the process.

> **Definitions**
>
> A histogram or line chart in which the horizontal axis represents *time* is called a **time-series graph**.
>
> A **stemplot** (or *stem-and-leaf plot*) is somewhat like a histogram turned sideways, except in place of bars we see a listing of data.

⏻ USING TECHNOLOGY—LINE CHARTS AND HISTOGRAMS

Excel Line charts are easy to create in Excel. The screen shot below shows the process for the binned data from Table 3.5 (p. 81). Follow these steps:

1. To get the dots in the centers of the bins for the scores, enter the center point of each bin in Column B. Then enter the frequencies in Column C.

2. Select the scores and frequencies, then choose the chart type "scatter" but with the option for connecting points with straight lines. You will get the graph as shown.

3. Use the chart options to improve design, labels, and more.

Note: You can also create a line chart with the "line" chart option in Excel. In that case, select *only* the frequencies when you begin the graphing process; then, in the "source data" dialog box, choose "series" and select the scores (Column B) as the "X values." The resulting graph should look the same as that created with the "scatter" option, except the data points will not have dots.

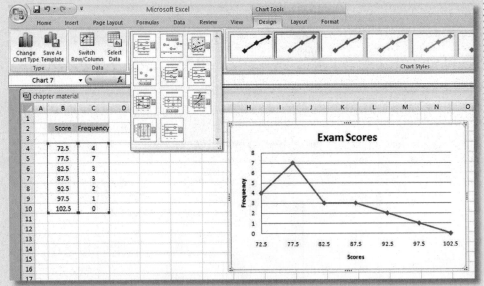

Microsoft Excel 2007.

Excel can also generate histograms with the use of add-ins, including the Data Analysis add-in that can be installed with some versions of Excel.

STATDISK To create a histogram in STATDISK, enter or open data in the STATDISK Data Window, click **Data,** click **Histogram,** and then click on the **Plot** button. To use your own settings, click on the "User defined" button before clicking on Plot. Click on the **Turn labels on** button to see the frequency for each class.

TI-83/84 PLUS To create a histogram using a TI-83/84 Plus calculator, enter or a list of data in L1 or use a list of values assigned to a name. Select the **STAT PLOT** function by pressing **2ND** **Y=**. Press **ENTER** and use the arrow keys to turn Plot1 to "On" and select the graph with bars. The screen display should be as shown here.

Image used with permission of Texas Instruments, Inc.

If you want to let the calculator determine the settings, press **ZOOM** **9** to get a histogram with default settings. (To use your own settings, press **WINDOW** and enter the maximum and minimum values. Press **GRAPH** to obtain the graph.)

EXAMPLE 4 Stemplot for Oscar-Winning Actresses

Make a stemplot for the data from Example 3. Discuss the pros and cons of the stemplot over a histogram.

SOLUTION   The data are the ages of the actresses (at time of award), which range from 21 to 80. We therefore choose to use the tens column of the ages as the stems and the ones column as the leaves. Because the ages range from the 20s to the 80s, the stem values range from 2 to 8; we put these on the left side of the stemplot shown in Figure 3.9. On the right, we place all the ones values that go with each tens value. For example, the first row of data from Example 3 shows the following ages:

21   22   24   24   25   25   25   25   26   26   26

All of these ages are in the 20s, so they all have the same stem of "2" in Figure 3.9 We then place each of these ages as one of the leaves on the right, which is why the first set of the leaves begins 1, 2, 4, 4, 5, 5, 5, 5, 6, 6, 6. You can continue through the data from Example 3 to see how all the leaves were created. Notice, for example, that there is only one age in the 50s, and it is 54; that is why the stem 5 (which represents the 50s) has only a single leaf with value 4.

The primary advantage of the stemplot over the histogram from Figure 3.8 is that the stemplot actually contains all of the original data. For example, the histogram shows only that there were 6 actresses with ages from 60–69; in contrast, by looking at the leaves that go with the stem "6" in the stemplot, you can see that the actual ages of these six actresses were 60, 61, 61, 61, 62, and 63. The primary drawback to the stemplot is that the extra information it contains makes it more visually complex than a histogram.

Stem (tens)	Leaves (ones)
2	1 2 4 4 5 5 5 5 6 6 6 6 6 7 7 7 7 8 8 8 8 9 9 9 9 9 9 9
3	0 0 0 1 1 1 2 2 2 2 3 3 3 3 3 3 4 4 4 5 5 5 5 5 6 6 7 7 8 8 8 8 9
4	0 1 1 1 1 1 2 2 5 5 8 9 9
5	4
6	0 1 1 1 2 3
7	7
8	0

Figure 3.9 Stemplot for the ages of Academy Award–winning actresses.

· · ●

EXAMPLE 5 Interpreting a Time-Series Graph

Figure 3.10 shows a time-series graph for the death rate (deaths per 1,000 people) in the United States since 1900. (For example, the 1905 death rate of 16 means that, for each 1,000 people living at the beginning of 1905, 16 people died during the year.) Discuss the general trend. Also consider the spike in 1918: If someone told you that this spike was due to battlefield deaths in World War I, would you believe it? Explain.

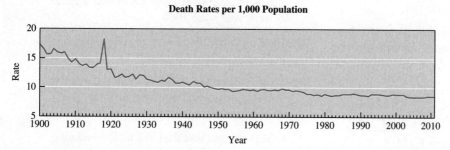

Figure 3.10 Historical U.S. death rates per 1,000 people.

Source: National Center for Health Statistics.

SOLUTION   The general trend in death rates is clearly downward, presumably because of improvements in medical science. For example, bacterial diseases such as pneumonia were major killers in the early 1900s, but are largely curable with antibiotics today. Regarding the spike in 1918: Although it coincides with the end of World War I, if the war were the cause of this spike, then we might expect to see a similar spike during World War II, but we don't. This suggests that there must have been some other reason for the spike. In fact, the spike reflects the effects of the deadly flu pandemic of 1918, which killed 850,000 people in the United States and an estimated 20 million people worldwide.   · · ●

TIME ⏱UT TO THINK

Do you get annual flu shots? Does knowing the tremendous impact of the 1918 flu epidemic affect your opinion as to the value of flu shots? Defend your opinion.

Section 3.2 Exercises

Statistical Literacy and Critical Thinking

1. **Distribution.** What do we mean by the distribution of data?

2. **Visualizing Data.** How is a histogram or line chart more helpful than a list of sample values for understanding a data distribution?

3. **Pareto Chart and Pie Chart.** What is an important advantage of a Pareto chart over a pie chart?

4. **Histogram and Stemplot.** Assume that a data set is used to construct a histogram and a stemplot. Using only the histogram, is it possible to re-create the original list of data values? Using only the stemplot, is it possible to re-create the original list of data values? What is an advantage of a stemplot over a histogram?

Does It Make Sense? For Exercises 5–8, decide whether the statement makes sense (or is clearly true) or does not make sense (or is clearly false). Explain clearly. Not all of these statements have definitive answers, so your explanation is more important than your chosen answer.

5. **Histogram.** I made a histogram to depict frequency counts of answers to the question "What political party do you belong to?"

6. **Pie Chart.** I used a pie chart to illustrate a data set showing how the cost of college has changed over time.

7. **Pareto Chart.** A quality control engineer wants to draw attention to the most serious causes of defects, so she uses a Pareto chart to illustrate the frequencies of the different causes of defects.

8. **Peak Values.** I made both a histogram and a line chart for a data set of the crime rate (number of crimes committed) each year from 1960 to 2010. The histogram showed that the crime rate peaked in 1982, while the line chart showed that it peaked in 1983.

Concepts and Applications

Most Appropriate Display. Exercises 9–12 describe data sets but do not give actual data. For each data set, state the type of graphic that you believe would be most appropriate for displaying the data, if they were available. Explain your choice.

9. **Incomes.** Incomes of college graduates who took a statistics course

10. **Political Party.** The political party affiliations of 1000 survey subjects

11. **Movie Theaters.** Number of movie theatres for each year since 1960

12. **Ages of Crash Victims.** Ages of people who died in car crashes last year

13. **What People Are Reading.** The pie chart in Figure 3.11 shows the results of a survey about what people are reading.

    a. Summarize these data in a table of relative frequencies.

    b. Construct a Pareto chart for these data.

    c. Which do you think is a better representation of the data: the pie chart or the Pareto chart? Why?

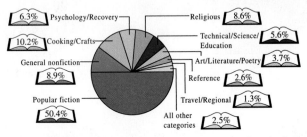

Figure 3.11   *Source:* Wall Street Journal Almanac, based on data from the Book Industry Study Group.

14. **Histogram.** The histogram in Figure 3.12 depicts cotinine levels (in milligrams per milliliter) of a sample of subjects who smoke cigarettes. Cotinine is a metabolite of nicotine, which means that cotinine is produced by the body when nicotine is absorbed. The data are from the Third National Health and Nutrition Examination Survey.

   a. How many subjects are represented in the histogram?

   b. How many subjects have cotinine levels below 400?

   c. How many subjects have cotinine levels above 150?

   d. What is the highest possible cotinine level of a subject represented in this histogram?

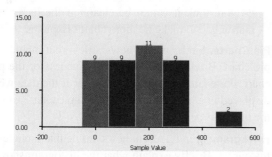

Figure 3.12

15. **Weights of Coke.** Exercise 11 in Section 3.1 required the construction of a frequency table from the weights (in pounds) of 36 cans of regular Coke. Use that frequency table to construct the corresponding histogram.

16. **Weights of Diet Coke.** Exercise 12 in Section 3.1 required the construction of a frequency table for the weights (in pounds) of 36 cans of Diet Coke. Use that frequency table to construct the corresponding histogram.

17. **Oscar-Winning Actors.** Exercise 13 in Section 3.1 required the construction of a frequency table for the ages of recent Academy Award–winning male actors at the time when they won their award. Use that frequency table to construct the corresponding histogram.

18. **Body Temperatures.** Exercise 14 in Section 3.1 required the construction of a frequency table for a list of body temperatures (°F) of randomly selected subjects. Use that frequency table to construct the corresponding histogram.

19. **Job Hunting.** A survey was conducted to determine how employees found their jobs. The table below lists the successful methods identified by 400 randomly selected employees. The data are based on results from the National Center for Career Strategies. Construct a Pareto chart that corresponds to the given data. Based on these results, what appears to be the best strategy for someone seeking employment?

Job sources of survey respondents	Frequency
Help-wanted ads	56
Executive search firms	44
Networking	280
Mass mailing	20

20. **Job Sources.** Refer to the data given in Exercise 19, and construct a pie chart. Compare the pie chart to the Pareto chart. Can you determine which graph is more effective in showing the relative importance of job sources?

21. **Job Application Mistakes** Chief financial officers of U.S. companies were surveyed about areas in which job applicants make mistakes. Here are the areas and the frequency of responses: interview (452); resume (297); cover letter (141); reference checks (143); interview follow-up (113); screening call (85). These results are based on data from Robert Half Finance and Accounting. Construct a pie chart representing the given data.

22. **Job Application Mistakes** Construct a Pareto chart of the data given in Exercise 21. Compare the Pareto chart to the pie chart. Which graph is more effective in showing the relative importance of the mistakes made by job applicants?

23. **Dotplot.** Refer to the QWERTY data in Exercise 19 in Section 3.1 and construct a dotplot.

24. **Dotplot.** Refer to the Dvorak data in Exercise 19 in Section 3.1 and construct a dotplot. Compare the result to the dotplot in Exercise 23. Based on the results, does either keyboard configuration appear to be better? Explain.

25. **Time-Series Graph for Cell Phone Subscriptions.** The following table shows the numbers of cell phone subscriptions (in thousands) in the United States for various years. Construct a time-series graph of the data. "Linear" growth would result in a graph that is approximately a straight line. Does the time-series graph appear to show linear growth?

Year	Number	Year	Number
1986	682	2000	109478
1988	2069	2002	140767
1990	5283	2004	182140
1992	11033	2006	233000
1994	24134	2008	262700
1996	44043	2010	302900
1998	69209		

26. **Time Series for Motor Vehicle Deaths.** The following values are numbers of motor vehicle deaths in the United States for years beginning with 1980. The data are arranged in order by row. Construct a time-series graph and then determine whether there appears to be a trend. If so, provide a possible explanation.

51,091	49,301	43,945	42,589	44,257	43,825
46,087	46,390	47,087	45,582	44,599	41,508
39,250	40,150	40,716	41,817	42,065	42,013
41,501	41,717	41,945	42,196	43,005	42,884
42,836	43,443	42,642	41,059	37,261	33,808
32,708					

27. **Stemplot.** Construct a stemplot of these test scores: 67, 72, 85, 75, 89, 89, 88, 90, 99, 100. How does the stemplot show the distribution of these data?

28. Stemplot. Listed below are the lengths (in minutes) of animated children's movies. Construct a stemplot. Does the stemplot show the distribution of the data? If so, how?

83	88	120	64	69	71	76	74	75	76	75
75	79	80	78	78	83	77	71	83	80	73
72	82	74	84	90	89	81	81	90	79	92
82	89	82	74	86	76	81	75	75	77	70
75	64	73	74	71	94					

PROJECTS FOR THE INTERNET & BEYOND

29. CO₂ Emissions. Look for updated data on international carbon dioxide emissions. Create a graph of the latest data and discuss any important features or trends that you notice.

30. Energy Table. Explore the energy tables at the U.S. Energy Information Administration (EIA) Web site. Choose a table that you find interesting and make a graph of its data. You may choose any of the graph types discussed in this section. Explain how you made your graph, and briefly discuss what can be learned from it.

31. Statistical Abstract. Go to the Web site for the *Statistical Abstract of the United States*. Explore the selection of "frequently requested tables." Choose one table of interest to you and make a graph from its data. You may choose any of the graph types discussed in this section. Explain how you made your graph and briefly discuss what can be learned from it.

32. Navel Data. Create an appropriate display of the navel data collected in Exercise 23 of Section 3.1. Discuss any special properties of this distribution.

IN THE NEWS

33. Bar Graphs. Find a recent news article that includes a bar graph with qualitative data categories.

   a. Briefly explain what the bar graph shows, and discuss whether it helps make the point of the news article. Are the labels clear?

   b. Briefly discuss whether the bar graph could be recast as a dotplot.

   c. Is the bar graph already a Pareto chart? If so, explain why you think it was drawn this way. If not, do you think it would be clearer if the bars were rearranged to make a Pareto chart? Explain.

34. Pie Charts. Find a recent news article that includes a pie chart. Briefly discuss the effectiveness of the pie chart. For example, would it be better if the data were displayed in a bar graph rather than a pie chart? Could the pie chart be improved in other ways?

35. Histograms. Find a recent news article that includes a histogram. Briefly explain what the histogram shows, and discuss whether it helps make the point of the news article. Are the labels clear? Is the histogram a time-series diagram? Explain.

36. Line Charts. Find a recent news article that includes a line chart. Briefly explain what the line chart shows, and discuss whether it helps make the point of the news article. Are the labels clear? Is the line chart a time-series diagram? Explain.

3.3 GRAPHICS IN THE MEDIA

The basic graphs we have studied so far are only the beginning of the many ways to depict data visually. In this section, we will explore some of the more complex types of graphics that are common in the media.

Multiple Bar Graphs and Line Charts

A **multiple bar graph** is a simple extension of a regular bar graph: It has two or more sets of bars that allow comparison between two or more data sets. All the data sets must have the same categories so that they can be displayed on the same graph. Figure 3.13 is a multiple bar graph with two sets of bars, one for men and one for women. (The median is a type of average; see Section 4.1.)

    Notice that the data categories (the different levels of educational attainment by gender) in Figure 3.13 are qualitative, which makes a bar chart the best choice for display. In cases for which data categories are quantitative, a **multiple line chart** is often a better choice. Figure 3.14 shows time-series data using four different lines for four different data sets. The multiple line chart works because all four data sets represent the same pair of variables: time and unemployment rate. They differ only in the groups for which the data are plotted, with each line representing data for a different level of educational attainment.

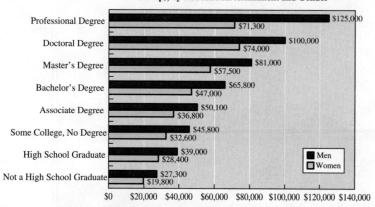

Figure 3.13 A multiple bar graph.

Source: The College Board, *Trends in Higher Education 2010*, based on 2008 data from the U.S. Census Bureau.

Figure 3.14 A multiple line chart.

Source: The College Board, *Trends in Higher Education 2010*, based on data through 2009 from the U.S. Bureau of Labor Statistics.

EXAMPLE 1 Education Pays

Study Figures 3.13 and 3.14. What general messages do they reveal? Comment on how the use of the multiple bar and line graphics helps convey these messages.

SOLUTION Figure 3.13 conveys two clear messages. First, by looking at the bars across all the categories, we see clearly that people with greater education have significantly higher median incomes. For example, the median earnings for men with a bachelor's degree are more than $25,000 higher than for men without a college education. This means that over a typical 40-year career (age 25 to 65), a college education is worth an extra *$1 million* in income, at least on average. The second message conveyed by the graph is that for equivalent levels of educational attainment, women still earn much less than men.

Figure 3.14 shows another added value of education: no matter what the unemployment rate (at least over the time period shown), unemployment has always been significantly lower for more highly educated people.

The graphic choices work well because of the easy comparisons they allow. For example, if the bar graphs for men and women were shown separately, it would be much more difficult to see the fact that women earn less than men with the same education. Similarly, if the unemployment line charts were shown separately, our eyes would be drawn more to the trends with time than to the more important differences in the unemployment rates for people with different levels of education.

··●

TIME OUT TO THINK

Together, Figures 3.13 and 3.14 make clear that education can be very valuable financially because it can easily raise your lifetime earnings by $1 million while dramatically reducing your risk of unemployment. Discuss whether these facts should alter the way you approach your own education or public policies regarding higher education.

BY THE WAY

More detailed data show that median earnings are higher for college graduates than non-graduates in nearly every field of work. However, the variations are quite significant. For example, students majoring in technical fields such as science or engineering generally earn more than students with other majors, and for any particular major, students who study more and get better grades tend to earn more than students who study less.

Stack Plots

Another way to show two or more related data sets simultaneously is with a **stack plot**, which shows different data sets in a vertical stack. Data can be stacked in both bar charts and line charts.

Figure 3.15 shows a stack plot of federal government spending categories as a percentage of the total, with data projected through 2016. Note that government spending is broken into five broad categories, with each category represented by a wedge in the figure. The

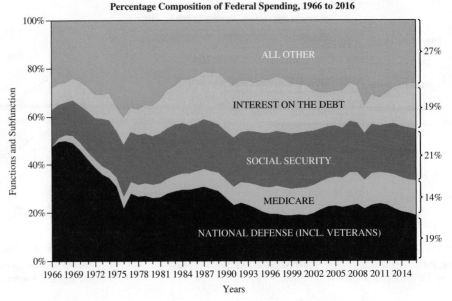

Figure 3.15   A stack plot for U.S. federal government spending by category as a percentage of total spending. Based on actual data through 2010, and projections made in 2010 for the years 2011 through 2016.

Source: U.S. Office of Management and Budget, Historical Tables published in 2011.

thickness of a wedge at a particular time tells you its value at that time. For example, the fact that the wedge for National Defense (which, in this chart, includes spending on Veterans) falls from about 48% in 1966 to about 19% in 2016 tells us that the percentage of the federal budget that goes to defense has declined dramatically in recent decades. The graph also makes clear that the trend over time is for Social Security, Medicare, and interest on the debt to make up a rapidly growing percentage of overall spending.

EXAMPLE 2   Interpreting the Stack Plot

As you are undoubtedly aware, the government currently spends much more money than it takes in; in fact, in recent years the revenue has been about one-third less than the spending. Therefore, balancing the budget requires either a dramatic reduction in spending, a substantial increase in tax revenue, or some combination of both. Suppose that you wanted to balance the budget with spending cuts only. How could you do it?

SOLUTION   Achieving a balanced budget through spending cuts alone would require cutting total spending by about one-third, which essentially means finding a way to take 33% out of the 2016 values shown at the right. In principle, there are many ways we could do this, but in practice there are difficult constraints. Interest on the debt is an obligation that cannot be reduced without a government default, and a default could potentially cripple the economy. In addition, many conservatives feel strongly that defense should not be cut any further, while many liberals tend to believe that there should be no cuts in Social Security and Medicare. Those constraints put heavy pressure on finding a way to cut spending without touching interest on the debt, Social Security, Medicare, or defense. However, because the remaining "all other" adds up to only about 27% of the budget in 2016, it's not possible to cut the budget by a third from this spending alone. The conclusion is that balancing the budget through spending cuts alone would require deep cuts not only to "all other" government programs but also to national defense, Social Security, and Medicare.   ·· ●

TIME (🕐)UT TO THINK

If *you* were a member of Congress, how would you propose to balance the budget? Do you think your proposal could pass Congress? Do you think you'd be able to be re-elected if it did? Defend your opinions.

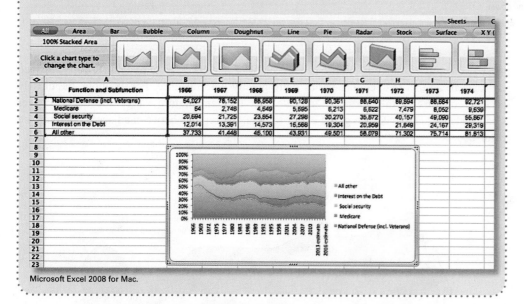

USING TECHNOLOGY—GRAPHS WITH MULTIPLE DATA SETS

Excel Excel provides a variety of options for making multiple bar graphs, multiple line graphs, stack plots, and more. The screen shot below shows some of the data that went into making Figure 3.15, along with a first attempt at a graph. Excel provides several ways of creating the desired graph, but here's a simple process: Notice that the years run along row 1, the different categories of spending are in column 1, and the data cells show the actual spending amounts in millions of dollars.

1. Enter all the data. Notice that the years run along row 1, the different categories of spending are in column 1, and the data cells show the actual spending amounts in millions of dollars.

2. Select all six rows, which means including both the data cells and the labels in row 1 and column 1. Then use the "Insert" menu to choose "Insert Chart…" That will bring up the chart selections that you see above row 1 in the screen shot below.

3. Choose your chart option. For this particular screen shot, we chose the third chart from the left; to the left of the chart options, you'll see that this chart type is identified as "100% stacked area," which means it makes the stack plot as percentages of each year's total rather than from the actual dollar values.

4. From this point, getting to a final graphic like that in Figure 3.15 is just a matter of choosing labeling options and other decorative features. Most of these features can be added within Excel, though sometimes it is easier to add them by importing the Excel graph into an art or photo editing program.

You should try your own similar data set and experiment with the various chart types. If you don't get the result that you expect, use Excel's "help" function to get more detailed instructions about making various types of graphs.

Microsoft Excel 2008 for Mac.

Geographical Data

There are many cases in which we are interested in geographical patterns in data. Figure 3.16 shows one common way of depicting **geographical data**. In this case, the map shows trends in energy use per capita (per person) in different states. The actual data values are shown in small print with each state, while the color coding shows the binned categories listed in the legend.

The display in Figure 3.16 works well for the energy data set because each state is associated with a unique energy usage per person. For data that vary continuously across geographical areas, a **contour map** is more convenient. Figure 3.17 shows a contour map of temperature over the United States at a particular time. Each of the *contours* (curvy lines) connects locations with the same temperature. For example, the temperature is 50° F everywhere along the contour labeled 50° and 60° F everywhere along the contour labeled 60°. Between

Per Person Energy Use by State (gallons of oil equivalent)

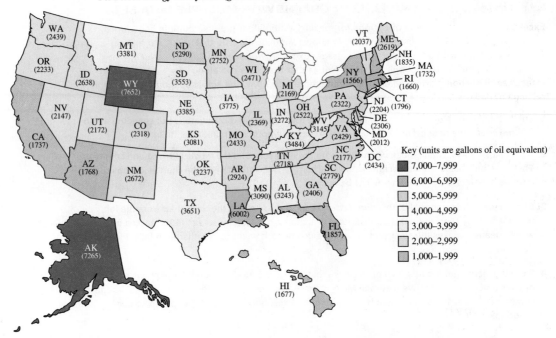

Figure 3.16 Geographical data can be displayed with a color-coded map. These data show per person energy usage by state, in units of "gallons of oil equivalent"; that is, the data represent the amount of oil that each person would use if all of the energy were generated by burning oil. In reality, about 40% of U.S. energy comes from oil, with most of the rest from coal, natural gas, nuclear, and hydroelectric power.

Source: U.S. Energy Information Administration, State Energy Data System (2009 data released 2011).

these two contours, the temperature is between 50°F and 60°F. Note that more closely spaced contours mean that temperature varies more greatly with distance. For example, the closely packed contours in the northeast indicate that the temperature varies substantially over small distances. To make the graph easier to read, the regions between adjacent contours are color-coded.

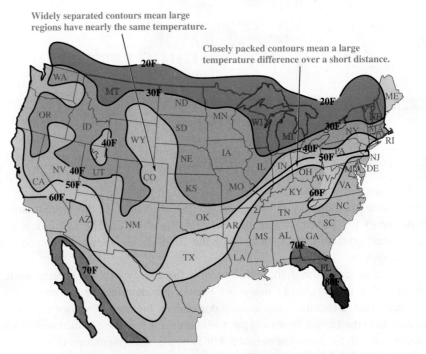

Figure 3.17 Geographical data that vary continuously, such as temperatures, can be displayed with a contour map.

EXAMPLE 3 Interpreting Geographical Data

Study Figures 3.16 and 3.17 and use them to answer the following questions.

a. Do you see any geographical trends that might explain the states with the lowest energy usage per person?

b. Were there any temperatures above 80°F in the United States on the date shown in Figure 3.17?

SOLUTION

a. The color coding shows that the states in the lowest category of energy use per person are all either warm-weather states (CA, AZ, FL, and HI) or states in the more densely populated regions of the northeast (NY, NH, CT, MA, and RI).

b. The 80°F contour passes through southern Florida, so the parts of Florida that lie south of this contour had temperatures above 80°F.   ⋯●

Three-Dimensional Graphics

Today, computer software makes it easy to give almost any graph a three-dimensional appearance. For example, Figure 3.18 shows the same bar graph as Figure 3.1, but "dressed up" with a three-dimensional look. It may look nice, but the three-dimensional effects are purely cosmetic; they do not provide any information that wasn't already shown in Figure 3.1.

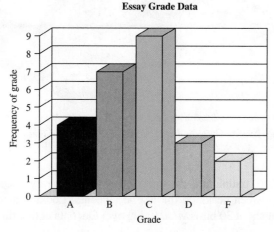

Essay Grade Data

Figure 3.18 This graph has a three-dimensional appearance, but it shows only two-dimensional data.

In contrast, each of the three axes in Figure 3.18 carries distinct information, making the graph a true three-dimensional graph. Researchers studying migration patterns of a bird species (the *Bobolink*) counted the number of birds flying over seven New York cities throughout the night. As shown on the inset map, the cities were aligned east-west so that the researchers would learn what parts of the state the birds flew over, and at what times of night, as they headed south for the winter. Notice that the three axes measure *number of birds, time of night,* and *east-west location.*

EXAMPLE 5 Bird Migration

Based on Figure 3.19, at about what time was the largest number of birds flying over the east-west line marked by the seven cities? Over what part of New York did most of the birds fly? Approximately how many birds passed over Oneonta around 12:00 midnight?

SOLUTION The number of birds detected in all the cities peaked between 3 and 5 hours after 8:30 p.m., or between about 11:30 p.m and 1:30 a.m. More birds flew over the two easternmost cities of Oneonta and Jefferson than over cities farther west, which means that most of the birds were flying over the eastern part of the state. To answer the specific question about

SONIC MAPPING TRACES BIRD MIGRATION

Sensors across New York State counted each occurrence of the nocturnal flight call of the bobolink to trace the fall migration on the night of Aug. 28–29, 1993. Computerized, the data showed the heaviest swath passing over the eastern part of the state.

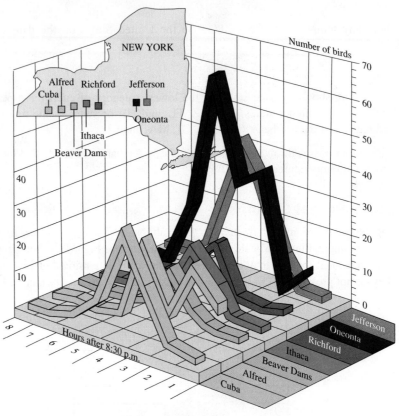

Figure 3.19   This graph shows true three-dimensional data.

Source: "Graph of Ornithology Data," by Bill Evans, from *New York Times.* © *New York Times.*

Oneonta, note that 12:00 midnight is the midpoint of time category 4. On the graph, this time aligns with the dip between peaks on the line at Oneonta. Looking across to the *number of birds* axis, we see that about 30 birds were flying over Oneonta at that time.   · · ●

Combination Graphics

All of the graphic types we have studied so far are common and fairly easy to create. But the media today are often filled with many varieties of even more complex graphics. For example, Figure 3.20 shows a graphic concerning the participation of women in the summer Olympics. This single graphic combines a line chart, many pie charts, and numerical data. It is certainly a case of a picture being worth far more than a thousand words.

EXAMPLE 6   Olympic Women

Describe three trends shown in Figure 3.20.

SOLUTION   The line chart shows that the total number of women competing in the summer Olympics has risen fairly steadily, especially since the 1960s, reaching nearly 5,000 in the 2012 games. The pie charts show that the percentage of women among all competitors has also increased, surpassing 44% in the 2012 games. The bold red numbers at the bottom show that the number of events in which women compete has also increased dramatically, reaching 140 in the 2012 games.

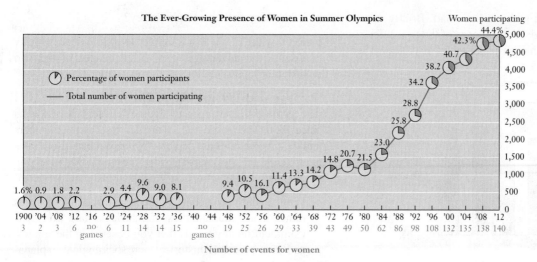

Figure 3.20 Women in the Olympics.

Source: Adapted from the *New York Times,* based on data from the International Olympic Committee.

TIME (🕐)UT TO THINK

Which of the trends shown in Figure 3.20 are likely to continue over the next few Olympic games? Which are not? Explain.

Section 3.3 Exercises

Statistical Literacy and Critical Thinking

1. **Three-Dimensional Histograms.** Can every histogram be converted to a three-dimensional graphic? Does the three-dimensional version of a histogram have more visual appeal? Does the three-dimensional version of a histogram provide any information not provided by the original histogram?

2. **Multiple Bar Graph.** What is a multiple bar graph, and how is it helpful?

3. **Geographical Data.** What are geographical data? Identify at least two ways to display geographical data.

4. **Contour Map.** What is a contour on a contour map? What does it mean when contours are close together? What does it mean when they are far apart?

Does It Make Sense? For Exercises 5–8, decide whether the statement makes sense (or is clearly true) or does not make sense (or is clearly false). Explain clearly. Not all of these statements have definitive answers, so your explanation is more important than your chosen answer.

5. **Three-Dimensional Graph.** A quality control engineer claims that because cars are three-dimensional objects, she needs a three-dimensional graph to display the production cost of a Corvette for each of the past 10 years.

6. **Contour Map.** A contour map could be used to display the ages of all full-time students at your college.

7. **Geographic Data.** A graphic artist for a magazine is depicting the populations of the 10 largest U.S. cities by using bars of different heights, with the bars positioned on the locations of the cities on a map of the United States.

8. **Stack Plot.** A multiple bar graph could be used to show the numbers of males, females, and total students at your college for each of the past 10 years.

Concepts and Applications

9. **Genders of Students.** The stack plot in Figure 3.21 shows the numbers of male and female higher education students for different years. Projections are from the U.S. National Center for Education Statistics.

   a. In words, discuss the trends revealed on this graphic.

b. Redraw the graph as a multiple line chart. Briefly discuss the advantages and disadvantages of the two different representations of this particular data set.

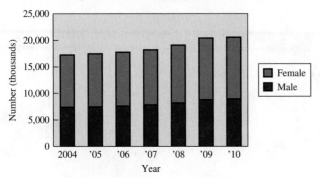

Figure 3.21   *Source:* National Center for Education Statistics.

10. Home Prices by Region. The graph in Figure 3.22 shows home prices in different regions of the United States. Note that the data have *not* been adjusted for the effects of inflation.

a. In words, describe the general trends that apply to the home price data for all regions.

b. In words, describe any differences that you notice among the different regions.

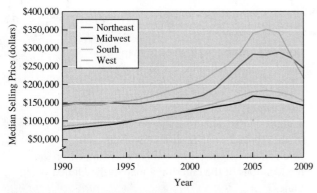

Figure 3.22   *Source:* National Association of Realtors.

11. Gender and Salary. Consider the display in Figure 3.23 of median salaries of males and females in recent years.

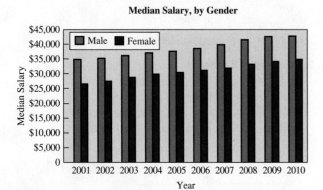

Figure 3.23   *Source:* U.S. Census Bureau.

a. What story does the graph convey?

b. Redraw the graph as a multiple (two) line chart. Briefly discuss the advantages and disadvantages of the two different representations of this particular data set.

12. Marriage and Divorce Rates. The graph in Figure 3.24 depicts the U.S. marriage and divorce rates for selected years since 1900. The marriage rates are depicted by the blue bars and the divorce rates are depicted by the red bars. Both rates are given in units of marriages/divorces per 1,000 people in the population (Department of Health and Human Services).

a. Why do these data consist of marriage and divorce *rates* rather than total numbers of marriages and divorces? Comment on any trends that you observe in these rates, and give plausible historical and sociological explanations for these trends.

b. Construct a stack plot of the marriage and divorce rate data. For each bar, place the divorce rate above the marriage rate. Which graph makes the comparisons easier: the multiple bar graph shown here or the stack plot? Explain.

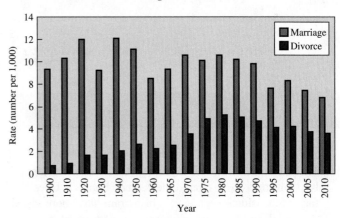

Figure 3.24   *Source:* National Center for Health Statistics.

13. Entitlement Spending. The stack plot in Figure 3.25 shows Congressional Budget Office data for actual (through 2011) and projected spending on entitlement programs through 2085 as percentages of the gross domestic product (GDP). Interpret the graph and summarize its message.

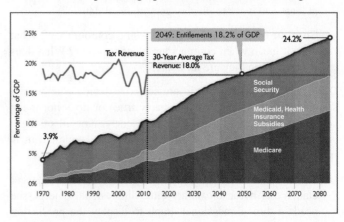

Figure 3.25   *Source:* "Tax Revenues Devoured By Medicare, Medicaid, and Social Security in 2045," from heritage.org, based on data from the Congressional Budget Office. Reprinted with permission.

14. College Degrees. The stacked line chart in Figure 3.26 shows the numbers of bachelor's degrees awarded to men and women since 1970.

Bachelor's Degrees Awarded

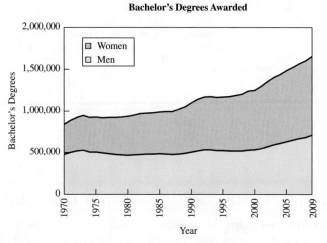

Figure 3.26 *Source:* National Center for Education Statistics.

a. Estimate the numbers of bachelor's degrees awarded to men and to women (separately) in 1970 and in 2010.

b. About when were the number of bachelor's degrees equal for males and females?

c. Comment on the overall trend.

d. Do you think the stacked line chart is an effective way to display these data? Briefly discuss other ways that might have been used instead.

15. Melanoma Mortality. Figure 3.27 shows how the mortality rate from *melanoma* (a form of skin cancer) varies on a county-by-county basis across the United States. The legend shows that the darker the shading in a county, the higher the mortality rate. Discuss a few of the trends revealed in the figure. If you were researching skin cancer, which regions might warrant special study? Why?

16. School Segregation. One way of measuring segregation is the likelihood that a black student will have classmates who are white. Figure 3.28 shows the probability that a black

Female Melanoma Mortality Rates by County

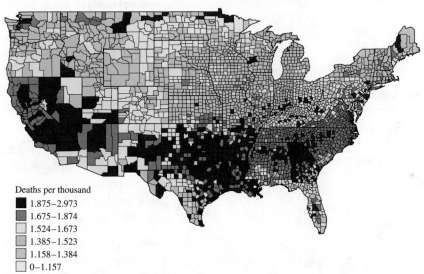

Deaths per thousand
- 1.875–2.973
- 1.675–1.874
- 1.524–1.673
- 1.385–1.523
- 1.158–1.384
- 0–1.157

Figure 3.27 *Source:* "Female Melanoma Mortality Rates by County," by Professor Karen Kafadar, Mathematics Department, University of Colorado at Denver. Reprinted with permission.

Probability That a Black Student Would Have White Classmates

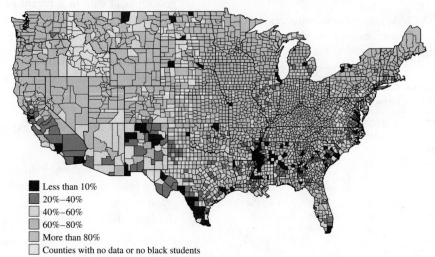

- Less than 10%
- 20%–40%
- 40%–60%
- 60%–80%
- More than 80%
- Counties with no data or no black students

Figure 3.28 *Source: New York Times.*

student would have white classmates, by county, during a recent academic year. Do there appear to be any significant regional differences? Can you pick out any differences between urban and rural areas? Discuss possible explanations for a few of the trends that you see in the figure.

Creating Graphics. Exercises 17–20 give tables of real data. For each table, make a graphical display of the data. You may choose any graphic type that you feel is appropriate to the data set. In addition to making the display, write a few sentences explaining why you chose this type of display and a few sentences describing interesting patterns in the data.

17. **Drinking and Driving.** The following table lists the numbers of persons killed in fatal car crashes for three different categories of blood alcohol content (BAC) of drivers. The data are from the U.S. Census Bureau.

Year	BAC of 0.0 (No alcohol)	BAC of 0.01 to 0.07	BAC of 0.08 and higher
1990	24,083	2,810	17,572
1995	25,968	2,300	13,423
2000	26,300	2,349	13,171
2005	27,542	2,350	13,532
2010 (est)	21,787	1,988	11,242

18. **Daily Newspapers.** The following table gives the number of daily newspapers and their total circulation (in millions) for selected years since 1920 (*Editor & Publisher*).

Year	Number of daily newspapers	Circulation (millions)
1920	2,042	27.8
1930	1,942	39.6
1940	1,878	41.1
1950	1,772	53.9
1960	1,763	58.8
1970	1,748	62.1
1980	1,747	62.2
1990	1,611	62.3
2000	1,485	56.1
2010 (est)	1,402	50.0

19. **Firearm Fatalities.** The following table summarizes deaths due to firearms in different nations in a recent year (Coalition to Stop Gun Violence).

Country	Total firearms deaths	Homicides by firearms	Suicides by firearms	Fatal accidents by firearms
United States	35,563	15,835	18,503	1,225
Germany	1,197	168	1,004	25
Canada	1,189	176	975	38
Australia	536	96	420	20
Spain	396	76	219	101
United Kingdom	277	72	193	12
Sweden	200	27	169	4
Vietnam	131	85	16	30
Japan	93	34	49	10

20. **Working Mothers.** The following table lists labor force participation rates (as percentages) of mothers categorized according to the age of their youngest child (based on data from the Bureau of Labor Statistics).

Year	Youngest child aged 6 to 17 years	Youngest child under 6 years of age
1980	64.3	46.8
1985	69.9	53.5
1990	74.7	58.2
1995	76.4	62.3
2000	79.0	65.3
2005	76.9	62.6
2010 (est)	77.5	63.6

PROJECTS FOR THE INTERNET & BEYOND

21. **Weather Maps.** Many Web sites offer contour maps with current weather data. Find at least two contour weather maps and discuss what they show.

22. **The Federal Budget.** Go to the Web site for the U.S. Office of Management and Budget (OMB) and look for some of its charts related to the federal budget. Pick two charts of particular interest to you and discuss the data they show.

IN THE NEWS

23. **Multiple Bar Graphs.** Find an example of a multiple bar graph or multiple line chart in a recent news report. Comment on the effectiveness of the display. Could another display have been used to depict the same data?

24. **Stack Plots.** Find an example of a stack plot in a recent news report. Comment on the effectiveness of the display. Could another display have been used to depict the same data?

25. **Geographical Data.** Find an example of a graph of geographical data in a recent news report. Comment on the effectiveness of the display. Could another display have been used to depict the same data?

26. **Three-Dimensional Displays.** Find an example of a three-dimensional display in a recent news report. Are three dimensions needed, or are they included for cosmetic reasons? Comment on the effectiveness of the display. Could another display have been used to depict the same data?

27. **Fancy News Graphics.** Find an example in the news of a graphic that combines two or more of the basic graphic types. Briefly explain what the graphic is showing, and discuss the effectiveness of the graphic.

3.4 A FEW CAUTIONS ABOUT GRAPHICS

As we have seen, graphics can offer clear and meaningful summaries of statistical data. However, even well-made graphics can be misleading if we are not careful in interpreting them, and poorly made graphics are almost always misleading. Moreover, some people use graphics in deliberately misleading ways. In this section, we discuss a few of the more common ways in which graphics can lead us astray.

Perceptual Distortions

Many graphics are drawn in a way that distorts our perception of them. Figure 3.29 shows one of the most common types of distortion. Dollar-shaped bars are used to show the declining value of the dollar over time. The problem is that the values are represented by the *lengths* of the dollar bills; for example, a 2010 dollar was worth $0.39 in 1980 dollars and therefore is drawn so that it is 39% as long as the 1980 dollar. However, our eyes tend to focus on the *areas* of the dollar bills, and the area of the 2010 dollar is only about 15% of the area of the 1980 dollar (because both the length and width of the 2010 dollar are reduced to 39% of the 1980 sizes, and $0.39^2 \approx 0.15$). This gives the perception that the value of the dollar shrank even more than it really did.

1980 = $1.00

1990 = $0.63

2010 = $0.39

Figure 3.29 The lengths of the dollars are proportional to their spending power, but our eyes are drawn to the areas, which decline more than the lengths.

> **TIME OUT TO THINK**
> Suppose the three dollars shown in Figure 3.29 were each represented by a three-dimensional stack of pennies, with the height of the stack proportional to the value of the dollar. Would that make the visual distortion of the data greater, less, or the same as the distortion in Figure 3.29? Explain.

Watch the Scales

Figure 3.30a shows the percentage of college students since 1910 who were women. At first glance, it appears that this percentage grew by a huge margin after about 1950. But the vertical axis scale does not begin at zero and does not end at 100%. The increase is still substantial but looks far less dramatic if we redraw the graph with the vertical axis covering the full range of 0 to 100% (Figure 3.30b). From a mathematical point of view, leaving out the zero point on a scale is perfectly honest and can make it easier to see small-scale trends in data. Nevertheless, as this example shows, it can be visually deceptive if you don't study the scale carefully.

BY THE WAY

German researchers in the latter part of the 19th century studied many types of graphics. The type of distortion shown in Figure 3.29 was so common that they gave it its own name, which translates roughly as "the old goosing up the effect by squaring the eyeball trick."

Women as a Percentage of All College Students

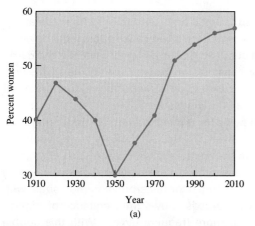

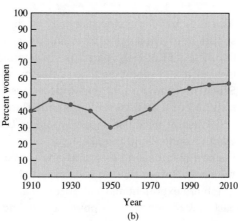

(a)   (b)

Figure 3.30 Both graphs show the same data, but they look very different because their vertical scales have different ranges.
Source: National Center for Education Statistics.

The easiest person to deceive is one's own self.

—Edward Bulwer-Lytton

Another issue can arise when graphs use nonlinear scales, meaning scales in which each increment does not always represent the same change in value. Consider Figure 3.31a, which shows how the speeds of the fastest computers have increased with time. At first glance, it appears that speeds have been increasing linearly. For example, it might look as if the speed increased by the same amount from 1990 to 2000 as it did from 1950 to 1960. However, if you look more closely, you'll see that each tick mark on the vertical scale represents a *tenfold* increase in speed. Now we see that computer speed grew from about 1 to 100 calculations per second from 1950 to 1960 and from about 100 million to 10 billion calculations per second between 1990 and 2000. This type of scale is called an **exponential scale** (or *logarithmic scale*) because it grows by powers of 10 and powers of 10 are *exponents*. (For example, 3 is the exponent in $10^3 = 1,000$.) It is always possible to convert an exponential scale back to an ordinary linear scale as shown in Figure 3.31b. However, comparing the two graphs should make clear why the exponential scale is so useful in this case: The exponential scale clearly shows the rapid gains in computer speeds, while the ordinary scale makes it impossible to see any detail in the early years shown on the graph. More generally, exponential scales are useful whenever data vary over a huge range of values.

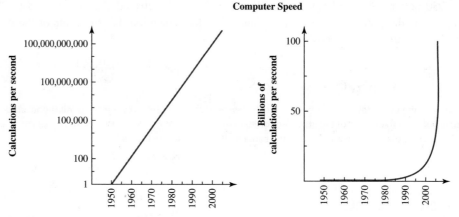

Figure 3.31   Both graphs show the same data, but the graph on the left uses an exponential scale.

TIME OUT TO THINK

Based on Figure 3.31a, can you predict the speed of the fastest computers in 2020? Could you make the same prediction with Figure 3.31b? Explain.

CASE STUDY   Asteroid Threat

Asteroids and comets occasionally hit the Earth. Small ones tend to burn up in the atmosphere or create small craters on impact, but larger ones can cause substantial devastation. About 65 million years ago, an asteroid about 10 kilometers in diameter hit the Earth, leaving a 200-kilometer-wide crater on the coast of the Yucatan peninsula in Mexico. Scientists estimate that this impact caused the extinction of about three-quarters of all species living on Earth at the time, including all the dinosaurs.

Clearly, a similar impact would be bad news for our civilization. We might therefore want to understand the likelihood of such an event. Figure 3.32 shows a graph relating the size of impacting asteroids and comets to the frequency with which such objects hit the Earth. Because of the wide range of sizes and time scales involved, *both* axes on this graph are exponential. The horizontal axis shows impactor (asteroid or comet) sizes, with each tick representing a power of 10. The vertical axis shows the frequency of impact; moving up on the vertical axis corresponds to more frequent events. With this double exponential graph, we can see trends clearly. For example, small objects of about 1 meter

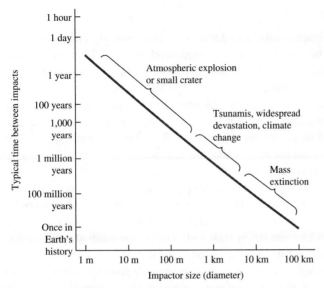

Figure 3.32 This graph shows how the frequency of impacts—and the magnitude of their effects—depends on the size of the impactor. Note that smaller impacts are much more frequent than larger ones.

in size strike the Earth every day, but cause little damage. At the other extreme, objects large enough to cause a mass extinction hit only about once every hundred million years.

The intermediate cases are probably the most worrisome. The graph indicates that objects that could cause "widespread devastation"—such as wiping out the population of a large city—can be expected as often as once every thousand years. This is often enough to warrant at least some preventive action. Currently, astronomers are trying to make more precise predictions about when an object might hit the Earth. If they discover an object that will hit the Earth, scientists will need to find a way to deflect it to prevent the impact.

Percentage Change Graphs

Is college getting more or less expensive? If you didn't look too carefully, Figure 3.33a might lead you to conclude that after peaking in the early 2000s, the cost of public colleges fell during the rest of that decade. But look more closely and you'll see that the vertical axis on this graph represents the *percentage change* in costs. The drop-off therefore means only that costs rose by smaller amounts, not that they fell. Actual college costs are shown in Figure 3.33b, which makes it clear that they rose every year. Graphs that show percentage change are very common; you'll find them in the financial news almost every day. But as you can see, they can be very misleading if you don't realize that they are showing change rather than actual values.

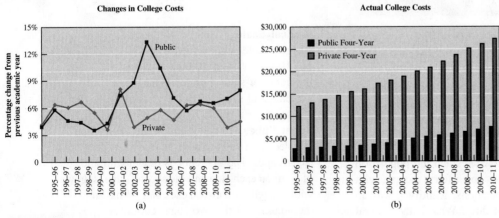

Figure 3.33 Trends in college costs: (a) annual percent change; (b) actual costs.
Source: The College Board.

Get your facts first, and then you can distort them as much as you please.

—Mark Twain

Pictographs

Pictographs are graphs embellished with additional artwork. The artwork may make the graph more appealing, but it can also distract or mislead. Figure 3.34 is a pictograph showing the rise in world population from 1804 to 2040 (numbers for future years are based on United Nations intermediate-case projections). The lengths of the bars correctly correspond to world population for the years listed. However, the artistic embellishments of this graph are deceptive in several ways. For example, your eye may be drawn to the figures of people lining the globe. Because this line of people rises from the left side of the pictograph to the center and then falls, it might give the impression that future world population will decline. In fact, the line of people is purely decorative and carries no information.

The more serious problem with this pictograph is that it makes it appear that world population has been rising linearly. However, notice that the time intervals on the horizontal axis are not the same in each case. For example, the interval between the bars for 1 billion and 2 billion people is 123 years (from 1804 to 1927), but the interval between the bars for 5 billion and 6 billion people is only 12 years (from 1987 to 1999).

Pictographs are very common. As this example shows, however, you have to study them carefully to extract the essential information and not be distracted by the cosmetic effects.

BY THE WAY

Demographers often characterize population growth by a doubling time—the time it takes the population to double. During the late 20th century, the doubling time for the human population was about 40 years. If the population continued to double at this rate, world population would reach more than 30 billion by 2100 and 190 billion by 2200. By about 2650, the human population would be so large that it would not fit on the Earth, even if everyone stood elbow to elbow everywhere.

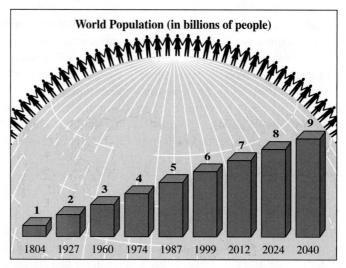

Figure 3.34   A pictograph of world population. The pictures add visual impact, but can also be misleading. Also notice that the horizontal scale (time) is not linear.

Source: United Nations Population Division, future projections based on intermediate case assumptions.

Section 3.4 Exercises

Statistical Literacy and Critical Thinking

1. **Exaggerating a Difference.** Under standard test conditions, the braking distance for a Honda Civic is 136 feet and the braking distance for a VW Jetta is 137 feet. Is that difference meaningful? How could a graph be constructed so that the difference is greatly exaggerated?

2. **Graph of Populations.** When constructing a graph showing the population of the United States and the population of Mexico, an illustrator draws two different people with heights proportional to the populations. Identify a way in which the graph might be misleading? What is the general name for such graphs that use drawings of people or objects?

3. **Vertical Scale.** A line chart has a vertical scale with values of 0, 1, 10, 100, 1000 and 10,000. What is the name of such a scale? What is an advantage of such a scale?

4. **Sugar Cubes.** To show how sugar production doubled from 1990 to now, an illustrator draws two sugar cubes. The first cube is drawn with a length of 1 cm on each side and the second cube is drawn with a length of 2 cm on each side. What are the volumes of the two sugar cubes? Is the illustration misleading? If so, how?

Concepts and Applications

5. Car Mileage. Figure 3.35 shows the highway fuel consumption (mi/gal) for the Chevrolet Aveo and the Honda Civic. How is the graph misleading? How could it be drawn so that it is not misleading?

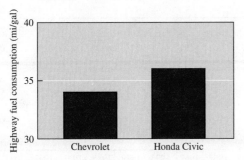

Figure 3.35 Car Mileage.

6. Comparing State Populations. Figure 3.36 depicts 2010 populations for California, Texas, and New York (based on data from the 2010 Census). How is this graph misleading? How could it be drawn so that it is not misleading?

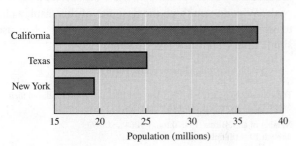

Figure 3.36 State Populations.

7. Pictograph. Figure 3.37 depicts the amounts of daily oil consumption in the United States and Japan. Does the illustration accurately depict the data? Why or why not?

Daily Oil Consumption
(millions of barrels)

Figure 3.37

8. Pictograph. Refer to Figure 3.37 used in Exercise 7 and construct a bar chart to depict the same data in a way that is fair and objective.

9. Three-Dimensional Pies. The pie charts in Figure 3.38 give the percentage of Americans in three age categories in 1990 and 2050 (projected).

a. Consider the 1990 age distribution. The actual percentages for the three categories for 1990 were 87.5% (others), 11.3% (60–84), and 1.2% (85+). Does the pie chart show these values accurately? Explain.

b. Consider the 2050 age distribution. The actual percentages for the three categories for 2050 were 80.0% (others), 15.4% (60–84), and 4.6% (85+). Does the pie chart show these values accurately? Explain.

c. Using the actual percentages given in parts a and b, draw flat (two-dimensional) pie charts to display these data. Explain why these pie charts give a more accurate picture than the three-dimensional pies.

d. Comment on the general trends shown in the two pie charts.

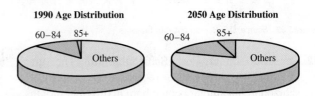

Figure 3.38   *Source:* U.S. Census Bureau.

10. Moore's Law. In 1965, Intel cofounder Gordon Moore initiated what has since become known as *Moore's law:* the number of transistors per square inch on integrated circuits will double approximately every 18 months. In the table below, the first row lists different years and the second row lists the number of transistors (in thousands) for different years.

Year	Transistors	Year	Transistors
1971	23	1997	7500
1974	5	2000	42,000
1978	29	2002	220,000
1982	120	2003	410,000
1985	275	2007	789,000
1989	1180	2011	2,600,000
1993	3100		

a. Construct a time-series graph of these data, using a uniform scale on both axes.

b. Make an exponential graph of these data in which the subdivisions on the vertical axis are 0, 1, 10, 100, 1,000, 10,000, 100,000, 1,000,000, and 100,000,000.

c. Compare the graphs in parts a and b.

11. Percentage Change in the CPI. The graph in Figure 3.39 shows the percentage change in the CPI over recent years. In what year (of the years displayed) was the change in the CPI the greatest? What happened in 2009? How do actual prices in 2010 compare to those in 1990? Based on this graph, what can you conclude about changes in prices during the period shown?

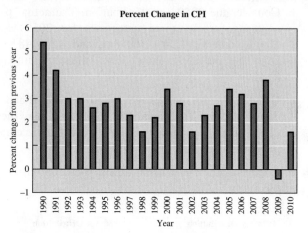

Figure 3.39   *Source:* U.S. Bureau of Labor Statistics.

12. Seasonal Effects on Schizophrenia? The graph in Figure 3.40 shows data regarding the relative risk of schizophrenia among people born in different months.

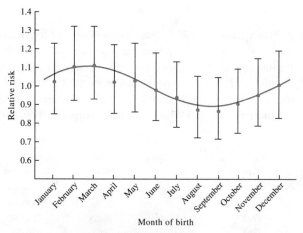

Figure 3.40   *Source: New England Journal of Medicine.*

a. Note that the scale of the vertical axis does not include zero. Sketch the same risk curve using an axis that includes zero. Comment on the effect of this change.

b. Each value of the relative risk is shown with a dot at its most likely value and with an "error bar" indicating the range in which the data value probably lies. The study concludes that "the risk was also significantly associated with the season of birth." Given the size of the error bars, does this claim appear justified? (Is it possible to draw a flat line that passes through all of the error bars?)

13. Constant Dollars. The graph in Figure 3.41 shows the minimum wage in the United States, together with its purchasing power, which is adjusted for inflation with 1996 used as the reference year. The graph represents the years from 1955 to 2011. Summarize what the graph shows.

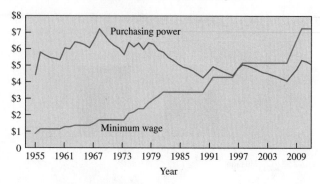

Figure 3.41   *Source:* U.S. Department of Labor.

14. Double Horizontal Scale. The graph in Figure 3.42 shows *simultaneously* the number of births in this country during two time periods: 1946–1964 and 1977–1994. When did the first baby boom peak? When did the second baby boom peak? Why do you think the designer of this display chose to superimpose the two time intervals, rather than use a single time scale from 1946 through 1994?

Baby Boomers and Their Babies

The baby-boom generation, born between 1946 and 1964, produced their own smaller boom between 1977 and 1994.

Number of U.S. Births, 1946 through 1964 and 1977 through 1994 (in millions)

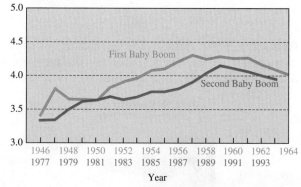

Figure 3.42   *Source:* Based on data from the National Center for Health Statistics.

PROJECTS FOR THE INTERNET & BEYOND

15. USA Snapshot. *USA Today* offers a daily pictograph for its "Snapshot." Find a snapshot from a recent issue of *USA Today*. Briefly discuss its purpose and effectiveness.

16. Image Search. Choose some topic that interests you for which you think that good statistical graphs should be available. Do an image search for the topic in Google, Bing, or other search engine. Does the search give you what you were looking for? Briefly discuss the value of the search results in terms of what you hoped to find.

IN THE NEWS

17. Distortions in the News. Find an example in a recent news report of a graph that involves some type of perceptual distortion. Explain the effects of the distortion, and describe how the graph could have been drawn more honestly.

18. Scale Problems in the News. Find an example in a recent news report of a graph in which the vertical scale does not start at zero. Suggest why the graph was drawn that way and also discuss any ways in which the graph might be misleading as a result.

19. Economic Graph in the News. Find an example in a recent news report of a graph that shows economic data over time. Are the data adjusted for inflation? Discuss the meaning of the graph and any ways in which it might be deceptive.

20. Pictograph in the News. Find an example of a pictograph in a recent news report. Discuss what the pictograph attempts to show, and discuss whether the artistic embellishments help or hinder this purpose.

21. Outstanding News Graph. Find a graph from a recent news report that, in your opinion, is truly outstanding in displaying data visually. Discuss what the graph shows, and explain why you think it is so outstanding.

22. Not-So-Outstanding News Graph. Find a graph from a recent news report that, in your opinion, fails in its attempt to display data visually in a meaningful way. Discuss what the graph was trying to show, explain why it failed, and explain how it could have been done better.

CHAPTER REVIEW EXERCISES

Listed below are measured weights (in pounds) of the contents in samples of cans of regular Pepsi and Diet Pepsi. Use these data for Exercises 1–3.

Regular Pepsi:

0.8258	0.8156	0.8211	0.8170	0.8216	0.8302
0.8192	0.8192	0.8271	0.8251	0.8227	0.8256
0.8139	0.8260	0.8227	0.8388	0.8260	0.8317
0.8247	0.8200	0.8172	0.8227	0.8244	0.8244
0.8319	0.8247	0.8214	0.8291	0.8227	0.8211
0.8401	0.8233	0.8291	0.8172	0.8233	0.8211

Diet Pepsi:

0.7925	0.7868	0.7846	0.7938	0.7861	0.7844
0.7795	0.7883	0.7879	0.7850	0.7899	0.7877
0.7852	0.7756	0.7837	0.7879	0.7839	0.7817
0.7822	0.7742	0.7833	0.7835	0.7855	0.7859
0.7775	0.7833	0.7835	0.7826	0.7815	0.7791
0.7866	0.7855	0.7848	0.7806	0.7773	0.7775

1. a. Construct a frequency table for the weights of regular Pepsi. Use bins of

   0.8130–0.8179
   0.8180–0.8229
   0.8230–0.8279
   0.8280–0.8329
   0.8330–0.8379
   0.8380–0.8429

b. Construct a frequency table for the weights of diet Pepsi. Use bins of

   0.7740–0.7779
   0.7780–0.7819
   0.7820–0.7859
   0.7860–0.7899
   0.7900–0.7939

c. Compare the frequency tables from parts a and b. What notable differences are there? How can those notable differences be explained?

2. a. Construct a relative frequency table for the weights of regular Pepsi. Use bins of

   0.8130–0.8179
   0.8180–0.8229
   0.8230–0.8279
   0.8280–0.8329
   0.8330–0.8379
   0.8380–0.8429

b. Construct a cumulative frequency table for the weights of regular Pepsi.

3. a. Use the result from Exercise 1a to construct a histogram for the weights of regular Pepsi.

b. Use the result from Exercise 1b to construct a histogram for the weights of Diet Pepsi.

c. Compare the histograms from parts a and b. How are they similar and how are they different?

4. **Pie Chart of Awful Sounds.** In a survey, 1,004 adults were asked to identify the most frustrating sound that they hear in a day. Two hundred seventy-nine chose jackhammers, 388 chose car alarms, 128 chose barking dogs, and 209 chose crying babies (based on data from Kelton Research). Construct a pie chart depicting these data.

5. **Pareto Chart.** Construct a Pareto chart from the data given in Exercise 4. Compare the Pareto chart to the pie chart. Which graph is more effective in showing the numbers of frustrating sounds? Explain.

6. **Bar Chart.** Figure 3.43 shows the numbers of U.S. adoptions from China in the years 2005 and 2010. What is wrong with this graph? Draw a graph that depicts the data in a fair and objective way.

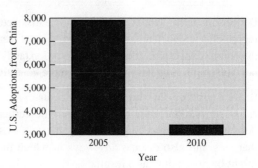

Figure 3.43

CHAPTER QUIZ

1. The IQ scores of 500 college football players are randomly selected. Which graph would be most appropriate for these data: histogram, bar chart, pie chart, multiple bar graph, or stack plot?

2. As a quality control manager at Sony, you find that defective CDs have various causes, including worn machinery, human error, bad supplies, and packaging mistreatment. Which of the following graphs would be best for describing the causes of defects: histogram, bar chart, Pareto chart, dotplot, or pie chart?

3. A stemplot is created from the intervals (min) between eruptions of the Old Faithful geyser in Yellowstone National Park, and one row of that stemplot is 6|1222279. Identify the values represented by that row.

4. The first class in a frequency table is 20–29 and the corresponding frequency is 15. What does the value of 15 indicate?

5. The first class in a relative frequency table is 20–29 and the corresponding relative frequency is 0.25. What does the value of 0.25 indicate?

6. The third class in a frequency table is 40–49 and the corresponding cumulative frequency is 80. What does the value of 80 indicate?

7. The bar chart in Figure 3.44 depicts the number of twin births in the United States in the years 2000 and 2008. In what way is this graph misleading?

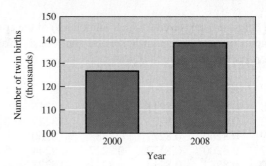

Figure 3.44

8. Construct a dotplot representing these six IQ scores: 90, 95, 95, 100, 105, 110.

9. Identify the values represented by the following stemplot:

```
0 | 0 1 1
1 | 0 1 2
3 |
4 | 9 9
```

10. In one county, the cost of snow removal tripled from 1980 to the current year. Why is it misleading to represent the amounts with images of two rectangular snowplows if the second is three times as wide and three times as tall as the first snowplow?

F CUS ON HISTORY

Can War Be Described with a Graph?

Can a war be described with a graph? Figure 3.45, created by Charles Joseph Minard in 1869, does so remarkably well. This graph tells the story of Napoleon's ill-fated Russian campaign of 1812, sometimes called Napoleon's death march.

The underlying map on Minard's graph shows a roughly 500-mile strip of land extending from the Niemen River on the Polish-Russian border to Moscow. The blue strip depicts the forward march of Napoleon's army. On Minard's original drawing, each millimeter of width represented 6,000 men; this reproduction is shown at a smaller size than the original. The march begins at the far left, where the strip is widest. Here, an army of 422,000 men triumphantly began a march toward Moscow on June 24, 1812. At the time, it was the largest army ever mobilized.

The narrowing of the strip as it approaches Moscow represents the unfolding decimation of the army. (The offshoots represent battalions that were sent off in other directions along the way.) Napoleon had brought only minimal food supplies, and hot summer weather accompanied by heavy rains brought rampant disease. Starvation, disease, and combat losses killed thousands of men each day. By the time the army entered Moscow on September 14, it had shrunk to 100,000 men. The worst was yet to come.

To Napoleon's dismay, the Russians evacuated Moscow prior to the French army's arrival. Deprived of the opportunity to engage the Russian troops and feeling that his army's condition was too poor to continue on to the Russian capital of St. Petersburg, Napoleon took his troops southward out of Moscow. The lower part of the strip on the graph (shown in maroon) represents the retreat, and the dark blue line near the bottom of the figure shows the nighttime temperatures as winter approached. We see that freezing temperatures had already set in by October 18.

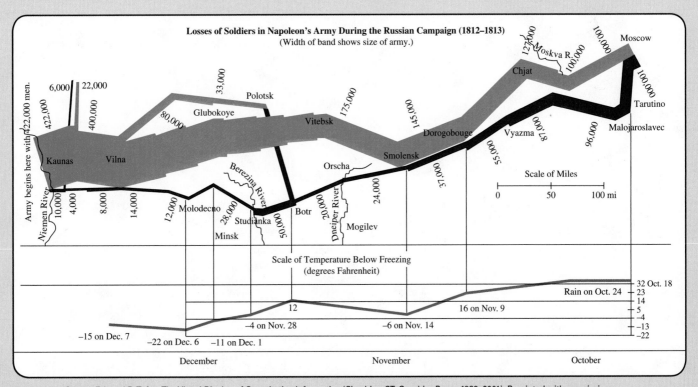

Figure 3.45 *Source:* Edward R. Tufte, *The Visual Display of Quantitative Information* (Cheshire, CT: Graphics Press, 1983, 2001). Reprinted with permission.

Temperatures plunged below 0°F in late November. The sudden narrowing of the lower strip around November 28 shows where 22,000 men perished on the banks of the Berezina River. Three-fourths of the survivors froze to death over the next few days, many on the bitter cold night of December 6. By the time the army reached Poland on December 14, only 10,000 of the original 422,000 remained.

In a famous analysis of graphical techniques, author Edward Tufte described Minard's graph as possibly "the best statistical graphic ever drawn." But a more dramatic statement came from a contemporary of Minard, E. J. Marey, who wrote that this graphic "brought tears to the eyes of all France."

QUESTIONS FOR DISCUSSION

1. Discuss how this graph helps to overcome the impersonal nature of the many deaths in a war. What kind of impact does it have on you personally?

2. Note that this graph plots six variables: two variables of direction (north-south and east-west), the size of the army, the location of the army, the direction of the army's movement, and temperatures during the retreat. Do you think Minard could have gotten the point across with fewer variables? Why or why not?

3. Discuss how you might make a similar graph for some other historical or political event.

• • • • • • • • • • • • • • • •

FOCUS ON
ENVIRONMENT

Are We Changing Earth's Atmosphere?

Sometimes a well-drawn figure can teach important lessons even before you understand everything about it. Figure 3.46 is a case in point.

As the label shows, the graph shows how the concentration of carbon dioxide in Earth's atmosphere varies with time. The units used for the concentration are *parts per million*, which means the number of carbon dioxide molecules among each one million molecules of air. For example, a concentration of 300 parts per million means that there are 300 molecules of carbon dioxide among each one million molecules of air, which is equivalent to a concentration of 3 in 10,000 or 0.03%.

Notice that the main graph shows how the carbon dioxide concentration has varied over the past 800,000 years. Scientists obtain these data by measuring the concentration of carbon dioxide trapped in bubbles in ancient ice, which they collect by drilling ice cores out of the ice sheets in Greenland

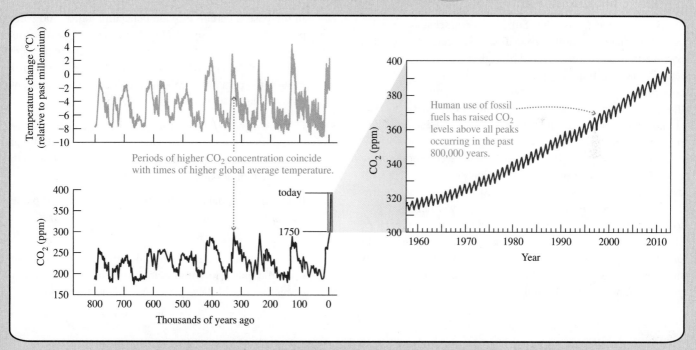

Figure 3.46   (Left) The atmospheric concentration of carbon dioxide and global average temperature reconstructed from ice core data for the past 800,000 years.
Source: European Project for Ice Coring in Antarctica. (Right) The concentration has been directly measured in the atmosphere since the late 1950s. Note: The wiggles in the right graph represent seasonal fluctuations; the line running through the wiggles represents the general trend. *Source:* National Oceanic and Atmospheric Administration.

or Antarctica. The deepest ice cores drilled to date reach down to ice that was deposited more than 800,000 years ago, which is why the graph can show data going back that far. The zoom-in on the right shows the carbon dioxide concentration directly measured in Earth's atmosphere since the late 1950s.

Even without knowing anything about the role of carbon dioxide, the graph should certainly grab your attention. For one thing, the many ups and downs visible in the graph for the past 800,000 years show that the carbon dioxide concentration varies substantially through natural processes. But notice that in all of those ups and downs, the concentration never rose above about 300 parts per million until just a couple of centuries ago. Now, the zoom-out shows the concentration rising by some 2 to 3 parts per million per year, a rate at which it will pass 400 parts per million by 2015 and 500 parts per million by about 2060. Clearly, something dramatic is happening to the carbon dioxide concentration.

The dramatic change should lead you to ask other questions. First, you might wonder whether the carbon dioxide concentration is important, and the answer is revealed by the notes on the main graph. The same ice core data that allow reconstruction of the past carbon dioxide concentration also allow reconstruction of past temperatures on Earth, and comparison reveals that the carbon dioxide concentration and the temperature tend to rise and fall in tandem. That is, past ice ages were marked by low carbon dioxide concentrations, while past warm periods were marked by higher concentrations. This leads to the question of whether the change in carbon dioxide concentration *causes* the changes in temperature. Although this graph alone does not answer that question, other statistical studies, which we'll discuss in Chapter 7 (see Focus, p. 273), provide strong evidence that there is indeed a cause and effect.

If the cause and effect are real, then the current dramatic rise in the carbon dioxide concentration should be cause for great concern. During many of the past warm periods, Earth's average temperature was several degrees warmer than it is today, so the fact that the carbon dioxide concentration is now skyrocketing suggests that the same might happen to our planet's temperature. Moreover, notice that the past data show that the changes in concentration (and temperature) often happen quite rapidly, suggesting that the process can feed back on itself. In that case, there's a strong risk that we have already started a process that may cause Earth's temperature to rise rapidly and dramatically over the coming decades.

The remaining question is whether the current rise in the carbon dioxide concentration is a natural phenomenon like the past rises or something being caused by humans. On this point, there is no doubt. By carefully studying the isotopic makeup of the carbon dioxide in the atmosphere and comparing it to that of various carbon dioxide sources, scientists have found that added carbon dioxide is coming primarily from the burning of fossil fuels.

Perhaps you've heard debate about whether global warming is a real or imagined threat. With a bit of extra information, such as knowing that the added carbon dioxide comes from human activity, Figure 3.46 makes clear that there's nothing imaginary about it.

QUESTIONS FOR DISCUSSION

1. Study Figure 3.46 carefully. How does the carbon dioxide concentration today compare to that of 1750? How does that of 1750 compare to that during the past 800,000 years? What conclusions can you draw from your answers to these questions?

2. In the past, a carbon dioxide concentration of 300 parts per million accompanied global average temperatures as much as about 9°F (5°C) higher than the global average temperature today. What do you think would happen if Earth's temperature rose that much over the next century? Is that a worst-case scenario? Explain.

3. Discuss some of the factors that will affect the future concentration of carbon dioxide in the atmosphere. What do *you* think should be done to slow or stop the growth in the carbon dioxide concentration?

4. Find the latest data for the atmospheric carbon dioxide concentration. Is the level still rising? How fast is it rising?

5. The graphs in Figure 3.46 summarize a great deal of scientific data. How important are clear graphics to our understanding of these data? Defend your opinion.

• • • • • • • • • • • • • • • • •

Describing Data

In Chapter 3, we discussed methods for displaying data distributions with tables and graphs. Now we are ready to study common methods for describing the center, shape, and variation of a collection of data. These methods are central to data analysis and, as you'll see, have applications to nearly every statistical study you encounter in the news. We'll conclude the chapter by studying a few surprises that occasionally turn up even when we look at data carefully.

It's no use trying to sum people up. One must follow hints, not exactly what is said, nor yet entirely what is done.

—Virginia Woolf

4.1 WHAT IS AVERAGE?

The term *average* comes up frequently in the news and other reports, but it does not always have the same meaning. As you will see in this section, the most appropriate definition of *average* depends on the situation.

Mean, Median, and Mode

Table 4.1 shows the number of movies (original and sequels or prequels) in each of five popular science fiction series. What is the average number of films in these series? One way to answer this question is to compute the **mean**. (The formal term of *arithmetic mean* is commonly referred to simply as the *mean*.) We find the mean by dividing the total number of movies by five (because there are five series listed in the data set):

$$\text{mean} = \frac{4 + 7 + 11 + 6 + 4}{5} = \frac{32}{5} = 6.4$$

In other words, these five series have a mean of 6.4 movies. More generally, we find the mean of any data set by dividing the sum of all the data values by the number of data values. The mean is what most people think of as the average. In essence, it represents the balance point for a quantitative data distribution, as shown in Figure 4.1.

We could also describe the average number of films by computing the **median,** or middle value, of the data set. To find a median, we arrange the data values in ascending (or descending) order, repeating data values that appear more than once. If the number of values is odd, there is exactly one value in the middle of the list, and this value is the median. If the number of values is even, there are two values in the middle of the list, and the median is the number that lies halfway between them. Putting the data in Table 4.1 in ascending order gives the list 4, 4, 6, 7, 11. The median number of movies is 6 because 6 is the middle number in the list.

The **mode** is the most common value or group of values in a data set. In the case of the movies, the mode is 4 because this value occurs twice in the data set, while the other values occur only once. A data set may have one mode, more than one mode, or no mode. Sometimes the mode refers to a group of closely spaced values rather than a single value. The mode is used more commonly for qualitative data than for quantitative data, as neither the mean nor the median can be used with qualitative data.

TABLE 4.1	Five Science Fiction Movie Series
Series	**Number of movies (as of 2012)**
Alien	4
Planet of the Apes	7
Star Trek	11
Star Wars	6
Terminator	4

Note: Counts only major releases in each series; does not include *Prometheus* in the *Alien* series.

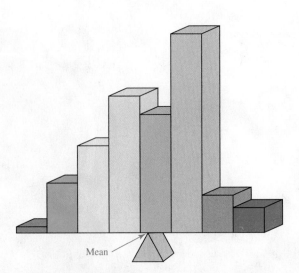

Figure 4.1 A histogram made from blocks would balance at the position of its mean.

Definitions—Measures of Center in a Distribution

The **mean** is what we most commonly call the average value. It is found as follows:

$$\text{mean} = \frac{\text{sum of all values}}{\text{total number of values}}$$

The **median** is the middle value in the sorted data set (or halfway between the two middle values if the number of values is even).

The **mode** is the most common value (or group of values) in a data set.

When rounding, we will use the following rule for all the calculations discussed in this chapter.

Rounding Rule for Statistical Calculations

In general, you should express answers with *one more* decimal place of precision than is found in the raw data. For example, if the data are given as whole numbers, you should round their mean to the nearest tenth; if the data are given to the nearest tenth, you should round their mean to the nearest hundredth; and so on. *As always, round only the final answer and not any intermediate values used in your calculations.*

Notice how we applied this rule in the movie example. The data in Table 4.1 consist of whole numbers, so we stated the mean as 6.4. If the calculation happens to end up with the same number of significant digits as the raw data, then keeping an extra decimal place is optional.

EXAMPLE 1 Price Data

Eight grocery stores sell the PR energy bar for the following prices:

$1.09 $1.29 $1.29 $1.35 $1.39 $1.49 $1.59 $1.79

Find the mean, median, and mode for these prices.

SOLUTION   The *mean* price is $1.41:

$$\text{mean} = \frac{\$1.09 + \$1.29 + \$1.29 + \$1.35 + \$1.39 + \$1.49 + \$1.59 + \$1.79}{8}$$

$$= \$1.41$$

To find the *median*, we first sort the data in ascending order:

$1.09, \ $1.29, \ $1.29, $1.35, \ $1.39, $1.49, \ $1.59, \ $1.79

3 values below   2 middle values   3 values above

Excel Excel provides the built-in function AVERAGE for calculating a mean and separate functions for MEDIAN and MODE. The screen shot below shows the use of these functions for the data from Example 1. Column B shows the functions and Column C shows the results.

◇	A	B	C
1	Data	1.09	
2		1.29	
3		1.29	
4		1.35	
5		1.39	
6		1.49	
7		1.59	
8		1.79	
9	Mean	=AVERAGE(B1:B8)	1.41
10	Median	=MEDIAN(B1:B8)	1.37
11	Mode	=MODE(B1:B8)	1.29
12			

Note: If you have a Windows version of Excel, you can get even more information by installing the Data Analysis Toolpak; to find it, use the Help feature and search for "Data Analysis." Once installed, click on **Data**, select **Data Analysis**, then select **Descriptive Statistics** in the pop-up window, and click **OK**. In the dialog box, enter the input range (such as A1:A8 for 8 values in column A), click on **Summary Statistics**, then click **OK**. Results will include the mean and median, as well as other statistics to be discussed in the following sections.

Alternatively, the XLSTAT add-in that is a supplement to this book can be used with both Windows and Mac computers. Enter the data as described above, click on XLSTAT, click on Visualizing Data, then select Descriptive Statistics. Enter the range of the data, such as A1:A8. (If the first row is the name of the data, be sure to click on the box next to "Sample labels."

STATDISK Enter the data in the Data Window or open an existing data set. Click on **Data** and select **Descriptive Statistics**. Now click on **Evaluate** to get the various descriptive statistics, including the mean and median, as well as other statistics to be discussed in the following sections.

TI-83/84 PLUS   First enter the data in list L1 by pressing `STAT`, then selecting **Edit** and pressing the `ENTER` key. After the data values have been entered, press `STAT` and select **CALC**, then select **1-Var Stats** and press the `ENTER` key twice. The display will include the mean and median, as well as other statistics to be discussed in the following sections. Use the down-arrow key `▽` to view the results that don't fit on the initial display.

Because there are eight prices (an even number), there are two values in the middle of the list: $1.35 and $1.39. Therefore the median lies halfway between these two values, which we calculate by adding them and dividing by 2:

$$\text{median} = \frac{\$1.35 + \$1.39}{2} = \$1.37$$

Using the rounding rule, we could express the mean and median as $1.410 and $1.370, respectively, though the extra zeros are optional in this case.

The *mode* is $1.29 because this price occurs more times than any other price.   ··●

Effects of Outliers

To explore the differences among the mean, median, and mode, imagine that the five graduating seniors on a college basketball team receive the following first-year contract offers to play in the National Basketball Association (zero indicates that the player did not receive a contract offer):

$$0 \; 0 \; 0 \; 0 \; \$10,000,000$$

The mean contract offer is

$$\text{mean} = \frac{0 + 0 + 0 + 0 + \$10,000,000}{5} = \$2,000,000$$

Is it therefore fair to say that the *average* senior on this basketball team received a $2 million contract offer?

TABLE 4.2	Comparison of Mean, Median, and Mode					
Measure	Definition	How common?	Existence	Takes every value into account?	Affected by outliers?	Advantages
Mean	$\dfrac{\text{sum of all values}}{\text{total number of values}}$	most familiar "average"	always exists	yes	yes	commonly understood; works well with many statistical methods
Median	middle value	common	always exists	no (aside from counting the total number of values)	no	when there are outliers, may be more representative of an "average" than the mean
Mode	most frequent value	sometimes used	may be no mode, one mode, or more than one mode	no	no	most appropriate for qualitative data (see Section 2.1)

Not really. The problem is that the single player receiving the large offer makes the mean much larger than it would be otherwise. If we ignore this one player and look only at the other four, the mean contract offer is zero. Because this one value of $10,000,000 is so extreme compared with the others, we say that it is an **outlier** (or *outlying value*). As our example shows, an outlier can pull the mean significantly upward (or downward), thereby making the mean unrepresentative of the data set as a whole.

> **Definition**
>
> An **outlier** in a data set is a value that is much higher or much lower than almost all other values.

While the outlier pulls the mean contract offer upward, it has no effect on the median contract offer, which remains zero for the five players. In general, the value of an outlier has no effect on the median, because outliers don't lie in the middle of a data set. Outliers do not affect the mode either. Table 4.2 summarizes the characteristics of the mean, median, and mode, including the effects of outliers on each measure.

> **TIME ◔UT TO THINK**
>
> Is it fair to use the median as the average contract offer for the five players? Why or why not?

Deciding how to deal with outliers is one of the more important issues in statistics. Sometimes, as in our basketball example, an outlier is a legitimate value that must be understood in order to interpret the mean and median properly. Other times, outliers may indicate mistakes in a data set. Deciding when outliers are important and when they may simply be mistakes can be very difficult.

EXAMPLE 2 Mistake?

A track coach wants to determine an appropriate heart rate for her athletes during their workouts. She chooses five of her best runners and asks them to wear heart monitors during a workout. In the middle of the workout, she reads the following heart rates for the five athletes: 130, 135, 140, 145, 325. Which is a better measure of the average in this case—the mean or the median? Why?

SOLUTION Four of the five values are fairly close together and seem reasonable for mid-workout heart rates. The high value of 325 is an outlier. This outlier seems likely to be a mistake (perhaps caused by a faulty heart monitor), because anyone with such a high heart rate should be in cardiac arrest. If the coach uses the mean as the average, she will be including this

outlier—which means she will be including any mistake made when it was recorded. If she uses the median as the average, she'll have a more reasonable value, because the median won't be affected by the outlier.   · · •

"Average" Confusion

The different meanings of *average* can lead to confusion. Sometimes this confusion arises because we are not told whether the average is the mean or the median, and other times because we are not given enough information about how the average was computed. The following examples illustrate two such situations.

EXAMPLE 3   Wage Dispute

A newspaper surveys wages for workers in regional high-tech companies and reports an average of $22 per hour. The workers at one large firm immediately request a pay raise, claiming that they work as hard as employees at other companies but their average wage is only $19. The management rejects their request, telling them that they are *overpaid* because their average wage, in fact, is $23. Can both sides be right? Explain.

SOLUTION   Both sides can be right if they are using different definitions of *average*. In this case, the workers may be using the median while the management uses the mean. For example, imagine that there are only five workers at the company and their wages are $19, $19, $19, $19, and $39. The median of these five wages is $19 (as the workers claimed), but the mean is $23 (as management claimed).   · · •

EXAMPLE 4   Which Mean?

All 100 first-year students at a small college take three courses in the Core Studies program. Two courses are taught in large lectures, with all 100 students in a single class. The third course is taught in 10 classes of 10 students each. Students and administrators get into an argument about whether classes are too large. The students claim that the mean size of their Core Studies classes is 70. The administrators claim that the mean class size is only 25. Can both sides be right? Explain

Figures won't lie, but liars will figure.

—Charles H. Grosvenor

SOLUTION   The students calculated the mean size of the classes in which each student is personally enrolled. Each student is taking two classes with enrollments of 100 and one class with an enrollment of 10, so the mean size of each student's classes is

$$\frac{\text{total enrollment in student's classes}}{\text{number of classes student is taking}} = \frac{100 + 100 + 10}{3} = 70$$

The administrators calculated the mean enrollment in all classes. There are two classes with 100 students and 10 classes with 10 students, making a total enrollment of 300 students in 12 classes. The mean enrollment per class is

$$\frac{\text{total enrollment}}{\text{number of classes}} = \frac{300}{12} = 25$$

The two claims about the mean are both correct, but the two sides are talking about different means. The students calculated the mean *class size per student*, while the administrators calculated the mean number of *students per class*.   · · •

> ### TIME ◷UT TO THINK
> In Example 4, could the administrators redistribute faculty assignments so that all classes have 25 students each? How? Discuss the advantages and disadvantages of such a change.

Weighted Mean

Suppose your course grade is based on four quizzes and one final exam. Each quiz counts as 15% of your final grade, and the final counts as 40%. Your quiz scores are 75, 80, 84, and 88, and your final exam score is 96. What is your overall score?

Because the final exam counts more than the quizzes, a simple mean of the five scores does not give your final score. Instead, we must assign a *weight* (indicating the relative importance) to each score. In this case, we assign weights of 15 (for the 15%) to each of the quizzes and 40 (for the 40%) to the final. We then find the **weighted mean** by adding the products of each score and its weight and then dividing by the sum of the weights:

$$\text{weighted mean} = \frac{(75 \times 15) + (80 \times 15) + (84 \times 15) + (88 \times 15) + (96 \times 40)}{15 + 15 + 15 + 15 + 40}$$

$$= \frac{8745}{100} = 87.45$$

The weighted mean of 87.45 properly accounts for the different weights of the quizzes and the exam. Following the rounding rule, we round this score to 87.5.

Weighted means are appropriate whenever the data values vary in their degree of importance. You can always find a weighted mean using the following formula.

BY THE WAY

Sports statistics that rate players or teams according to their performance in many different categories are usually weighted means. Examples include the earned run average (ERA) and slugging percentage in baseball, the quarterback rating in football, and computerized rankings of college teams.

Definitions

A **weighted mean** accounts for variations in the relative importance of data values. Each data value is assigned a weight and the weighted mean is

$$\text{weighted mean} = \frac{\text{sum of (each data value} \times \text{its weight)}}{\text{sum of all weights}}$$

TIME ⏱ OUT TO THINK

Because the weights are percentages in the course grade example, we could think of the weights as 0.15 and 0.40 rather than 15 and 40. Calculate the weighted mean by using the weights of 0.15 and 0.40. Do you still find the same answer? Why or why not?

EXAMPLE 5 GPA

Randall has 38 credits with a grade of A, 22 credits with a grade of B, and 7 credits with a grade of C. What is his grade point average (GPA)? Base the GPA on values of 4.0 points for an A, 3.0 points for a B, and 2.0 points for a C.

SOLUTION The grades of A, B, and C represent data values of 4.0, 3.0, and 2.0, respectively. The numbers of credits are the weights. The As represent a data value of 4 with a weight of 38, the Bs represent a data value of 3 with a weight of 22, and the Cs represent a data value of 2 with a weight of 7. The weighted mean is

$$\text{weighted mean} = \frac{(4 \times 38) + (3 \times 22) + (2 \times 7)}{38 + 22 + 7} = \frac{232}{67} = 3.46$$

Following our rounding rule, we round Randall's GPA from 3.46 to 3.5.

EXAMPLE 6   Stock Voting

Voting in corporate elections is usually weighted by the amount of stock owned by each voter. Suppose a company has five stockholders who vote on whether the company should embark on a new advertising campaign. The votes (Y = yes, N = no) are as follows:

Stockholder	Shares owned	Vote
A	225	Y
B	170	Y
C	275	Y
D	500	N
E	90	N

According to the company's bylaws, the measure needs 60% of the vote to pass. Does it pass?

SOLUTION   We can regard a yes vote as a value of 1 and a no vote as a value of 0. The number of shares is the weight for the vote of each stockholder, so Stockholder A's vote represents a value of 1 with a weight of 225, stockholder B's vote represents a value of 1 with a weight of 170, and so on. The weighted mean vote is

$$\text{weighted mean} = \frac{(1 \times 225) + (1 \times 170) + (1 \times 275) + (0 \times 500) + (0 \times 90)}{225 + 170 + 275 + 500 + 90}$$

$$= \frac{670}{1260} \approx 0.53$$

The weighted vote is 53% (or 0.53) in favor, which is short of the required 60%, so the measure does not pass.   •••

Means with Summation Notation (Optional Section)

Many statistical formulas, including the formula for the mean, can be written compactly with a mathematical notation called *summation notation*. The symbol Σ (the Greek capital letter *sigma*) is called the *summation sign* and indicates that a set of numbers should be added. We use the symbol x to represent *each* value in a data set, so we write the sum of all the data values as

$$\text{sum of all values} = \Sigma x$$

For example, if a sample consists of 25 exam scores, Σx represents the sum of all 25 scores. Similarly, if a sample consists of the incomes of 10,000 families, Σx represents the total dollar value of all 10,000 incomes.

We use n to represent the total number of values in the sample. Thus, the general formula for the mean is

$$\bar{x} = \text{sample mean} = \frac{\text{sum of all values}}{\text{total number of values}} = \frac{\Sigma x}{n}$$

The symbol $\bar{x}$ is the standard symbol for the mean of a sample. When dealing with the mean of a population rather than a sample, statisticians instead use the Greek letter μ (*mu*).

Summation notation also makes it easy to express a general formula for the weighted mean. Again we use the symbol x to represent each data value, and we let w represent the weight of each data value. The sum of the products of each data value and its corresponding weight is $\Sigma(x \times w)$. The sum of the weights is Σw. Therefore, the formula for the weighted mean is

$$\text{weighted mean} = \frac{\Sigma(x \times w)}{\Sigma w}$$

TECHNICAL NOTE

Summations are often written with the use of an *index* that specifies how to step through the sum. For example, the symbol x_i indicates the *i*th data value in the set; the letter *i* is the index. We then write the sum of all values as

$$\sum_{i=1}^{n} x_i$$

We read this expression as "the sum of the x_i values, starting with $i = 1$ and continuing to $i = n$, where n is the total number of data values in the set." With this notation, the mean is written

$$\bar{x} = \frac{1}{n} \sum_{i=1}^{n} x_i$$

Means and Medians with Binned Data (Optional Section)

The ideas of this section can be extended to binned data simply by assuming that the middle value in the bin represents all the data values in the bin. For example, consider the following table of 50 binned integer data values:

Bin	Frequency
0–6	10
7–13	10
14–20	10
21–27	20

The middle value of the first bin is 3, so we assume that the value of 3 occurs 10 times. Continuing this way, the sum of the 50 values in the table is

$$(3 \times 10) + (10 \times 10) + (17 \times 10) + (24 \times 20) = 780$$

The mean is therefore 780/50 = 15.6. With 50 values, the median is between the 25th and 26th values. These values fall within the bin 14–20, so we call this bin the **median class** for the data. The mode is the bin with the highest frequency—the bin 21–27 in this case.

Section 4.1 Exercises

Statistical Literacy and Critical Thinking

1. **Average and Mean.** Do the terms "average" and "mean" have the same meaning? Explain.

2. **Mean and Median.** A statistics class consists of 24 students, all but one of whom are unemployed or are employed in low-paying part-time jobs. One student works as an executive secretary earning $75,000 per year. Which does a better job of describing the income of a typical student in the class: the mean or the median? Why?

3. **Outlier.** For the same statistics class described in Exercise 2, is the executive secretary's salary an outlier? Why or why not? In general, is an outlier defined in an exact way so that it can be clearly and objectively identified?

4. **Mean Income.** An economist wants to find the mean annual income for all adults in the United States. She knows that it is not practical to survey each member of the adult population, so she refers to an almanac and finds the mean income listed for each of the 50 states. She adds the 50 state means and divides by 50. Is the result the mean income for the United States? Why or why not?

Does It Make Sense? For Exercises 5–8, decide whether the statement makes sense (or is clearly true) or does not make sense (or is clearly false). Explain clearly; not all of these statements have definitive answers, so your explanation is more important than your chosen answer.

5. **Mean.** The number on the jersey of each New York Giants football player is recorded, then the mean of those numbers is computed.

6. **Mode.** A data set of incomes has modes of $50,000 and $80,000.

7. **Mean, Median, and Mode.** A researcher studying an income distribution obtains the same value of $75,000 for the mean, median, and mode.

8. **Weighted Mean.** A professor calculates final grades using a weighted mean in which the final exam counts twice as much as the midterm.

Appropriate Average. Exercises 9–12 list "averages" that someone might want to know. In each case, state whether the mean or median would give a better description of the "average." Explain your reasoning.

9. **Height.** The average height of all active professional basketball players

10. **Salary.** The average salary of all active professional basketball players

11. **Ages.** The following ages (years) of survey respondents: 22, 19, 21, 27, "over 65," "over 80"

12. **Pulse Rates.** The average resting pulse rate of 500 randomly selected female statistics students

Concepts and Applications

Mean, Median, and Mode. Exercises 13–20 each list a set of numbers. In each case, find the mean, median, and mode of the listed numbers.

13. **Number of Words.** Pages from *Merriam-Webster's Collegiate Dictionary*, 11th edition, were randomly selected. Here are the numbers of words defined on those pages:

   51 63 36 43 34 62 73 39 53 79

14. **Space Shuttle Flights.** Listed below are the durations (in hours) of a sample of all flights of NASA's Space Transport System (space shuttle):

   73   95   235   192   165   262   191   376
   259   235   381   331   221   244   0

15. **Perception of Time.** Actual times (in seconds) recorded when statistics students participated in an experiment to test their ability to determine when one minute (60 seconds) had passed:

   53 52 75 62 68 58 49 49

16. **Body Temperatures.** Body temperatures (in degrees Fahrenheit) of randomly selected normal and healthy adults:

   98.6   98.6   98.0   98.0   99.0
   98.4   98.4   98.4   98.4   98.6

17. **Blood Alcohol.** Blood alcohol concentrations of drivers involved in fatal crashes and then given jail sentences (based on data from the U.S. Department of Justice):

   0.27   0.17   0.17   0.16   0.13   0.24
   0.29   0.24   0.14   0.16   0.12   0.16

18. **Old Faithful Geyser.** Time intervals (in minutes) between eruptions of Old Faithful geyser in Yellowstone National Park:

   98   92   95   87   96   90
   65   92   95   93   98   94

19. **Weights of M&Ms.** Weights (in grams) of randomly selected M&M plain candies:

   0.957   0.912   0.842   0.925   0.939   0.886
   0.914   0.913   0.958   0.947   0.920

20. **Quarters.** Weights (in grams) of quarters in circulation:

   5.60   5.63   5.58   5.56   5.66   5.58   5.57   5.59
   5.67   5.61   5.84   5.73   5.53   5.58   5.52   5.65
   5.57   5.71   5.59   5.53   5.63   5.68

21. **Alphabetic States.** The following table gives the total area in square miles (land and water) of the seven states with names beginning with the letters A through C.

State	Area
Alabama	52,200
Alaska	615,200
Arizona	114,000
Arkansas	53,200
California	158,900
Colorado	104,100
Connecticut	5,500

   a. Find the mean area and median area for these states.

   b. Which state is an outlier on the high end? If you eliminate this state, what are the new mean and median areas for this data set?

c. Which state is an outlier on the low end? If you eliminate this state, what are the new mean and median areas for this data set?

22. **Outlier Coke.** Cans of regular Coca-Cola vary slightly in weight. Here are the measured weights of seven cans, in pounds:

   0.8161   0.8194   0.8165   0.8176
   0.7901   0.8143   0.8126

   a. Find the mean and median of these weights.

   b. Which, if any, of these weights would you consider to be an outlier? Explain.

   c. What are the mean and median weights if the outlier is excluded?

23. **Raising Your Grade.** Suppose you have scores of 80, 84, 87, and 89 on quizzes in a mathematics class.

   a. What is the mean of these scores?

   b. What score would you need on the next quiz to have an overall mean of 88?

   c. If the maximum score on a quiz is 100, is it possible to have a mean of 90 after the fifth quiz? Explain.

24. **Raising Your Grade.** Suppose you have scores of 60, 70, 65, 85, and 85 on exams in a sociology class.

   a. What is the mean of these scores?

   b. What score would you need on the next exam to have an overall mean of 75?

   c. If the maximum score on an exam is 100, what is the maximum mean score that you could possibly have after the next exam? Explain.

25. **Comparing Averages.** Suppose that school district officials claim that the average reading score for fourth-graders in the district is 73 (out of a possible 100). As a principal, you know that your fourth-graders had the following scores: 55, 60, 68, 70, 87, 88, 95. Would you be justified in claiming that your students scored above the district average? Explain.

26. **Comparing Averages.** Suppose the National Basketball Association (NBA) reports that the average height of basketball players is 6'8". As a coach, you know that the players on your starting lineup have heights of 6'5", 6'6", 6'6", 7'0", and 7'2". Would you be justified in claiming that your starting lineup has above average height for the NBA? Explain.

27. **Average Peaches.** A grocer has three baskets of peaches. One holds 50 peaches and weighs 18 pounds, one holds 55 peaches and weighs 22 pounds, and the third holds 60 peaches and weighs 24 pounds. What is the mean weight of all of the peaches combined? Explain.

28. **Average Confusion.** An instructor has a first-period class with 25 students, and they had a mean score of 86% on the midterm exam. The second-period class has 30 students, and they had a mean score of 84% on the same exam. Does it follow that the mean score for both classes combined is 85%? Explain.

29. Different Means? Each of the 300 students at a high school takes the same four courses. Three of the courses are each taught in 15 classes of 20 students each. The fourth course is taught in 3 classes of 100 students each. Find the mean class size experienced by each student. Find the mean class size in the 48 courses. Are the two means the same?

30. Final Grade. Your course grade is based on one midterm that counts as 15% of your final grade, one class project that counts as 20% of your final grade, a set of homework assignments that counts as 40% of your final grade, and a final exam that counts as 25% of your final grade. Your midterm score is 75, your project score is 90, your homework score is 85, and your final exam score is 72. What is your overall final score?

31. Batting Average. A batting average in baseball is determined by dividing the total number of hits by the total number of at-bats (neglecting walks, sacrifices, and a few other special cases). A player goes 2 for 4 (2 hits in 4 at-bats) in the first game, 0 for 3 in the second game, and 3 for 5 in the third game. What is his batting average? In what way is this number an "average"?

32. Averaging Averages. Suppose a player has a batting average over many games of 0.200 (he's not very good). In his next game, he goes 2 for 4, which is a batting average of 0.500 for the game. Does it follow that his new batting average is $(0.200 + 0.500)/2 = 0.350$? Explain.

33. Batteries. A manufacturer uses two different production sites to make batteries for cell phones. There is a defect rate of 2% at one of the sites, and the defect rate at the other site is 4%. Does it follow that when the batteries from the two sites are combined, the overall rate of defects is 3%? Explain.

34. Slugging Average. In addition to the batting average, another measure of hitting performance in baseball is called the slugging average. In finding a slugging average, a single is worth 1 point, a double is worth 2 points, a triple is worth 3 points, and a home run is worth 4 points. A player's slugging average is the total number of points divided by the total number of at-bats (neglecting walks, sacrifices, and a few other special cases). A player has three singles in five at-bats in the first game, a triple and a single in four at-bats in the second game, and a double and a home run in five at-bats in the third game.

a. What is his batting average?

b. What is his slugging average?

c. Is it possible for a slugging average to be more than 1? Explain.

35. Stockholder Voting. A small company has four stockholders. One stockholder has 400 shares, a second stockholder has 300 shares, a third stockholder has 200 shares, and the fourth stockholder has 100 shares. In a vote on a new advertising campaign, the first stockholder votes yes, and the other three stockholders vote no. Explain how the outcome of the vote can be expressed as a weighted mean. What is the outcome of the vote?

36. GPA. One common system for computing a grade point average (GPA) assigns 4 points to an A, 3 points to a B, 2 points to a C, 1 point to a D, and 0 points to an F. What is the GPA of a student who gets an A in a 4-credit course, a B in each of two 3-credit courses, and a C in a 1-credit course?

37. Phenotypes of Peas. An experiment was conducted to determine whether a deficiency of carbon dioxide in soil affects the phenotypes of peas. Listed below are the phenotype codes, where 1 = smooth yellow, 2 = smooth green, 3 = wrinkled yellow, and 4 = wrinkled green. Can the measures of center be obtained for these values? Do the results make sense?

2	1	1	1	1	1	1	4	1	2	2	1	2
3	3	2	3	1	3	1	3	1	3	2	2

38. U.S. Population Center. Imagine taking a huge flat map of the United States and placing weights on it to represent where people live. The point at which the map would balance is called the mean center of population. Figure 4.2

Figure 4.2 Mean center of population.

Source: Statistical Abstract of the United States.

shows how the location of the mean center of population has shifted from 1790 to 2010. Briefly explain the pattern shown on this map.

PROJECTS FOR THE INTERNET & BEYOND

39. **Salary Data.** Many Web sites offer data on salaries in different careers. Find salary data for a career you are considering. What are the mean and median salaries for this career? How do these salaries compare with those of other careers that interest you?

40. **Is the Median the Message?** Read the article "The Median Isn't the Message," by Stephen Jay Gould, which is posted on the Web. Write a few paragraphs in which you describe the message that Gould was trying to get across. How is this message important to other patients diagnosed with cancer?

41. **Navel Data.** Bin the data collected in Exercise 23 of Section 3.1. Then make a frequency table, and draw a histogram of the distribution. What is the mean of the distribution? What is the median of the distribution? An old theory says that, on average, the navel ratio of humans is the golden ratio: $\left(1 + \sqrt{5}\right)/2$. Does this theory seem accurate based on your observations?

IN THE NEWS

42. **Daily Averages.** Cite three examples of averages that you deal with in your own life (such as grade point average or batting average). In each case, explain whether the average is a mean, a median, or some other type of average. Briefly describe how the average is useful to you.

43. **Averages in the News.** Find three recent news articles that refer to some type of average. In each case, explain whether the average is a mean, a median, or some other type of average.

4.2   SHAPES OF DISTRIBUTIONS

In the previous section we discussed how to describe the center of a quantitative data distribution with measures such as the mean and median. We now turn our attention to the overall *shape* of a distribution, which we often describe with three characteristics: its number of modes, its symmetry or skewness, and its variation. Although these three characteristics carry less information than a complete graph of the distribution, they are still useful.

Note that, because we are interested in the *general* shapes of distributions, it's often easier to examine graphs that show smooth curves to fit the original data sets. Figure 4.3 shows three examples of this idea, two in which the distributions are shown as histograms and one in which the distribution is shown as a line chart. In each case, the smooth curves make good approximations to the original distributions.

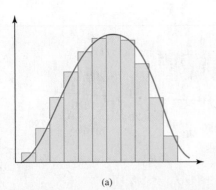
(a)

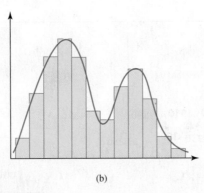
(b)

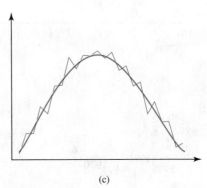
(c)

Figure 4.3   The smooth curves approximate the shapes of the distributions.

Number of Modes

One simple way to describe the shape of a distribution is by its number of peaks, or modes. Figure 4.4a shows a distribution, called a **uniform distribution**, that has no mode because all data values have the same frequency. Figure 4.4b shows a distribution with a single peak as its mode. It is called a **single-peaked**, or **unimodal**, distribution. By convention, any peak in a distribution is considered a mode, even if not all peaks have the same height. For example,

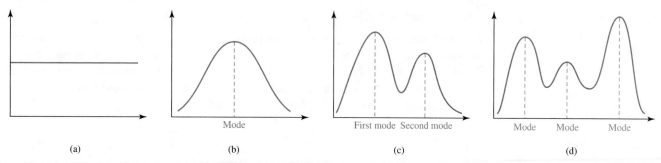

Figure 4.4 (a) A uniform distribution has no mode. (b) A single-peaked distribution has one mode. (c) A bimodal distribution has two modes. (d) A trimodal distribution has three modes.

the distribution in Figure 4.4c is said to have two modes, even though the second peak is lower than the first; it is a *bimodal* distribution. Similarly, the distribution in Figure 4.4d is said to have three modes; it is a *trimodal* distribution.

EXAMPLE 1 Number of Modes

How many modes would you expect for each of the following distributions? Why? Make a rough sketch for each distribution, with clearly labeled axes.

a. Heights of 1,000 randomly selected adult women

b. Hours spent watching football on TV in January for 1,000 randomly selected adult Americans

c. Weekly sales throughout the year at a retail clothing store for children

d. The number of people with particular last digits (0 through 9) in their Social Security numbers

SOLUTION   Figure 4.5 shows sketches of the distributions.

a. The distribution of heights of women is single-peaked because many women are at or near the mean height, with fewer and fewer women at heights much greater or less than the mean.

b. The distribution of times spent watching football on TV for 1,000 randomly selected adult Americans is likely to be bimodal (two modes). One mode represents the mean watching time of men, and the other represents the mean watching time of women.

c. The distribution of weekly sales throughout the year at a retail clothing store for children is likely to have several modes. For example, it will probably have a mode in spring for sales of summer clothing, a mode in late summer for back-to-school sales, and another mode in winter for holiday sales.

d. The last digits of Social Security numbers are essentially random, so the number of people with each different last digit (0 through 9) should be about the same. That is, about 10% of all Social Security numbers end in 0, 10% end in 1, and so on. It is therefore a uniform distribution with no mode.

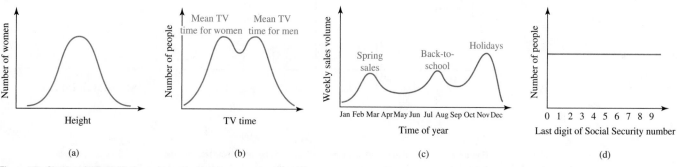

Figure 4.5   Sketches for Example 1.

Symmetry or Skewness

A second simple way to describe the shape of a distribution is in terms of its symmetry or skewness. A distribution is **symmetric** if its left half is a mirror image of its right half. The distributions in Figure 4.6 are all symmetric. The symmetric distribution in Figure 4.6a, with a single peak and a characteristic bell shape, is known in statistics as a *normal distribution*; it is so important that we will devote Chapter 5 to its study.

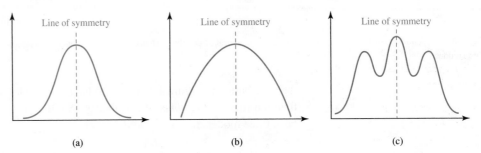

Figure 4.6 These distributions are all symmetric because their left halves are mirror images of their right halves. Note that (a) and (b) are single-peaked (unimodal), whereas (c) is triple-peaked (trimodal).

A distribution that is not symmetric must have values that tend to be more spread out on one side than on the other. In this case, we say that the distribution is **skewed.** Figure 4.7a shows a distribution in which the values are more spread out on the left, meaning that some values are outliers at low values. We say that such a distribution is **left-skewed** (or *negatively skewed*), because it looks as if it has a tail that has been pulled toward the left. Figure 4.7b shows a distribution in which the values are more spread out on a tail extending to the right, making it **right-skewed** (or *positively skewed*).

Figure 4.7 also shows how skewness affects the relative positions of the mean, median, and mode. By definition, the mode is at the peak in a single-peaked distribution. A left-skewed distribution pulls both the mean and median to the left of the mode, meaning to values less than the mode. In addition, outliers at the low end of the data set make the mean less than the median (see Table 4.2 on page 149). Similarly, a right-skewed distribution pulls the mean and median to the right of the mode), and the outliers at the high end of the data set make the mean greater than the median. When the distribution is symmetric and single-peaked, both the mean and the median are equal to the mode.

TECHNICAL NOTE

A left-skewed distribution is also called *negatively skewed,* and a right-skewed distribution is also called *positively skewed.* A symmetric distribution has *zero skewness.*

Definitions

A distribution is **symmetric** if its left half is a mirror image of its right half.

A distribution is **left-skewed** if its values are more spread out on the left side.

A distribution is **right-skewed** if its values are more spread out on the right side.

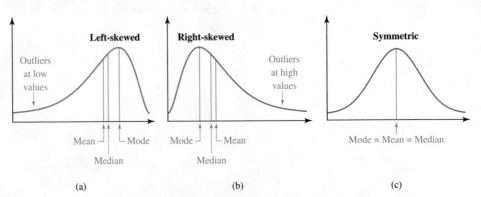

Figure 4.7 (a) Skewed to the left (left-skewed): The mean and median are less than the mode. (b) Skewed to the right (right-skewed). The mean and median are greater than the mode. (c) Symmetric distribution: The mean, median, and mode are the same.

EXAMPLE ② Skewness

For each of the following situations, state whether you expect the distribution to be symmetric, left-skewed, or right-skewed. Explain.

a. Heights of a sample of 100 women

b. Family income in the United States

c. Speeds of cars on a road where a visible patrol car is using radar to detect speeders

SOLUTION

a. The distribution of heights of women is symmetric, because roughly equal numbers of women are shorter and taller than the mean and extremes of height are rare on either side of the mean.

b. The distribution of family incomes is right-skewed. Most families are middle-class, so the mode of this distribution is a middle-class income. But a few very high-income families pull the mean to a considerably higher value, stretching the distribution to the right (high-income) side.

c. Drivers usually slow down when they are aware of a patrol car looking for speeders. Few if any drivers will be exceeding the speed limit, but some drivers tend to slow to well below the speed limit. Thus, the distribution of speeds is therefore left-skewed, with a mode near the speed limit but a few cars going well below the speed limit.

BY THE WAY

Median family income in the United States is about $50,000, which is substantially lower than the *mean* family income of about $68,000 (Data for 2011).

BY THE WAY

Speed kills. On average, in the United States, someone is killed in an auto accident about every 12 minutes. About one-third of these fatalities involve a speeding driver.

Variation

A third way to describe a distribution is by its **variation**, which is a measure of how much the data values are spread out. A distribution in which most data are clustered together has a low variation. As shown in Figure 4.8a, such a distribution has a fairly sharp peak. The variation is higher when the data are distributed more widely around the center, which makes the peak broader. Figure 4.8b shows a distribution with a moderate variation and Figure 4.8c shows a distribution with a high variation. We'll discuss methods for describing the variation quantitatively in the next section.

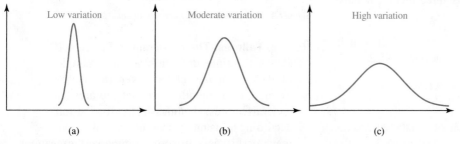

(a)   (b)   (c)

Figure 4.8   From left to right, these three distributions have increasing variation.

> **Definitions**
> **Variation** describes how widely data are spread out about the center of a data set.

EXAMPLE **3** Variation in Marathon Times

How would you expect the variation to differ between times in the Olympic marathon and times in the New York City marathon? Explain.

SOLUTION The Olympic marathon invites only elite runners, whose times are likely to be clustered relatively near world-record times. The New York City marathon allows runners of all abilities, whose times are spread over a very wide range (from near the world record of just over two hours to many hours). Therefore, the variation among the times should be greater in the New York City marathon than in the Olympic marathon. $\quad\cdots\bullet$

Section 4.2 Exercises

Statistical Literacy and Critical Thinking

1. **Symmetry.** In the United States, there are many people with little or no accumulated wealth, and there are a few people with very large amounts of wealth. What does this suggest about the symmetry of the distribution of wealth?

2. **Distribution.** When the digits 0 through 9 are selected for a state lottery, the digits are selected in a way that they are all equally likely. Which term best describes the distribution of selected digits: skewed, bimodal, uniform, or unimodal?

3. **IQ Scores.** Consider the IQ scores of professors who teach statistics courses compared with the IQ scores of adults randomly selected from the general population. Which of these two sets of IQ scores has less variation? What effect does the lower variation have on a graph of the distribution of those IQ scores?

4. **Skewness.** What is skewness in a graph?

Does It Make Sense? For Exercises 5–8, decide whether the statement makes sense (or is clearly true) or does not make sense (or is clearly false). Explain clearly; not all of these statements have definitive answers, so your explanation is more important than your chosen answer.

5. **Symmetry.** Because a data set has three modes, it must have a skewed distribution.

6. **Symmetry.** Examination of the data set reveals that it is symmetric with a mean of 98.2 and a median of 98.2.

7. **Distribution.** Examination of a data set reveals that its distribution is left-skewed and unimodal.

8. **Uniform Distribution.** Examination of a data set reveals that the mean and median are both equal to 98.2, so the distribution must be uniform.

Concepts and Applications

9. **Old Faithful.** The histogram in Figure 4.9 shows the times between eruptions of Old Faithful geyser in Yellowstone National Park for a sample of 300 eruptions (with 299 times between eruptions). Over the histogram, draw a smooth curve that captures its general features. Then classify the distribution according to its number of modes and its symmetry or skewness. In words, summarize the meaning of your results.

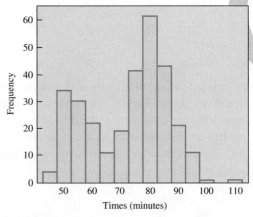

Times Between Eruptions of Old Faithful

Figure 4.9 *Source:* Hand et al., *Handbook of Small Data Sets.*

10. **Chip Failures.** The histogram in Figure 4.10 shows the time until failure for a sample of 108 computer chips. Over the histogram, draw a smooth curve that captures its general features. Then classify the distribution according to its number of modes and its symmetry or skewness. In words, summarize the meaning of your results.

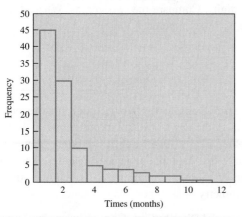

Failure Time of Computer Chips

Figure 4.10 *Source:* Hand et al., *Handbook of Small Data Sets.*

11. Rugby Weights. The histogram in Figure 4.11 shows the weights of a sample of 391 rugby players. Over the histogram, draw a smooth curve that captures its general features. Then classify the distribution according to its number of modes and its symmetry or skewness. In words, summarize the meaning of your results.

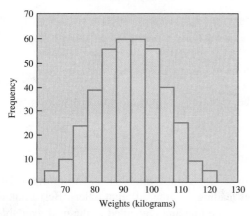

Weights of Rugby Players

Figure 4.11 *Source:* Hand et al., *Handbook of Small Data Sets.*

12. Penny Weights. The histogram in Figure 4.12 shows the weights (in grams) of 72 pennies. Over the histogram, draw a smooth curve that captures its general features. Then classify the distribution according to its number of modes and its symmetry or skewness. What feature of the graph reflects the fact that 35 of the pennies were made before 1983 and consist of 95% copper and 5% zinc whereas the other 37 pennies were made after 1983 and are 2.5% copper and 97.5% zinc?

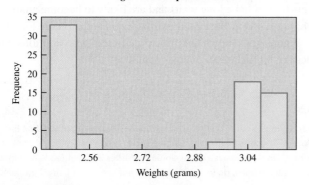

Weights of a Sample of Pennies

Figure 4.12 *Source: Measurements* by Mario F. Triola.

13. Baseball Salaries. In a recent year, the 817 professional baseball players had salaries with the following characteristics:

- The mean was $3,250,178.
- The median was $1,152,000.
- The salaries ranged from a low of $400,000 to a high of $33,000,000.

  a. Describe the shape of the distribution of salaries. Is the distribution symmetric? Is it left-skewed? Is it right-skewed?

  b. About how many players had salaries of $1,152,000 or higher?

14. Boston Rainfall. The daily rainfall amounts (in inches) for Boston in a recent year consist of 365 values with these properties:

- The mean daily rainfall amount is 0.083 inch.
- The median of the daily rainfall amounts is 0 inches.
- The minimum daily rainfall amount is 0 inches and the maximum is 1.48 inches.

  a. How is it possible that the minimum of the 365 values is 0 inches and the median is also 0 inches?

  b. Describe the distribution as symmetric, left-skewed, or right-skewed.

  c. Can you determine the exact number of days that it rained? Can you conclude anything about the number of days that it rained? Explain.

Describing Distributions. For each distribution described in Exercises 15–26, answer the following questions:

a. How many modes would you expect for the distribution?

b. Would you expect the distribution to be symmetric, left-skewed, or right-skewed?

15. Incomes. The annual incomes of all those in a statistics class, including the instructor

16. Reaction Times. The reaction times of 500 randomly selected drivers, measured under standard conditions

17. Heights. The heights of 250 randomly selected male attorneys

18. Heights. The heights of 500 male students, half of whom are adults while the other half are eight years of age

19. Weights of Cola. The weights of the cola in 1000 randomly selected cans of Coke

20. Vehicle Weights. The weights of cars in a fleet consisting of 50 compact cars and 50 delivery trucks

21. Patients. The ages of 1,000 randomly selected patients being treated for dementia

22. Speeds. The speeds of drivers on a highway in Montana

23. Patron Ages. The ages of adults who visit the National Air and Space Museum

24. **Patron Ages.** The ages of people who visit Disneyworld

25. **Incomes.** The incomes of people sitting in luxury boxes at the Super Bowl

26. **Times.** The amounts of time that 5000 randomly selected individual taxpayers used to prepare their federal tax returns

PROJECTS FOR THE INTERNET & BEYOND

27. **New York Marathon.** The Web site for the New York City marathon gives frequency data for finish times in the most recent marathon. Study the data, make a rough sketch of the distribution, and describe the shape of the distribution in words.

28. **Tax Stats.** The IRS Web site provides statistics collected from tax returns on income, refunds, and much more. Choose a set of statistics from this Web site and study the distribution. Describe the distribution in words, and discuss anything you learn that is relevant to national tax policies.

29. **Social Security Data.** Survey a sample of fellow students, asking each to indicate the last digit of her or his Social Security number. Also ask each participant to indicate the fifth digit. Draw one graph showing the distribution of the last digits and another graph showing the distribution of the fifth digits. Compare the two graphs. What notable difference becomes apparent?

IN THE NEWS

30. **Distributions in the News.** Find three recent examples in the news of distributions shown as histograms or line charts. Over each distribution, draw a smooth curve that captures its general features. Then classify the distribution according to its number of modes, symmetry or skewness, and variation.

31. **Trimodal Distribution.** Give an example of a real distribution that you expect to have *three* modes. Make a rough sketch of the distribution; be sure to label the axes on your sketch.

32. **Skewed Distribution.** Give an example of a real distribution that you would expect to be either right- or left-skewed. Make a rough sketch of the distribution; be sure to label the axes on your sketch.

4.3 MEASURES OF VARIATION

In Section 4.2, we saw how to describe variation qualitatively. In this section we describe quantitative measures of variation.

Why Variation Matters

We mortals cross the ocean of this world,
Each in his average cabin of a life.

—Robert Browning

Imagine customers waiting in line for tellers at two different banks. Customers at Big Bank can enter any one of three different lines leading to three different tellers. Best Bank also has three tellers, but all customers wait in a single line and are called to the next available teller. The following values are waiting times, in minutes, for 11 customers at each bank. The times are arranged in ascending order.

Big Bank (three lines): 4.1 5.2 5.6 6.2 6.7 7.2 7.7 7.7 8.5 9.3 11.0

Best Bank (one lines): 6.6 6.7 6.7 6.9 7.1 7.2 7.3 7.4 7.7 7.8 7.8

You'll probably find more unhappy customers at Big Bank than at Best Bank, but this is *not* because the average wait is any longer. In fact, you should verify for yourself that the mean and median waiting times are 7.2 minutes at both banks. The difference in customer satisfaction comes from the *variation* at the two banks. The waiting times at Big Bank vary over a fairly wide range, so a few customers have long waits and are likely to become annoyed. In contrast, the variation of the waiting times at Best Bank is small, so all customers feel they are being treated roughly equally. Figure 4.13 shows the difference in the two variations with histograms in which the data values are binned to the nearest minute.

TIME OUT TO THINK

Explain *why* Big Bank, with three separate lines, should have a greater variation in waiting times than Best Bank. Then consider several places where you commonly wait in lines, such as a grocery store, a bank, a theme park ride, or a fast food restaurant. Do these places use a single customer line that feeds multiple clerks or multiple lines? If a place uses multiple lines, do you think a single line would be better? Explain.

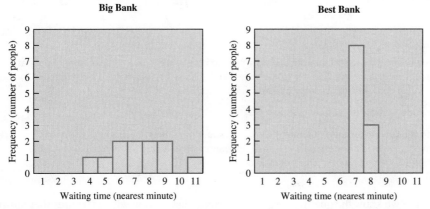

Figure 4.13 Histograms for the waiting times at Big Bank and Best Bank, shown with data binned to the nearest minute.

BY THE WAY

The idea of waiting in line (or *queuing*) is important not only for people but also for data, particularly for data streaming through the Internet. Major corporations often employ statisticians to help them make sure that data move smoothly and without bottlenecks through their servers and Web pages.

Range

The simplest way to describe the variation of a data set is to compute its **range**, defined as the difference between the highest (maximum) and lowest (minimum) values. For the example of the two banks, the waiting times for Big Bank vary from 4.1 to 11.0 minutes, so the range is $11.0 - 4.1 = 6.9$ minutes. The waiting times for Best Bank vary from 6.6 to 7.8 minutes, so the range is $7.8 - 6.6 = 1.2$ minutes. The range for Big Bank is much larger, reflecting its greater variation.

Definitions

The **range** of a set of data values is the difference between its highest and lowest data values:

$$\text{range} = \text{highest value (max)} - \text{lowest value (min)}$$

Although the range is easy to compute and can be useful, it occasionally can be misleading, as the next example shows.

EXAMPLE 1 Misleading Range

Consider the following two sets of quiz scores for nine students. Which set has the greater range? Would you also say that this set has the greater variation?

$$Quiz\ 1: \quad 1 \quad 10 \quad 10 \quad 10 \quad 10 \quad 10 \quad 10 \quad 10 \quad 10$$
$$Quiz\ 2: \quad 2 \quad 3 \quad 4 \quad 5 \quad 6 \quad 7 \quad 8 \quad 9 \quad 10$$

SOLUTION   The range for Quiz 1 is $10 - 1 = 9$ points, which is greater than the range for Quiz 2 of $10 - 2 = 8$ points. However, aside from a single low score (an outlier), Quiz 1 has no variation at all because every other student got a 10. In contrast, no two students got the same score on Quiz 2, and the scores are spread throughout the list of possible scores. Quiz 2 therefore has greater variation even though Quiz 1 has greater range.   ··●

Quartiles and the Five-Number Summary

A better way to describe variation is to consider a few intermediate data values in addition to the high and low values. A common way involves looking at the **quartiles**, or values that divide the data distribution into quarters. The following list repeats the waiting times at the two banks, with the quartiles shown in bold. Note that the middle quartile, which divides the data set in half, is simply the median.

	Lower quartile (Q_1) ↓		Median (Q_2) ↓		Upper quartile (Q_3) ↓	

Big Bank: 4.1   5.2   **5.6**   6.2   6.7   **7.2**   7.7   7.7   **8.5**   9.3   11.0

Best Bank: 6.6   6.7   **6.7**   6.9   7.1   **7.2**   7.3   7.4   **7.7**   7.8   7.8

Definitions

The **lower quartile** (or **first quartile** or Q_1 divides the lowest fourth of a data set from the upper three-fourths. It is the median of the data values in the *lower half* of a data set. (Exclude the middle value in the data set if the number of data points is odd.)

The **middle quartile** (or **second quartile** or Q_2 is the overall median.

The **upper quartile** (or **third quartile** or Q_3 divides the lowest three-fourths of a data set from the upper fourth. It is the median of the data values in the *upper half* of a data set. (Exclude the middle value in the data set if the number of data points is odd.)

Once we know the quartiles, we can describe a distribution with a **five-number summary**, consisting of the low value, the lower quartile, the median, the upper quartile, and the high value. For the waiting times at the two banks, the five-number summaries are as follows:

Big Bank:		*Best Bank:*	
low	= 4.1	low	= 6.6
lower quartile	= 5.6	lower quartile	= 6.7
median	= 7.2	median	= 7.2
upper quartile	= 8.5	upper quartile	= 7.7
high	= 11.0	high	= 7.8

The Five-Number Summary

The **five-number summary** for a data distribution consists of the following five numbers:

low value   lower quartile   median   upper quartile   high value

We can display the five-number summary with a graph called a **boxplot** (or *box-and-whisker plot*). Using a number line for reference, we enclose the values from the lower to the upper quartiles in a box. We then draw a line through the box at the median and add two "whiskers," extending from the box to the low and high values. Figure 4.14 shows boxplots for the bank waiting times. Both the box and the whiskers for Big Bank are broader than those for Best Bank, indicating that the waiting times have greater variation at Big Bank.

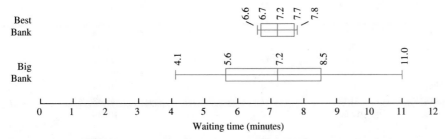

Figure 4.14   Boxplots show that the variation of the waiting times is greater at Big Bank than at Best Bank.

Drawing a Boxplot

Step 1.   Draw a number line that spans all the values in the data set.

Step 2.   Enclose the values from the lower to the upper quartile in a box. (The thickness of the box has no meaning.)

Step 3.   Draw a line through the box at the median.

Step 4.   Add "whiskers" extending to the low and high values.

TECHNICAL NOTE

The boxplots shown in this text are called *skeletal boxplots*. Some boxplots are drawn with outliers marked by an asterisk ($*$) or a dot and the whiskers extending only to the smallest and largest *non*outliers; these types of boxplots are called *modified boxplots*.

⏻ USING TECHNOLOGY—**BOXPLOTS**

Excel Although Excel itself is not designed to generate a boxplot, it can be generated using XLSTAT that is a supplement to this book. Load XLSTAT (the Excel add-in), then enter or copy the data into a column of the spreadsheet. Click on **XLSTAT,** click on **Visualizing Data,** then select **Univariate plots.** Enter the range of cells containing the data, such as A1:A11. (If the first cell includes the name of the data, click on the box next to "Sample labels.") Click **OK** to continue. The result will be a boxplot with two features: (1) The boxplot will be vertical, and (2) the exact values of the quartiles are likely to be somewhat different from those found using the procedure described above.

STATDISK Enter the data in the Data Window, then click on **Data,** then **Boxplot.** Click on the columns that you want to include, then click on **Plot.**

TI-83/84 PLUS Enter the sample data in list L1 or enter the data and assign them to a list name. Now select **STAT PLOT** by pressing `2ND` `Y=`. Press `ENTER`, then select the option of **ON.** For a simple boxplot as described in Part 1 of this section, select the boxplot type that is positioned in the middle of the second row; for a modified boxplot as described in Part 2 of this section, select the boxplot that is positioned at the far left of the second row. The Xlist should indicate L1 and the Freq value should be 1. Now press `ZOOM` and select option 9 for **ZoomStat.** Press `ENTER` and the boxplot should be displayed. You can use the arrow keys to move right or left so that values can be read from the horizontal scale.

EXAMPLE 2   **Passive and Active Smoke**

One way to study exposure to cigarette smoke is by measuring blood levels of *serum cotinine*, a metabolic product of nicotine that the body absorbs from cigarette smoke. Table 4.3 lists serum cotinine levels from samples of 50 smokers ("active smoke") and 50 nonsmokers who are exposed to cigarette smoke at home or at work ("passive smoke"). Compare the two data sets (smokers and nonsmokers) with five-number summaries and boxplots, and discuss your results.

BY THE WAY

Passive smoke is particularly harmful to young children. Apparently, the toxins in cigarette smoke have a greater effect on developing bodies than on full-grown adults. A similar effect is found for most other toxins, which is why it is especially important to limit children's exposure to toxic chemicals.

TABLE 4.3	Serum Cotinine Levels (nanograms per milliliter of blood) in Samples of 50 Smokers and 50 Nonsmokers Exposed to Passive Smoke, with Data Values Listed in Ascending Order				
Order number	Smokers	Nonsmokers	Order number	Smokers	Nonsmokers
1	0.08	0.03	26	34.21	0.82
2	0.14	0.07	27	36.73	0.97
3	0.27	0.08	28	37.73	1.12
4	0.44	0.08	29	39.48	1.23
5	0.51	0.09	30	48.58	1.37
6	1.78	0.09	31	51.21	1.40
7	2.55	0.10	32	56.74	1.67
8	3.03	0.11	33	58.69	1.98
9	3.44	0.12	34	72.37	2.33
10	4.98	0.12	35	104.54	2.42
11	6.87	0.14	36	114.49	2.66
12	11.12	0.17	37	145.43	2.87
13	12.58	0.20	38	187.34	3.13
14	13.73	0.23	39	226.82	3.54
15	14.42	0.27	40	267.83	3.76
16	18.22	0.28	41	328.46	4.58
17	19.28	0.30	42	388.74	5.31
18	20.16	0.33	43	405.28	6.20
19	23.67	0.37	44	415.38	7.14
20	25.00	0.38	45	417.82	7.25
21	25.39	0.44	46	539.62	10.23
22	29.41	0.49	47	592.79	10.83
23	30.71	0.51	48	688.36	17.11
24	32.54	0.51	49	692.51	37.44
25	32.56	0.68	50	983.41	61.33

Note: The column "Order number" is included to make it easier to read the table.
Source: National Health and Nutrition Examination Survey, National Institutes of Health.

SOLUTION The two data sets are already in ascending order, making it easy to construct the five-number summary. Each has 50 data points, so the median lies halfway between the 25th and 26th values. For the smokers, the 25th and 26th values are 32.56 and 34.21, respectively, so the median is

$$\frac{32.56 + 34.21}{2} = 33.385$$

For the nonsmokers, the 25th and 26th values are 0.68 and 0.82, respectively, so the median is

$$\frac{0.68 + 0.82}{2} = 0.75$$

The lower quartile is the median of the *lower half* of the values, which is the 13th value in each set. The upper quartile is the median of the *upper half* of the values, which is the 38th value in each set. The five-number summaries for the two data sets are as follows:

Active smoke:

low $\doteq$ 0.08 ng/ml

lower quartile = 12.58 ng/ml

median = 33.385 ng/ml

upper quartile = 187.34 ng/ml

high = 983.41 ng/ml

Passive smoke:

low = 0.03 ng/ml

lower quartile = 0.20 ng/ml

median = 0.75 ng/ml

upper quartile = 3.13 ng/ml

high = 61.33 ng/ml

Figure 4.15 shows boxplots for the two data sets. The boxplots make it easy to see some key features of the data sets. For example, it is immediately clear that the active smokers have a higher median level of serum cotinine, as well as a greater variation in levels. We conclude that smokers absorb considerably more nicotine than do nonsmokers exposed to passive smoke. Nevertheless, the levels in the passive smokers are much higher than those found in people who had no exposure to cigarette smoke (as demonstrated by other data, not shown here). Indeed, the nonsmoker with the high value for passive smoke has absorbed more nicotine than

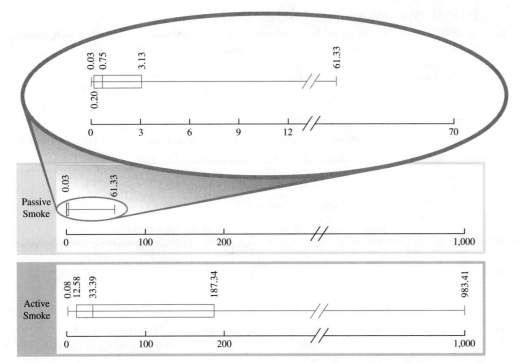

Figure 4.15 Boxplots for the data in Table 4.3.

the median smoker. We conclude that passive smoke can expose nonsmokers to significant amounts of nicotine. Given the known dangers of cigarette smoke, these results give us reason to be concerned about possible health effects from passive smoke.

Percentiles

Quartiles divide a data set into 4 segments. It is possible to divide a data set even more. For example, *quintiles* divide a data set into 5 segments, and *deciles* divide a data set into 10 segments. It is particularly common to divide data sets into 100 segments using **percentiles**. Roughly speaking, the 35th percentile, for example, is a value that separates the bottom 35% of the data values from the top 65%. (More precisely, the 35th percentile is greater than or equal to at least 35% of the data values and less than or equal to at least 65% of the data values.)

If a data value lies between two percentiles, it is common to say that the data value lies *in* the lower percentile. For example, if you score higher than 84.7% of all people taking a college entrance examination, we say that your score is in the 84th percentile.

> **TECHNICAL NOTE**
>
> As with quartiles, statisticians and various statistics software packages may use slightly different procedures to calculate percentiles, resulting in slightly different values.

> **Definitions**
>
> The **nth percentile** of a data set divides the bottom n% of data values from the top (100 − n)%. A data value that lies between two percentiles is often said to lie *in* the lower percentile. You can approximate the percentile of any data value with the following formula:
>
> $$\text{percentile of data value} = \frac{\text{number of values less than this data value}}{\text{total number of values in data set}} \times 100$$

There are different procedures for finding a data value corresponding to a given percentile, but one approximate approach is to find the Lth value, where L is the product of the percentile (in decimal form) and the sample size. For example, with 50 sample values, the 12th percentile is around the $0.12 \times 50 = 6th$ value.

EXAMPLE 3 Smoke Exposure Percentiles

Answer the following questions concerning the data in Table 4.3.

a. What is the percentile for the data value of 104.54 ng/ml for smokers?

b. What is the percentile for the data value of 61.33 ng/ml for nonsmokers?

c. What data value marks the 36th percentile for the smokers? For the nonsmokers?

SOLUTION The following results are approximate.

a. The data value of 104.54 ng/ml for smokers is the 35th data value in the set, which means that 34 data values lie below it. Thus, its percentile is

$$\frac{\text{number of values less than 104.54 ng/ml}}{\text{total number of values in data set}} = \times 100 = \frac{34}{50} \times 100 = 68$$

In other words, the 35th data value marks the 68th percentile.

b. The data value of 61.33 ng/ml for nonsmokers is the 50th and highest data value in the set, which means that 49 data values lie below it. Thus, its percentile is

$$\frac{\text{number of values less than 61.33 ng/ml}}{\text{total number of values in data set}} \times 100 = \frac{49}{50} \times 100 = 98$$

In other words, the highest data value in this set lies in the 98th percentile.

c. Because there are 50 data values in the set, the 36th percentile is around the $0.36 \times 50 = 18$th value. For smokers this value is 20.16 ng/ml, and for nonsmokers it is 0.33 ng/ml.   ⋅⋅●

Standard Deviation

The five-number summary characterizes variation well, but statisticians often prefer to describe variation with a single number. The single number most commonly used to describe variation is called the **standard deviation.**

The standard deviation is a measure of how widely data values are spread around the mean of a data set. To calculate a standard deviation, we first find the mean and then find how much each data value "deviates" from the mean. Consider our bank data sets, in which the mean waiting time was 7.2 minutes for both Big Bank and Best Bank. For a waiting time of 8.2 minutes, the **deviation** from the mean is equal to 8.2 minutes − 7.2 minutes = 1.0 minute, meaning that it is 1.0 minute greater than the mean. For a waiting time of 5.2 minutes, the deviation from the mean is equal to 5.2 minutes − 7.2 minutes = −2 minutes (*negative* 2 minutes), because it is 2.0 minutes *less* than the mean.

In essence, the standard deviation is a measure of the average of all the deviations from the mean. However, because the mean of the deviations is always zero (because the positive deviations exactly balance the negative deviations), we calculate the standard deviation by first finding a mean of the *squares* of the deviations (because squares are always positive) and taking a square root in the end. For technical reasons, we divide the sum of the squares by the total number of data values *minus* 1.

> **Calculating the Standard Deviation**
>
> To calculate the standard deviation for any data set:
>
> Step 1. Compute the mean of the data set. Then find the deviation from the mean for every data value by subtracting the mean from the data value. That is, for every data value,
>
> $$\text{deviation from mean} = \text{data value} - \text{mean}$$
>
> Step 2. Find the squares (second power) of all the deviations from the mean.
>
> Step 3. Add all the squares of the deviations from the mean.
>
> Step 4. Divide this sum by the total number of data values *minus* 1.
>
> Step 5. The standard deviation is the square root of this quotient. Overall, these steps produce the standard deviation formula:
>
> $$\text{standard deviation} = \sqrt{\frac{\text{sum of (deviations from the mean)}^2}{\text{total number of data values} - 1}}$$

Note that, because we square the deviations in Step 3 and then take the square root in Step 5, the units of the standard deviation are the same as the units of the data values. For example, if the data values have units of minutes, the standard deviation also has units of minutes.

The standard deviation formula is easy to use in principle, but the calculations become tedious for all but the smallest data sets. As a result, it is usually calculated with the aid of a calculator or computer. Nevertheless, you'll find the standard deviation formula easier to understand if you try a few examples in which you work through the calculations in detail.

EXAMPLE 4 Calculating Standard Deviation

Calculate the standard deviations for the waiting times at Big Bank and Best Bank.

SOLUTION   We follow the five steps to calculate the standard deviations. Table 4.4 shows how to organize the work in the first three steps. The first column for each bank lists the waiting times (in minutes). The second column lists the deviations from the mean (Step 1), which we already know to be 7.2 minutes for both banks. The third column lists the squares of the deviations (Step 2). We add all the squared deviations to find the sum at the bottom of the third column (Step 3). For Step 4, we divide the sums from Step 3 by the total number of data values *minus* 1. Because there are 11 data values, we divide by 10:

$$Big\ Bank: \quad \frac{38.46}{10} = 3.846$$

$$Best\ Bank: \quad \frac{1.98}{10} = 0.198$$

Finally, Step 5 tells us that the standard deviations are the square roots of the numbers from Step 4:

$$Big\ Bank: \quad standard\ deviation = \sqrt{3.846} \approx 1.96\ minutes$$

$$Best\ Bank: \quad standard\ deviation = \sqrt{0.198} \approx 0.44\ minutes$$

We conclude that the standard deviation of the waiting times is about 1.96 minutes at Big Bank and 0.44 minute at Best Bank. As we expected, the waiting times showed greater variation at Big Bank, which is why the lines at Big Bank annoyed more customers than did those at Best Bank.

TABLE 4.4	**Calculating Standard Deviation**					
	Big Bank			**Best Bank**		
Time	**Deviation (Time − Mean)**	**(Deviation)²**	**Time**	**Deviation (Time − Mean)**	**(Deviation)²**	
4.1	$4.1 - 7.2 = -3.1$	$(-3.1)^2 = 9.61$	6.6	$6.6 - 7.2 = -0.6$	$(-0.6)^2 = 0.36$	
5.2	$5.2 - 7.2 = -2.0$	$(-2.0)^2 = 4.00$	6.7	$6.7 - 7.2 = -0.5$	$(-0.5)^2 = 0.25$	
5.6	$5.6 - 7.2 = -1.6$	$(-1.6)^2 = 2.56$	6.7	$6.7 - 7.2 = -0.5$	$(-0.5)^2 = 0.25$	
6.2	$6.2 - 7.2 = -1.0$	$(-1.0)^2 = 1.00$	6.9	$6.9 - 7.2 = -0.3$	$(-0.3)^2 = 0.09$	
6.7	$6.7 - 7.2 = -0.5$	$(-0.5)^2 = 0.25$	7.1	$7.1 - 7.2 = -0.1$	$(-0.1)^2 = 0.01$	
7.2	$7.2 - 7.2 = 0.0$	$(0.0)^2 = 0.0$	7.2	$7.2 - 7.2 = 0.0$	$(0.0)^2 = 0.0$	
7.7	$7.7 - 7.2 = 0.5$	$(0.5)^2 = 0.25$	7.3	$7.3 - 7.2 = 0.1$	$(0.1)^2 = 0.01$	
7.7	$7.7 - 7.2 = 0.5$	$(0.5)^2 = 0.25$	7.4	$7.4 - 7.2 = 0.2$	$(0.2)^2 = 0.04$	
8.5	$8.5 - 7.2 = 1.3$	$(1.3)^2 = 1.69$	7.7	$7.7 - 7.2 = 0.5$	$(0.5)^2 = 0.25$	
9.3	$9.3 - 7.2 = 2.1$	$(2.1)^2 = 4.41$	7.8	$7.8 - 7.2 = 0.6$	$(0.6)^2 = 0.36$	
11.0	$11.0 - 7.2 = 3.8$	$(3.8)^2 = 14.44$	7.8	$7.8 - 7.2 = 0.6$	$(0.6)^2 = 0.36$	
		Sum = 38.46			**Sum = 1.98**	

Excel The built-in Excel function STDEV automates the calculation of standard deviation, so that all you have to do is enter the data and then use the function. The screen shot below shows the process for the Big Bank data, with the functions shown in Column B and the results in Column C.

◇	A	B	C
1		**Big Bank**	
2	Data	4.1	
3		5.2	
4		5.6	
5		6.2	
6		6.7	
7		7.2	
8		7.7	
9		7.7	
10		8.5	
11		9.3	
12		11	
13	Mean	=AVERAGE(B2:B12)	7.2
14	St. Dev.	=STDEV(B2:B12)	1.96
15			

Note: Alternatively, the procedures given in Section 4.1 using the Analysis Tookit or XLSTAT will give results that include the value of the standard deviation

Statdisk or **TI 83/84**: Use the same procedures given in Section 4.1, and the results will include the value of the standard deviation.

TIME ⏱ UT TO THINK

Look closely at the individual deviations in Table 4.4 in Example 4. Do the standard deviations for the two data sets seem like reasonable "averages" for the deviations? Explain.

Interpreting the Standard Deviation

A good way to develop a deeper understanding of the standard deviation is to consider an approximation called the **range rule of thumb**, summarized in the following box.

TECHNICAL NOTE

Another way of interpreting the standard deviation uses a mathematical rule called *Chebyshev's Theorem*. It states that, for any data distribution, at least 75% of all data values lie within two standard deviations of the mean, and at least 89% of all data values lie within three deviations of the mean.

The Range Rule of Thumb

The standard deviation is *approximately* related to the range of a distribution by the **range rule of thumb**:

$$\text{standard deviation} \approx \frac{\text{range}}{4}$$

If we know the range of a distribution (range = high − low), we can use this rule to estimate the standard deviation. Alternatively, if we know the standard deviation, we can use this rule to estimate the low and high values as follows:

$$\text{low value} \approx \text{mean} - (2 \times \text{standard deviation})$$

$$\text{high value} \approx \text{mean} + (2 \times \text{standard deviation})$$

The range rule of thumb does not work well when the high or low values are outliers.

The range rule of thumb works reasonably well for data sets in which values are distributed fairly evenly. It does not work well when the high or low values are extreme outliers. You must therefore use judgment in deciding whether the range rule of thumb is applicable in a particular case, and in all cases remember that the range rule of thumb yields rough approximations, not exact results.

EXAMPLE ⑤ Using the Range Rule of Thumb

Use the range rule of thumb to estimate the standard deviations for the waiting times at Big Bank and Best Bank. Compare the estimates to the actual values found in Example 4.

SOLUTION   The waiting times for Big Bank vary from 4.1 to 11.0 minutes, which means a range of $11.0 - 4.1 = 6.9$ minutes. The waiting times for Best Bank vary from 6.6 to 7.8 minutes, for a range of $7.8 - 6.6 = 1.2$ minutes. Thus, the range rule of thumb gives the following estimates for the standard deviations:

$$\textit{Big Bank:} \quad \text{standard derivation} \approx \frac{6.9}{4} = 1.7$$

$$\textit{Best Bank:} \quad \text{standard derivation} \approx \frac{1.2}{4} = 0.3$$

The actual standard deviations calculated in Example 4 are 1.96 and 0.44, respectively. For these two cases, the estimates from the range rule of thumb slightly underestimate the actual standard deviations. Nevertheless, the estimates put us in the right ballpark, showing that the rule is useful.

$\cdots \bullet$

EXAMPLE 6   Estimating a Range

Studies of the gas mileage of a Prius under varying driving conditions show that it gets a mean of 45 miles per gallon with a standard deviation of 4 miles per gallon. Estimate the minimum and maximum typical gas mileage amounts that you can expect under ordinary driving conditions.

SOLUTION   From the range rule of thumb, the low and high values for gas mileage are approximately

$$\text{low value} \approx \text{mean} - (2 \times \text{standard deviation}) = 45 - (2 \times 4) = 37$$

$$\text{high value} \approx \text{mean} + (2 \times \text{standard deviation}) = 45 + (2 \times 4) = 53$$

The range of gas mileage for the car is roughly from a minimum of 37 miles per gallon to a maximum of 53 miles per gallon.

$\cdots \bullet$

Standard Deviation with Summation Notation (Optional Section)

The summation notation introduced earlier makes it easy to write the standard deviation formula in a compact form. Recall that x represents the individual values in a data set and $\bar{x}$ represents the mean of the data set. We can therefore write the deviation from the mean for any data value as

$$\text{deviation} = \text{data value} - \text{mean} = x - \bar{x}$$

We can now write the sum of all squared deviations as

$$\text{sum of all squared deviations} = \sum (x - \bar{x})^2$$

The remaining steps in the calculation of the standard deviation are to divide this sum by $n - 1$ and then take the square root. You should confirm that the following formula summarizes the five steps in the earlier box:

$$s = \text{standard deviation} = \sqrt{\frac{\sum (x - \bar{x})^2}{n - 1}}$$

The symbol s is the conventional symbol for the standard deviation of a sample. For the standard deviation of a population, statisticians use the Greek letter σ (sigma), and the term $n - 1$ in the formula is replaced by N (the population size). Consequently, you will get slightly different results for the standard deviation depending on whether you assume the data represent a sample or a population.

TECHNICAL NOTE

The formula for the *variance* is

$$s^2 = \frac{\sum (x - \bar{x})^2}{n - 1}$$

The standard symbol for the variance, s^2, reflects the fact that it is the square of the standard deviation.

Section 4.3 Exercises

Statistical Literacy and Critical Thinking

1. **Range.** How is the range computed for a set of sample data? Is the range a measure of variation? What is a major disadvantage of the range?

2. **Standard Deviation.** Assume that you are manufacturing aspirin tablets that are supposed to contain 325 mg of aspirin. If the standard deviation of the amounts of aspirin is calculated, which value would you prefer: a standard deviation of 20 mg or a standard deviation of 10 mg? Why?

3. **Correct Statement?** In the book *How to Lie with Charts,* the author writes that "the standard deviation is usually shown as plus or minus the difference between the high and the mean, and the low and the mean. For example, if the mean is 1, the high is 3, and the low is −1, the standard deviation is ±2." Is that statement correct? Why or why not?

4. **Quartiles.** For the salaries paid to the 817 professional baseball players in a recent year, the first quartile is $4,355,101. What do we mean when we say that $4,355,101 is the first quartile?

Does It Make Sense? For Exercises 5–8, decide whether the statement makes sense (or is clearly true) or does not make sense (or is clearly false). Explain clearly; not all of these statements have definitive answers, so your explanation is more important than your chosen answer.

5. **SAT Score.** Jennifer received an SAT score that was equal to the first quartile and the 35th percentile.

6. **Lengths.** The house key lengths of 15 statistics students are measured and rounded to the nearest centimeter, and all 15 values are the same, so the standard deviation is 0 cm.

7. **Baseball Salaries.** For a recent year, the 817 salaries paid to professional baseball players have a median of $1,152,000 and a second quartile with the same value of $1,152,000.

8. **Baseball Salaries.** If the range of salaries paid to baseball players in the American League is less than the range of salaries paid to baseball players in the National League, then the American League salaries must have a smaller standard deviation than the National League salaries.

Concepts and Applications

Range and Standard Deviation. Exercises 9–16 each list a set of numbers. In each case, find the range and standard deviation. (The same sets of numbers were used in Exercises 13–20 in Section 4.1.)

9. **Number of Words.** Pages from *Merriam-Webster's Collegiate Dictionary*, 11th edition, were randomly selected. Here are the numbers of words defined on those pages:

   51   63   36   43   34   62   73   39   53   79

10. **Space Shuttle Flights.** Listed below are the durations (in hours) of a sample of all flights of NASA's Space Transport System (space shuttle):

    73   95 235 192 165 262 191 376
    259 235 381 331 221 244   0

11. **Perception of Time.** Actual times (in seconds) recorded when statistics students participated in an experiment to test their ability to determine when one minute (60 seconds) had passed:

    53   52   75   62   68   58   49   49

12. **Body Temperatures.** Body temperatures (in degrees Fahrenheit) of randomly selected normal and healthy adults:

    98.6   98.6   98.0   98.0   99.0
    98.4   98.4   98.4   98.4   98.6

13. **Blood Alcohol.** Blood alcohol concentrations of drivers involved in fatal crashes and then given jail sentences (based on data from the U.S. Department of Justice):

    0.27   0.17   0.17   0.16   0.13   0.24
    0.29   0.24   0.14   0.16   0.12   0.16

14. **Old Faithful Geyser.** Time intervals (in minutes) between eruptions of Old Faithful geyser in Yellowstone National Park:

    98   92   95   87   96   90
    65   92   95   93   98   94

15. **Weights of M&Ms.** Weights (in grams) of randomly selected M&M plain candies:

    0.957   0.912   0.842   0.925   0.939   0.886
    0.914   0.913   0.958   0.947   0.920

16. **Quarters.** Weights (in grams) of quarters in circulation:

    5.60   5.63   5.58   5.56   5.66   5.58   5.57   5.59
    5.67   5.61   5.84   5.73   5.53   5.58   5.52   5.65
    5.57   5.71   5.59   5.53   5.63   5.68

Comparing Variation. In Exercises 17–20, find the range and standard deviation for each of the two samples and then compare the two sets of results.

17. **It's Raining Cats.** Statistics are sometimes used to compare or identify authors of different works. The lengths of the first 20 words in the foreword by Tennessee Williams in *Cat on a Hot Tin Roof* are listed along with the lengths of the first 20 words in *The Cat in the Hat* by Dr. Seuss. Does there appear to be a difference in variation?

Cat on a Hot Tin Roof:

2	6	2	2	1	4	4	2	4	2
3	8	4	2	2	7	7	2	3	11

The Cat in the Hat:

3	3	3	3	5	2	3	3	3	2
4	2	2	3	2	3	5	3	4	4

18. BMI for Miss America. The trend of thinner Miss America winners has generated charges that the contest encourages unhealthy diet habits among young women. Listed below are body mass indexes (BMI) for Miss America winners from two different time periods. Does there appear to be a difference in variation?

BMI (from the 1920s and 1930s):

20.4 21.9 22.1 22.3 20.3 18.8 18.9 19.4 18.4 19.1

BMI (from recent winners):

19.5 20.3 19.6 20.2 17.8 17.9 19.1 18.8 17.6 16.8

19. Weather Forecast Accuracy. In an analysis of the accuracy of weather forecasts, the actual high temperatures are compared with the high temperatures predicted one day earlier and the high temperatures predicted five days earlier. Listed below are the errors between the predicted temperatures and the actual high temperatures for consecutive days in Dutchess County, New York. Do the standard deviations suggest that the temperatures predicted one day in advance are more accurate than those predicted five days in advance, as we might expect?

(*actual high*) − (*high predicted one day earlier*):

2	2	0	0	−3	−2	1
−2	8	1	0	−1	0	1

(*actual high*) − (*high predicted five days earlier*):

0	−3	2	5	−6	−9	4
−1	6	−2	−2	−1	6	−4

20. Treatment Effect. Researchers at Pennsylvania State University conducted experiments with poplar trees. Listed below are weights (in kilograms) of poplar trees given no treatment and poplar trees treated with fertilizer and irrigation. Does there appear to be a difference between the two standard deviations?

No treatment:

0.15   0.02   0.16   0.37   0.22

Fertilizer and irrigation:

2.03   0.27   0.92   1.07   2.38

21. Calculating Percentiles. A statistics professor with too much time on his hands weighed each M&M candy in a bag of 465 plain M&M candies.

a. One of the M&Ms weighed 0.776 gram and it was heavier than 25 of the other M&Ms. What is the percentile of this particular value?

b. One of the M&Ms weighed 0.876 gram and it was heavier than 322 of the other M&Ms. What is the percentile of this particular value?

c. One of the M&Ms weighed 0.856 gram and it was heavier than 224 of the other M&Ms. What is the percentile of this particular value?

22. Calculating Percentiles. A data set consists of the 85 ages of women at the time that they won an Oscar in the category of best actress.

a. One of the actresses was 40 years of age, and she was older than 63 of the other actresses at the time that they won Oscars. What is the percentile of the age of 40?

b. One of the actresses was 54 years of age, and she was older than 76 of the other actresses at the time that they won Oscars. What is the percentile of the age of 54?

c. One of the actresses was 60 years of age, and she was older than 77 of the other actresses at the time that they won Oscars. What is the percentile of the age of 60?

23. Understanding Standard Deviation. The following four sets of 7 numbers all have a mean of 9.

$$\{9, 9, 9, 9, 9, 9, 9,\} \quad \{8, 8, 9, 9, 9, 10, 10\}$$

$$\{8, 8, 8, 9, 10, 10, 10\} \quad \{6, 6, 6, 9, 12, 12, 12\}$$

a. Make a histogram for each set.

b. Give the five-number summary and draw a boxplot for each set.

c. Compute the standard deviation for each set.

d. Based on your results, briefly explain how the standard deviation provides a useful single-number summary of the variation in these data sets.

24. Understanding Standard Deviation. The following four sets of 7 numbers all have a mean of 6.

$$\{6, 6, 6, 6, 6, 6, 6\}, \quad \{5, 5, 6, 6, 6, 7, 7\}$$

$$\{5, 5, 5, 6, 7, 7, 7\}, \quad \{3, 3, 3, 6, 9, 9, 9\}$$

a. Make a histogram for each set.

b. Give the five-number summary and draw a boxplot for each set.

c. Compute the standard deviation for each set.

d. Based on your results, briefly explain how the standard deviation provides a useful single-number summary of the variation in these data sets.

Comparing Variations. For each of Exercises 25–28, do the following:

a. Find the mean, median, and range for each of the two data sets.

b. Give the five-number summary and draw a boxplot for each of the two data sets.

c. Find the standard deviation for each of the two data sets.

d. Apply the range rule of thumb to estimate the standard deviation of each of the two data sets. How well does the rule work in each case? Briefly discuss why it does or does not work well.

e. Based on all your results, compare and discuss the two data sets in terms of their center and variation.

25. The following data sets give the ages in years of a sample of cars in a faculty parking lot and a student parking lot at the College of Portland.

Faculty:

2   3   1   0   1   2   4   3   3   2   1

Student:

5   6   8   2   7   10   1   4   6   10   9

26. The following data sets give the driving speeds in miles per hour of the first nine cars to pass through a school zone and the first nine cars to pass through a downtown intersection.

School:

20   18   23   21   19   18   17   24   25

Downtown:

29   31   35   24   31   26   36   31   28

27. The following data sets show the ages of the first seven U.S. Presidents (Washington through Jackson) and seven recent U.S. Presidents (Ford through Obama) at the time of inauguration.

First 7:

57   61   57   57   58   57   61

Last 7:

61   52   69   64   46   54   47

28. The following data sets give the approximate lengths of Beethoven's nine symphonies and Mahler's nine symphonies (in minutes).

Beethoven:

28   36   50   33   30   40   38   26   68

Mahler:

52   85   94   50   72   72   80   90   80

29. Manufacturing. You are in charge of a manufacturing process that produces car batteries that are supposed to provide 12 volts of power. Manufacturing occurs at two different sites. The first site produces batteries with a mean of 12.1 volts and a standard deviation of 0.5 volt, while the second site produces batteries with a mean of 12.2 volts and a standard deviation of 0.1 volt. Which site has better quality? Why?

30. Managing Complaints. You manage a small ice cream shop in which your employees scoop the ice cream by hand. Each night, you total your sales and the total volume of ice cream sold. You find that on nights when an employee named Ben is working, the mean price of the ice cream sold is $1.75 per pint with a standard deviation of $0.05. On nights when an employee named Jerry is working, the mean price of the ice cream sold is $1.70 per pint with a standard deviation of $0.35. Which employee is more likely to be generating complaints of "too small" servings? Explain.

31. Portfolio Standard Deviation. The book *Investments*, by Zvi Bodie, Alex Kane, and Alan Marcus, claims that the annual percentage returns for investment portfolios with a single stock have a standard deviation of 0.55, while the annual percentage returns for portfolios with 32 stocks have a standard deviation of 0.325. Explain how the standard deviation measures the risk in these two types of portfolios.

32. Batting Standard Deviation. For the past 100 years, the mean batting average in the major leagues has remained fairly constant at about 0.260. However, the standard deviation of batting averages has decreased from about 0.049 in the 1870s to 0.031 in the present. What does this tell us about the batting averages of players? Based on these facts, would you expect batting averages above 0.350 to be more or less common today than in the past? Explain.

PROJECTS FOR THE INTERNET & BEYOND

33. Secondhand Smoke. At the Web sites of the American Lung Association and the U.S. Environmental Protection Agency, find statistical data concerning the health effects of secondhand (passive) smoke. Write a short summary of your findings and your opinions about whether and how this health issue should be addressed by government.

34. Kids and the Media. A recent study by the Kaiser Family Foundation looked at the role of media (for example, television, books, computers) in the lives of children. The report, which is on the Kaiser Family Foundation Web site, gives many data distributions concerning, for example, how much time children spend daily with each medium. Study at least three of the distributions in the report that you find particularly interesting. Summarize each distribution in words, and discuss your opinions of the social consequences of the findings.

35. Measuring Variation. The range and standard deviation use different approaches to measure variation in a data set. Construct two different data sets configured so that the range of the first set is *greater than* the range of the second set (suggesting that the first set has more variation) but the standard deviation of the first set is *less than* the standard deviation of the second set (suggesting that the first set has less variation).

IN THE NEWS

36. Ranges in the News. Find two examples of data distributions in recent news reports; they may be given either as tables or as graphs. In each case, state the range of the distribution and explain its meaning in the context of the news report. Estimate the standard deviation by applying the range rule of thumb.

37. Summarizing a News Data Set. Find an example of a data distribution given in the form of a table in a recent news report. Make a five-number summary and a boxplot for the distribution.

4.4 STATISTICAL PARADOXES

The government administers polygraph tests ("lie detectors") to new applicants for sensitive security jobs. The polygraph tests are reputed to be 90% accurate; that is, they catch 90% of the people who are lying and validate 90% of the people who are truthful. Most people therefore guess that only 10% of the people who fail a polygraph test have been falsely identified as lying. In fact, the actual percentage of false accusations can be *much* higher — more than 90% in some cases. How can this be?

We'll discuss the answer soon, but the moral of this story should already be clear: Even when we describe data carefully according to the principles discussed in the first three sections of this chapter, we may still be led to very surprising conclusions. Before we get to the polygraph issue, let's start with a couple of other statistical surprises.

Better in Each Case, but Worse Overall

Suppose a pharmaceutical company creates a new treatment for acne. To decide whether the new treatment is better than an older treatment, the company gives the old treatment to 90 patients and gives the new treatment to 110 patients. Some patients had mild acne and others had severe acne. Table 4.5 summarizes the results after four weeks of treatment, broken down by which treatment was given and whether the patient's acne was mild or severe. If you study the table carefully, you will notice these key facts:

- Among patients with *mild* acne:
  10 received the old treatment and 2 were cured, for a 20% cure rate.
  90 received the new treatment and 30 were cured, for a 33% cure rate.
- Among patients with *severe* acne:
  80 received the old treatment and 40 were cured, for a 50% cure rate.
  20 received the new treatment and 12 were cured, for a 60% cure rate.

TABLE 4.5	Results of Acne Treatments			
	Mild acne		Severe acne	
	Cured	Not cured	Cured	Not cured
Old treatment	2	8	40	40
New treatment	30	60	12	8

Notice that the new treatment had a higher cure rate *both* for patients with mild acne (33% for the new treatment vs. 20% for the old) and for patients with severe acne (60% for the new treatment vs. 50% for the old). Is it therefore fair for the company to claim that their new treatment is better than the old treatment?

At first, this might seem to make sense. But instead of looking at the data for the mild and severe acne patients separately, let's look at the *overall* results:

- A total of 90 patients received the old treatment and 42 were cured (2 out of 10 with mild acne and 40 out of 80 with severe acne), for an overall cure rate of 42/90 = 46.7%.

- A total of 110 patients received the new treatment and 42 were cured (30 out of 90 with mild acne and 12 out of 20 with severe acne), for an overall cure rate of 42/110 = 38.2%.

Overall, the *old* treatment had the higher cure rate, despite the fact that the new treatment had a higher rate for both mild and severe acne cases.

This example illustrates that it is possible for something to appear better in each of two or more group comparisons but actually be worse overall. If you look carefully, you'll see that this occurs because of the way in which the overall results are divided into unequally sized groups (in this case, mild acne patients and severe acne patients).

BY THE WAY

The general case in which a set of data gives different results for each of several group comparisons than it does when the groups are taken together is known as *Simpson's paradox*, so named because it was described by Edward Simpson in 1951. However, the same idea was actually described around 1900 by Scottish statistician George Yule.

EXAMPLE **1** **Who Played Better?**

Table 4.6 gives the shooting performance of two players in each half of a basketball game. Shaq had a higher shooting percentage in both the first half (40% to 25%) and the second half (75% to 70%). Can Shaq claim that he had the better game?

TABLE 4.6	Basketball Shots					
	First half			Second half		
Player	Baskets	Attempts	Percent	Baskets	Attempts	Percent
Shaq	4	10	40%	3	4	75%
Vince	1	4	25%	7	10	70%

SOLUTION   No, and we can see why by looking at the overall game statistics. Shaq made a total of 7 baskets (4 in the first half and 3 in the second half) on 14 shots (10 in the first half and 4 in the second half), for an overall shooting percentage of 7/14 = 50%. Vince made a total of 8 baskets on 14 shots, for an overall shooting percentage of 8/14 = 57.1% Surprisingly, even though Shaq had a higher shooting percentage in both halves, Vince had a better overall shooting percentage for the game.   ·· ●

Does a Positive Mammogram Mean Cancer?

We often associate tumors with cancers, but most tumors are not cancers. Medically, any kind of abnormal swelling or tissue growth is considered a tumor. A tumor caused by cancer is said to be *malignant* (or *cancerous*); all others are said to be *benign*.

Imagine you are a doctor or nurse treating a patient who has a breast tumor. The patient will be understandably nervous, but you can give her some comfort by telling her that only about 1 in 100 breast tumors turns out to be malignant. But, just to be safe, you order a mammogram to determine whether her tumor is one of the 1% that are malignant.

Now, suppose the mammogram comes back positive, suggesting that the tumor is malignant. Mammograms are not perfect, so the positive result does not necessarily mean that your patient has breast cancer. More specifically, let's assume that the mammogram screening is 85% accurate: It will correctly identify 85% of malignant tumors as malignant and 85% of benign tumors as benign. When you tell your patient that her mammogram was positive, what should you tell her about the chance that she actually has cancer?

Because the mammogram screening is 85% accurate, most people guess that the positive result means that the patient probably has cancer. Studies have shown that most doctors also believe this to be the case and would tell the patient to be prepared for cancer treatment. But a more careful analysis shows otherwise. In fact, the chance that the patient has cancer is still quite small—about 5%. We can see why by analyzing some numbers.

Consider a study in which mammograms are given to 10,000 women with breast tumors. Assuming that 1% of tumors are malignant, 1% × 10,000 = 100 of the women actually have cancer; the remaining 9,900 women have benign tumors. Table 4.7 summarizes the mammogram results. Notice the following:

- The mammogram screening correctly identifies 85% of the 100 malignant tumors as malignant. Thus, it gives positive (malignant) results for 85 of the malignant tumors; these cases are called **true positives**. In the other 15 malignant cases, the result is negative, even though the women actually have cancer; these cases are **false negatives**.

TABLE 4.7	Summary of Results for 10,000 Mammograms (when in fact 100 tumors are malignant and 9,900 are benign)		
	Tumor is malignant	Tumor is benign	Total
Positive mammogram	85 true positives	1,485 false positives	1,570
Negative mammogram	15 false negatives	8,415 true negatives	8,430
Total	100	9,900	10,000

BY THE WAY

This mammogram example and the polygraph example that follows it illustrate cases in which conditional probabilities (discussed in Section 6.5) lead to confusion. The proper way of handling conditional probabilities was discovered by the Reverend Thomas Bayes (1702–1761) and is often called *Bayes rule.*

- The mammogram screening correctly identifies 85% of the 9,900 benign tumors as benign. Thus, it gives negative (benign) results for 85% × 9,900 = 8,415 of the benign tumors; these cases are **true negatives**. The remaining 9,900 − 8,485 = 1,485 women get positive results in which the mammogram incorrectly identifies their tumors as malignant; these cases are **false positives**.

Overall, the mammogram screening gives positive results to 85 women who actually have cancer and to 1,485 women who do *not* have cancer. The total number of positive results is 85 + 1,485 = 1,570. Because only 85 of these are true positives (the rest are false positives), the chance that a positive result really means cancer is only 85/1,570 = 0.054, or 5.4%. Therefore, when your patient's mammogram comes back positive, you should reassure her that there's still only a small chance that she has cancer.

EXAMPLE 2 False Negatives

Suppose you are a doctor seeing a patient with a breast tumor. Her mammogram comes back negative. Based on the numbers in Table 4.7, what is the chance that she has cancer?

SOLUTION   For the 10,000 cases summarized in Table 4.7, the mammograms are negative for 15 women with cancer and for 8,415 women with benign tumors. The total number of negative results is 15 + 8,415 = 8,430. Thus, the fraction of women with cancer who have false negatives is 15/8,430 = 0.0018, or slightly less than 2 in 1,000. In other words, the chance that a woman with a negative mammogram has cancer is only about 2 in 1,000.   ·· •

> **TIME ◷UT TO THINK**
> While the chance of cancer with a negative mammogram is small, it is not zero. Therefore, it might seem like a good idea to biopsy all tumors, just to be sure. However, biopsies involve surgery, which means they can be painful and expensive, among other things. Given these facts, do you think that biopsies should be routine for all tumors? Should they be routine for cases of positive mammograms? Defend your opinion.

Polygraphs and Drug Tests

We're now ready to return to the question asked at the beginning of this section, about how a 90% accurate polygraph test can lead to a surprising number of false accusations. The explanation is very similar to that used in the case of the mammograms.

Suppose the government gives the polygraph test to 1,000 applicants for sensitive security jobs. Further suppose that 990 of these 1,000 people tell the truth on their polygraph test, while only 10 people lie. For a test that is 90% accurate, we find the following results:

- Of the 10 people who lie, the polygraph correctly identifies 90%, meaning that 9 fail the test (they are identified as liars) and 1 passes.
- Of the 990 people who tell the truth, the polygraph correctly identifies 90%, meaning that 90% × 990 = 891 truthful people pass the test and the other 10% × 990 = 99 truthful people fail the test.

Figure 4.16 summarizes these results. The total number of people who fail the test is 9 + 99 = 108. Of these, only 9 were actually liars; the other 99 were falsely accused of lying. That is, 99 out of 108, or 99/108 = 91.7%, of the people who fail the test were actually telling the truth.

The percentage of people who are falsely accused in any real situation depends on both the accuracy of the test and the proportion of people who are lying. Nevertheless, for the numbers given here, we have an astounding result: Assuming the government rejects applicants who fail the polygraph test, then almost 92% of the rejected applicants were actually being truthful and may have been highly qualified for the jobs.

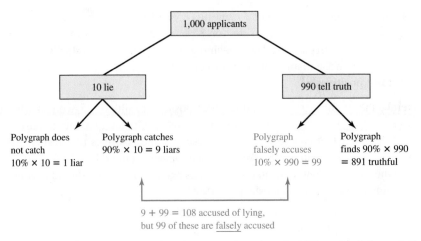

Figure 4.16 A tree diagram summarizes results of a 90% accurate polygraph test for 1,000 people, of whom only 10 are lying.

TIME OUT TO THINK

Imagine that you are falsely accused of a crime. The police suggest that, if you are truly innocent, you should agree to take a polygraph test. Would you do it? Why or why not?

EXAMPLE 3 High School Drug Testing

All athletes participating in a regional high school track and field championship must provide a urine sample for a drug test. Those who fail are eliminated from the meet and suspended from competition for the following year. Studies show that, at the laboratory selected, the drug tests are 95% accurate. Assume that 4% of the athletes actually use drugs. What fraction of the athletes who fail the test are falsely accused and therefore suspended without cause?

SOLUTION The easiest way to answer this question is by using some sample numbers. Suppose there are 1,000 athletes in the meet. Then 4%, or 40 athletes, actually use drugs; the remaining 960 athletes do not use drugs. In that case, the 95% accurate drug test should return the following results:

- 95% of the 40 athletes who use drugs, or 0.95 × 40 = 38 athletes, fail the test. The other 2 athletes who use drugs pass the test.
- 95% of the 960 athletes who do not use drugs pass the test, but 5% of these 960, or 0.05 × 960 = 48 athletes, fail.

The total number of athletes who fail the test is 38 + 48 = 86. But 48 of these athletes who fail the test, or 48/86 = 56%, are actually nonusers. Despite the 95% accuracy of the drug test, more than half of the suspended students are innocent of drug use. ···•

Section 4.4 Exercises

Statistical Literacy and Critical Thinking

1. **False Positive and False Negative.** A baseball player is given a test for banned substances. For this test, what is a false positive? What is a false negative? What is a true positive? What is a true negative?

2. **Positive Test Result.** Jennifer is given a pregnancy test. What does it mean when she is told that the result is positive?

3. **Test Result.** If you apply a test for the presence of alcohol, what is the result called if the test correctly indicates that the subject has not consumed alcohol?

4. **Better in Each Half, Worse Overall.** When the Giants and Patriots football teams play each other, can one of the quarterbacks have a higher passing

percentage in each half while having a lower passing percentage for the entire game?

Does It Make Sense? For Exercises 5–8, decide whether the statement makes sense (or is clearly true) or does not make sense (or is clearly false). Explain clearly; not all of these statements have definitive answers, so your explanation is more important than your chosen answer.

5. **Course Average.** Ann and Bret are taking the same statistics course, in which the final grade is determined by assignments and exams. Ann's mean score on the assignments is higher than Bret's, and Ann's mean score on the exams is higher than Bret's. It follows that Ann's overall mean score in the course is higher than Bret's.

6. **Batting Average.** Ann's batting average for the first half of the softball season is higher than Bret's, and Ann's batting average for the second half of the season is higher than Bret's. It follows that Ann's batting average for the entire season is higher than Bret's.

7. **Test Results.** After taking a test for the presence of a disease, a patient is happy because the physician announces that the test results are positive.

8. **Test Accuracy.** If a drug test is 90% accurate, it follows that 90% of those who test positive are actual drug users.

Concepts and Applications

9. **Batting Percentages.** The table below shows the batting records of two baseball players in the first half (first 81 games) and last half of a season.

	First half		
Player	Hits	At-bats	Batting average
Josh	50	150	0.333
Jude	10	50	0.200
	Second half		
Player	Hits	At-bats	Batting average
Josh	35	70	0.500
Jude	70	150	0.467

Who had the higher batting average in the first half of the season? Who had the higher batting average in the second half? Who had the higher overall batting average? Explain how these results illustrate Simpson's paradox.

10. **Passing Percentages.** The table below shows the passing records of two rival quarterbacks in the first half and second half of a football game.

	First half		
Player	Completions	Attempts	Percent
Allan	8	20	40%
Abner	2	6	33%
	Second half		
Player	Completions	Attempts	Percent
Allan	3	6	50%
Abner	12	25	48%

Who had the higher completion percentage in the first half? Who had the higher completion percentage in the second half? Who had the higher overall completion percentage? Explain how these results illustrate Simpson's paradox.

11. **Test Scores.** The table below shows eighth-grade mathematics test scores in Nebraska and New Jersey. The scores are separated according to the race of the student. Also shown are the state averages for all races.

	White	Nonwhite	Average for all races
Nebraska	281	250	277
New Jersey	283	252	272

Source: National Assessment of Educational Progress, from *Chance* magazine.

a. Which state had the higher scores in both racial categories? Which state had the higher overall average across both racial categories?

b. Explain how a state could score lower in both categories and still have a higher overall average.

c. Now consider the table below, which gives the percentages of whites and nonwhites in each state. Use these percentages to verify that the overall average test score in Nebraska is 277, as claimed in the first table.

	White	Nonwhite
Nebraska	87%	13%
New Jersey	66%	34%

d. Use the racial percentages to verify that the overall average test score in New Jersey is 272, as claimed in the first table.

e. Explain briefly, in your own words, how Simpson's paradox appeared in this case.

12. **Test Scores.** Consider the following table comparing the grade point averages (GPAs) and mathematics SAT scores of high school students in 1988 and 1998 (before the SAT test format was revised).

	% students		SAT score		
GPA	1988	1998	1988	1998	Change
A+	4	7	632	629	−3
A	11	15	586	582	−4
A−	13	16	556	554	−2
B	53	48	490	487	−3
C	19	14	431	428	−3
Overall average			504	514	+10

Source: Cited in *Chance*, Vol. 12, No. 2, 1999, from data in *New York Times*, September 2, 1999.

a. In general terms, how did the SAT scores of the students in the five grade categories change between 1988 and 1998?

b. How did the overall average SAT score change between 1988 and 1998?

c. How is this an example of Simpson's paradox?

13. Tuberculosis Deaths. The following table shows deaths due to tuberculosis (TB) in New York City and Richmond, Virginia, in 1910.

Race	New York	
	Population	TB deaths
White	4,675,000	8400
Nonwhite	92,000	500
Total	4,767,000	8900

Race	Richmond	
	Population	TB deaths
White	81,000	130
Nonwhite	47,000	160
Total	128,000	290

Source: Cohen and Nagel, *An Introduction to Logic and Scientific Method,* Harcourt, Brace and World, 1934.

a. Compute the death rates for whites, nonwhites, and all residents in New York City.

b. Compute the death rates for whites, nonwhites, and all residents in Richmond.

c. Explain why this is an example of Simpson's paradox and explain how the paradox arises.

14. Weight Training. Two cross-country running teams participated in a (hypothetical) study in which a fraction of each team used weight training to supplement a running workout. The remaining runners did not use weight training. At the end of the season, the mean improvement in race times (in seconds) was recorded in the table below.

	Mean improvement (seconds)		
	Weight training	No weight training	Team average
Gazelles	10	2	6.0
Cheetahs	9	1	6.2

Describe how Simpson's paradox arises in this table. Resolve the paradox by finding the percentage of each team that used weight training.

15. Basketball Records. Consider the following (hypothetical) basketball records for Spelman and Morehouse Colleges.

	Spelman College	Morehouse College
Home games	10 wins, 19 losses	9 wins, 19 losses
Away games	12 wins, 4 losses	56 wins, 20 losses

a. Give numerical evidence to support the claim that Spelman College has a better team than Morehouse College.

b. Give numerical evidence to support the claim that Morehouse College has a better team than Spelman College.

c. Which claim do you think makes more sense? Why?

16. Better Drug. Two drugs, A and B, were tested on a total of 2,000 patients, half of whom were women and half of whom were men. Drug A was given to 900 patients and Drug B to 1,100 patients. The results appear in the table below.

	Women	Men
Drug A	5 of 100 cured	400 of 800 cured
Drug B	101 of 900 cured	196 of 200 cured

a. Give numerical evidence to support the claim that Drug B is more effective than Drug A.

b. Give numerical evidence to support the claim that Drug A is more effective than Drug B.

c. Which claim do you think makes more sense? Why?

17. Polygraph Test. The results in the table below are from experiments conducted by researchers Charles R. Honts (Boise State University) and Gordon H. Barland (Department of Defense Polygraph Institute). In each case, it was known whether the subject lied, so the table indicates when the polygraph test was correct.

	Did the Subject Actually Lie?	
	No	Yes
Polygraph test indicated that the subject *lied*.	15	42
Polygraph test indicated that the subject did *not lie*.	32	9

a. Based on the test results, how many subjects appeared to be lying? Of these, how many were actually lying and how many were telling the truth? What percentage of those who appear to be lying were not actually lying?

b. Based on the test results, how many subjects appeared to be telling the truth? Of those, how many were actually telling the truth? What percentage of those who appeared to be telling the truth are actually truthful?

18. Disease Test. Suppose a test for a disease is 80% accurate for those who have the disease (true positives) and 80% accurate for those who do not have the disease (true negatives). Within a sample of 4,000 patients, the incidence rate of the disease matches the national average, which is 1.5%.

	Disease	No disease	Total
Test positive	48	788	836
Test negative	12	3,152	3,164
Total	60	3,940	4,000

a. Of those with the disease, what percentage test positive?

b. Of those who test positive, what percentage have the disease? Compare this result to the one in part a and explain why they are different.

c. Suppose a patient tests positive for the disease. As a doctor using this table, how would you describe the patient's chance of actually having the disease? Compare this figure to the overall incidence rate of the disease.

Further Applications

19. Hiring Statistics. (This problem is based on an example in "Ask Marilyn" column in *Parade Magazine*.) A company decided to expand, so it opened a factory, generating 455 jobs. For the 70 white-collar positions, 200 males and 200 females applied. Of the females who applied, 20% were hired, while only 15% of the males were hired. Of the 400 males applying for the blue-collar positions, 75% were hired, while 85% of the 100 females who applied were hired. How does looking at the white-collar and blue-collar positions separately suggest a hiring preference for women? Do the overall data support the idea that the company hires women preferentially? Explain why this is an example of Simpson's paradox and how the paradox can be resolved.

20. Drug Trials. (This problem is based on an example from the "Ask Marilyn" column in *Parade Magazine*.) A company runs two trials of two treatments for an illness. In the first trial, Treatment A cures 20% of the cases (40 out of 200) and Treatment B cures 15% of the cases (30 out of 200). In the second trial, Treatment A cures 85% of the cases (85 out of 100) and Treatment B cures 75% of the cases (300 out of 400). Which treatment had the better cure rate in the two trials individually? Which treatment had the better overall cure rate? Explain why this is an example of Simpson's paradox and how the paradox can be resolved.

21. HIV Risks. The New York State Department of Health estimates a 10% rate of HIV for the at-risk population and a 0.3% rate for the general population. Tests for HIV are 95% accurate in detecting both true negatives and true positives. Random selection and testing of 5,000 at-risk people and 20,000 people from the general population results in the following table.

	At-risk population	
	Test positive	Test negative
Infected	475	25
Not infected	225	4,275
	General population	
	Test positive	Test negative
Infected	57	3
Not infected	997	18,943

a. Verify that incidence rates for the general and at-risk populations are 0.3% and 10%, respectively. Also verify that detection rates for the general and at-risk populations are 95%.

b. Consider the at-risk population. Of those with HIV, what percentage test positive? Of those who test positive, what percentage have HIV? Explain why these two percentages are different.

c. Suppose a patient in the at-risk category tests positive for the disease. As a doctor using this table, how would you describe the patient's chance of actually having the disease? Compare this figure with the overall incidence rate of the disease.

d. Consider the general population. Of those with HIV, what percentage test positive? Of those who test positive, what percentage have HIV? Explain why these two percentages are different.

e. Suppose a patient in the general population tests positive for the disease. As a doctor using this table, how would you describe the patient's chance of actually having the disease? Compare this figure to the overall incidence rate of the disease.

PROJECTS FOR THE INTERNET & BEYOND

22. Polygraph Arguments. Visit Web sites devoted to either opposing or supporting the use of polygraph tests. Summarize the arguments on both sides, specifically noting the role that false negative rates play in the discussion.

23. Drug Testing. Explore the issue of drug testing either in the workplace or in athletic competitions. Discuss the legality of drug testing in these settings and the accuracy of the tests that are commonly conducted.

24. Cancer Screening. Investigate recommendations concerning routine screening for some type of cancer (for example, breast cancer, prostate cancer, or colon cancer). Explain how the accuracy of the screening test is measured. How is the test useful? How can its results be misleading?

IN THE NEWS

25. Polygraphs. Find a recent article in which someone or some group proposes a polygraph test to determine whether a person is being truthful. In light of what you know about polygraph tests, do you think the results will be meaningful? Why or why not?

26. Drug Testing and Athletes. Find a news report concerning drug testing of athletes. Summarize how the testing is being used, and discuss whether the testing is reliable.

CHAPTER REVIEW EXERCISES

1. Nicotine in Cigarettes. Listed below are the nicotine amounts (in mg per cigarette) for samples of filtered and non-filtered cigarettes. Do filters appear to be effective in reducing the amount of nicotine?

Non-filtered: 1.1 1.7 1.7 1.1 1.1 1.4 1.1 1.4 1.0 1.2

Filtered: 0.4 1.0 1.2 0.8 0.8 1.0 1.1 1.1 1.1 0.78

a. Find the mean and median for each of the two data sets.

b. Find the range and standard deviation for each of the two data sets.

c. Give the five-number summary and construct a boxplot for each of the two data sets.

d. Apply the range rule of thumb to estimate the standard deviation of each of the two data sets. How well does the rule work in each case? Briefly discuss why it does or does not work well.

e. Based on all your results, compare and discuss the two data sets in terms of their center and variation. Does there appear to be a difference between the amounts of nicotine in non-filtered cigarettes and filtered cigarettes?

2. Combine the two samples from Review Exercise 1 and find the following:

a. The percentile for the amount of 1.4 mg

b. The mode

3. a. What is the standard deviation for a sample of 50 values, all of which are the same?

b. Which of the following two car batteries would you prefer to buy, and why?

- One taken from a population with a mean life of 48 months and a standard deviation of 2 months

- One taken from a population with a mean life of 48 months and a standard deviation of 6 months

c. If an outlier is included with a sample of 50 values, what is the effect of the outlier on the mean?

d. If an outlier is included with a sample of 50 values, what is the effect of the outlier on the median?

e. If an outlier is included with a sample of 50 values, what is the effect of the outlier on the range?

f. If an outlier is included with a sample of 50 values, what is the effect of the outlier on the standard deviation?

CHAPTER QUIZ

1. When you add the pulse rates of 65, 74, 88, 77, and 92, then divide by the number of values, the result is 79.2. Which term best describes this value: average, mean, median, mode, or standard deviation?

2. Find the median of the pulse rates given in Exercise 1.

3. What is the range of the pulse rates given in Exercise 1?

4. The standard deviation of the pulse rates given in Exercise 1 is 10.9. What characteristic does that value measure?

5. A histogram is constructed for a large set of pulse rates of adult males, and it is found that the distribution is symmetric and unimodal. What does this imply about the values of the mean and median?

6. Indicate whether the given statement could apply to a data set consisting of 1,000 values that are all different.

a. The 20th percentile is greater than the 30th percentile.

b. The median is greater than the first quartile.

c. The third quartile is greater than the first quartile.

d. The mean is equal to the median.

e. The range is zero.

7. A standard test for braking reaction time of drivers is designed so that the mean is 2.00 sec and the standard deviation is 0.25 sec. Based on the range rule of thumb, what are the likely low and high values?

8. Use the range rule of thumb to estimate the standard deviation of the pulse rates given in Exercise 1. How does the result compare to the actual standard deviation of 10.9?

9. Find the standard deviation of these body temperatures (°F): 98.2, 98.2, 98.2, 98.2, and 98.2.

10. Identify the components that constitute the five-number summary for a data set.

FOCUS ON
THE STOCK MARKET

What's Average About the Dow?

As "averages" go, this one is extraordinary. You can't watch the news without hearing what happened to it, and many people spend hours tracking it each day. It is by far the most famous indicator of stock market performance. We are talking, of course, about the Dow Jones Industrial Average, or DJIA for short. But what exactly is it?

The easiest way to understand the DJIA is by looking at its history. As the modern industrial era got under way in the late 19th century, most people considered stocks to be dangerous and highly speculative investments. One reason was a lack of regulation that made it easy for wealthy speculators, unscrupulous managers, and corporate raiders to manipulate stock prices. But another reason was that, given the complexities of daily stock trading, even Wall Street professionals had a hard time figuring out whether stocks in general were going up (a "bull market") or down (a "bear market"). Charles H. Dow, the founder (along with Edward D. Jones) and first editor of the *Wall Street Journal*, believed he could rectify this problem by creating an "average" for the stock market as a whole. If the average was up, the market was up, and if the average was down, the market was down.

To keep the average simple, Dow chose 12 large corporations to include in his average. On May 26, 1896, he added the stock prices of these 12 companies and divided by 12, finding a mean stock price of $40.94. This was the first value for the DJIA. As Dow had hoped, it suddenly became easy for the public to follow the market's direction just by comparing his average from day to day, month to month, or year to year.

The basic idea behind the DJIA is still the same, although the list now includes 30 stocks rather than 12; the list is selected by the editors of the *Wall Street Journal*, who occasionally change the stocks on the list. However, the DJIA is no longer the mean price of its 30 stocks. Instead, it is calculated by adding the prices of its 30 stocks and dividing by a special divisor. Because of this divisor, we now think of the DJIA as an index that helps us keep track of stock values, rather than as an actual average of stock prices.

The divisor is designed to preserve continuity in the underlying value represented by the DJIA, and it therefore must change whenever the list of 30 stocks changes or when a company on the list has a stock split. A simple example shows why the divisor must change when the list changes. Suppose the DJIA consisted of only 2 stocks (rather than 30): Stock A with a price of $100 and Stock B with a price of $50. The mean price of these two stocks is ($100 + $50)/2 = $75. Now, suppose that we change the list by replacing Stock B with Stock C and that Stock C's price is $200. The new mean is ($100 + $200)/2 = $150, so merely replacing one stock on the list would raise the mean price from $75 to $150. Therefore, to keep the "value" of the DJIA constant when we change this list, we must divide the new mean of $150 by 2. In this way, the DJIA remains 75 both before and after the list change, but we can no longer think of this 75 as a mean price in dollars.

To see why a stock split changes the divisor, again suppose the index consists of just two stocks: Stock X at $100 and Stock Y at $50, for a mean price of $75. Now, suppose Stock X undergoes a 2-for-1 stock split, so that its new price is $50. With both stocks now priced at $50, the mean price after the stock split would also be $50. In other words, even though a stock split does not affect a company's total value (it only changes the number and prices of its shares), we'd find a drop in the mean price from $75 to $50. In this case, we can preserve continuity by dividing the new mean of 50 by 2/3 (which is equivalent to multiplying by 3/2) so that the DJIA holds at 75 both before and after the stock split.

Just as in these simple examples, the real divisor changes with every list change or stock split, so it has changed many times since Charles Dow first calculated the DJIA as an actual mean. The current value of the divisor is published daily in the *Wall Street Journal*.

Given that there are now well over 10,000 actively traded stocks, it might seem remarkable that a sample of only 30 could reflect overall market activity. But today, when computers make it easy to calculate stock market "averages" in many other ways, we can look at historical data and see that the DJIA has indeed been a reliable indicator of overall market performance. Figure 4.17 shows the historical performance of the DJIA.

If you study Figure 4.17 carefully, you may be tempted to think that you can see patterns that would allow you to forecast precise values of the market in the future. Unfortunately, no

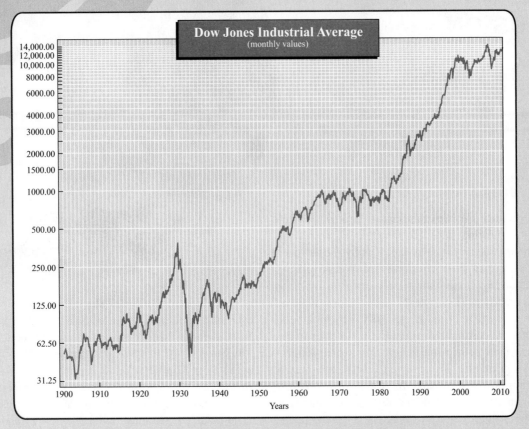

Figure 4.17 Historical values of the Dow Jones Industrial Average, 1900 through 2011 Note that the vertical axis uses an exponential scale in which the value doubles with each equivalent increment in height; this makes it easier to see the changes that occurred when the DJIA was low compared with its value today. Chart courtesy of StockCharts.com.

one has ever found a way to make reliable forecasts, and most economists now believe that such forecasts are impossible.

The futility of trying to forecast the market is illustrated by the story of the esteemed Professor Benjamin Graham, often called the father of "value investing." In the spring of 1951, one of his students came to him for some investment advice. Professor Graham noted that the DJIA then stood at 250, but that it had fallen below 200 at least once during every year since its inception in 1896. Because it had not yet fallen below 200 in 1951, Professor Graham advised his student to hold off on buying until it did. Professor Graham presumably followed his own advice, but the student did not. Instead, the student invested his "about 10 thousand bucks" in the market right away. As it turned out, the market never did fall below 200 in 1951 or any time thereafter. And the student, named Warren Buffet, became a billionaire many times over.

QUESTIONS FOR DISCUSSION

1. The stock market is still considered a riskier investment than, say, bank savings accounts or bonds. Nevertheless, financial advisors almost universally recommend holding at least some stocks, which is quite different from the situation that prevailed a century ago. What role do you think the DJIA played in building investors' confidence in the stock market?

2. The DJIA is only one of many different stock market indices in wide use today. Briefly look up a few other indices, such as the S&P 500, the Russell 2000, and the NASDAQ. How do these indices differ from the DJIA? Do you think that any of them should be considered more reliable indicators of the overall market than the DJIA? Why or why not?

3. The 30 stocks in the DJIA represent a sample of the more than 10,000 actively traded stocks, but it is not a *random* sample because it is chosen by particular editors for particular reasons that may include personal biases. Suppose that you chose a random sample of 30 stocks and tracked their prices. Do you think that such a random sample would track the market as well as the stocks in the DJIA? Why or why not?

4. Create your own "portfolio" of 10 stocks that you'd like to own, and assume you own 100 shares of each. Calculate the total value of your portfolio today, and track price changes over the next month. At the end of the month, calculate the percent change in the value of your portfolio. How did the performance of your portfolio compare to the performance of the DJIA during the month? If you really owned these stocks, would you continue to hold them or would you sell? Explain.

F CUS ON
ECONOMICS

Are the Rich Getting Richer?

The media love to report on the lavish spending of the super-rich, making it seem that the rich keep getting richer while the rest of us are left behind. But is it true?

If we wish to draw general conclusions about how the average person is faring compared to the rich, we must look at the overall income distribution. Economists have developed a number, called the **Gini Index,** that is used to describe the level of equality or inequality in the income distribution. The Gini Index is defined so that it can range only between 0 and 1. A Gini Index of 0 indicates perfect income equality, in which every person has precisely the same income. A Gini Index of 1 indicates perfect inequality, in which a single person has all the income and no one else has anything. Figure 4.18 shows the Gini Index in the United States since 1947. Note that the Gini Index fell from 1947 to 1968, indicating that the income distribution became more uniform during this period. The Gini Index has generally risen ever since, indicating that the rich are, indeed, getting richer.

Although the Gini Index provides a simple single-number summary of income inequality, the number itself is fairly difficult to interpret (and to calculate). An alternative way to look at the income distribution is to study income quintiles, which divide the population into fifths by income. Often, the highest quintile is further broken down to show how the top 5% of income earners compares with others.

Figure 4.19 shows the share of total income received by each quintile and the top 5% in the United States in different decades. The height of each bar (the number on top of it) represents the share of total income. For example, the 3.3 on the bar for the lowest quintile in 2010 means that the poorest 20% of the population received only 3.3% of the total income in the United States. Similarly, the 50.2 on the bar for the top quintile in 2010 means that the richest 20% of the population received 50.2% of the income. Note also that the richest 5% received 21.3% of the income—nearly double the total income of the poorest 40% of the population. If you study this graph carefully, you'll see that the share of income earned by the first four quintiles—which means all but the richest 20% of the population—dropped since 1970. Meanwhile, the share earned by the richest 20% rose substantially, as did the share of the top 5%. In other words, this graph also confirms that the rich have been getting richer compared to most of the population.

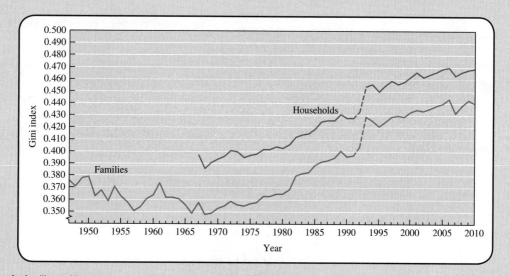

Figure 4.18 Gini Index for families and households, 1947–2010. Household data, which include single people and households in which the members are not part of the same family, have been taken only since 1967. The dashed segments in 1993 indicate a change in the methodology for data collection, so the corresponding rise in the Gini Index may be partially or wholly due to this change rather than a real change in income inequality. *Source:* Adapted from data by the U.S. Census Bureau.

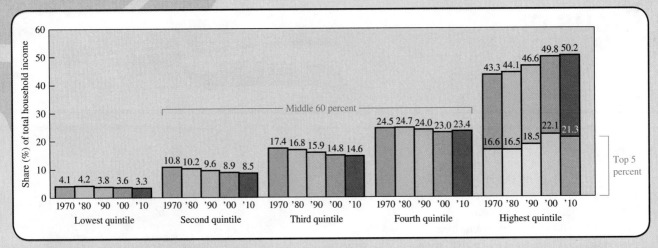

Figure 4.19 Share of total household income by quintile (and top 5%) at 10-year intervals. *Source:* U.S. Census Bureau.

Now that we've established that the rich are getting richer, the next question is whether it matters. Most people, including most economists, have traditionally assumed that rising income inequality is bad for democracies. But a few economists from both the left and the right of the political spectrum argue that the change in recent decades is different. For one thing, the change meets a widely accepted ethical condition called the *Pareto criterion*, after the Italian economist Vilfredo Pareto (for whom Pareto charts are also named): Any change is good if it makes someone better off without making anyone else worse off. The Pareto criterion appears to be satisfied because overall growth in the U.S. economy has been helping nearly everyone. In other words, most people may have a smaller percentage of total income than they had in the past, but they still have more absolute income and therefore are living better than they did in the past.

Secondly, today's rich differ from the rich in the past. For example, as recently as 1980, 60% of the "Forbes 400" (the richest 400 people) had inherited most of their wealth. Today, less than 20% of the Forbes 400 represents old money. The implication is that while you had to be born rich in the past, today you can *become* rich by getting educated and working hard. Surely, it is a good thing to encourage education and hard work.

Finally, while overall income inequality has increased, the income inequality among different races and between men and women has decreased. In other words, it is now easier than it was in the past for African Americans, Hispanics, and women to earn as much as white males. Again, this is surely a good thing for our democratic values, even if we still have a long way to go before the inequalities are completely eliminated.

QUESTIONS FOR DISCUSSION

1. Compare several different ways of looking at the data shown in Figure 4.18 and Figure 4.19. For example, does one seem to indicate a larger change in income inequality than the other? Can you think of other possible ways to display income data that might give a different picture than those shown here?

2. Do you agree that the Pareto criterion is a good way to evaluate the ethics of economic change? Why or why not?

3. Overall, do you think the increase in income inequality has been a good or bad thing for the United States? Will it be good if the trend continues? Defend your opinion.

4. Although economic data suggest that the vast majority of Americans are better off today than they were a few decades ago, the poorest Americans still live in difficult economic conditions. What do you think can or should be done to help improve the lives of the poor? Can your suggestion be implemented without harming the overall economy? Explain.

5

A Normal World

When you walk into a store, how do you know if a sale price is really a good price? When you exercise and your heart rate rises, how do you know if it has risen enough, but not too much, for a good workout? If your 12-year-old daughter runs a mile in 5 minutes, is she a future Olympic hopeful? These questions seem very different, but from a statistical standpoint they are very similar: Each one asks whether a particular number (price, heart rate, running time) is somehow unusual. In this chapter, we will discuss how we can answer such questions with the aid of the bell-shaped *normal distribution*.

Nothing in life is to be feared.
It is only to be understood.

—Marie Curie

5.1   WHAT IS NORMAL?

Suppose a friend is pregnant and due to give birth on June 30. Would you advise her to schedule an important business meeting for June 16, two weeks before the due date? Answering this question requires knowing whether the baby is likely to arrive more than 14 days before the due date. For that, we need to examine data concerning due dates and actual birth dates.

Figure 5.1 is a histogram for a distribution of 300 natural births at Providence Memorial Hospital; the data are hypothetical, but based on how births would be distributed without medical intervention. The horizontal axis shows how many days before or after the due date a baby was born. Negative numbers represent births *prior* to the due date, zero represents a birth on the due date, and positive numbers represent births *after* the due date. The left vertical axis shows the number of births for each 4-day bin. For example, the frequency of 35 for the highest bar corresponds to the bin from −2 days to 2 days; it shows that out of the 300 total births in the sample, 35 births occurred within 2 days of the due date.

To answer our question about whether a birth is likely to occur more than 14 days early, it is more useful to look at the *relative frequencies*. Recall that the relative frequency of any data value is its frequency divided by the total number of data values (see Section 3.1). Figure 5.1 shows relative frequencies on the right vertical axis. For example, the bin for −14 days to −10 days (shaded dark blue) has a relative frequency of about 0.07, or 7%. That is, about 7% of the 300 births occurred between 14 days and 10 days before the due date.

We can find the proportion of births that occurred more than 14 days before the due date simply by adding the relative frequencies for the bins to the left of −14; you can measure the graph to confirm that these bins have a total relative frequency of about 0.21, which means that about 21% of the births in this data set occurred more than 14 days before the due date. Based on these data, your friend has about a 1 in 5 chance of her baby being born on or before the date of the business meeting. If the meeting is important, it might be good to schedule it earlier.

BY THE WAY

The Scottish politician John Sinclair (1754–1835) was one of the first collectors of economic, demographic, and agricultural data. He is credited with introducing the words *statistics* and *statistical* into the English language, having heard them used in Germany to refer to matters of state.

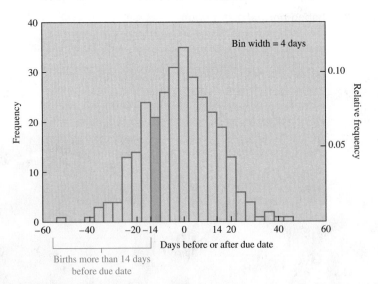

Figure 5.1 Histogram of frequencies (left vertical axis) and relative frequencies (right vertical axis) for birth dates relative to due date. (These data are hypothetical.) Negative numbers refer to births before the due date; positive numbers refer to births after the due date. The width of each bin is 4 days. For example, the bin shaded dark blue represents births occurring between 10 and 14 days early.

The Normal Distribution

The distribution of the birth data has a fairly distinctive shape, which is easier to see if we overlay the histogram with a smooth curve (Figure 5.2). For our present purposes, the shape of this smooth distribution has three very important characteristics:

- The distribution is *single-peaked.* Its mode, or most common birth date, is the due date.
- The distribution is *symmetric* around its single peak; therefore, its median and mean are the same as its mode. The median is the due date because equal numbers of births occur before and after this date. The mean is also the due date because, for every birth before the due date, there is a birth the same number of days after the due date.
- The distribution is spread out in a way that makes it resemble the shape of a bell, so we call it a "bell-shaped" distribution.

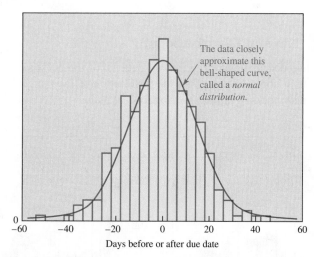

The data closely approximate this bell-shaped curve, called a *normal distribution.*

Days before or after due date

Figure 5.2 A smooth normal distribution curve is drawn over the histogram of Figure 5.1.

The smooth distribution in Figure 5.2, with these three characteristics, is called a **normal distribution**. All normal distributions have the same characteristic bell shape, but they can differ in their mean and in their variation. Figure 5.3 shows two different normal distributions. Both have the same mean, but distribution (a) has greater variation. As we'll discuss in the next section, knowing the *standard deviation* (see Section 4.3) of a normal distribution tells us everything we need to know about its variation. Therefore, a normal distribution can be fully described with just two numbers: its mean and its standard deviation.

TECHNICAL NOTE

Although we will not use it in this text, the following algebraic function describes a normal distribution with mean μ and standard deviation σ:

$$y = \frac{e^{-\frac{1}{2}[(x-\mu)/\sigma]^2}}{\sigma\sqrt{2\pi}}$$

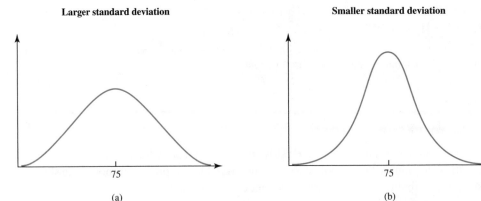

Larger standard deviation

Smaller standard deviation

75

75

(a)

(b)

Figure 5.3 Both distributions are normal and have the same mean of 75, but the distribution on the left has a larger standard deviation.

> **Definition**
>
> The **normal distribution** is a symmetric, bell-shaped distribution with a single peak. Its peak corresponds to the mean, median, and mode of the distribution. Its variation can be characterized by the standard deviation of the distribution.

BY THE WAY

Using data taken from French and Scottish soldiers, the Belgian social scientist Adolphe Quetelet realized in the 1830s that human characteristics such as height and chest circumference are normally distributed. This observation led him to coin the term "the average man." Quetelet was the first foreign member of the American Statistical Association.

EXAMPLE **1** **Normal Distributions?**

Figure 5.4 shows two distributions: (a) a famous data set of the chest sizes of 5,738 Scottish militiamen collected in about 1846 and (b) the distribution of the population densities of the 50 states. Is either distribution a normal distribution? Explain.

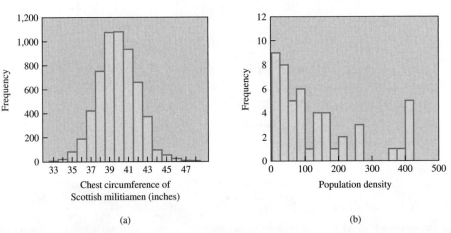

Chest circumference of Scottish militiamen (inches)

Population density

(a)

(b)

Figure 5.4 *Source of* (a): Adolphe Quetelet, *Lettres à S. A. R. le Duc Régnant de Saxe-Cobourg et Gotha*, 1846.

SOLUTION The distribution in Figure 5.4a is nearly symmetric, with a mean between 39 and 40 inches. Values far from the mean are less common, giving it the bell shape of a normal distribution. The distribution in Figure 5.4b shows that most states have low population densities, but a few have much higher densities. This fact makes the distribution right-skewed, so it is not a normal distribution. $\cdots\bullet$

The Normal Distribution and Relative Frequencies

Recall that the total relative frequency for any data set must be 1 (see Section 3.1). Now consider the smooth curve for the normal distribution in Figure 5.2, which is repeated in Figure 5.5. Although we no longer show individual bars, we can still associate the height of the normal curve with the relative frequency. The fact that the relative frequencies must sum to 1 becomes the condition that the area under the normal curve must be 1.

The key idea is this: *The relative frequency for any range of data values is the area under the curve covering that range of values.* For example, a precise calculation shows that the shaded region to the left of −14 days in Figure 5.5 represents about 18% of the total area under the curve. We therefore conclude that the relative frequency is about 0.18 for data values less than −14 days, which means that about 18% of births are more than 14 days early. Similarly, the shaded region to the right of 18 days represents about 12% of the total area under the curve. We therefore conclude that the relative frequency is about 0.12 for data values greater than 18 days, which means that about 12% of births are more than 18 days late. Altogether, we see that 18% + 12% = 30% of all births are either more than 14 days early or more than 18 days late.

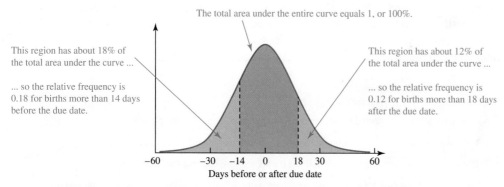

Figure 5.5 The percentage of the total area in any region under the normal curve tells us the relative frequency of data values in that region.

Relative Frequencies and the Normal Distribution

- The area that lies under the normal distribution curve corresponding to a range of values on the horizontal axis is the relative frequency of those values.

- Because the total relative frequency must be 1, the total area under the normal distribution curve must equal 1, or 100%.

EXAMPLE 2   Estimating Areas

Look again at the normal distribution in Figure 5.5.

a. Estimate the percentage of births occurring between 0 and 60 days after the due date.

b. Estimate the percentage of births occurring between 14 days before and 14 days after the due date.

SOLUTION

a. About half of the total area under the curve lies in the region between 0 days and 60 days. This means that about 50% of the births in the sample occur between 0 and 60 days after the due date.

b. Figure 5.5 shows that about 18% of the births occur more than 14 days before the due date. Because the distribution is symmetric, about 18% must also occur more than 14 days after the due date. Therefore, a total of about 18% + 18% = 36% of births occur either more than 14 days before or more than 14 days after the due date. The question asked about the remaining region, which means *between* 14 days before and 14 days after the due date, so this region must represent 100% − 36% = 64% of the births.   · · ●

TIME ⏱ OUT TO THINK

About what percentage of births in Figure 5.5 occur between 14 days early and 18 days late? Explain. (Hint: Remember that the total area under the curve is 100%.)

BY THE WAY

The normal distribution curve is often called a *Gaussian curve* in honor of the 19th-century German mathematician Carl Friedrich Gauss. The American logician Charles Peirce introduced the term *normal distribution* in about 1870.

When Can We Expect a Normal Distribution?

We can better appreciate the importance of the normal distribution if we understand why it is so common. Consider a human characteristic such as height, which closely approximates a normal distribution. Most men or women have heights clustered near the mean height (for their sex), so a data set of heights has a peak at the mean height. But as we consider heights increasingly far from the mean on either side, we find fewer and fewer people. This "tailing off" of heights far from the mean produces the two tails of the normal distribution.

On a deeper level, any quantity that is influenced by many factors is likely to follow a normal distribution. Adult heights are the result of many genetic and environmental factors. Scores on SAT tests or IQ tests tend to be normally distributed because each test score is determined from many individual test questions. Sports statistics, such as batting averages, tend to be normally distributed because they involve many people with many different levels of skill. More generally, we can expect a data set to have a nearly normal distribution if it meets the following conditions.

> **Conditions for a Normal Distribution**
>
> A data set that satisfies the following four criteria is likely to have a nearly normal distribution:
>
> 1. Most data values are clustered near the mean, giving the distribution a well-defined single peak.
>
> 2. Data values are spread evenly around the mean, making the distribution symmetric.
>
> 3. Larger deviations from the mean become increasingly rare, producing the tapering tails of the distribution.
>
> 4. Individual data values result from a combination of many different factors, such as genetic and environmental factors.

EXAMPLE 3   Is It a Normal Distribution?

Which of the following variables would you expect to have a normal or nearly normal distribution?

a. Scores on a very easy test

b. Shoe sizes of a random sample of adult women

c. The number of apples in each of 100 full bushel baskets

SOLUTION

a. Tests have a maximum possible score (100%) that limits the size of data values. If the test is very easy, the mean will be high and many scores will be near the maximum. The fewer lower scores can be spread out well below the mean. We therefore expect the distribution to be left-skewed and non-normal.

b. Foot length is a human trait determined by many genetic and environmental factors. We therefore expect lengths of women's feet to cluster near a mean and become less common farther from the mean, giving the distribution a bell shape, so the lengths have a nearly normal distribution.

c. The number of apples in a bushel basket varies with the size of the apples. We expect that in the distribution there will be a single mode that should be close to the mean number of apples per basket. The number of baskets with more than the mean number of apples should be close to the number of baskets with fewer than the mean number of apples. We therefore expect the number of apples per basket to have a nearly normal distribution.   ⋅⋅●

> TIME 🕐UT TO THINK
>
> Would you expect scores on a moderately difficult exam to have a normal distribution? Suggest two more quantities that you would expect to be normally distributed.

Section 5.1 Exercises

Statistical Literacy and Critical Thinking

1. **Normal Distribution.** When we refer to a "normal" distribution, does the word *normal* have the same meaning as in ordinary language, or does it have a special meaning in statistics? What exactly is a normal distribution?

2. **Normal Distribution.** A normal distribution is informally and loosely described as a probability distribution that is "bell-shaped" when graphed. Draw a rough sketch of the bell shape that characterizes normal distributions.

3. **Random Digits.** Many states have lotteries that involve the random selection of digits 0, 1, 2, ..., 9. Is the distribution of those digits a normal distribution? Why or why not?

4. **Areas.** Birth weights in the United States are normally distributed with a mean (in grams) of 3420 g and a standard deviation of 495 g. If you graph this normal distribution, the area to the right of 4000 g is 0.12. What is the area to the left of 4000 g?

Does It Make Sense? For Exercises 5–8, decide whether the statement makes sense (or is clearly true) or does not make sense (or is clearly false). Explain clearly; not all of these have definitive answers, so your explanation is more important than your chosen answer.

5. **Heights of Women.** A sample of 2,000 women is randomly selected, and it is found that the heights of the women are normally distributed with a mean of 63.6 in.

6. **Brains.** As part of a study of the relationship between brain size and IQ, a random sample of 250 adult males is obtained and their brain volumes are measured and found to be normally distributed.

7. **IQ Scores.** The mean of a normally distributed set of IQ scores is 100, and 60% of the scores are over 105.

8. **Salaries.** An economist plans to obtain the current salaries of all professional football players, and she predicts that those salaries will have a normal distribution.

Concepts and Applications

9. **What Is Normal?** Identify the distribution in Figure 5.6 that is not normal. Of the two normal distributions, which has the larger standard deviation?

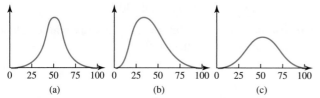

Figure 5.6

10. **What Is Normal?** Identify the distribution in Figure 5.7 that is not normal. Of the two normal distributions, which has the larger standard deviation?

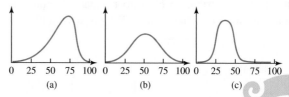

Figure 5.7

Normal Variables. For each of the data sets in Exercises 11–18, state whether you would expect it to be normally distributed. Explain your reasoning.

11. **Weights of Quarters.** The exact weights of a random sample of quarters manufactured in 2012 by the U.S. Mint.

12. **Incomes.** The incomes of randomly selected adults in the United States.

13. **Lottery.** The numbers selected in the Pennsylvania "Match 6" lottery, in which players attempt to match six randomly selected numbers between 1 and 49.

14. **SAT Scores.** All of the SAT scores from last year.

15. **White Blood Cell Counts.** The measured white blood cell counts of 500 randomly selected adult women.

16. **Flight Delays.** The lengths of time that commercial aircraft are delayed before departing.

17. **Waiting Times.** The waiting times at a bus stop if the bus comes once every 10 minutes and you arrive at random times.

18. **Parking Ticket Fines.** The amounts of the fines from parking tickets found on a random sample of 1,000 parked cars.

19. **Movie Lengths.** Figure 5.8 shows a histogram for the lengths of 60 movies. The mean movie length is 110.5 minutes. Is this distribution close to normal? Should this variable have a normal distribution? Why or why not?

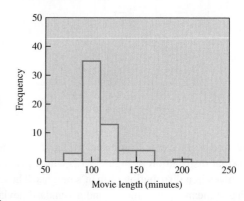

Figure 5.8

20. Pulse Rates. Figure 5.9 shows a histogram for the pulse rates of 98 students. The mean pulse rate is 71.2 beats per minute. Is this distribution close to normal? Should this variable have a normal distribution? Why or why not?

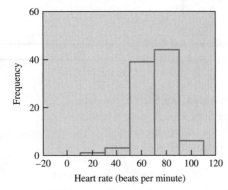

Figure 5.9

21. Quarter Weights. Figure 5.10 shows a histogram for the weights of 50 randomly selected quarters. The mean weight is 5.62 grams. Is this distribution close to normal? Should this variable have a normal distribution? Why or why not?

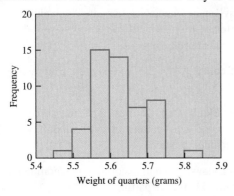

Figure 5.10

22. Aspirin Weights. Figure 5.11 shows a histogram for the weights of 30 randomly selected aspirin tablets. The mean weight is 665.4 milligrams. Is this distribution close to normal? Should this variable have a normal distribution? Why or why not?

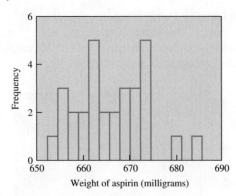

Figure 5.11

23. Areas and Relative Frequencies. Consider the graph of the normal distribution in Figure 5.12, which gives relative frequencies in a distribution of men's heights. The distribution has a mean of 69.6 inches and a standard deviation of 2.8 inches.

a. What is the total area under the curve?

b. Estimate (using area) the relative frequency of values less than 67.

c. Estimate the relative frequency of values greater than 67.

d. Estimate the relative frequency of values between 67 and 70.

e. Estimate the relative frequency of values greater than 70.

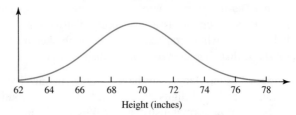

Figure 5.12

24. Areas and Relative Frequencies. Consider the graph of the normal distribution in Figure 5.13, which shows the relative frequencies in a distribution of IQ scores. The distribution has a mean of 100 and a standard deviation of 16.

a. What is the total area under the curve?

b. Estimate (using area) the relative frequency of values less than 100.

c. Estimate the relative frequency of values greater than 110.

d. Estimate the relative frequency of values less than 110.

e. Estimate the relative frequency of values between 100 and 110.

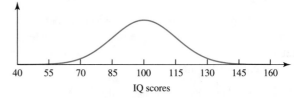

Figure 5.13

25. Estimating Areas. Consider the graph of the normal distribution in Figure 5.14, which illustrates the relative frequencies in a distribution of systolic blood pressures for a sample of female students. The distribution has a standard deviation of 14.

a. What is the mean of the distribution?

b. Estimate (using area) the percentage of students whose blood pressure is less than 100.

c. Estimate the percentage of students whose blood pressure is between 110 and 130.

d. Estimate the percentage of students whose blood pressure is greater than 130.

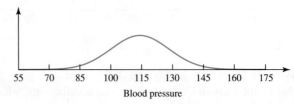

Figure 5.14

26. Estimating Areas. Consider the graph of the normal distribution in Figure 5.15, which gives the relative frequencies in a distribution of body weights for a sample of male students.

  a. What is the mean of the distribution?

  b. Estimate (using area) the percentage of students whose weight is less than 140.

  c. Estimate the percentage of students whose weight is greater than 170.

  d. Estimate the percentage of students whose weight is between 140 and 160.

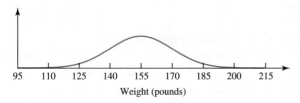

Weight (pounds)

Figure 5.15

PROJECTS FOR THE INTERNET & BEYOND

27. SAT Score Distributions. The College Board Web site gives the distribution of SAT scores (usually in 50-point bins). Collect these data and construct a histogram for each part of the test. Discuss the validity of the claim that SAT scores are normally distributed.

28. Finding Normal Distributions. Using the guidelines given in the text, choose a variable that you think should be nearly normally distributed. Collect at least 30 data values for the variable and make a histogram. Comment on how closely the distribution fits a normal distribution. In what ways does it differ from a normal distribution? Try to explain the differences.

29. Movie Lengths. Collect data to support or refute the claim that movies have gotten shorter over the decades. Specifically, make a histogram of movie lengths for each decade from the 1940s through the present, find the mean movie length for each sample, and comment on whether these distributions are normal. Discuss your results and give plausible reasons for any trends that you observe.

IN THE NEWS

30. Normal Distributions. Rarely does a news article refer to the actual distribution of a variable or state that a variable is normally distributed. Nevertheless, variables mentioned in news reports must have some distribution. Find two variables in news reports that you suspect have nearly normal distributions. Explain your reasoning.

31. Non-normal Distributions. Find two variables in news reports that you suspect *do not* have nearly normal distributions. Explain your reasoning.

5.2   PROPERTIES OF THE NORMAL DISTRIBUTION

Consider a *Consumer Reports* survey in which participants were asked how long they owned their last TV set before they replaced it. The variable of interest in this survey is *replacement time for television sets*. Based on the survey, the distribution of replacement times has a mean of about 8.2 years, which we denote as μ (the Greek letter *mu*). The standard deviation of the distribution is about 1.1 years, which we denote as σ (the Greek letter *sigma*). Making the reasonable assumption that the distribution of TV replacement times is approximately normal, we can picture it as shown in Figure 5.16.

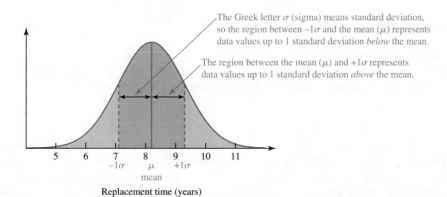

The Greek letter σ (sigma) means standard deviation, so the region between -1σ and the mean (μ) represents data values up to 1 standard deviation *below* the mean.

The region between the mean (μ) and $+1\sigma$ represents data values up to 1 standard deviation *above* the mean.

Replacement time (years)

Figure 5.16 Normal distribution for replacement times for TV sets with a mean of $\mu = 8.2$ years and a standard deviation of $\sigma = 1.1$ years.

TECHNICAL NOTE

A normal distribution can have any value for the mean and any positive value for the standard deviation. The term *standard normal distribution* specifically refers to a normal distribution with a mean of 0 and a standard deviation of 1.

Because all normal distributions have the same bell shape, knowing the mean and standard deviation of a distribution allows us to know much about where the data values lie. For example, if we measure areas under the curve in Figure 5.16, we find that about two-thirds of the area lies within 1 standard deviation of the mean, which in this case is between $8.2 - 1.1 = 7.1$ years and $8.2 + 1.1 = 9.3$ years. Therefore, the TV replacement time is between 7.1 and 9.3 years for about two-thirds of the people surveyed. Similarly, about 95% of the area lies within 2 standard deviations of the mean, which in this case is between $8.2 - 2.2 = 6.0$ years and $8.2 + 2.2 = 10.4$ years. We conclude that the TV replacement time is between 6.0 and 10.4 years for about 95% of the people surveyed.

A simple rule, called the **68-95-99.7 rule**, gives precise guidelines for the percentage of data values that lie within 1, 2, and 3 standard deviations of the mean for any normal distribution. The following box states the rule in words, and Figure 5.17 shows it visually.

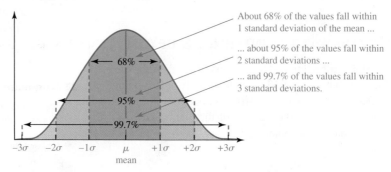

Figure 5.17 Normal distribution illustrating the 68-95-99.7 rule.

The 68-95-99.7 Rule for a Normal Distribution

• About 68% (more precisely, 68.3%), or just over two-thirds, of the data values fall within 1 standard deviation of the mean.

• About 95% (more precisely, 95.4%) of the data values fall within 2 standard deviations of the mean.

• About 99.7% of the data values fall within 3 standard deviations of the mean.

EXAMPLE 1   SAT Scores

The tests that make up the verbal (critical reading) and mathematics parts of the SAT (and the GRE, LSAT, and GMAT) are designed so that their scores are normally distributed with a mean of $\mu = 500$ and a standard deviation of $\sigma = 100$. Interpret this statement.

TECHNICAL NOTE

As discussed in the "Focus on Psychology" section at the end of the chapter, the mean score on a particular component of the SAT may differ from 500 depending on the test and the year it is given.

SOLUTION From the 68-95-99.7 rule, about 68% of students have scores within 1 standard deviation (100 points) of the mean of 500 points; that is, about 68% of students score between 400 and 600. About 95% of students score within 2 standard deviations (200 points) of the mean, or between 300 and 700. And about 99.7% of students score within 3 standard deviations (300 points) of the mean, or between 200 and 800. Figure 5.18 shows this interpretation graphically; note that the horizontal axis shows both actual scores and distance from the mean in standard deviations.

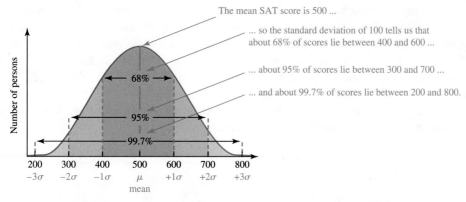

Figure 5.18 Normal distribution for scores on components of the SAT, showing the percentages associated with 1, 2, and 3 standard deviations.

EXAMPLE 2 Detecting Counterfeits

Vending machines can be adjusted to reject coins above and below certain weights. The weights of legal U.S. quarters have a normal distribution with a mean of 5.67 grams and a standard deviation of 0.0700 gram. If a vending machine is adjusted to reject quarters that weigh more than 5.81 grams and less than 5.53 grams, what percentage of legal quarters will be rejected by the machine?

SOLUTION A weight of 5.81 is 0.14 gram, or 2 standard deviations, above the mean. A weight of 5.53 is 0.14 gram, or 2 standard deviations, below the mean. Therefore, by accepting only quarters within the weight range 5.53 to 5.81 grams, the machine accepts quarters that are within 2 standard deviations of the mean and rejects those that are more than 2 standard deviations from the mean. By the 68-95-99.7 rule, 95% of legal quarters will be accepted and 5% of legal quarters will be rejected.

Applying the 68-95-99.7 Rule

We can apply the 68-95-99.7 rule to answer many questions about the frequencies (or relative frequencies) of data values in a normal distribution. Consider an exam taken by 1,000 students for which the scores are normally distributed with a mean of $\mu = 75$ and a standard deviation of $\sigma = 7$. How many students scored above 82?

A score of 82 is 7 points, or 1 standard deviation, above the mean of 75. The 68-95-99.7 rule tells us that about 68% of the scores are *within* 1 standard deviation of the mean. Therefore, about $100\% - 68\% = 32\%$ of the scores are *more than* 1 standard deviation from the mean. Half of this 32%, or 16%, of the scores are more than 1 standard deviation *below* the mean; the other 16% of the scores are more than 1 standard deviation *above* the mean (Figure 5.19a). We conclude that about 16% of 1,000 students, or 160 students, scored above 82.

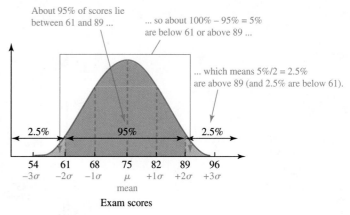

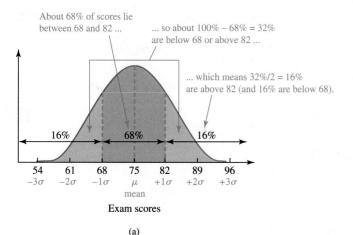

Figure 5.19 A normal distribution of test scores with a mean of 75 and a standard deviation of 7. (a) 68% of the scores lie within 1 standard deviation of the mean. (b) 95% of the scores lie within 2 standard deviations of the mean.

Similarly, suppose we want to know how many students scored below 61. A score of 61 is 14 points, or 2 standard deviations, below the mean of 75. The 68-95-99.7 rule tells us that about 95% of the scores are *within* 2 standard deviations of the mean, so about 5% of the scores are *more than* 2 standard deviations from the mean. Half of this 5%, or 2.5%, of the scores are more than 2 standard deviations *below* the mean (Figure 5.19b), so we conclude that about 2.5% of 1,000 students, or 25 students, scored below 61.

Because 95% of the scores fall between 61 and 89, we sometimes say the scores outside this range are *unusual* because they are relatively rare.

Identifying Unusual Results

In statistics, we often need to distinguish values that are typical, or "usual," from values that are "unusual." By applying the 68-95-99.7 rule, we find that about 95% of all values from a normal distribution lie within 2 standard deviations of the mean. This implies that, among all values, 5% lie more than 2 standard deviations away from the mean. We can use this property to identify values that are relatively "unusual": **Unusual values** are values that are more than 2 standard deviations away from the mean.

EXAMPLE 3   Traveling and Pregnancy

Consider again the question of whether you should advise a pregnant friend to schedule an important business meeting 2 weeks before her due date. Actual data suggest that the number of days between the birth date and the due date is normally distributed with a mean of $\mu = 0$ days and a standard deviation of $\sigma = 15$ days. How would you help your friend make the decision? Would a birth 2 weeks before the due date be considered "unusual"?

SOLUTION   Your friend is assuming that she will *not* have given birth 14 days, or roughly 1 standard deviation, before her due date. But because this outcome is well within 2 standard deviations of the mean, it is *not* unusual. By the 68-95-99.7 rule, the day of birth for 68% of pregnancies is within 1 standard deviation of the mean, or between -15 days and 15 days from the due date. This means that about $100\% - 68\% = 32\%$ of births occur either more than 15 days early or more than 15 days late. Therefore, half of this 32%, or 16%, of all births are more than 15 days early (Figure 5.20). You should tell your friend that about 16% of all births occur more than 15 days before the due date. If your friend likes to think in terms of probability, you could say that there is a 0.16 (about 1 in 6) chance that she will give birth on or before her meeting date.

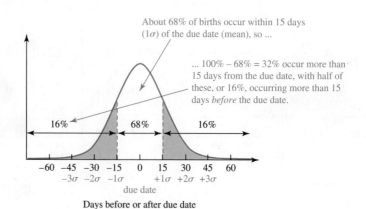

Figure 5.20   About 16% of births occur more than 15 days before the due date.

EXAMPLE 4 Normal Heart Rate

You measure your resting heart rate at noon every day for a year and record the data. You discover that the data have a normal distribution with a mean of 66 and a standard deviation of 4. On how many days was your heart rate below 58 beats per minute?

SOLUTION A heart rate of 58 is 8 (or 2 standard deviations) below the mean. According to the 68-95-99.7 rule, about 95% of the data values are within 2 standard deviations of the mean. Therefore, 2.5% of the data values are more than 2 standard deviations *below* the mean, and 2.5% of the data values are more than 2 standard deviations *above* the mean. On 2.5% of 365 days, or about 9 days, your measured heart rate was below 58 beats per minute.   ·· •

> ### TIME (L)UT TO THINK
> As Example 4 suggests, multiple measurements of the resting heart rate of a *single* individual are normally distributed. Would you expect the average resting heart rates of *many* individuals to be normally distributed? Which distribution would you expect to have the larger standard deviation? Why?

Standard Scores

The 68-95-99.7 rule applies only to data values that are exactly 1, 2, or 3 standard deviations from the mean. For other cases, we can generalize this rule if we know precisely how many standard deviations from the mean a particular data value lies. The number of standard deviations a data value lies above or below the mean is called its **standard score** (or *z-score*), often abbreviated by the letter z. For example:

- The standard score of the mean is $z = 0$, because it is 0 standard deviations from the mean.
- The standard score of a data value 1.5 standard deviations *above* the mean is $z = 1.5$.
- The standard score of a data value 2.4 standard deviations *below* the mean is $z = -2.4$.

The following box summarizes the computation of standard scores.

> ### Computing Standard Scores
> The number of standard deviations a data value lies above or below the mean is called its standard score (or z-score), defined by
>
> $$z = \text{standard score} = \frac{\text{data value} - \text{mean}}{\text{standard deviation}}$$
>
> The standard score is positive for data values above the mean and negative for data values below the mean.

EXAMPLE 5 Standard IQ Scores

The Stanford-Binet IQ test is scaled so that scores have a mean of 100 and a standard deviation of 16. Find the standard scores for IQs of 85, 100, and 125.

SOLUTION We calculate the standard scores for these IQs by using the standard score formula with a mean of 100 and standard deviation of 16.

$$\text{standard score for 85: } z = \frac{85 - 100}{16} = -0.94$$

$$\text{standard score for 100: } z = \frac{100 - 100}{16} = 0.00$$

$$\text{standard score for 125: } z = \frac{125 - 100}{16} = 1.56$$

We can interpret these standard scores as follows: 85 is 0.94 standard deviation *below* the mean, 100 is equal to the mean, and 125 is 1.56 standard deviations *above* the mean. Figure 5.21 shows these values on the distribution of IQ scores.

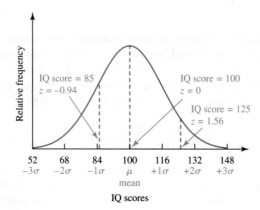

Figure 5.21 Standard scores for IQ scores of 85, 100, and 125.

Standard Scores and Percentiles

Once we know the standard score of a data value, the properties of the normal distribution allow us to find its **percentile** in the distribution. You are probably familiar with the idea of percentiles. For example, if you scored in the 45th percentile on the SAT, 45% of the SAT scores were lower than yours.

> **Percentiles**
>
> The *n*th percentile of a data set is the smallest value in the set with the property that n% of the data values are less than or equal to it. A data value that lies between two percentiles is said to lie in the lower percentile.

We can convert standard scores to percentiles with a *standard score table*, such as Table 5.1 (or see Appendix A for a more detailed table), or with computer software. For each of many standard scores in a normal distribution, the table gives the percentage of values in the distribution *less than or equal to* that value. For example, the table shows that 55.96% of the values in a normal distribution have a standard score less than or equal to 0.15. In other words, a data value with a standard score of 0.15 lies in the 55th percentile.

> ⏻ USING TECHNOLOGY—**STANDARD SCORES IN EXCEL**
>
> Excel's built-in function STANDARDIZE returns the standard score for any data value, given the mean and standard deviation of the distribution. Enter the data in the form "=STANDARDIZE(data value, mean, standard deviation)." The screen shot below shows this function with a table for the data value, mean, and standard deviation from Example 5; the function references the cells (B1, B2, B3) containing those data. Column C shows the results.
>
>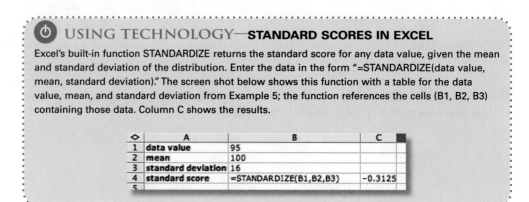

TABLE 5.1	Standard Scores and Percentiles for a Normal Distribution (cumulative values from the *left*)								
Standard score	%	Standard score	%	Standard score	%	Standard score	%
−3.5	0.02	−1.0	15.87	0.0	50.00	1.1	86.43
−3.0	0.13	−0.95	17.11	0.05	51.99	1.2	88.49
−2.9	0.19	−0.90	18.41	0.10	53.98	1.3	90.32
−2.8	0.26	−0.85	19.77	0.15	55.96	1.4	91.92
−2.7	0.35	−0.80	21.19	0.20	57.93	1.5	93.32
−2.6	0.47	−0.75	22.66	0.25	59.87	1.6	94.52
−2.5	0.62	−0.70	24.20	0.30	61.79	1.7	95.54
−2.4	0.82	−0.65	25.78	0.35	63.68	1.8	96.41
−2.3	1.07	−0.60	27.43	0.40	65.54	1.9	97.13
−2.2	1.39	−0.55	29.12	0.45	67.36	2.0	97.72
−2.1	1.79	−0.50	30.85	0.50	69.15	2.1	98.21
−2.0	2.28	−0.45	32.64	0.55	70.88	2.2	98.61
−1.9	2.87	−0.40	34.46	0.60	72.57	2.3	98.93
−1.8	3.59	−0.35	36.32	0.65	74.22	2.4	99.18
−1.7	4.46	−0.30	38.21	0.70	75.80	2.5	99.38
−1.6	5.48	−0.25	40.13	0.75	77.34	2.6	99.53
−1.5	6.68	−0.20	42.07	0.80	78.81	2.7	99.65
−1.4	8.08	−0.15	44.04	0.85	80.23	2.8	99.74
−1.3	9.68	−0.10	46.02	0.90	81.59	2.9	99.81
−1.2	11.51	−0.05	48.01	0.95	82.89	3.0	99.87
−1.1	13.57	0.0	50.00	1.0	84.13	3.5	99.98

Note: The table shows percentiles for standard scores between −3.5 and +3.5, though much lower and higher standard scores are possible. (Appendix A has a more detailed standard score table.) The % column gives the percentage of values in the distribution less than the corresponding standard score.

EXAMPLE 6 Cholesterol Levels

Cholesterol levels in men 18 to 24 years of age are normally distributed with a mean of 178 and a standard deviation of 41.

a. What is the percentile for a 20-year-old man with a cholesterol level of 190?

b. What cholesterol level corresponds to the 90th percentile, the level at which treatment may be necessary?

SOLUTION

a. The *standard score* for a cholesterol level of 190 is

$$z = \text{standard score} = \frac{\text{data value} - \text{mean}}{\text{standard deviation}} = \frac{190 - 178}{41} \approx 0.29$$

Table 5.1 shows that a standard score of 0.29 corresponds to about the 61st percentile.

b. Table 5.1 shows that 90.32% of all data values have a standard score less than 1.3. That is, the 90th percentile is about 1.3 standard deviations above the mean. Given the mean cholesterol level of 178 and the standard deviation of 41, a cholesterol level 1.3 standard deviations above the mean is

$$\underbrace{178}_{\text{mean}} + \underbrace{(1.3 \times 41)}_{\substack{1.3 \text{ standard} \\ \text{deviations}}} = 231.3$$

The 90th percentile begins at a cholesterol level of about 231, so anyone with this level or higher may be in need of treatment.   ·· ●

⏻ **USING TECHNOLOGY—STANDARD SCORES AND PERCENTILES**

Excel The built-in function NORMDIST saves you the work of using a table like Table 5.1. Find a data value's percentile in a normal distribution by entering "=NORMDIST (data value, mean, standard deviation, TRUE)"; the fourth input (TRUE) is necessary to get a percentile result. The screen shot below shows the calculation of the standard score and percentile for the data in Example 6a. Column C shows the results; note that the percentile is displayed as 0.615, so you need to multiply by 100 to express it as a percentage (61.5%).

◇	A	B	C
1	data value	190	
2	mean	178	
3	standard deviation	41	
4	standard score	=STANDARDIZE(B1,B2,B3)	0.29
5	percentile	=NORMDIST(B1,B2,B3,TRUE)	0.615

STATDISK STATDISK can be used in place of Table 5.1. Select **Analysis, Probability Distributions, Normal Distribution.** Either enter the z score to find corresponding areas, or enter the cumulative area from the left to find the z score. After entering a value, click on the **Evaluate** button. The display will include the area to the left of the z score as well as the area to the right of the z score.

TI-83/84 Plus

- *Finding Area:* To find the area between two values, press **2ND** , then press **VARS** to get to the **DISTR** (distribution) menu. Select **normalcdf.** Enter the two values, the mean, and the standard deviation, all separated by commas, as in this format: (left value, right value, mean, standard deviation). *Hint:* If there is no left value, enter the left value as −999999, and if there is no right value, enter the right value as 999999. For example, **normalcdf(80, 85, 100, 15)** returns a value of 0.0674 (rounded), indicating that for a normally distributed population with mean 100 and standard deviation 15, 6.74% of the values are between 80 and 85.

- *Finding x Value:* To find a value corresponding to a known area, press **2ND** , then press **VARS** to get to the **DISTR** (distribution) menu. Select **invNorm,** and proceed to enter the total area to the left of the value, the mean, and the standard deviation in this format with the commas included: (total area to the left, mean, standard deviation). For example, **invNorm(0.4, 100, 15)** returns a value of 96.2 (rounded), indicating that for a normally distributed population with mean 100 and standard deviation 15, the value of 96.2 has an area of 0.4 (or 40%) to its left.

EXAMPLE 7 IQ Scores

IQ scores are normally distributed with a mean of 100 and a standard deviation of 16 (see Example 5). What are the IQ scores for people in the 75th and 40th percentiles on IQ tests?

SOLUTION Table 5.1 shows that the 75th percentile falls *between* standard scores of 0.65 and 0.70; we can estimate that it has a standard score of about 0.67. This corresponds to an IQ that is 0.67 standard deviation, or about $0.67 \times 16 = 11$ points, above the mean of 100. Therefore, a person in the 75th percentile has an IQ of 111. The 40th percentile corresponds to a standard score of approximately −0.25, or a score that is $0.25 \times 16 = 4$ points *below* the mean of 100. So a person in the 40th percentile has an IQ of 96. ⋯●

EXAMPLE 8 Women in the Army

The heights of American women ages 18 to 24 are normally distributed with a mean of 65 inches and a standard deviation of 2.5 inches. In order to serve in the U.S. Army, women must be between 58 inches and 80 inches tall. What percentage of women are ineligible to serve based on their height?

SOLUTION The standard scores for the army's minimum and maximum heights of 58 inches and 80 inches are

$$\text{For 58 inches: } z = \frac{58 - 65}{2.5} = -2.8$$

$$\text{For 80 inches: } z = \frac{80 - 65}{2.5} = 6.0$$

Table 5.1 shows that a standard score of −2.8 corresponds to the 0.26 percentile. A standard score of 6.0 does not appear in Table 5.1, which means it is above the 99.98th percentile

(the highest percentile shown in the table). We conclude that 0.26% of all women are too short to serve in the army and fewer than 0.02% of all women are too tall to serve in the army. Altogether, fewer than about 0.28% of all women, or about 1 out of every 400 women, are ineligible to serve in the army based on their height. $\cdots\bullet$

Toward Probability

Suppose you pick a baby at random and ask whether the baby was born more than 15 days prior to his or her due date. Because births are normally distributed around the due date with a standard deviation of 15 days, we know that 16% of all births occur more than 15 days prior to the due date (see Example 3). For an individual baby chosen at random, we can therefore say that there's a 0.16 chance (about 1 in 6) that the baby was born more than 15 days early. In other words, the properties of the normal distribution allow us to make a *probability statement* about an individual. In this case, our statement is that the probability of a birth occurring more than 15 days early is 0.16.

This example shows that the properties of the normal distribution can be restated in terms of ideas of probability. In fact, much of the work we will do throughout the rest of this text is closely tied to ideas of probability. For this reason, we will devote the next chapter to studying fundamental ideas of probability. But first, in the next section, we will use the basic ideas we have discussed so far to introduce one of the most important concepts in statistics.

Section 5.2 Exercises

Statistical Literacy and Critical Thinking

1. **Standard Score.** Men's heights are normally distributed with a mean of 69.0 in. and a standard deviation of 2.8 in. What is the standard z-score for a man with a height of 69.0 in.?

2. **Standard Score.** The standard score for the height of a male is $z = -2$. Is the height of this male above or below the mean height of all males? How many standard deviations away from the mean is this?

3. **Distributions.** For rolling a die, the mean outcome is 3.5. Can we apply the 68-95-99.7 rule and conclude that 95% of all outcomes fall within 2 standard deviations of 3.5? Why or why not?

4. **z-Scores and Percentages.** Table 5.1 includes standard scores and percentiles. Can a z-score be a negative value? Can a percentile be a negative value? Explain.

Does It Make Sense? For Exercises 5–8, decide whether the statement makes sense (or is clearly true) or does not make sense (or is clearly false). Explain clearly; not all of these have definitive answers, so your explanation is more important than your chosen answer.

5. **Test Scores.** Scores on a statistics test are normally distributed with a mean of 75 and a standard deviation of 75.

6. **Birth Weights**. Birth weights (in grams) in Lichtenstein are normally distributed with a mean of 3,420 g and a standard deviation of 0 g.

7. **Depth Perception Scores.** Scores on a standard test of depth perception are normally distributed with two different modes.

8. **SAT Scores.** SAT scores are normally distributed with a mean of 1518 and a standard deviation of 325.

Concepts and Applications

9. **Using the 68-95-99.7 Rule.** A test of depth perception is designed so that scores are normally distributed with a mean of 50 and a standard deviation of 10. Use the 68-95-99.7 rule to find the following values.

    a. Percentage of scores less than 50

    b. Percentage of scores less than 60

    c. Percentage of scores greater than 70

    d. Percentage of scores greater than 40

    e. Percentage of scores between 40 and 70

10. **Using the 68-95-99.7 Rule.** Assume the resting pulse rates for a sample of individuals are normally distributed with a mean of 70 and a standard deviation of 15. Use the 68-95-99.7 rule to find the following quantities.

    a. Percentage of pulse rates less than 70

    b. Percentage of pulse rates greater than 55

    c. Percentage of pulse rates between 55 and 70

    d. Percentage of pulse rates between 55 and 100

    e. Percentage pulse of rates between 70 and 100

11. **Applying the 68-95-99.7 Rule.** In a study of facial behavior, people in a control group are timed for eye contact in a 5-minute period. Their times are normally distributed with a mean of 184.0 seconds and a standard deviation of 55.0 seconds (based on data from "Ethological Study of Facial Behavior

in Nonparanoid and Paranoid Schizophrenic Patients," by Pittman, Olk, Orr, and Singh, *Psychiatry*, Vol. 144, No. 1). Use the 68-95-99.7 rule to find the indicated quantity.

a. Find the percentage of times within 55.0 seconds of the mean of 184.0 seconds.

b. Find the percentage of times within 110.0 seconds of the mean of 184.0 seconds.

c. Find the percentage of times within 165.0 seconds of the mean of 184.0 seconds.

d. Find the percentage of times between 184.0 seconds and 294 seconds.

12. Applying the 68-95-99.7 Rule. When designing the placement of a CD player in a new model car, engineers must consider the forward grip reach of the driver. Women have forward grip reaches that are normally distributed with a mean of 27.0 inches and a standard deviation of 1.3 inches (based on anthropometric survey data from Gordon, Churchill et al.). Use the 68-95-99.7 rule to find the indicated quantity.

a. Find the percentage of women with forward grip reaches between 24.4 inches and 29.6 inches.

b. Find the percentage of women with forward grip reaches less than 30.9 inches.

c. Find the percentage of women with forward grip reaches between 27.0 inches and 28.3 inches.

IQ Scores. For Exercises 13–24, use the normal distribution of IQ scores, which has a mean of 100 and a standard deviation of 16. Use Table 5.1 (on page 175) to find the indicated quantities. Note: Table 5.1 shows standard scores from -3.5 to $+3.5$. In these problems, for standard scores above 3.5, use a percentile of 99.99%; for all standard scores below -3.5, use a percentile of 0.01%.

13. Percentage of scores greater than 100

14. Percentage of scores less than 116

15. Percentage of scores less than 68

16. Percentage of scores greater than 108

17. Percentage of scores less than 132

18. Percentage of scores less than 92

19. Percentage of scores greater than 76

20. Percentage of scores greater than 148

21. Percentage of scores between 84 and 116

22. Percentage of scores between 68 and 132

23. Percentage of scores between 52 and 116

24. Percentage of scores between 76 and 108

Heights of Women. For Exercises 25–36, use the normal distribution of heights of adult women, which has a mean of 162 centimeters and a standard deviation of 6 centimeters.

Use Table 5.1 to find the indicated quantities. Note: Table 5.1 shows standard scores from -3.5 to $+3.5$. In these problems, for standard scores above 3.5, use a percentile of 99.99%; for all standard scores below -3.5, use a percentile of 0.01%.

25. The percentage of heights greater than 162 centimeters

26. The percentage of heights less than 168 centimeters

27. The percentage of heights greater than 156 centimeters

28. The percentage of heights greater than 171 centimeters

29. The percentage of heights less than 177 centimeters

30. The percentage of heights less than 147 centimeters

31. The percentage of heights less than 144 centimeters

32. The percentage of heights greater than 179 centimeters

33. The percentage of heights between 156 centimeters and 168 centimeters

34. The percentage of heights between 159 centimeters and 165 centimeters

35. The percentage of heights between 148 centimeters and 170 centimeters

36. The percentage of heights between 146 centimeters and 156 centimeters

37. Coin Weights. Consider the following table, showing the official mean weight and estimated standard deviation for five U.S. coins. Suppose a vending machine is designed to reject all coins with weights more than 2 standard deviations above or below the mean. For each coin, find the range of weights that is acceptable to the vending machine. In each case, what percentage of legal coins is rejected by the machine?

Coin	Weight (grams)	Estimated standard deviation (grams)
Cent	2.500	0.03
Nickel	5.000	0.06
Dime	2.268	0.03
Quarter	5.670	0.07
Half dollar	11.340	0.14

38. Pregnancy Lengths. Lengths of pregnancies are normally distributed with a mean of 268 days and a standard deviation of 15 days.

a. What is the percentage of pregnancies that last less than 250 days?

b. What is the percentage of pregnancies that last more than 300 days?

c. If a birth is considered premature if the pregnancy lasts less than 238 days, what is the percentage of premature births?

39. SAT Scores. Based on data from the College Board, SAT scores are normally distributed with a mean of 1518 and a standard deviation of 325.

a. Find the percentage of SAT scores greater than 2000.

b. Find the percentage of SAT scores less than 1500.

c. Find the percentage of SAT scores between 1600 and 2100.

40. GRE Scores. Assume that the scores on the Graduate Record Exam (GRE) are normally distributed with a mean of 497 and a standard deviation of 115.

    a. A graduate school requires a GRE score of 650 for admission. To what percentile does this correspond?

    b. A graduate school requires a GRE score in the 95th percentile for admission. To what actual score does this correspond?

41. Calibrating Barometers. Researchers for a manufacturer of barometers (devices to measure atmospheric pressure) read each of 50 barometers at the same time of day. The mean of the readings is 30.4 (inches of mercury) with a standard deviation of 0.23 inch, and the readings appear to be normally distributed.

    a. What percentage of the barometers read over 31?

    b. What percentage of the barometers read less than 30?

    c. The company decides to reject barometers that read more than 1.5 standard deviations above or below the mean. What is the critical reading below which barometers will be rejected? What is the critical reading above which barometers will be rejected?

    d. What would you take as the actual atmospheric pressure at the time the barometers were read? Explain.

42. Spelling Bee Scores. At the district spelling bee, the 60 girls have a mean score of 71 points with a standard deviation of 6, while the 50 boys have a mean score of 66 points with a standard deviation of 5 points. Those students with a score greater than 75 are eligible to go to the state spelling bee. What percentage of those going to the state bee will be girls?

43. Being a Marine. According to data from the National Health Survey, the heights of adult men are normally distributed with a mean of 69.0 inches and a standard deviation of 2.8 inches. The U.S. Marine Corps requires that men have heights between 64 inches and 78 inches. What percentage of American men are eligible for the Marines based on height?

44. Movie Lengths. Based on a random sample of movie lengths, the mean length is 110.5 minutes with a standard deviation of 22.4 minutes. Assume that movie lengths are normally distributed.

    a. What fraction of movies are more than 2 hours long?

    b. What fraction of movies are less than $1\frac{1}{2}$ hours long?

    c. What is the probability that a randomly selected movie will be less than $2\frac{1}{2}$ hours long?

PROJECTS FOR THE INTERNET & BEYOND

45. Normal Distribution Demonstrations on the Web. Do an Internet search on the keywords "normal distribution" and find an animated demonstration of the normal distribution. Describe how the demonstration works and the useful features that you observed.

46. Estimating a Minute. Ask survey subjects to estimate 1 minute without looking at a watch or clock. Each subject should say "go" at the beginning of the minute and then "stop" when he or she thinks that 1 minute has passed. (Alternatively, you could repeatedly time yourself by looking away from any watch or clock and then noting the correct time when you think that 1 minute has passed.) Use a watch to record the actual times. Construct a graph of the estimates. Is the graph approximately normal? Does the mean appear to be close to 1 minute?

5.3   THE CENTRAL LIMIT THEOREM

A high school English teacher has 100 seniors taking a college placement test. The test is designed to have a mean score of 500 and a standard deviation of 100. Using the methods of this chapter, she can determine the percentage of individual scores that are likely to be above, say, 600. But can she predict anything about the performance of her *group* of 100 students? For example, what is the likelihood that the mean score of the group will be above 600? This type of question, in which we ask about the mean score for a group or sample drawn from a much larger population, can be answered with the *Central Limit Theorem.*

    Before we get to the theorem itself, we can develop some insight by thinking about dice rolling. Suppose we roll *one* die 1,000 times and record the outcome of each roll, which can be the number 1, 2, 3, 4, 5, or 6. Figure 5.22 shows a histogram of outcomes. All six outcomes have roughly the same relative frequency, because the die is equally likely to land in each of the six possible ways. That is, the histogram shows a (nearly) *uniform distribution* (see Section 4.2). Using the methods described in Chapter 4, we can compute that the mean is 3.41 and the standard deviation is 1.73 for this distribution.

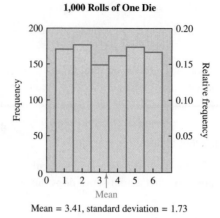

1,000 Rolls of One Die

Mean = 3.41, standard deviation = 1.73

Figure 5.22 Frequency and relative frequency distribution of outcomes from rolling one die 1,000 times.

Now suppose we roll *two* dice 1,000 times and record the *mean* of the two numbers that appear on each roll (Figure 5.23). To find the mean for a single roll, we add the two numbers and divide by 2. For example, if the two dice come up 3 and 5, the mean for the roll is $(3 + 5)/2 = 4$. The possible values of the mean on a roll of two dice are 1.0, 1.5, 2.0, ..., 5.0, 5.5, 6.0.

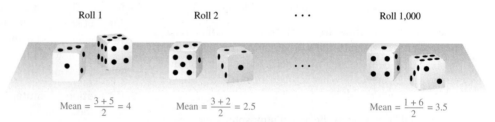

Roll 1   Roll 2   · · ·   Roll 1,000

Mean $= \dfrac{3+5}{2} = 4$ Mean $= \dfrac{3+2}{2} = 2.5$ Mean $= \dfrac{1+6}{2} = 3.5$

Figure 5.23 This diagram represents the idea of rolling two dice 1,000 times and recording the *mean* on each roll. The mean of the values on two dice is their sum divided by 2. This mean can range from $(1 + 1)/2 = 1$ to $(6 + 6)/2 = 6$.

Figure 5.24a shows a typical result of rolling two dice 1,000 times. The most common values in this distribution are the central values 3.0, 3.5, and 4.0. These values are common because they can occur in several ways. For example, a mean of 3.5 can occur if the two dice land as 1 and 6, 2 and 5, 3 and 4, 4 and 3, 5 and 2, or 6 and 1. High and low values occur less frequently because they can occur in fewer ways. For example, a roll can have a mean of 1.0 only if both dice land showing a 1. Again, it is possible to compute the mean and standard deviation for this distribution, and they turn out to be 3.43 and 1.21, respectively.

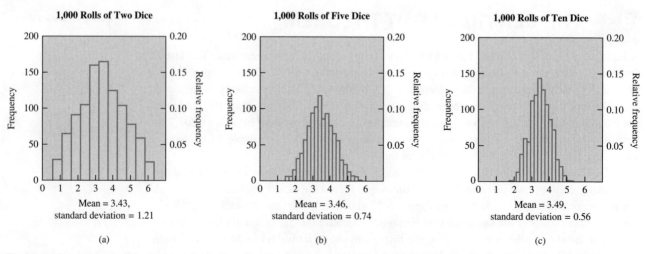

1,000 Rolls of Two Dice

Mean = 3.43,
standard deviation = 1.21

(a)

1,000 Rolls of Five Dice

Mean = 3.46,
standard deviation = 0.74

(b)

1,000 Rolls of Ten Dice

Mean = 3.49,
standard deviation = 0.56

(c)

Figure 5.24 Frequency and relative frequency distributions of sample means from rolling (a) two dice 1,000 times, (b) five dice 1,000 times, and (c) ten dice 1,000 times.

What happens if we increase the number of dice we roll? Suppose we roll five dice 1,000 times and record the mean of the five numbers on each roll. A histogram for this experiment is shown in Figure 5.24b. Once again we see that the central values around 3.5 occur most frequently, but the spread of the distribution is narrower than in the two previous cases. Computing the mean and standard deviation of this distribution, we get values of 3.46 and 0.74, respectively.

If we further increase the number of dice to 10 on each of 1,000 rolls, we find the histogram in Figure 5.24c, which is even narrower. In this case, the mean is 3.49 and standard deviation is 0.56.

Table 5.2 summarizes the four experiments we've described. Columns 2 and 3 of the table refer to a *distribution of means* because, in each of the dice rolling experiments, we recorded the mean of each of 1,000 rolls (that is, the mean of one die on each roll, of two dice on each roll, of five dice on each roll, or of 10 dice on each roll). In other words, the mean for all 1,000 rolls in an experiment is *a mean of the distribution of means* (Column 2). Similarly, the standard deviation for all 1,000 rolls in an experiment is *standard deviation of the distribution of means* (Column 3).

TABLE 5.2	Summary of Dice Rolling Experiments	
Number of dice rolled each time	Mean of the distribution of means	Standard deviation of the distribution of means
1	3.41	1.73
2	3.43	1.21
5	3.46	0.74
10	3.49	0.56

A remarkable insight emerges from these four experiments. Rolling $n = 1$ die 1,000 times can be regarded as taking 1,000 samples of size $n = 1$ from the population of all possible dice rolls. Rolling $n = 2$ dice 1,000 times can be viewed as taking 1,000 samples of size $n = 2$. Likewise, rolling $n = 5$ and $n = 10$ dice 1,000 times is like taking 1,000 samples of size $n = 5$ and $n = 10$, respectively. Table 5.2 shows that as the sample size increases, the mean of the distribution of means approaches the value 3.5 and the standard deviation becomes smaller (making the distribution narrower). More important, the distribution looks more and more like a normal distribution as the sample size increases. This latter fact may seem surprising because we have taken samples from a *uniform* distribution (the outcomes of rolling a single die shown in Figure 5.22), *not* from a normal distribution. Nevertheless, the distribution of means clearly approaches a normal distribution for large sample sizes. This fact is a consequence of the **Central Limit Theorem**.

TECHNICAL NOTE

(1) For practical purposes, the distribution of means will be nearly normal if the sample size is larger than 30. (2) If the original population is normally distributed, then the sample means will be normally distributed for *any* sample size n. (3) In the ideal case, where the distribution of means is formed from *all* possible samples, the mean of the distribution of means *equals* μ and the standard deviation of the distribution of means *equals* $\sigma/\sqrt{n}$.

The Central Limit Theorem

Suppose we take many random samples of size n for a variable with any distribution (not necessarily a normal distribution) and record the distribution of the *means* of each sample. Then,

1. The distribution of means will be approximately a normal distribution for large sample sizes.

2. The mean of the distribution of means approaches the population mean, μ, for large sample sizes.

3. The standard deviation of the distribution of means approaches $\sigma/\sqrt{n}$ for large sample sizes, where σ is the standard deviation of the population.

Be sure to note the third point above: The standard deviation of the distribution of sample means is not the standard deviation of the population, σ, but rather $\sigma/\sqrt{n}$, where n is the size of the samples.

TIME ⏲ OUT TO THINK

Confirm that the standard deviations of the distributions of means given in Table 5.2 for $n = 2, 5, 10$ agree with the prediction of the Central Limit Theorem, given that $\sigma = 1.73$ (the population standard deviation found in Figure 5.22). For example, with $n = 2$, $\sigma/\sqrt{2} = 1.22 \approx 1.21$.

Let's summarize the ingredients of the Central Limit Theorem. We always start with a particular variable, such as the outcomes of rolling a die or the weights of people, that varies randomly over a population. The variable has a certain mean, μ, and standard deviation, σ, which we may or may not know. This variable can have a distribution of *any* shape, not necessarily the shape of a normal distribution. Now we take many samples of that variable, with n items in each sample, and find the mean of each sample (such as the mean value of n dice or the mean weight of a sample of n people). If we then make a histogram of the means from the many samples, we will see a distribution that is close to a normal distribution. The larger the sample size, n, the more closely the distribution of means approximates a normal distribution. Careful study of Figure 5.25 should help solidify these important ideas.

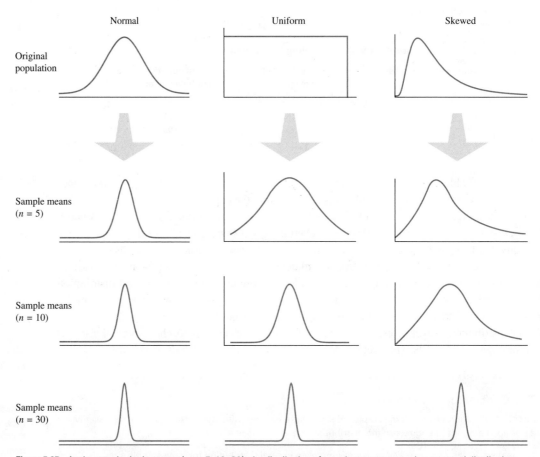

Figure 5.25 As the sample size increases ($n = 5, 10, 30$), the distribution of sample means approaches a normal distribution, regardless of the shape of the original distribution. The larger the sample size, the smaller is the standard deviation of the distribution of sample means.

EXAMPLE ❶ Predicting Test Scores

You are a middle school principal and your 100 eighth-graders are about to take a national standardized test. The test is designed so that the mean score is $\mu = 400$ with a standard deviation of $\sigma = 70$. Assume the scores are normally distributed.

a. What is the likelihood that *one* of your eighth-graders, selected at random, will score below 375 on the exam?

b. Your performance as a principal depends on how well your entire *group* of eighth-graders scores on the exam. What is the likelihood that your group of 100 eighth-graders will have a *mean* score below 375?

SOLUTION

a. In dealing with an individual score, we use the method of standard scores discussed in Section 5.2. Given the mean of 400 and standard deviation of 70, a score of 375 has a standard score of

$$z = \frac{\text{data value} - \text{mean}}{\text{standard deviation}} = \frac{375 - 400}{70} = -0.36$$

According to Table 5.1, a standard score of -0.36 corresponds to about the 36th percentile—that is, 36% of all students can be expected to score below 375. In other words, there is about a 0.36 chance that a randomly selected student will score below 375.

b. The question about the mean of a *group* of students must be handled with the Central Limit Theorem. According to this theorem, if we take random samples of size $n = 100$ students and compute the mean test score of each group, the distribution of means is approximately normal. Moreover, the mean of this distribution is $\mu = 400$ and its standard deviation is $\sigma/\sqrt{n} = 70/\sqrt{100} = 7$. With these values for the mean and standard deviation, the standard score for a mean test score of 375 is

$$z = \frac{\text{data value} - \text{mean}}{\text{standard deviation}} = \frac{375 - 400}{7} = -3.57$$

When you are listening to corn pop, are you hearing the Central Limit Theorem?

—William A. Massey

Table 5.1 shows that a standard score of -3.5 corresponds to the 0.02th percentile, and the standard score in this case is even lower. In other words, fewer than 0.02% of all random samples of 100 students will have a mean score of less than 375. Therefore, the chance that a randomly selected group of 100 students will have a mean score below 375 is less than 0.0002, or about 1 in 5,000.

This example has an important lesson. The likelihood of an *individual* scoring below 375 is more than 1 in 3 (36%), but the likelihood of a *group* of 100 students having a mean score below 375 is less than 1 in 5,000 (0.02%). In other words, there is much more variation in the scores of individuals than in the means of groups of individuals.   ••●

EXAMPLE 2 Salary Equity

The mean salary of the 9,000 employees at Holley Inc. is $\mu = \$26,400$ with a standard deviation of $\sigma = \$2,420$. A pollster samples 400 randomly selected employees and finds that the mean salary of the sample is $26,650. Is it likely that the pollster would get these results by chance, or does the discrepancy suggest that the pollster's results are suspect?

SOLUTION   The question deals with the mean of a *group* of 400 individuals, which is a case for the Central Limit Theorem. The theorem tells us that if we select many groups of 400 individuals and compute the mean of each group, the distribution of means will be close to normal with a mean of $\mu = \$26,400$ and a standard deviation of $\sigma/\sqrt{n} = \$2,420/\sqrt{400} = \$121$. Within the distribution of means, a mean salary of $26,650 has a *standard score* of

$$z = \frac{\text{data value} - \text{mean}}{\text{standard deviation}} = \frac{\$26,650 - \$26,400}{\$121} = 2.07$$

In other words, if we assume that the sample is randomly selected, its mean salary is more than 2 standard deviations above the mean salary of the entire company. According to Table 5.1, a standard score of 2.07 lies near the 98th percentile. Thus, the mean salary of *this* sample is greater than the mean salary we would find in 98% of the possible samples of 400 workers. That is, the likelihood of selecting a group of 400 workers with a mean salary above $26,650 is about 2%, or 0.02. The mean salary of the sample is surprisingly high; perhaps the survey was flawed.   ••●

TIME OUT TO THINK

Would a salary of $26,650 for an *individual* worker lie above or below the 98th percentile? Explain.

The Value of the Central Limit Theorem

The Central Limit Theorem allows us to say something about the mean of a group if we know the mean, μ, and the standard deviation, σ, of the entire population. This can be useful, but it turns out that the opposite application is far more important.

Two major activities of statistics are making estimates of population means and testing claims about population means. Suppose we do *not* know the mean of a variable for the entire population. Is it possible to make a good estimate of the population mean (such as the mean income of all Internet users) knowing only the mean of a much smaller sample? As you can probably guess, being able to answer this type of question lies at the heart of statistical sampling, especially in polls and surveys. The Central Limit Theorem provides the key to answering such questions. We will return to this topic in Chapter 8.

Section 5.3 Exercises

Statistical Literacy and Critical Thinking

1. **Texas Lottery.** In each drawing for the Texas Pick 3 lottery, 3 digits between 0 and 9 (inclusive) are randomly selected. What is the distribution of the selected digits? If the mean is calculated for each drawing, can the distribution of the sample means be treated as a normal distribution?

2. **Notation.** In this section, it was noted that the standard deviation of sample means is $\sigma/\sqrt{n}$. In that expression, what does σ represent and what does n represent?

3. **Central Limit Theorem.** A process consists of repeating this operation: Randomly select two values from a normally distributed population and then find the mean of the two values. Will the sample means be normally distributed, even though each sample has only two values?

4. **Central Limit Theorem.** The population of unemployed adults has ages with mean μ and standard deviation σ. Samples of unemployed adults are randomly selected so that there are exactly 100 in each sample. For each sample, the mean age is computed. What does the Central Limit Theorem tell us about the distribution of those mean ages?

Concepts and Applications

5. **IQ Scores and the Central Limit Theorem.** IQ scores are normally distributed with a mean of 100 and a standard deviation of 16. Assume that many samples of size n are taken from a large population of people and the mean IQ score is computed for each sample.

   a. If the sample size is $n = 64$, find the mean and standard deviation of the distribution of sample means.

   b. If the sample size is $n = 100$, find the mean and standard deviation of the distribution of sample means.

   c. Why is the standard deviation in part a different from the standard deviation in part b?

6. **SAT Scores and the Central Limit Theorem.** Based on data from the College Board, assume that SAT scores are normally distributed with a mean of 1518 and a standard deviation of 325. Assume that many samples of size n are taken from a large population of students and the mean SAT score is computed for each sample.

   a. If the sample size is $n = 100$, find the mean and standard deviation of the distribution of sample means.

   b. If the sample size is $n = 2,500$, find the mean and standard deviation of the distribution of sample means.

   c. Why is the standard deviation in part a different from the standard deviation in part b?

7. **Twelve-Sided Dice and the Central Limit Theorem.** Rolling a fair *12-sided* die produces a uniformly distributed set of numbers between 1 and 12 with a mean of 6.5 and a standard deviation of 3.452. Assume that n 12-sided dice are rolled many times and the mean of the n outcomes is computed each time.

   a. Find the mean and standard deviation of the resulting distribution of sample means for $n = 81$.

   b. Find the mean and standard deviation of the resulting distribution of sample means for $n = 100$.

   c. Why is the standard deviation in part a different from the standard deviation in part b?

8. **Ten-Sided Dice and the Central Limit Theorem.** Rolling a fair *10-sided* die produces a uniformly distributed set of numbers between 1 and 10 with a mean of 5.5 and a standard deviation of 2.872. Assume that n 10-sided dice are rolled many times and the mean of the n outcomes is computed each time.

   a. Find the mean and standard deviation of the resulting distribution of sample means for $n = 49$.

   b. Find the mean and standard deviation of the resulting distribution of sample means for $n = 400$.

   c. Why is the standard deviation in part a different from the standard deviation in part b?

Male Weights. In Exercises 9–12, assume that weights of men are normally distributed with a mean of 170 lb and a standard deviation of 30 lb (approximate values based on data from the National Health and Nutrition Examination Survey).

9. What percentage of individual men have weights less than 185 lb? If samples of 36 men are randomly selected and the mean weight is computed for each sample, what percentage of the sample means are less than 185 lb?

10. What percentage of individual men have weights greater than 167 lb? If samples of 100 men are randomly selected and the mean weight is computed for each sample, what percentage of the sample means are greater than 167 lb?

11. What percentage of individual men have weights between 164 lb and 176 lb? If samples of 100 men are randomly selected and the mean weight is computed for each sample, what percentage of sample means is between 164 lb and 176 lb?

12. What percentage of individual men have weights between 176 lb and 185 lb? If samples of 25 men are randomly selected and the mean weight is computed for each sample, what percentage of sample means is between 176 lb and 185 lb?

13. **Amounts of Cola.** Assume that cans of cola are filled so that the actual amounts are normally distributed with a mean of 12.00 ounces and a standard deviation of 0.11 ounce.

   a. What is the likelihood that a sample of 36 cans will have a mean amount of at least 12.05 ounces?

   b. Given the result in part a, is it reasonable to believe that the cans are actually filled with a mean of 12.00 ounces? If the mean is not 12.00 ounces, are consumers being cheated?

14. **Designing Strobe Lights.** An aircraft strobe light is designed so that the times between flashes are normally distributed with a mean of 3.00 seconds and a standard deviation of 0.40 second.

   a. What is the likelihood that an individual time is greater than 4.00 seconds?

   b. What is the likelihood that the mean for 60 randomly selected times is greater than 4.00 seconds?

   c. Given that the strobe light is intended to help other pilots see an aircraft, which result is more relevant for assessing the safety of the strobe light: the result in part a or the result in part b? Why?

15. **Designing Motorcycle Helmets.** Engineers must consider the breadths of male heads when designing motorcycle helmets for men. Men have head breadths that are normally distributed with a mean of 6.0 inches and a standard deviation of 1.0 inch (based on anthropometric survey data from Gordon, Churchill et al.).

   a. If one male is randomly selected, what is the likelihood that his head breadth is less than 6.2 inches?

   b. The Safeguard Helmet company plans an initial production run of 100 helmets. How likely is it that 100 randomly selected men have a mean head breadth of less than 6.2 inches?

   c. The production manager sees the result in part b and reasons that all helmets should be made for men with head breadths of less than 6.2 inches, because they would fit all but a few men. What is wrong with that reasoning?

16. **Staying Out of Hot Water.** In planning for hot water requirements, the manager of the Luxurion Hotel finds that guests spend a mean of 11.4 minutes each day in the shower (based on data from the Opinion Research Corporation). Assume that the shower times are normally distributed with a standard deviation of 2.7 minutes.

   a. Find the percentage of guests who shower for more than 12 minutes.

   b. The hotel has installed a system that can provide enough hot water provided that the mean shower time for 84 guests is less than 12 minutes. If the hotel currently has 84 guests, how likely is it that there will not be enough hot water? Does the current system appear to be effective?

17. **Redesign of Ejection Seats.** When women were allowed to become pilots of fighter jets, engineers needed to redesign the ejection seats because they had been designed for men only. The ACES-II ejection seats were designed for men weighing between 140 pounds and 211 pounds. The population of women has normally distributed weights with a mean of 143 pounds and a standard deviation of 29 pounds (based on data from the National Health Survey).

   a. What percentage of women have weights between 140 pounds and 211 pounds?

   b. If 36 women are randomly selected, how likely is it that their mean weight is between 140 pounds and 211 pounds?

   c. For redesigning the fighter jet ejection seats to better accommodate women, which probability is more relevant: the result in part a or the result in part b? Why?

18. **Labeling of M&M Packages.** M&M plain candies have weights that are normally distributed with a mean weight

of 0.8565 gram and a standard deviation of 0.0518 gram (based on measurements from one of the authors). A random sample of 100 M&M candies is obtained from a package containing 465 candies; the package label states that the net weight is 396.9 grams. (If every package has 465 candies, the mean weight of the candies must exceed $396.9/465 = 0.8535$ for the net contents to weigh at least 396.9 grams.)

a. If 1 M&M plain candy is randomly selected, how likely is it that it weighs more than 0.8535 gram?

b. If 465 M&M plain candies are randomly selected, how likely is it that their mean weight is at least 0.8535 gram?

c. Given these results, does it seem that the Mars Company is providing M&M consumers with the amount claimed on the label?

19. **Vending Machines.** Currently, quarters have weights that are normally distributed with a mean of 5.670 grams and a standard deviation of 0.062 gram. A vending machine is configured to accept only those quarters with weights between 5.550 grams and 5.790 grams.

a. If 280 different quarters are inserted into the vending machine, what is the expected number of rejected quarters?

b. If 280 different quarters are inserted into the vending machine, how likely is it that the mean falls between the limits of 5.550 grams and 5.790 grams?

c. If you owned the vending machine, which result would concern you more: the result in part a or the result in part b? Why?

20. **Aircraft Safety Standards.** Federal Aviation Administration rules require airlines to estimate the weight of a passenger as 195 pounds, including carry-on baggage. Men have weights (without baggage) that are normally distributed with a mean of 172 pounds and a standard deviation of 29 pounds.

a. If one adult male is randomly selected and is assumed to have 20 pounds of carry-on baggage, how likely is it that his total weight is greater than 195 pounds?

b. If a Boeing 767-300 aircraft is full of 213 adult male passengers and each is assumed to have 20 pounds of carry-on baggage, how likely is it that the mean passenger weight (including carry-on baggage) is greater than 195 pounds? Does a pilot have to be concerned about exceeding this weight limit?

21. **Kindergarten Desks.** In designing new desks for an incoming class of 25 kindergarten girls, an important characteristic of the desks is that they must accommodate the sitting heights of those students. (The sitting height is the height

of a seated student measured from the bottom of the feet to the top of the knee.) Kindergarten girls have normally distributed sitting heights with a mean of 61.2 cm and a standard deviation of 3.1 cm.

a. What percentage of individual kindergarten girls have sitting heights greater than 63 cm?

b. For randomly selected groups of 25 kindergarten girls, what percentage of the group means is greater than 63 cm?

c. Which result is more relevant for the design of the desks: the result from part a or part b? Why?

22. **Blood Pressure in Women.** Systolic blood pressure for women between the ages of 18 and 24 is normally distributed with a mean of 114.8 (millimeters of mercury) and a standard deviation of 13.1.

a. What is the likelihood that an individual woman has a blood pressure above 125?

b. Suppose a random sample of $n = 300$ women is selected and the mean blood pressure for the sample is computed. What is the likelihood that the mean blood pressure for the sample will be above 125?

c. Suppose a random sample of $n = 300$ women is selected and the mean blood pressure for the sample is computed. What is the likelihood that the mean blood pressure for the sample will be below 114?

PROJECTS FOR THE INTERNET & BEYOND

23. **Central Limit Theorem on the Internet.** Doing an Internet search on "central limit theorem" will uncover many sites devoted to this subject. Find a site that has animated demonstrations of the Central Limit Theorem. Describe in your own words what you observed and how it illustrates the Central Limit Theorem.

24. **The Quincunx on the Internet.** Do an Internet search on "central limit theorem" or "quincunx" and find a site that has an animated demonstration of the quincunx (or Galton's board). Describe the quincunx and explain how it illustrates the Central Limit Theorem.

25. **Dice Rolling.** Demonstrate the Central Limit Theorem using dice, as discussed in this section. Give each person in your class as many dice as possible. Begin by rolling one die and making a histogram of the outcomes. Then let every person roll two dice and make a histogram of the mean for each roll. Increase the number of dice in each roll as long as dice and time allow. Comment on the appearance of the histogram at each stage.

CHAPTER REVIEW EXERCISES

1. For each of the following situations, state whether the distribution of values is likely to be a normal distribution. Give a brief explanation justifying your choice.

   a. Numbers resulting from spins of a roulette wheel. (There are 38 equally likely slots with numbers 0, 00, 1, 2, 3, ..., 36.)

   b. Weights of 12-year old girls

   c. Scores on a test designed to measure knowledge of current events

2. The College of Portland uses a diagnostics test to correctly place students in math courses. Scores on the test are normally distributed with a mean of 400 and a standard deviation of 50.

   a. Using the 68-95-99.7 rule, find the percentage of scores within 50 of the mean of 400.

   b. Using the 68-95-99.7 rule, find the percentage of scores between 300 and 500.

   c. Is a score of 550 unusual? Why or why not?

3. Assume that body temperatures of healthy adults are normally distributed with a mean of 98.20°F and a standard deviation of 0.62°F (based on data from University of Maryland researchers).

   a. If you have a body temperature of 99.00°F, what is your percentile score?

   b. Convert 99.00°F to a standard score (or z-score).

   c. Is a body temperature of 99.00°F "unusual"? Why or why not?

   d. Fifty adults are randomly selected. What is the likelihood that the mean of their body temperatures is 97.98°F or lower?

   e. A person's body temperature is found to be 101.00°F. Is this result "unusual"? Why or why not? What should you conclude?

   f. What body temperature is the 95th percentile?

   g. What body temperature is the 5th percentile?

   h. Bellevue Hospital in New York City uses 100.6°F as the lowest temperature considered to indicate a fever. What percentage of normal and healthy adults would be considered to have a fever? Does this percentage suggest that a cutoff of 100.6°F is appropriate?

   i. If, instead of assuming that the mean body temperature is 98.20°F, we assume that the mean is 98.60°F (as many people believe), what is the chance of randomly selecting 106 people and getting a mean of 98.20°F or lower? (Continue to assume that the standard deviation is 0.62°F.) University of Maryland researchers did get such a result. What should we conclude?

1. Which of the following statements are correct?

   a. A normal distribution is any distribution that typically occurs.

   b. The graph of a normal distribution is bell-shaped.

   c. The graph of a normal distribution has one mode.

   d. In a normal distribution, the mean and median are equal.

   e. In a normal distribution, the standard deviation is 0.

2. Car seat belts are tested for strength, and it is found that the results are normally distributed with a mean of 2,400 lb and a standard deviation of 400 lb. Using the 68-95-99.7 rule, find the percentage of scores between 1,600 lb and 3,200 lb.

3. Car seat belts are tested for strength, and it is found that the results are normally distributed with a mean of 2,400 lb and a standard deviation of 400 lb. Use Table 5.1 to find the percentage of scores between 1,600 lb and 3,200 lb.

4. A population is normally distributed with a mean of 850 and a standard deviation of 49, and samples of size 100 are randomly selected. What is the mean of the sample means?

5. A population is normally distributed with a mean of 850 and a standard deviation of 49, and samples of size 100 are randomly selected. What is the standard deviation of the sample means?

6. A population is normally distributed with a mean of 850 and a standard deviation of 49. What is the standard z-score corresponding to 801?

7. A population is normally distributed with a mean of 850 and a standard deviation of 49. What percentage of scores are greater than 801?

8. A population is normally distributed with a mean of 850 and a standard deviation of 49. If 2.28% of the scores are less than 752, what percentage of scores are greater than 752?

9. A population is normally distributed with a mean of 850 and a standard deviation of 49. Find the percentage of scores between 801 and 948.

10. Which of the following is likely to have a distribution that is closest to a normal distribution?

    a. The outcomes that occur when a single die is rolled many times

    b. The outcomes that occur when two dice are rolled many times and the mean is computed each time

    c. The outcomes that occur when five dice are rolled many times and the mean is computed each time

FOCUS ON EDUCATION

What Can We Learn from SAT Trends?

The Scholastic Aptitude Test (SAT) has been taken by college-bound high school students since 1941, when 11,000 students took the first test. Today it is taken by more than 2 million students each year. Until 2006, there were only two parts to the SAT: a "verbal" test and a mathematics test. In 2006, the verbal test was changed to become the "critical reading" test and writing was added as a third test.

The scores on each part of the SAT were originally scaled to have a mean of 500 and a standard deviation of 100. The maximum and minimum scores on each part are 800 and 200, respectively, which are 3 standard deviations above and below the mean. (Scores more than 3 standard deviations above or below the mean are assigned the maximum 800 or minimum 200.)

The mean does not stay at 500 from year to year, however. Each year's test shares some common questions with prior years' tests, and the College Board (which designs the test) uses these common questions to compare each year's test results to those from prior years. The intent is for a score of 500 to represent the same level of achievement at all times, so scores can be compared from one year to the next. For example, if the mean score in a particular year is higher than 500, it implies students performed better on average than did students at the time that the average was set to 500.

Trends in SAT scores are widely used to assess the general state of American education. Figure 5.26 shows the average scores on the verbal/critical reading (meaning the verbal test prior to 2006 and the critical reading test since) and math parts of the SAT between 1972 and 2010; results for the writing test are not shown, because there are not yet enough data to look for trends over time.

The trends show that if the scores from year to year are truly comparable, then students taking the SAT in recent years have significantly poorer verbal skills than students of a few decades ago, though somewhat better mathematical skills. But before we accept this conclusion, we must answer two important questions:

1. Are trends among the *sample* of high school students who take the SAT representative of trends for the *population* of all high school students?

2. Aside from the small number of common questions that link one year to the next, the test changes every year. Can test scores from one year legitimately be compared to scores in other years?

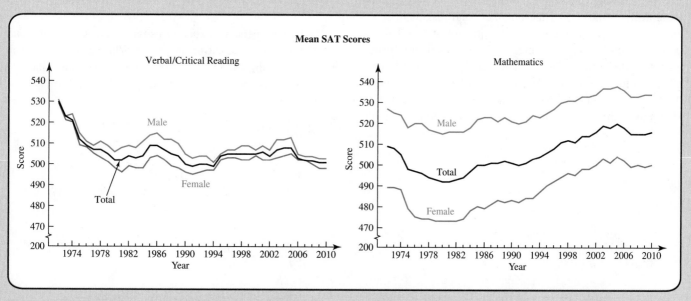

Figure 5.26 Verbal/critical reading and mathematics SAT scores, 1972–2010. Scores for years before 1996 are shown at their "recentered" values, rather than their original values. *Source:* The College Board.

Unfortunately, neither question can be answered with a clear yes. For example, only about a third of high school graduates took the SAT in the 1970s, while more than 50% of high school graduates take it today. If these samples represent the top tier of high school students, the decline in verbal scores might simply reflect the fact that a greater range of student abilities is represented in later samples.

The second question is even more difficult. Even if some test questions stay the same from one year to the next, different material may be emphasized in high schools, thereby changing the likelihood that students will answer the same questions correctly from one year to the next.

Further complications arise when the test undergoes major changes. For example, in 1994 the use of calculators was allowed for the first time on the mathematics section, and additional time was allotted for the entire test (the verbal test also underwent significant changes that year). Is it a coincidence that mathematics scores began an upward trend starting that year, or could it be that the test became "easier" as a result of the changes? Similarly, in 2006 the combined critical reading and mathematics scores suffered their largest decline in more than 30 years. Critics attributed the decline to the fact that the SAT became longer and harder as a result of adding the writing test, while the College Board attributed the decline to a decrease in the number of repeat test-takers. (Repeat test-takers tend to raise the average score, because individual students make significant gains in their scores when they take the test a second or third time.) Clearly, changes in the structure of the test make it difficult to determine whether changes in average scores reflect real changes in education.

Perhaps even more significantly, the calibration of the SAT test underwent a major change in 1996. By that time, scores had declined significantly from the originally planned mean of 500: The mean verbal score had fallen to about 420, while the mean mathematics score had fallen to about 470. Because the range of possible scores goes only from 200 to 800, these declines meant that the minimum and maximum scores no longer represented the same number of standard deviations from the mean, which created difficulties with statistical analysis. The College Board therefore decided to "recenter" all scores to a mean of 500 in 1996. The recentering affected different percentiles differently, but it effectively added about 80 points to the mean of the verbal scores and 30 points to the mean of the mathematics scores. The scores shown in Figure 5.26 are the "recentered" scores. For example, Figure 5.26 shows a mean verbal score of almost exactly 500 in 1994, but if you look back at news reports from that time, you'll see that the actual mean verbal score was about 420 in 1994.

The College Board argues that its statistical analysis has been done with great care, and that despite recentering and changes to the test, trends in the SAT reflect real changes in education. Critics generally agree that the statistical analysis has been done well, but argue that other factors, such as the change in the test-taking population and changes in our education system, make the comparisons invalid. For students, however, the key fact is that many colleges continue to use SAT scores as a part of their decision on whether to admit applicants. As long as that remains the case, the SAT will remain one of the nation's most important tests—and the debate over its merits will surely continue.

QUESTIONS FOR DISCUSSION

1. Do you think that comparisons of SAT scores over two years are meaningful? Over 10 years? Do you agree that the long-term trends indicate that students today have poorer verbal skills but better mathematical skills than those of a few decades ago? Defend your opinions.

2. The downward trend in verbal skills was a major reason why the College Board added the writing test in 2006. The hope was that adding the new test would cause students to focus more on writing in high school and therefore improve both in writing and in overall verbal abilities. What do you think of this rationale? Do you think it will be successful?

3. Notice that scores for males have been consistently higher than scores for females. Why do you think this is the case? Do you think that changes in our education system could eliminate this gap? Defend your opinions.

4. Discuss personal experiences with the SAT among your classmates. Based on these personal experiences, what do you think the SAT is measuring? Do you think the test is a reasonable way to predict students' performance in college? Why or why not?

FOCUS ON PSYCHOLOGY

Are We Smarter than Our Parents?

Most kids tend to think that they're smarter than their parents, but is it possible that they're right? If you believe the results of IQ tests, not only are we smarter than our parents, on average, but our parents are smarter than our grandparents. In fact, almost all of us would have ranked as geniuses if we'd lived a hundred years ago. Of course, before any of us start touting our Einstein-like abilities, it would be good to investigate what lies behind this startling claim.

The idea of an IQ, which stands for *intelligence quotient*, was invented by French psychologist Alfred Binet (1857–1911). Binet created a test that he hoped would identify children in need of special help in school.[*] He gave his test to many children and then calculated each child's IQ by dividing the child's "mental age" by his or her physical age (and multiplying by 100). For example, a 5-year-old child who scored as well as an average 6-year-old was said to have a mental age of 6, and therefore an IQ of $(6 \div 5) \times 100$, or 120. Note that, by this definition, IQ tests make sense only for children. However, later researchers, especially psychologists for the U.S. Army, extended the idea of IQ so that it could be applied to adults as well.

Today, IQ is defined by a normal distribution with a mean of 100 and a standard deviation of 16. Traditionally, psychologists have classified people with an IQ below 70 (about 2 standard deviations below the mean of 100) as "intellectually deficient," and people who score above 130 (about 2 standard deviations above the mean) as "intellectually superior."

You're probably aware of the controversy that surrounds IQ tests, which boils down to two key issues:

- Do IQ tests measure intelligence or something else?
- If they do measure intelligence, is it something that is innate and determined by heredity or something that can be molded by environment and education?

A full discussion of these issues is too involved to cover here, but a surprising trend in IQ scores sheds light on these issues. As we'll see shortly, the trend is quite pronounced, but it was long hidden because of the way IQ tests are scored. There are several different, competing versions of IQ tests, and most of them are regularly changed and updated. But in all cases, the scores are adjusted to fit a normal distribution with a mean of 100 and standard deviation of 16. In other words, the scoring of

[*]Binet himself assumed that intelligence could be molded and warned against taking his tests as a measure of any innate or inherited abilities. However, many later psychologists concluded that IQ tests could measure innate intelligence, which led to their being used for separating school children, military recruits, and many other groups of people according to supposed intellectual ability.

an IQ test is essentially done in the same way that an instructor might grade an exam "on a curve." Because of this adjustment, the mean on IQ tests is *always* 100, which makes it impossible for measured IQ scores to rise and fall with time.

Nevertheless, a few IQ tests have not been changed and updated substantially over time, and even tests that have changed considerably often still repeat some old questions. In the early 1980s, a political science professor named Dr. James Flynn began to look at the raw, unadjusted scores on unchanged tests and questions. The results were astounding.

Dr. Flynn found that raw scores have been steadily rising, although the precise amount of the rise varies somewhat with the type of IQ test. The highest rates of increase are found on tests that purport to measure abstract reasoning abilities (such as the "Raven's" tests). For these tests, Dr. Flynn found that the unadjusted IQ scores of people in industrialized countries have been rising at a rate of about 6 points per decade. In other words, a person who scored 100 on a test given in 2010 would have scored about 106 on a test in 2000, 112 on a test in 1990, and so on. Over a hundred years, this would imply a rise of some 60 points, suggesting that someone who scores an "intellectually deficient" IQ of 70 today would have rated an "intellectually superior" IQ of 130 a century ago.

This long-term trend toward rising scores on IQ tests is now called the *Flynn effect*. It is present for all types of IQ tests, though not always to the same degree as with abstract reasoning tests. For example, Figure 5.27 shows how results changed on one of the most widely used IQ tests (the Stanford-Binet test) between 1932 and 1997. Note that, in terms of unadjusted scores, the mean rose by 20 points during that time period. In other words, if people who scored an IQ of 100 on a 1997 test were instead scored on a 1932 test, they would rate an IQ of 120. As the figure shows, about one-fourth of the 1997 test-takers would have rated "intellectually superior" on the 1932 test. There is some evidence that the rise in scores may have begun to slow or halt in more recent years, though the data are still subject to debate.

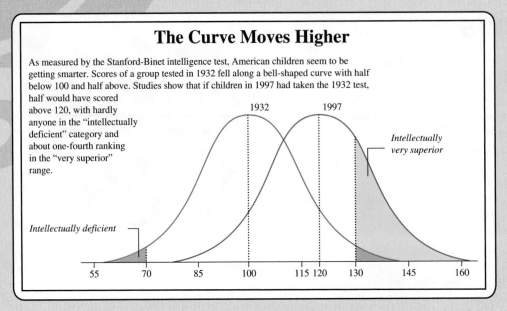

The Curve Moves Higher

As measured by the Stanford-Binet intelligence test, American children seem to be getting smarter. Scores of a group tested in 1932 fell along a bell-shaped curve with half below 100 and half above. Studies show that if children in 1997 had taken the 1932 test, half would have scored above 120, with hardly anyone in the "intellectually deficient" category and about one-fourth ranking in the "very superior" range.

1932   1997

Intellectually very superior

Intellectually deficient

55   70   85   100   115 120   130   145   160

Figure 5.27 *Source:* Adapted from the *New York Times,* based on data from Neisser, Ulric (ed.). *The Rising Curve: Long-Term Gains in IQ and Related Measures.* American Psychological Association, 1998.

Many other scientists have investigated the Flynn effect, and there is general agreement that the long-term trend is real. The implication is clear: Whatever IQ tests measure, people today really *do* have more of it than people just a few decades ago. If IQ tests measure intelligence, then it means we really are smarter than our parents (on average), who in turn are smarter than our grandparents (on average).

Of course, if IQ tests don't measure "intelligence" but only measure some type of skill, then the rise in scores may indicate only that today's children have more practice at that skill than past children. The fact that the greatest rise is seen on tests of abstract thinking lends some support to this idea. These tests often involve such problems as solving puzzles and looking for patterns among sets of shapes, and these types of problems are now much more common in games than they were in the past.

While the Flynn effect does not answer the question of whether IQ tests measure intelligence, it may tell us one important thing: If IQs really have been rising as the Flynn effect suggests, then IQ cannot be an entirely inherited trait, because inherited traits cannot change that much in just a few decades. That is, if IQ tests are measuring intelligence, then intelligence can be molded by environmental as well as hereditary factors.

Dr. Flynn's discovery has already changed the way psychologists look at IQ tests, and it is sure to remain an active topic of research. Moreover, given the many uses to which modern society has put IQ tests, the Flynn effect is likely to have profound social and political consequences as well. So back to our starting question: Are we smarter than our parents? We really can't say, but we can certainly hope so, because it will take a lot of brainpower to solve the problems of the future.

QUESTIONS FOR DISCUSSION

1. Which explanation do you favor for the Flynn effect: that people are getting smarter or that people are getting more practice at the skills measured on IQ tests? Defend your opinion.

2. The rise in performance on IQ tests contrasts sharply with a steady decline in performance over the past few decades on many tests that measure factual knowledge, such as the SAT. Think of several possible ways to explain these contrasting results, and form an opinion as to the most likely explanation.

3. Results on IQ tests tend to differ among different ethnic groups. Some people have used this fact to argue that some ethnic groups tend to be intellectually superior to others. Can such an argument still be supported in light of the Flynn effect? Defend your opinion.

4. Discuss some of the common uses of IQ tests. Do you think that IQ tests *should* be used for these purposes? Does the Flynn effect alter your thoughts about the uses of IQ tests? Explain.

• • • • • • • • • • • • • • • •

Probability in Statistics

Most statistical studies seek to learn something about a *population* from a much smaller *sample*. Therefore, a key question in any statistical study is whether it is valid to generalize from a sample to the population. To answer this question, we must understand the likelihood, or probability, that what we've learned about the sample also applies to the population. In this chapter, we will focus on a few basic ideas of probability that are commonly used in statistics. As you will see, these ideas of probability also have many applications in their own right.

Probability is the very guide of life.

—Cicero (106–43 BC)

6.1 THE ROLE OF PROBABILITY IN STATISTICS: STATISTICAL SIGNIFICANCE

To see why probability is so important in statistics, imagine that you are trying to test whether a coin is fair—that is, whether it is equally likely to land on heads or tails. If you toss the coin 100 times and get 52 heads and 48 tails, should you conclude that the coin is unfair? No. While we should expect to see *roughly* 50 heads and 50 tails in every 100 tosses of a fair coin, we also expect some variation from one sample of 100 tosses to another. We're not surprised to observe a small deviation from a perfect 50-50 split between heads and tails, because we expect small deviations to occur *by chance*.

In contrast, suppose you toss a coin 100 times and the results are 20 heads and 80 tails. This is such a substantial deviation from a 50-50 split that you'd probably conclude that the coin is unfair. In other words, while it's *possible* that you observed a rare set of 100 tosses with a fair coin, it's more likely that the coin is unfair. When the difference between what is observed and what is expected seems unlikely to be explained by chance alone, we say the difference is **statistically significant**.

> **Definition**
>
> A set of measurements or observations in a statistical study is said to be **statistically significant** if it is unlikely to have occurred by chance.

EXAMPLE 1 Likely Stories?

a. *A detective in Detroit finds that 25 of the 62 guns used in crimes during the past week were sold by the same gun shop.* This finding is statistically significant. Because there are many gun shops in the Detroit area, having 25 out of 62 guns come from the same shop seems unlikely to have occurred by chance.

b. *In terms of the global average temperature, the 10 years from 2001–2010 were 10 of the 11 hottest years on record to that time, using a record going back to 1880.* Having 10 of the 11 hottest years on record occur in a single decade is statistically significant. Indeed, having such a streak of hot years is very unlikely to have occurred by chance alone, and therefore provides significant evidence of a warming world.

c. *The team with the worst win-loss record in basketball wins one game against the defending league champions.* This one win is *not* statistically significant because although we expect a team with a poor win-loss record to lose most of its games, we also expect it to win occasionally, even against the defending league champions. ··●

From Sample to Population

Let's look at the idea of statistical significance in an opinion poll. Suppose that in a poll of 1,000 randomly selected people, 51% support the President. A week later, in another poll with a different randomly selected sample of 1,000 people, only 49% support the President. Should you conclude that the opinions of Americans changed during the one week between the polls?

You can probably guess that the answer is *no*. The poll results are *sample statistics* (see Section 1.1): 51% of the people in the first sample support the President. We can use this result to estimate the *population parameter*, which is the percentage of *all* Americans who support the President. At best, if the first poll were conducted well, it would say that the percentage of Americans who support the President is *close* to 51%. Similarly, the 49% result in the second poll means that the percentage of Americans supporting the President is *close* to 49%. Because the two sample statistics differed only slightly (51% versus 49%), it's quite possible that the

TECHNICAL NOTE

The difference between 49% and 51% is not statistically significant for typical polls, but it can be for a poll involving a very large sample size. In general, any difference can be significant if the sample size is large enough.

real percentage of Americans supporting the President did not change at all. Instead, the two polls reflect expected and reasonable differences between the two samples.

In contrast, suppose the first poll found that 75% of the sample supported the President, and the second poll found that only 30% supported the President. Assuming the polls were carefully conducted, it's highly unlikely that two groups of 1,000 randomly chosen people could differ so much by chance alone. In this case, we would look for another explanation. Perhaps Americans' opinions about the President really did change in the week between the polls.

In terms of statistical significance, the change from 51% to 49% in the first set of polls is *not* statistically significant, because we can reasonably attribute this change to chance variations between the two samples. However, in the second set of polls, the change from 75% to 30% is statistically significant, because it is unlikely to have occurred by chance.

EXAMPLE 2 Statistical Significance in Experiments

A researcher conducts a double-blind experiment that tests whether a new herbal formula is effective in preventing colds. During a 3-month period, the 100 randomly selected people in a treatment group take the herbal formula while the 100 randomly selected people in a control group take a placebo. The results show that 30 people in the treatment group get colds, compared with 32 people in the control group. Can we conclude that the herbal formula is effective in preventing colds?

SOLUTION   Whether a person gets a cold during any 3-month period depends on many unpredictable factors. Therefore, we should not expect the number of people with colds in any two groups of 100 people to be exactly the same. In this case, the difference between 30 people getting colds in the treatment group and 32 people getting colds in the control group is small enough to be explainable by chance. That is, the difference is not statistically significant, and we should not conclude that the treatment is effective.   ·· ●

Quantifying Statistical Significance

In Example 2, we said that the difference between 30 colds in the treatment group and 32 colds in the control group was not statistically significant. This conclusion was fairly obvious because the difference was so small. But suppose that 24 people in the treatment group had colds, compared with the 32 in the control group. Would the difference between 24 and 32 be large enough to be considered statistically significant? The definition of statistical significance that we've been using so far is too vague to answer this question. We need a way to quantify the idea of statistical significance.

In general, we determine statistical significance by using probability to quantify the likelihood that a result may have occurred by chance. We ask a question like this one: *Is the probability that the observed difference occurred by chance less than or equal to 0.05 (or 1 in 20)?* If the answer is *yes* (the probability is less than or equal to 0.05), then we say that the difference is *statistically significant at the 0.05 level*. If the answer is *no*, the observed difference is reasonably likely to have occurred by chance, so we say that it is not statistically significant.

The choice of 0.05 is somewhat arbitrary, but it's a figure that statisticians frequently use. Nevertheless, other probabilities are sometimes used, such as 0.1 or 0.01. Statistical significance at the 0.01 level is stronger than significance at the 0.05 level, which is stronger than significance at the 0.1 level.

He that leaves nothing to chance will do few things ill, but will do very few things.

—George Savile Halifax

> **Quantifying Statistical Significance**
>
> • If the probability of an observed difference occurring by chance is 0.05 (or 1 in 20) or less, the difference is statistically significant at the 0.05 level.
>
> • If the probability of an observed difference occurring by chance is 0.01 (or 1 in 100) or less, the difference is statistically significant at the 0.01 level.

You can probably see that caution is in order when working with statistical significance. We would expect roughly 1 in 20 trials to give results that are statistically significant at the 0.05 level even when the results actually occurred by chance. In other words, statistical significance at the 0.05 level—or at almost any level, for that matter—is *no guarantee* that an important effect or difference is present.

TIME OUT TO THINK

Suppose an experiment finds that people taking a new herbal remedy get fewer colds than people taking a placebo, and the results are statistically significant at the 0.01 level. Has the experiment *proven* that the herbal remedy works? Explain.

EXAMPLE 3 Polio Vaccine Significance

In the test of the Salk polio vaccine (see Section 1.1), 33 of the 200,000 children in the treatment group got paralytic polio, while 115 of the 200,000 in the control group got paralytic polio. Calculations show that the probability of this difference between the groups occurring by chance is less than 0.01. Describe the implications of this result.

SOLUTION The results of the polio vaccine test are statistically significant at the 0.01 level, meaning that there is a 0.01 chance (or less) that the difference between the control and treatment groups occurred by chance. Therefore, we can be fairly confident that the vaccine really was responsible for the fewer cases of polio in the treatment group. (In fact, the probability of the Salk results occurring by chance is *much* less than 0.01, so researchers were quite convinced that the vaccine worked; as we'll discuss in Chapter 9, this probability is called a "*P*-value.") ⋯●

Section 6.1 Exercises

Statistical Literacy and Critical Thinking

1. **Statistical Significance.** In an experiment testing a method of gender selection intended to increase the likelihood that a baby is a girl, 1,000 couples give birth to 501 girls and 499 boys. A company representative argues that this is evidence that the method is effective, because the probability of getting 501 girls in 1,000 births by chance is only 0.025, which is less than 0.05. Do you agree with this argument? Why or why not?

2. **Statistical Significance.** Does the term *statistical significance* refer to results that are significant in the sense that they have great importance? Explain.

3. **Statistical Significance.** If a particular result is statistically significant at the 0.05 level, must it also be statistically significant at the 0.01 level? Why or why not?

4. **Statistical Significance.** If a particular result is statistically significant at the 0.01 level, must it also be statistically significant at the 0.05 level? Why or why not?

Does It Make Sense? For Exercises 5–8, decide whether the statement makes sense (or is clearly true) or does not make sense (or is clearly false). Explain clearly; not all of these have definitive answers, so your explanation is more important than your chosen answer.

5. **Unemployment.** The rate of unemployment has statistical significance because it has such a strong effect on the economy.

6. **Statistical Significance.** In an experiment testing a method of gender selection, 1,000 couples give birth to 550 girls and 450 boys. Because the probability of getting such extreme results by chance is only 0.0009, the results are statistically significant.

7. **Gender Selection.** In a test of a technique of gender selection, the 100 babies born consist of at least 80 girls. Because there is about 1 chance in a billion of getting at least 80 girls among 100 babies, the results are statistically significant.

8. **Clinical Trial.** In a clinical trial of a treatment for reducing back pain, the difference between the treatment group and the control group (with no treatment) was found to be statistically significant. This means that the treatment will definitely ease back pain for everyone.

Concepts and Applications

Subjective Significance. For each event in Exercises 9–16, state whether the difference between what occurred and what you would have expected by chance is statistically significant. Discuss any implications of the statistical significance.

9. Statistics Class Survey. In a survey, 50 students are randomly selected among all college students currently taking a statistics course, and it is found that they are all females.

10. Voter Survey. In a pre-election survey, 100 likely voters are randomly selected from adults in the United States, and it is found that 6% of them are Democrats.

11. Rolls of a Die. In 6 rolls of a six-sided die, the outcome of 6 never occurs.

12. Lottery. A common lottery game is to select four digits, each between 0 and 9 (inclusive). Winning requires that you get the same four digits that are drawn, and they must be in the same order. You select four digits for one ticket, and none of your selections matches the numbers drawn.

13. Survey. In conducting a survey of adults in the United States, a pollster claims that he randomly selected 20 subjects and all of them were women.

14. Subway Riders. A commuter enters a New York City subway car near Times Square and finds that it is occupied by 50 men, all of whom are bald.

15. Jury Composition. For a trial on a charge of failure to pay child support, the jury consists of exactly 6 men and 6 women.

16. Clinical Trial. In a clinical trial of a new drug intended to treat allergies, 5 of the 80 subjects in the treatment group experienced headaches, and 8 of the 160 subjects in the control group experienced headaches.

17. Fuel Tests. Thirty identical cars are selected for a fuel test. Half of the cars are filled with regular gasoline, and the other half are filled with a new experimental fuel. The cars in the first group average 29.3 miles per gallon, while the cars in the second group average 35.5 miles per gallon. Discuss whether this difference seems statistically significant.

18. Carpal Tunnel Syndrome Treatments. An experiment was conducted to determine whether there is a difference between the success rates from treating carpal tunnel syndrome with surgery and with splinting. The success rate for 73 patients treated with surgery was 92%, and the success rate for 83 patients treated with splints was 72% (based on data from "Splinting vs. Surgery in the Treatment of Carpal Tunnel Syndrome" by Gerritsen et al., *Journal of the American Medical Association*, Vol. 288, No. 10). Discuss whether this difference appears to be statistically significant.

19. Gender Selection. The Genetics and IVF Institute conducted a clinical trial of its method for gender selection. The latest actual results showed that among 945 babies born to couples using the XSORT method of gender selection, 879 were girls.

20. Bednets and Malaria. In a randomized controlled trial in Kenya, insecticide-treated bednets were tested as a way to reduce malaria. Among 343 infants who used the bednets, 15 developed malaria. Among 294 infants not using bednets, 27 developed malaria (based on data from "Sustainability of Reductions in Malaria Transmission and Infant Mortality in Western Kenya with Use of Insecticide-Treated Bednets" by Lindblade et al., *Journal of the American Medical Association*, Vol. 291, No. 21). Assuming that the bednets have no effect, there is a probability of 0.015 of getting these results by chance. Do the results appear to have statistical significance? Do the bednets appear to be effective?

21. Human Body Temperature. In a study by researchers at the University of Maryland, the body temperatures of 106 individuals were measured; the mean for the sample was 98.20°F. It is commonly believed that the mean body temperature is 98.60°F. The difference between the sample mean and the accepted value is significant at the 0.05 level.

a. Discuss the meaning of the significance level in this case.

b. If we assume that the mean body temperature is actually 98.6°F, the probability of getting a sample with a mean of 98.20°F or less is 0.000000001. Interpret this probability value.

22. Seat Belts and Children. In a study of children injured in automobile crashes (*American Journal of Public Health*, Vol. 82, No. 3), those wearing seat belts had a mean stay of 0.83 day in an intensive care unit. Those not wearing seat belts had a mean stay of 1.39 days. The difference in means between the two groups is significant at the 0.0001 level. Interpret this result.

23. SAT Preparation. A study of 75 students who took an SAT preparation course (*American Education Research Journal*, Vol. 19, No. 3) concluded that the mean improvement on the SAT was 0.6 point. If we assume that the preparation course has no effect, the probability of getting a mean improvement of 0.6 point by chance is 0.08. Discuss whether this preparation course results in statistically significant improvement.

24. Weight by Age. A National Health Survey determined that the mean weight of a sample of 804 men ages 25 to 34 years was 176 pounds, while the mean weight of a sample of 1,657 men ages 65 to 74 years was 164 pounds. The difference is significant at the 0.01 level. Interpret this result.

PROJECTS FOR THE INTERNET & BEYOND

25. Significance in Vital Statistics. Visit a Web site that has vital statistics (for example, the U.S. Census Bureau or the National Center for Health Statistics). Choose a question such as one of the following:

- Are there significant differences in numbers of births among months?
- Are there significant differences in numbers of natural deaths among days of the week?
- Are there significant differences in infant mortality rates among selected states?
- Are there significant differences in incidences of a particular disease among selected states?
- Are there significant differences among the marriage rates in various states?

Collect the relevant data and determine subjectively whether you think the observed differences are significant. Explain why, and provide possible explanations for the significance or lack of significance.

26. Lengths of Rivers. Using an almanac or the Internet, find the lengths of the principal rivers of the world. Construct a list of the leading digits only. Does any particular digit occur more often than the others? Does that digit occur significantly more often? Explain.

IN THE NEWS

27. Statistical Significance. Find a recent newspaper article on a statistical study in which the idea of statistical significance is used. Write a one-page summary of the study and the result that is considered to be statistically significant. Also include a brief discussion of whether you believe the result, given its statistical significance.

28. Significant Experiment? Find a recent news story about a statistical study that used an experiment to determine whether some new treatment was effective. Based on the available information, briefly discuss what you can conclude about the statistical significance of the results. Given this significance (or lack thereof), do you think the new treatment is useful? Explain.

29. Personal Statistical Significance. Describe an incident in your own life that did not meet your expectation, defied the odds, or seemed unlikely to have occurred by chance. Would you call this incident statistically significant? To what did you attribute the event?

6.2   BASICS OF PROBABILITY

In this world, nothing is certain but death and taxes.

—Benjamin Franklin

In Section 6.1, we saw that ideas of probability are fundamental to statistics, in part through the concept of statistical significance. We will return to the topic of statistical significance in Chapters 7 through 10. First, however, we need to explore a few essential ideas of probability and their applications in everyday life.

Let's begin by considering a toss of two coins. Figure 6.1 shows that there are four different ways the coins could fall. We say that each of these four ways is a different **outcome** of the coin toss. The outcomes are the most basic possible results of the coin toss. But suppose we are interested only in the number of heads. Because the two middle outcomes in Figure 6.1 each have 1 head, we say that these two outcomes represent the same **event**. That is, an event describes one or more possible outcomes that all have the same property of interest—in this case, the same number of heads. Figure 6.1 shows that there are four possible outcomes for the two-coin toss, but only three possible events: 0 heads, 1 head, and 2 heads.

> **Definition**
>
> **Outcomes** are the most basic possible results of observations or experiments.
>
> An **event** is a collection of one or more outcomes that share a property of interest.

Outcome	🪙 🪙	🪙 🪙	🪙 🪙	🪙 🪙
Event	0 heads	1 head	1 head	2 heads

Figure 6.1   The four possible outcomes for a toss of two coins. The two middle outcomes both represent the same *event* of 1 head.

Mathematically, we express probabilities as numbers between 0 and 1. For example, the probability of a coin landing on heads is one-half, or 0.5. If an event is impossible, such as the event of being in two places at the same time, we assign it a probability of 0. At the other extreme, an event that is certain to occur, such as the event of the Sun being above the horizon in the daytime, is given a probability of 1. Figure 6.2 shows the scale of probability values, along with common expressions of likelihood.

It's helpful to have some special notation for probability. We write P(event) to mean the probability of an event, and often denote events by letters or symbols. For example, if we use H to represent the event of a head on a coin toss, then we write the probability of heads as $P(\text{H}) = 0.5$.

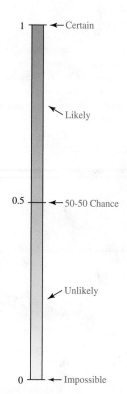

Figure 6.2 The scale shows various degrees of certainty as expressed by probabilities.

Expressing Probability

The probability of an event, expressed as P(event), is always between 0 and 1 (inclusive). A probability of 0 means that the event is impossible and a probability of 1 means that the event is certain.

Theoretical Probabilities

There are three basic techniques for finding probabilities, called the *theoretical method*, the *relative frequency method*, and the *subjective method*. We'll begin with the theoretical method.

When we say that the probability of heads on a coin toss is 1/2, we are assuming that the coin is fair and is equally likely to land on heads or tails. In essence, the probability is based on a theory of how the coin behaves, so we say that the probability of 1/2 comes from the theoretical method. As another example, consider rolling a single die. Because there are six equally likely outcomes (Figure 6.3), the theoretical probability for each outcome is 1/6.

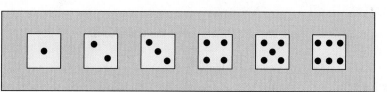

Figure 6.3 The six possible outcomes for a roll of one six-sided die.

As long as all outcomes are equally likely, we can use the following procedure to calculate theoretical probabilities.

Theoretical Method for Equally Likely Outcomes

Step 1. Count the total number of possible outcomes.

Step 2. Among all the possible outcomes, count the number of ways the event of interest, A, can occur.

Step 3. Determine the probability, $P(A)$, from

$$P(A) = \frac{\text{number of ways } A \text{ can occur}}{\text{total number of outcomes}}$$

EXAMPLE 1 Guessing Birthdays

You select a person at random from a large group at a conference. What is the probability that the person selected has a birthday in July? Assume 365 days in a year.

SOLUTION If we assume that all birthdays are equally likely, we can use the three-step theoretical method.

Step 1. Each possible birthday represents an outcome, so there are 365 possible outcomes.

Step 2. July has 31 days, so 31 of the 365 possible outcomes represent the event of a July birthday.

Step 3. The probability that a randomly selected person has a birthday in July is

$$P(\text{July birthday}) = \frac{31}{365} \approx 0.0849$$

which is slightly more than 1 in 12.   ...●

Counting Outcomes

Suppose we toss two coins and want to count the total number of outcomes. The toss of the first coin has two possible outcomes: heads (H) or tails (T). The toss of the second coin also has two possible outcomes. The two outcomes for the first coin can each occur with either of the two outcomes for the second coin, making a total of $2 \times 2 = 4$ possible outcomes for the two coins together. The tree diagram of Figure 6.4a shows an easy way to identify the four outcomes: HH, HT, TH, and TT.

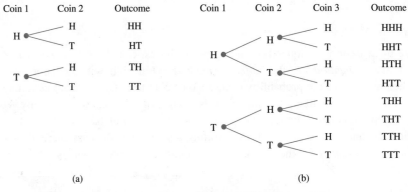

Figure 6.4   Tree diagrams showing the outcomes of tossing (a) two and (b) three coins.

> ### TIME ⏲ OUT TO THINK
> Are the outcomes for tossing one coin twice in a row the same as those for tossing two coins at the same time? Explain.

We can now extend this thinking. If we toss three coins, there are $2 \times 2 \times 2 = 8$ possible outcomes, all shown in Figure 6.4b. This idea is the basis for the following counting rule.

> ### Counting Outcomes
> Suppose process A has *a* possible outcomes and process B has *b* possible outcomes. Assuming the outcomes of the processes do not affect each other, the number of different outcomes for the two processes combined is $a \times b$. This idea extends to any number of processes. For example, if a third process C has *c* possible outcomes, the number of possible outcomes for the three processes combined is $a \times b \times c$.

EXAMPLE ❷   Dice Counting

a. How many outcomes are there if you roll two fair dice?

b. What is the probability of rolling two 1's (snake eyes) when two fair dice are rolled?

SOLUTION

a. Rolling a single die has six equally likely outcomes (see Figure 6.3). Therefore, when two fair dice are rolled, there are $6 \times 6 = 36$ different outcomes.

b. Of the 36 possible outcomes for rolling two fair dice, only one is the event of interest (two 1's). Therefore, the probability of rolling two 1's is

$$P(\text{two 1's}) = \frac{\text{number of ways two 1's can occur}}{\text{total number of outcomes}} = \frac{1}{36} = 0.0278$$

··●

EXAMPLE 3 Counting Children

What is the probability that a randomly selected family with three children has two girls and one boy? Assume that births of boys and girls are equally likely.

SOLUTION We apply the three-step theoretical method.

Step 1. There are two possible outcomes for each birth: boy (B) or girl (G). For a family with three children, the total number of possible outcomes (birth orders) is $2 \times 2 \times 2 = 8$: BBB, BBG, BGB, BGG, GBB, GBG, GGB, GGG.

Step 2. Of these eight possible outcomes, three of them have two girls and one boy: BGG, GBG, and GGB.

Step 3. Therefore, the probability that a family with three children has two girls and one boy is

$$P(\text{two girls}) = \frac{\text{number of outcomes with two girls}}{\text{total number of outcomes}} = \frac{3}{8} = 0.375$$

··●

BY THE WAY

Births of boys and girls are *not* equally likely. Naturally, there are approximately 105 male births for every 100 female births. However, male death rates are higher than female death rates, so female adults outnumber male adults.

TIME ◷UT TO THINK

How many different four-child families are possible if birth order is taken into account? What is the probability of a couple having a four-child family with four girls?

Relative Frequency Probabilities

A second way to determine probabilities is to *approximate* the probability of an event A by making many observations and counting the number of times event A occurs. This approach is called the **relative frequency method** (or *empirical* method). For example, if we observe that it rains an average of 100 days per year, we might say that the probability of rain on a randomly selected day is 100/365. We apply this method as follows.

Relative Frequency Method

Step 1. Repeat or observe a process many times and count the number of times the event of interest, A, occurs.

Step 2. Estimate $P(A)$ by

$$P(A) = \frac{\text{number of times } A \text{ occurred}}{\text{total number of observations}}$$

EXAMPLE 4 500-Year Flood

Geological records indicate that a river has crested above a particular high flood level four times in the past 2,000 years. What is the relative frequency probability that the river will crest above the high flood level next year?

SOLUTION   Based on the data, the probability of the river cresting above this flood level in any single year is

$$\frac{\text{number of years with flood}}{\text{total number of years}} = \frac{4}{2,000} = \frac{1}{500}$$

Because a flood of this magnitude occurs on average once every 500 years, it is called a "500-year flood." The probability of having a flood of this magnitude in any given year is 1/500, or 0.002.

··●

Subjective Probabilities

The third method for determining probabilities is to estimate a **subjective probability** using experience or intuition. For example, you could make a subjective estimate of the probability that a friend will be married in the next year or of the probability that a good grade in statistics will help you get the job you want.

BY THE WAY

Another approach to finding probabilities, called the *Monte Carlo method*, uses computer simulations. This technique essentially finds relative frequency probabilities; in this case, observations are made by causing a computer to behave in a way that is essentially the same as the actual event.

> **Three Approaches to Finding Probability**
>
> A **theoretical probability** is based on assuming that all outcomes are equally likely. It is determined by dividing the number of ways an event can occur by the total number of possible outcomes.
>
> A **relative frequency probability** is based on observations or experiments. It is the relative frequency of the event of interest.
>
> A **subjective probability** is an estimate based on experience or intuition.

EXAMPLE 5   Which Method?

Identify the method that resulted in the following statements.

a. The chance that you'll get married in the next year is zero.

b. Based on government data, the chance of dying in an automobile accident is about 1 in 8,000 (per year).

c. The chance of rolling a 7 with a 12-sided die is 1/12.

SOLUTION

a. This is a subjective probability because it is based on a feeling at the current moment.

b. This is a relative frequency probability because it is based on observed data on past automobile accidents.

BY THE WAY

Theoretical methods are also called *a priori* methods. The words *a priori* are Latin for "before the fact" or "before experience."

c. This is a theoretical probability because it is based on assuming that a fair twelve-sided die is equally likely to land on any of its 12 sides.

··●

Probability of an Event *Not* Occurring

Suppose we are interested in the probability that a particular event or outcome does *not* occur. For example, consider the probability of a wrong answer on a multiple-choice question with five possible answers. The probability of answering correctly with a random guess is 1/5, so the probability of *not* answering correctly is 4/5. Notice that the sum of the two probabilities must be 1, because the answer must be either right or wrong. We can generalize this idea.

> **Probability of an Event *Not* Occurring**
>
> Suppose the probability of an event A is $P(A)$. Then the probability that event A does *not* occur is $P(\text{not } A) = 1 - P(A)$. Note: The event *not A* is called the **complement** of the event A; the "not" is often designated by a bar, so $\overline{A}$ means *not A*.

EXAMPLE ⑥ Is Scanner Accuracy the Same for Specials?

In a study of checkout scanning systems, samples of purchases were used to compare the scanned prices to the posted prices. Table 6.1 summarizes results for a sample of 819 items. Based on these data, what is the probability that a regular-priced item has a scanning error? What is the probability that an advertised-special item has a scanning error?

TABLE 6.1	Scanner Accuracy	
	Regular-priced items	**Advertised-special items**
Undercharge	20	7
Overcharge	15	29
Correct price	384	364

Source: Ronald Goodstein, "UPC Scanner Pricing Systems: Are They Accurate?" *Journal of Marketing,* Vol. 58.

SOLUTION   We can let R represent a regular-priced item being scanned correctly. Because 384 of the 419 regular-priced items are correctly scanned,

$$P(R) = \frac{384}{419} = 0.916$$

The event of a scanning error is the complement of the event of a correct scan, so the probability of a regular-priced item being subject to a scanning error (either undercharged or overcharged) is

$$P(\text{not } R) = 1 - 0.916 = 0.084$$

Now let A represent an advertised-special item being scanned correctly. Of the 400 advertised-special items in the sample, 364 are scanned correctly. Therefore,

$$P(A) = \frac{364}{400} = 0.910$$

The probability of an advertised-special item being scanned incorrectly is

$$P(\text{not } A) = 1 - 0.910 = 0.090$$

The error rates are nearly equal. However, Table 6.1 also shows that the errors differed in their effects on customers: Most of the errors made with advertised-special items were overcharges, while most of the errors with regular-priced items were undercharges.   •·●

Probability Distributions

In Chapters 3 through 5, we worked with frequency and relative frequency distributions—for example, distributions of age or income. One of the most fundamental ideas in probability and statistics is that of a *probability distribution*. As the name suggests, a probability distribution is a distribution in which the variable of interest is associated with a probability.

　Suppose you toss two coins simultaneously. As we found earlier (see Figure 6.4a), there are a total of four possible outcomes for the two coins: HH, HT, TH, and TT. Notice, however, that these four outcomes represent only three different events: one outcome represents 2 heads (HH), one outcome represents 2 tails (TT), and two outcomes (HT, TH) represent 1 head and 1 tail. Therefore, the probability of two heads is $P(\text{HH}) = 1/4 = 0.25$; the probability of two tails is $P(\text{TT}) = 1/4 = 0.25$; and the probability of one head and one tail is $P(\text{H and T}) = 2/4 = 0.50$. We can display this **probability distribution** as a table (Table 6.2) or a histogram (Figure 6.5). Note that the sum of all the probabilities must be 1 (because exactly one of the possible results must occur).

BY THE WAY

A *lot* is a person's portion or allocation. The term came to be used for any object or marker that identified a person in a selection by chance. The casting of lots was used in the Trojan Wars to determine who would throw the first spear and in the Old Testament to divide conquered lands.

TABLE 6.2	Tossing Two Coins
Event	**Probability**
2 heads, 0 tails	0.25
1 head, 1 tail	0.50
0 heads, 2 tails	0.25
Total	**1**

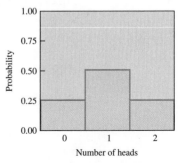

Figure 6.5 Histogram showing the probability distribution for the results of tossing two coins.

BY THE WAY

Another common way to express likelihood is to use odds. The odds *against* an event are the ratio of the probability that the event does not occur to the probability that it does occur. For example, the odds against rolling a 6 with a fair die are (5/6)/(1/6), or 5 to 1. The odds used in gambling are called *payoff odds*; they express your net gain on a winning bet. For example, suppose that the payoff odds on a particular horse at a horse race are 3 to 1. This means that for each $1 you bet on this horse, you will gain $3 if the horse wins (and get your original $1 back).

Making a Probability Distribution

A **probability distribution** represents the probabilities of all possible events. Do the following to make a display of a probability distribution:

Step 1. List all possible *outcomes*. Use a table or figure if it is helpful.

Step 2. Identify outcomes that represent the same *event*. Find the probability of each event.

Step 3. Make a table in which one column lists each event and another column lists each probability. The sum of all the probabilities must be 1.

EXAMPLE 7   Tossing Three Coins

Make a probability distribution for the number of heads that occurs when three coins are tossed simultaneously.

SOLUTION   We apply the three-step process.

Step 1. The number of different outcomes when three coins are tossed is $2 \times 2 \times 2 = 8$ (see Figure 6.4b): HHH, HHT, HTH, HTT, THH, THT, TTH, and TTT.

Step 2. The eight outcomes represent four possible events: 0 heads, 1 head, 2 heads, and 3 heads. Notice that only one outcome represents the event of 0 heads, so its probability is 1/8; the same is true for the event of three heads (0 tails). The remaining events of 1 head (and 2 tails) or 2 heads (and 1 tail) each occur three times, so their probabilities are each 3/8.

Step 3. Table 6.3 shows the probability distribution, with the four events listed in the left column and their probabilities in the right column.

TABLE 6.3	Tossing Three Coins
Result	**Probability**
3 heads (0 tails)	1/8
2 heads (1 tail)	3/8
1 head (2 tails)	3/8
0 heads (3 tails)	1/8
Total	1

TIME OUT TO THINK

How many different *outcomes* are possible when you toss four coins? If you are interested in the number of heads, how many different *events* are possible?

EXAMPLE 8   Two Dice Distribution

Make a probability distribution for the sum of the dice when two dice are rolled.

SOLUTION   There are six ways for each die to land (see Figure 6.3), so there are $6 \times 6 = 36$ possible outcomes for a roll of two dice. Table 6.4 shows all 36 outcomes by listing one die along the rows and the other along the columns, with each cell showing for the two dice.

TABLE 6.4	Outcomes and Sums for the Roll of Two Dice					
	1	**2**	**3**	**4**	**5**	**6**
1	1 + 1 = 2	1 + 2 = 3	1 + 3 = 4	1 + 4 = 5	1 + 5 = 6	1 + 6 = 7
2	2 + 1 = 3	2 + 2 = 4	2 + 3 = 5	2 + 4 = 6	2 + 5 = 7	2 + 6 = 8
3	3 + 1 = 4	3 + 2 = 5	3 + 3 = 6	3 + 4 = 7	3 + 5 = 8	3 + 6 = 9
4	4 + 1 = 5	4 + 2 = 6	4 + 3 = 7	4 + 4 = 8	4 + 5 = 9	4 + 6 = 10
5	5 + 1 = 6	5 + 2 = 7	5 + 3 = 8	5 + 4 = 9	5 + 5 = 10	5 + 6 = 11
6	6 + 1 = 7	6 + 2 = 8	6 + 3 = 9	6 + 4 = 10	6 + 5 = 11	6 + 6 = 12

Notice that the possible sums, which are the *events* in this case, are 2 through 12. We find the probability of each sum by counting the number of times it occurs and dividing by the total of 36 possible outcomes. For example, the five highlighted outcomes in the table have a sum of 8, so the probability of a sum of 8 is 5/36. Table 6.5 shows the complete probability distribution, which is also shown as a histogram in Figure 6.6.

TABLE 6.5	Probability Distribution for the Sum of Two Dice											
Event (sum)	2	3	4	5	6	7	8	9	10	11	12	**Total**
Probability	$\frac{1}{36}$	$\frac{2}{36}$	$\frac{3}{36}$	$\frac{4}{36}$	$\frac{5}{36}$	$\frac{6}{36}$	$\frac{5}{36}$	$\frac{4}{36}$	$\frac{3}{36}$	$\frac{2}{36}$	$\frac{1}{36}$	**1**

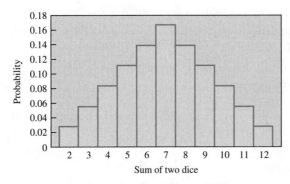

Figure 6.6 Histogram showing the probability distribution for the sum of two dice.

Section 6.2 Exercises

Statistical Literacy and Critical Thinking

1. **Notation.** If *A* denotes the event that you answer a particular true/false test question correctly, what do each of the following represent: $P(A)$, $P(not\ A)$, $P(\overline{A})$, and what are their values?

2. **Probability of Life.** A student reasons that there is a probability of 1/2 that there is life on Neptune, because there are two possible events: There is life or there is not life. Is this reasoning correct? Why or why not?

3. **Interpreting Probability.** What do we mean when we say that "the probability of getting 20 babies of the same gender when 20 random babies are born is 1/524,288? Is such an event *unusual*? Why or why not?

4. **Subjective Probability.** Use subjective judgment to estimate the probability that the next time you ride an elevator, it gets stuck between floors.

Does It Make Sense? For Exercises 5–10, decide whether the statement makes sense (or is clearly true) or does not make sense (or is clearly false). Explain clearly; not all of these have definitive answers, so your explanation is more important than your chosen answer.

5. **Certain Event.** When randomly selecting a day of the week, it is certain that you will select a day containing the letter *y*, so $P(y) = 1$.

6. **Impossible Event.** Because it is impossible for Thanksgiving to fall on Tuesday, the probability of Thanksgiving falling on Tuesday is 0.

7. **Complementary Events.** If there is a 0.9 probability that it will rain

sometime today, then there is a probability of 0.1 that it will not rain sometime today.

8. **Car Crash.** An insurance company states that the probability that a particular car will be involved in a car crash this year is 0.6 and the probability that the car will not be involved in a car crash this year is 0.3.

9. **Lightning.** Jack estimates that the subjective probability of his being struck by lightning sometime next year is 1/2.

10. **Lightning.** Jill estimates that the subjective probability of her being struck by lightning sometime next year is 1/1,000,000.

Concepts and Applications

Theoretical Probabilities. For Exercises 11–18, use the theoretical method to determine the probability of the given outcome or event. State any assumptions that you need to make.

11. **Die.** Rolling a die and getting an outcome that is greater than 2.

12. **Test Question.** Making a correct random guess for an answer to a particular multiple-choice question with possible answers of a, b, c, d, e, one of which is correct.

13. **Roulette.** Getting an outcome of a red slot when a roulette wheel is spun (A roulette wheel has slots of 0, 00, 1, 2, 3, ..., 36, and 18 of those slots are red.).

14. **Birthday.** Finding that the next President of the United States was born on Saturday.

15. **Birthday.** Finding that the next person you meet has the same birthday as yours (Ignore leap years.).

16. **Births.** Finding that the next baby born in Alaska is a girl.

17. **Births.** Finding that the next baby born to a couple is a girl, given that the couple already has two children and they are both boys.

18. **Dice.** Rolling a pair of dice and getting an outcome (sum) of 12.

Complementary Events. Exercises 19–26 involve complementary events. In each exercise, find the probability of the given event. State any assumptions that you use.

19. **Die.** What is the probability of rolling a fair die and not getting an outcome less than 7?

20. **Die.** What is the probability of rolling a fair die and not getting an outcome that is greater than 6?

21. **Week Days.** What is the probability of randomly selecting a day of the week and not getting Monday?

22. **Birthday.** What is the probability of finding that the next President of the United States was not born on Saturday?

23. **Basketball.** What is the probability that a 55% free-throw shooter will miss her next free throw?

24. **Testing.** What is the probability of guessing incorrectly when making a random guess on a multiple-choice test

question with possible answers of a, b, c, d, and e, one of which is correct?

25. **Baseball.** What is the probability that a 0.280 hitter in baseball will not get a hit on his next at-bat?

26. **Defects.** What is the probability of not getting a defective fuse when one fuse is randomly selected from an assembly line and 1% of the fuses are defective?

Theoretical Probabilities. For Exercises 27–30, use the theoretical method to determine the probability of the given outcome or event. State any assumptions that you need to make.

27. **M&Ms.** A bag contains 10 red M&Ms, 15 blue M&Ms, and 20 yellow M&Ms. What is the probability of drawing a red M&M? A blue M&M? A yellow M&M? Something besides a yellow M&M?

28. **Test Questions.** The New England College of Medicine uses an admissions test with multiple-choice questions, each with five possible answers, only one of which is correct. If you guess randomly on every question, what score might you expect to get? (Express the answer as a percentage.)

29. **Three-Child Family.** Suppose you randomly select a family with three children. Assume that births of boys and girls are equally likely. What is the probability that the family has each of the following?

   a. Three girls

   b. Two boys and a girl

   c. A girl, a boy, and a boy, in that order

   d. At least one girl

   e. At least two boys

30. **Four-Child Family.** Suppose you randomly select a family with four children. Assume that births of boys and girls are equally likely.

   a. How many birth orders are possible? List all of them.

   b. What is the probability that the family has four boys? Four girls?

   c. What is the probability that the family has a boy, a girl, a boy, and a girl, in that order?

   d. What is the probability that the family has two girls and two boys in any order?

Relative Frequency Probabilities. Use the relative frequency method to estimate the probabilities in Exercises 31–34.

31. **Weather Forecast.** After recording the forecasts of your local weatherman for 30 days, you conclude that he gave a correct forecast 18 times. What is the probability that his next forecast will be correct?

32. **Flood.** What is the probability of a 100-year flood this year?

33. **Basketball.** Halfway through the season, a basketball player has hit 86% of her free throws. What is the probability that her next free throw will be successful?

34. Surgery. In a clinical trial of 73 carpal tunnel syndrome patients treated with surgery, 67 had successful treatments (based on data from "Splinting vs. Surgery in the Treatment of Carpal Tunnel Syndrome" by Gerritsen et al., *Journal of the American Medical Association*, Vol. 288, No. 10). What is the probability that the next surgery treatment will be successful?

35. Senior Citizen Probabilities. In the year 2000, there were 34.7 million people over 65 years of age out of a U.S. population of 281 million. In the year 2050, it is estimated that there will be 78.9 million people over 65 years of age out of a U.S. population of 394 million. Would your chances of meeting a person over 65 at random be greater in 2000 or in 2050? Explain.

36. Age at First Marriage. The following table gives percentages of women and men married for the first time in several age categories (U.S. Census Bureau).

	Under 20	20–24	25–29	30–34	35–44	45–64	Over 65
Women	16.6	40.8	27.2	10.1	4.5	0.7	0.1
Men	6.6	36.0	34.3	14.8	7.1	1.1	0.1

a. What is the probability that a randomly encountered married woman was married, for the first time, between the ages of 35 and 44?

b. What is the probability that a randomly encountered married man was married, for the first time, before he was 20 years old?

c. Construct a bar chart consisting of side-by-side bars representing men and women.

37. Four-Coin Probability Distribution.

a. Construct a table similar to Table 6.2, showing all possible outcomes of tossing four coins at once.

b. Construct a table similar to Table 6.3, showing the probability distribution for the events 4 heads, 3 heads, 2 heads, 1 head, and 0 heads when you toss four coins at once.

c. What is the probability of getting 2 heads and 2 tails when you toss four coins at once?

d. What is the probability of tossing anything except 4 heads when you toss four coins at once?

e. Which of the five possible events (4 heads, 3 heads, 2 heads, 1 head, 0 heads) is most likely to occur?

38. Colorado Lottery Distribution. The histogram in Figure 6.7 shows the distribution of 5,964 Colorado lottery numbers (possible values range from 1 to 42).

a. Assuming the lottery drawings are random, what would you expect the probability of any number to be?

b. Based on the histogram, what is the relative frequency probability of the most frequently appearing number?

c. Based on the histogram, what is the relative frequency probability of the least frequently appearing number?

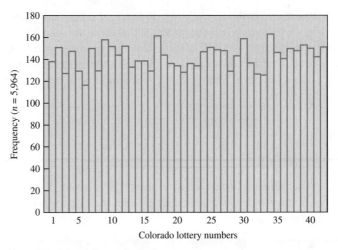

Figure 6.7 Colorado lottery distribution.

d. Comment on the deviations of the empirical probabilities from the expected probabilities. Would you say these deviations are significant?

PROJECTS FOR THE INTERNET & BEYOND

39. Blood Groups. The four major blood groups are designated A, B, AB, and O. Within each group there are two Rh types: positive and negative. Using library resources or the Internet, find data on the relative frequency of blood groups, including the Rh types. Construct a table showing the probability of meeting someone in each of the eight combinations of blood group and Rh type.

40. Age and Gender. The proportions of men and women in the population change with age. Using current data from a Web site, construct a table showing the probability of meeting a male or a female in each of these age categories: 0–5, 6–10, 11–20, 21–30, 31–40, 41–50, 51–60, 61–70, 71–80, over 80.

41. Thumb Tack Probabilities. Find a standard thumb tack and practice tossing it onto a flat surface. Notice that there are two different outcomes: The tack can land point down or point up.

a. Toss the tack 50 times and record the outcomes.

b. Give the relative frequency probabilities of the two outcomes based on these results.

c. If possible, ask several other people to repeat the process. How well do your probabilities agree?

42. Three-Coin Experiment. Toss three coins at once 50 times and record the outcomes in terms of the number of heads. Based on your observations, give the relative frequency probabilities of the outcomes. Do they agree with the theoretical probabilities? Explain and discuss your results.

43. Randomizing a Survey. Suppose you want to conduct a survey involving a sensitive question that not all participants may choose to answer honestly (for example, a question involving cheating on taxes or drug use). Here is a way to conduct the survey and protect the identity of respondents. We will assume that the sensitive question requires a *yes* or *no* answer. First, ask all respondents to toss a fair coin. Then give the following instructions:

- If you toss a head, then answer the decoy question (yes/no): Were you born on an even day of the month?

- If you toss a tail, then answer the real survey question (yes/no).

After all participants have answered *yes* or *no* to the question they were assigned, count the total numbers of *yes* and *no* responses.

a. Choose a question that may not produce totally honest responses and conduct a survey in your class using this technique.

b. Given only the total numbers of *yes* and *no* responses to both questions, explain how you can estimate the number of people who answered *yes* and *no* to the real question.

c. Will the results computed in part b be exact? Explain.

d. Suppose the decoy question was replaced by these instructions: If you toss a head, then answer *yes*. Can you still determine the number of people who answered *yes* and *no* to the real question?

IN THE NEWS

44. Theoretical Probabilities. Find a news article or research report that cites a theoretical probability. Provide a one-paragraph discussion.

45. Relative Frequency Probabilities. Find a news article or research report that makes use of a relative frequency (or empirical) probability. Provide a one-paragraph discussion.

46. Subjective Probabilities. Find a news article or research report that refers to a subjective probability. Provide a one-paragraph discussion.

47. Probability Distributions. Find a news article or research report that cites or makes use of a probability distribution. Provide a one-paragraph discussion.

6.3 THE LAW OF LARGE NUMBERS

The false ideas prevalent among all classes of the community, cultured as well as uncultured, respecting chance and luck, illustrate the truth that common consent argues almost of necessity [from] error.

—Richard Proctor,
Chance and Luck (1887 textbook)

If you toss a coin once, you cannot predict exactly how it will land; you can state only that the probability of a head is 0.5. If you toss the coin 100 times, you still cannot predict precisely how many heads will occur. However, you can reasonably expect to get heads *close to* 50% of the time. If you toss the coin 1,000 times, you can expect the proportion of heads to be even closer to 50%. In general, the more times you toss the coin, the closer the percentage of heads will be to exactly 50%. The idea that large numbers of events may show some pattern even while individual events are unpredictable is called the **law of large numbers** (sometimes called the *law of averages*). This law plays a very important role in statistics.

> **The Law of Large Numbers**
>
> The **law of large numbers** applies to a process for which the probability of an event A is $P(A)$ and the results of repeated trials do not depend on results of earlier trials (they are *independent*). It states:
> *If the process is repeated through many trials, the proportion of the trials in which event A occurs will be close to the probability $P(A)$. The larger the number of trials, the closer the proportion should be to $P(A)$.*

We can illustrate the law of large numbers with a die-rolling experiment. The probability of a 1 on a single roll is $P(1) = 1/6 = 0.167$. To avoid the tedium of rolling the die many times, we can let a computer *simulate* random rolls of the die. Figure 6.8 shows the results of a computer simulation of rolling a single die 5,000 times. The horizontal axis gives the number of rolls, and the height of the curve gives the proportion of 1's. Although the curve bounces around when the number of rolls is small, for larger numbers of rolls the proportion of 1's approaches the probability of 0.167—just as predicted by the law of large numbers.

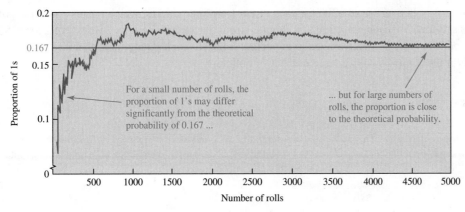

Figure 6.8 Results of a computer simulation of rolling a die. As the number of rolls grows large, the proportion of 1's gets close to the theoretical probability of a 1 on a single roll, which is 0.167.

EXAMPLE 1   Roulette

A roulette wheel has 38 numbers: 18 black numbers, 18 red numbers, and the numbers 0 and 00 in green. (Assume that all outcomes—the 38 numbers—have equal probability.)

a. What is the probability of getting a red number on any spin?

b. If patrons in a casino spin the wheel 100,000 times, how many times should you expect a red number?

SOLUTION

a. The theoretical probability of getting a red number on any spin is

$$P(A) = \frac{\text{number of ways red can occur}}{\text{total number of outcomes}} = \frac{18}{38} = 0.474$$

b. The law of large numbers tells us that as the game is played more and more times, the proportion of times that a red number appears should get closer to 0.474. In 100,000 tries, the wheel should come up red close to 47.4% of the time, or about 47,400 times. $\cdot\cdot\bullet$

Expected Value

Suppose the InsureAll Company sells a special type of insurance in which it promises to pay you $100,000 in the event that you must quit your job because of serious illness. Based on data from past claims, the probability that a policyholder will be paid for loss of job is 1 in 500. Should the insurance company expect to earn a profit if it sells the policies for $250 each?

   If InsureAll sells only a few policies, the profit or loss is unpredictable. For example, selling 100 policies for $250 each would generate revenue of 100 × $250 = $25,000. If none of the 100 policyholders files a claim, the company will make a tidy profit. On the other hand, if InsureAll must pay a $100,000 claim to even one policyholder, it will face a huge loss.

   In contrast, if InsureAll sells a large number of policies, the law of large numbers tells us that the proportion of policies for which claims will have to be paid should be very close to the 1 in 500 probability for a single policy. For example, if the company sells 1 million policies, it should expect that the number of policyholders collecting on a $100,000 claim will be close to

$$\underbrace{1,000,000}_{\substack{\text{number of}\\\text{policies}}} \times \underbrace{\frac{1}{500}}_{\substack{\text{probability of}\\\$100,000\text{ claim}}} = 2,000 \text{ claims}$$

Paying these 2,000 claims will cost

$$2,000 \times \$100,000 = \$200 \text{ million}$$

This cost is an *average* of $200 for each of the 1 million policies, which means that if the policies sell for $250 each, the company should expect to earn an average of $250 − $200 = $50 per policy. We call this average the **expected value** for each policy; note that it is "expected" only if the company sells a large number of policies.

> ### Definition
>
> The **expected value** of a variable is the weighted average of all its possible values. Because it is an average, we should expect to find the "expected value" only when there are a large number of events, so that the law of large numbers comes into play.

We can find the same expected value with a more formal procedure. The insurance example involves two distinct events, each with a particular *probability* and *value* for the company:

1. In the event that a person buys a policy, the value to the company is the $250 price of the policy. The probability of this event is 1 because everyone who buys a policy pays $250.

2. In the event that a person is paid for a claim, the value to the company is −$100,000; it is negative because the company loses $100,000 in this case. The probability of this event is 1/500.

We now multiply the value of each event by its probability and add the results to find the expected value of each insurance policy:

$$\text{expected value} = \underbrace{\$250}_{\substack{\text{value of} \\ \text{policy sale}}} \times \underbrace{1}_{\substack{\text{probability of} \\ \text{earning 250 on sale}}} + \underbrace{(-\$100,000)}_{\substack{\text{value of} \\ \text{claim}}} \times \underbrace{\frac{1}{500}}_{\substack{\text{probability of} \\ \text{paying claim}}}$$

$$= \$250 - \$200 = \$50$$

This expected profit of $50 per policy is the same answer we found earlier. Note that it amounts to a profit of $50 million on sales of 1 million policies.

> ### Calculating Expected Value
>
> Consider two events, each with its own value and probability. The **expected value** is
>
> $$\text{expected value} = \binom{\text{value of}}{\text{event 1}} \times \binom{\text{probability of}}{\text{event 1}} + \binom{\text{value of}}{\text{event 2}} \times \binom{\text{probability of}}{\text{event 2}}$$
>
> This formula can be extended to any number of events by including more terms in the sum.

> ### TIME OUT TO THINK
>
> Should the insurance company expect to see a profit of $50 on each individual policy? Should it expect a profit of $50,000 on 1,000 policies? Explain.

EXAMPLE 2 Lottery Expectations

Suppose that $1 lottery tickets have the following probabilities: 1 in 5 to win a free ticket (worth $1), 1 in 100 to win $5, 1 in 100,000 to win $1,000, and 1 in 10 million to win $1 million. What is the expected value of a lottery ticket? Discuss the implications. (Note: Winners do *not* get back the $1 they spend on the ticket.)

SOLUTION   The easiest way to proceed is to make a table (below) of all the relevant events with their values and probabilities. We are calculating the expected value of a lottery ticket to *you;* thus, the ticket price has a negative value because it costs you money, while the values of the winnings are positive.

Event	Value	Probability	Value × probability
Ticket purchase	−$1	1	$(-\$1) \times 1 = -\$1.00$
Win free ticket	$1	$\frac{1}{5}$	$\$1 \times \frac{1}{5} = \$0.20$
Win $5	$5	$\frac{1}{100}$	$\$5 \times \frac{1}{100} = \$0.05$
Win $1,000	$1,000	$\frac{1}{100,000}$	$\$1,000 \times \frac{1}{100,000} = \$0.01$
Win $1 million	$1,000,000	$\frac{1}{10,000,000}$	$\$1,000,000 \times \frac{1}{10,000,000} = \$0.10$
			Sum of last column: −$0.64

The expected value is the sum of all the products *value × probability*, which the final column of the table shows to be −$0.64. In other words, averaged over many tickets, you should expect to lose 64¢ for each lottery ticket that you buy. If you buy, say, 1,000 tickets, you should expect to *lose* about $1,000 \times \$0.64 = \$640$.   ·•●

© 1994 Scott Adams, Inc. Distributed by United Feature Syndicate. Reprinted with permission. All rights reserved.

The Gambler's Fallacy

Consider a simple game involving a coin toss: You win $1 if the coin lands heads, and you lose $1 if it lands tails. Suppose you toss the coin 100 times and get 45 heads and 55 tails, putting you $10 in the hole. Are you "due" for a streak of better luck?

You probably recognize that the answer is *no*: Your past bad luck has no bearing on your future chances. However, many gamblers—especially compulsive gamblers—guess just the opposite. They believe that when their luck has been bad, it's due for a change. This mistaken belief is often called the **gambler's fallacy** (or the *gambler's ruin*).

> **Definition**
>
> The **gambler's fallacy** is the mistaken belief that a streak of bad luck makes a person "due" for a streak of good luck.

One reason people succumb to the gambler's fallacy is a misunderstanding of the law of large numbers. In the coin-toss game, the law of large numbers tells us that the proportion of heads tends to be closer to 50% for larger numbers of tosses. But this does *not* mean that you are likely to recover early losses. To see why, study Table 6.6, which shows results from computer simulations of large numbers of coin tosses. Note that as the number of tosses increases, the percentage of heads gets closer to exactly 50%, just as the law of large numbers predicts. However, the last column shows that the *difference* between the number of heads and the number of tails continues to grow—meaning that the losses (the difference between the numbers of heads and tails) grow larger even as the proportion of heads approaches 50%.

Everyone who bets any part of his fortune, however small, on a mathematically unfair game of chance acts irrationally... The imprudence of a gambler will be the greater the larger part of his fortune which he exposes to a game of chance.

—Daniel Bernoulli, 18th-century mathematician

TABLE 6.6	Outcomes of Coin Tossing Trials		
Number of tosses	Number of heads	Percentage of heads	Difference between numbers of heads and tails
100	45	45%	10
1,000	470	47%	60
10,000	4,950	49.5%	100
100,000	49,900	49.9%	200

EXAMPLE 3 Continued Losses

You are playing the coin-toss game in which you win $1 for heads and lose $1 for tails. After 100 tosses, you are $10 in the hole because you have 45 heads and 55 tails. You continue playing until you've tossed the coin 1,000 times, at which point you've gotten 480 heads and 520 tails. Is this result consistent with what we expect from the law of large numbers? Have you gained back any of your losses? Explain.

SOLUTION The proportion of heads in your first 100 tosses was 45%. After 1,000 tosses, the proportion of heads has increased to 480 out of 1,000, or 48%. Because the proportion of heads moved closer to 50%, the results are consistent with what we expect from the law of large numbers. However, you've now won $480 (for the 480 heads) and lost $520 (for the 520 tails), for a net loss of $40. Thus, your losses *increased*, despite the fact that the proportion of heads grew closer to 50%.   ··●

Streaks

Another common misunderstanding that contributes to the gambler's fallacy involves expectations about streaks. Suppose you toss a coin six times and see the outcome HHHHHH (all heads). Then you toss it six more times and see the outcome HTTHTH. Most people would say that the latter outcome is "natural" while the streak of all heads is surprising. But, in fact, both outcomes are equally likely. The total number of possible outcomes for six coins is $2 \times 2 \times 2 \times 2 \times 2 \times 2 = 64$, and every individual outcome has the same probability of 1/64.

Moreover, suppose you just tossed six heads and had to bet on the outcome of the next toss. You might think that, given the run of heads, a tail is "due" on the next toss. But the probability of a head or a tail on the next toss is still 0.50; the coin has no memory of previous tosses.

TIME OUT TO THINK

Is a family with six boys more or less likely to have a boy for the next child? Is a basketball player who has hit 20 consecutive free throws more or less likely to hit her next free throw? Is the weather on one day independent of the weather on the next (as assumed in the next example)? Explain.

EXAMPLE 4 Planning for Rain

A farmer knows that at this time of year in his part of the country, the probability of rain on a given day is 0.5. It hasn't rained in 10 days, and he needs to decide whether to start irrigating. Is he justified in postponing irrigation because he is due for a rainy day?

SOLUTION The 10-day dry spell is unexpected, and, like a gambler, the farmer is having a "losing streak." However, if we assume that weather events are independent from one day to the next, then the probability of rain is still 0.5 on any given day, so he is not "due" for rain.   ··●

Section 6.3 Exercises

Statistical Literacy and Critical Thinking

1. **Law of Large Numbers.** In your own words, describe the law of large numbers.

2. **Expected Value.** A geneticist computes the expected number of girls in 5 births and obtains the result of 2.5 girls. He rounds the result to 3 girls, reasoning that it is impossible to get 2.5 girls in 5 births. Is that reasoning correct? Why or why not?

3. **Gambling Strategy.** A professional gambler playing blackjack in the Venetian casino has lost each of his first 10 bets. He begins to place larger bets, reasoning that his current proportion of wins (which is 0) will increase to get closer to the average number of wins. Is his betting strategy sound? Is his reasoning correct? Explain.

4. **Gambler's Fallacy.** In your own words, describe the gambler's fallacy.

Does It Make Sense? For Exercises 5–8, decide whether the statement makes sense (or is clearly true) or does not make sense (or is clearly false). Explain clearly; not all of these have definitive answers, so your explanation is more important than your chosen answer.

5. **Gambling Strategy.** Steve learns that for a $5 bet on a number in roulette, the expected return is $4.74, but for a $5 bet on the pass line in the game of craps, the expected return is $4.93, so it is better to play this particular craps bet than the roulette bet.

6. **Lottery.** Kelly studies the Wisconsin Pick 3 lottery and finds that for a $1 bet, the expected return is 50¢, so she reasons that this is bad bet and she does not play.

7. **Lottery.** When considering the chances of winning a lottery, Kim reasons that the number combination of 1, 2, 3, 4, 5, and 6 is less likely to occur because of the obvious pattern.

8. **Lottery.** Jennifer purchases a state lottery ticket and she avoids the combination of 1, 2, 3, 4, 5, and 6. She reasons that this combination has the same chance as any other combination, but the jackpot is divided among the winners, and many people are likely to select 1, 2, 3, 4, 5, and 6, so the jackpot will be much smaller, so it is better to select some other combination of numbers that will not be chosen by many other people.

Concepts and Applications

9. **Gender Selection.** In planning an experiment designed to test the effectiveness of a gender selection method, assume that boys and girls are equally likely to be born. Among 500 births, should we expect to get exactly 250 boys and 250 girls? As the number of births increases, what does the law of large numbers tell us about the proportion of girls?

10. **Speedy Driver.** A person who has a habit of driving fast has never had an accident or traffic citation. What does it mean to say that "the law of averages will catch up with him"? Is it true? Explain.

11. **Should You Play?** Suppose someone offers you this opportunity: You can place a bet of $10 and roll a single die once. You win twice the outcome of the die. For example, if you roll a 6, you win $12 for a net profit of $2. What is the expected value of this game? Should you play?

12. **Kentucky's Pick 4 Lottery.** If you bet $1 in Kentucky's Pick 4 lottery game, you either lose $1 or gain $4,999. (The winning prize is $5,000, but your $1 bet is not returned, so the net gain is $4,999.) The game is played by selecting a four-digit number between 0000 and 9999. What is the probability of winning? If you bet $1 on 1234, what is the expected value of your gain or loss?

13. **Extra Points in Football.** Football teams have the option of trying to score either 1 or 2 extra points after a touchdown. They get 1 point by kicking the ball through the goal posts or 2 points by running or passing the ball across the goal line. For a recent year in the NFL, 1-point kicks were successful 94% of the time, while 2-point attempts were successful only 37% of the time. In either case, failure means 0 points. Calculate the expected values of the 1-point and 2-point attempts. Based on these expected values, which option makes more sense in most cases? Can you think of any circumstances in which a team should make a decision different from what the expected values suggest? Explain.

14. **Insurance Claims.** An actuary at an insurance company estimates from existing data that on a $1,000 policy, an average of 1 in 100 policyholders will file a $20,000 claim, an average of 1 in 200 policyholders will file a $50,000 claim, and an average of 1 in 500 policyholders will file a $100,000 claim.

    a. What is the expected value to the company for each policy sold?

    b. If the company sells 100,000 policies, can it expect a profit? Explain the assumptions of this calculation.

15. **Expected Waiting Time.** Suppose that you arrive at a bus stop randomly, so all arrival times are equally likely. The bus arrives regularly every 30 minutes without delay (say, on the hour and on the half hour). What is the expected value of your waiting time? Explain how you got your answer.

16. Powerball Lottery. The multi-state Powerball lottery advertises the following prizes and probabilities of winning for a single $1 ticket. Assume the jackpot has a value of $30 million one week. Note that there is more than one way to win some of the monetary prizes (for example, two ways to win $100), so the table gives the probability for each way. What is the expected value of the winnings for a single lottery ticket? If you spend $365 per year on the lottery, how much can you expect to win or lose?

Prize	Probability
Jackpot	1 in 80,089,128
$100,000	1 in 1,953,393
$5,000	1 in 364,042
$100	1 in 8,879
$100	1 in 8,466
$7	1 in 207
$7	1 in 605
$4	1 in 188
$3	1 in 74

17. Big Game. The Multi-State Big Game lottery advertises the following prizes and probabilities of winning for a single $1 ticket. The jackpot is variable, but assume it has an average value of $3 million. Note that the same prize can be given to two outcomes with different probabilities. What is the expected value of a single lottery ticket to you? If you spend $365 per year on the lottery, how much can you expect to win or lose?

Prize	Probability
Jackpot	1 in 76,275,360
$150,000	1 in 2,179,296
$5,000	1 in 339,002
$150	1 in 9,686
$100	1 in 7,705
$5	1 in 220
$5	1 in 538
$2	1 in 102
$1	1 in 62

18. Expected Value in Roulette. When you give the Venetian casino in Las Vegas $5 for a bet on the number 7 in roulette, you have a 37/38 probability of losing $5 and you have a 1/38 probability of making a net gain of $175. (The prize is $180, but your $5 bet is not returned, so the net gain is $175.) If you bet $5 that the outcome is an odd number, the probability of losing $5 is 20/38 and the probability of making a net gain of $5 is 18/38. (If you bet $5 on an odd number and win, you are given $10 that includes your bet, so the net gain is $5.)

  a. If you bet $5 on the number 7, what is your expected value?

  b. If you bet $5 that the outcome is an odd number, what is your expected value?

  c. Which of these options is best: bet on 7, bet on odd, or don't bet? Why?

19. Expected Value in Casino Dice. When you give a casino $5 for a bet on the "pass line" in a casino game of dice,

there is a 251/495 probability that you will lose $5 and there is a 244/495 probability that you will make a net gain of $5. (If you win, the casino gives you $5 and you get to keep your $5 bet, so the net gain is $5.) What is your expected value? In the long run, how much do you lose for each dollar bet?

20. New Jersey Pick 4. In New Jersey's Pick 4 lottery game, you pay 50¢ to select a sequence of four digits, such as 2273, from the 10,000 different possible four-digit sequences. If you select the same sequence of four digits that are drawn, you win and collect $2788. What is your expected value? In the long run, how much do you lose for each 50¢ bet?

21. Psychology of Expected Values. In 1953, a French economist named Maurice Allais conducted a survey of how people assess risk. Here are two scenarios that he used, each of which required people to choose between two options.

Decision 1:

Option A: 100% chance of gaining $1,000,000

Option B: 10% chance of gaining $2,500,000; 89% chance of gaining $1,000,000; and 1% chance of gaining nothing

Decision 2:

Option A: 11% chance of gaining $1,000,000 and 89% chance of gaining nothing

Option B: 10% chance of gaining $2,500,000 and 90% chance of gaining nothing

Allais discovered that for decision 1, most people chose option A, while for decision 2, most people chose option B.

  a. For each decision, find the expected value of each option.

  b. Are the responses given in the survey consistent with the expected values?

  c. Give a possible explanation for the responses in Allais's survey.

22. Expected Value for a Magazine Sweepstakes. *Reader's Digest* ran a sweepstakes in which prizes were listed along with the chances of winning: $1,000,000 (1 chance in 90,000,000), $100,000 (1 chance in 110,000,000), $25,000 (1 chance in 110,000,000), $5,000 (1 chance in 36,667,000), and $2,500 (1 chance in 27,500,000).

  a. Assuming that there is no cost to enter the sweepstakes, find the expected value of the amount won for one entry.

  b. Find the expected value if the cost of entering this sweepstakes is the cost of a postage stamp. Is it worth entering this contest?

23. Gambler's Fallacy and Dice. Suppose you roll a die with a friend, with the following rules: For every even number you roll, you win $1 from your friend; for every odd number you roll, you pay $1 to your friend.

a. What are the chances of rolling an even number on one roll of a fair die? An odd number?

b. Suppose that on the first 100 rolls, you roll 45 even numbers. How much money have you won or lost?

c. Suppose that on the second 100 rolls, your luck improves and you roll 47 even numbers. How much money have you won or lost over 200 rolls?

d. Suppose that over the next 300 rolls, your luck again improves and you roll 148 even numbers. How much money have you won or lost over 500 rolls?

e. What was the percentage of even numbers after 100, 200, and 500 rolls? Explain why this game illustrates the gambler's fallacy.

f. How many even numbers would you have to roll in the next 100 rolls to break even?

24. **Behind in Coin Tossing: Can You Catch Up?** Suppose that you toss a fair coin 100 times, getting 38 heads and 62 tails, which is 24 more tails than heads.

a. Explain why, on your next toss, the *difference* in the numbers of heads and tails is as likely to grow to 25 as it is to shrink to 23.

b. Extend your explanation from part a to explain why, if you toss the coin 1,000 more times, the final difference in the numbers of heads and tails is as likely to be larger than 24 as it is to be smaller than 24.

c. Suppose that you continue tossing the coin. Explain why the following statement is true: If you stop at any random time, you always are more likely to have fewer heads than tails, in total.

d. Suppose that you are betting on heads with each coin toss. After the first 100 tosses, you are well on the losing side (having lost the bet 62 times while winning only 38 times). Explain why, if you continue to bet, you will most likely remain on the losing side. How is this answer related to the gambler's fallacy?

PROJECTS FOR THE INTERNET & BEYOND

25. **Analyzing Lotteries on the Web.** Go to the Web site for all U.S. lotteries and study the summary of state and multi-state lottery odds and prizes. Pick five lotteries and determine the expected value for winnings in each case. Discuss your results.

26. **Law of Large Numbers.** Use a coin to simulate 100 births: Flip the coin 100 times, recording the results, and then convert the outcomes to genders of babies (tail = boy and head = girl). Use the results to fill in the following table. What happens to the proportion of girls as the sample size increases? How does this illustrate the law of large numbers?

Number of births	10	20	30	40	50	60	70	80	90	100
Proportion of girls										

IN THE NEWS

27. **Personal Law of Large Numbers.** Describe a situation in which you personally have made use of the law of large numbers, either correctly or incorrectly. Why did you use the law of large numbers in this situation? Was it helpful?

28. **Gambler's Fallacy in Life.** Describe a situation in which you or someone you know has fallen victim to the gambler's fallacy. How could the situation have been dealt with correctly?

29. **Gambler's Ruin.** Describe a situation you know of in which someone lost nearly everything through gambling. Did his or her strategy appear to be rational, or did it appear to be the result of a destructive addiction?

6.4 IDEAS OF RISK AND LIFE EXPECTANCY

A smooth-talking but honest salesman comes to you, offering a new product:

I can't reveal the details yet, but you will love this product! It will improve your life in more ways than you can count. Its only downside is that it will eventually kill everyone who uses it. Will you buy one?

Not likely. After all, could any product be so great that you would die for it? A few weeks later, the salesman shows up again:

No one was buying, so we've made some improvements. Your chance of being killed by the product is now only 1 in 10. Ready to buy?

Despite the improvement, most people still would send the salesman home and wait for his inevitable return:

Okay, this time we've really perfected it. We've made it so safe that it would take 25 years for it to kill as many people as live in San Francisco. And it can be yours for a mere $30,000.

You may be surprised to realize that, if you are like most Americans, you'll jump at this offer. The product is, after all, the automobile. It does indeed improve our lives in many ways, and $30,000 is a typical price. And, given that more than 30,000 Americans are killed in auto accidents each year, it kills about the equivalent of the population of San Francisco (roughly 800,000) in about 25 years.

As this example shows, we frequently make tradeoffs between benefits and risks. In this section, we'll see how ideas of probability can help us quantify risk, thereby allowing us to make informed decisions about such tradeoffs.

Risk and Travel

Are you safer in a small car or in a sport utility vehicle? Are cars today safer than those 30 years ago? If you need to travel across country, are you safer flying or driving? To answer these and many similar questions, we must quantify the risk involved in travel. We can then make decisions appropriate for our own personal circumstances.

Travel risk is often expressed in terms of an **accident rate** or **death rate**. For example, suppose an annual accident rate is 750 accidents per 100,000 people. This means that, within a group of 100,000 people, on average 750 will have an accident over the period of a year. The statement is in essence an expected value, which means it also represents a probability: It tells us that the probability of a person being involved in an accident (in one year) is 750 in 100,000, or 0.0075.

This concept of travel risk is straightforward, but we must still interpret the numbers with care. For example, travel risks are sometimes stated *per 100,000 people,* as above, but other times they are stated *per trip* or *per mile.* If we use death rates *per trip* to compare the risks of flying and driving, we neglect the fact that airplane trips are typically much longer than automobile trips. Similarly, if we use accident rates *per person,* we neglect the fact that most automobile accidents involve only minor injuries.

EXAMPLE ❶ Is Driving Getting Safer?

Figure 6.9 shows the number of automobile fatalities and the total number of miles driven (among all Americans) for each year over a period of more than four decades. In terms of death rate per mile driven, how has the risk of driving changed?

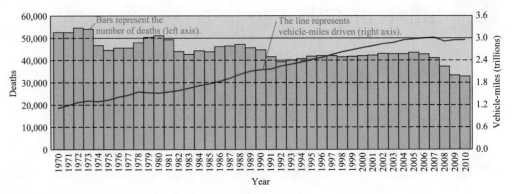

Figure 6.9 This graph shows the annual number of deaths (bars) and the annual number of vehicle-miles driven (line) in the United States from 1970 to 2010. *Source:* National Transportation Safety Board.

SOLUTION   We start by comparing the approximate death rates per mile for 1970 and 2010.

$$1970: \quad \frac{52{,}000 \text{ deaths}}{1 \times 10^{12} \text{ miles}} \approx 5.2 \times 10^{-8} \text{ death per mile}$$

$$2010: \quad \frac{33{,}000 \text{ deaths}}{3 \times 10^{12} \text{ miles}} \approx 1.1 \times 10^{-8} \text{ death per mile}$$

Note that 5.2×10^{-8} death per mile is equivalent to 5.2 deaths per 100 million miles. We conclude that, over the approximately four decade period, the death rate per 100 million miles dropped from about 5.2 to about 1.1. By this measure, driving became much safer over the period. Most researchers believe the improvements resulted from better automobile design and from safety features, such as shoulder belts and air bags. $\cdots\bullet$

EXAMPLE ❷ **Which Is Safer: Flying or Driving?**

For the period 1990 through 2010, the average (mean) number of deaths in commercial airplane accidents in the United States was roughly 60 per year. (The actual number varies significantly from year to year.) As of 2010, airplane passengers in the United States travel a total of about 8 billion miles per year. Use these numbers to calculate the death rate per mile of air travel. Compare the risk of flying to the risk of driving.

SOLUTION   Assuming 60 deaths and 8 billion miles in an average year, the risk of air travel is

$$\frac{60 \text{ deaths}}{8 \times 10^{9} \text{miles}} \approx 0.75 \times 10^{-8} \text{ death per mile}$$

This risk of about 0.75 deaths per 100 million miles is about two thirds of the 2010 risk of 1.1 deaths per 100 million miles for driving (see Example 1). Note that, because the average air trip covers a considerably longer distance than the average driving trip, the risk *per trip* is higher for air travel, even though the risk *per mile* is higher for car travel. $\cdots\bullet$

> *The cost of living is going up and the chance of living is going down.*
>
> —Flip Wilson,
> comedian

> **TIME** 🕐 **UT TO THINK**
> Suppose you need to make the 800-mile trip from Atlanta to Houston. Do you think it is safer to fly or to drive? Defend your opinion.

Vital Statistics

Data concerning births and deaths of citizens, often called *vital statistics*, are very important to understanding risk-benefit tradeoffs. For example, insurance companies use vital statistics to assess risks and set rates. Health professionals study vital statistics to assess medical progress and decide where research resources should be concentrated. Demographers use birth and death rates to predict future population trends.

One important set of vital statistics, shown in Table 6.7, concerns causes of death. These data are extremely general; a more complete table would categorize data by age, sex, and race. Vital statistics are often expressed in terms of deaths per 100,000 people, which makes it easier to compare the rates for different years and for different states or countries.

> *Only those who risk going too far can possibly find out how far one can go.*
>
> —T. S. Eliot

TABLE 6.7	**Leading Causes of Death in the United States (in a single recent year)**		
Cause	**Deaths**	**Cause**	**Deaths**
Heart disease	684,462	Diabetes	73,249
Cancer	554,643	Pneumonia/Influenza	65,681
Stroke	157,803	Alzheimer's disease	63,343
Pulmonary disease	126,128	Kidney disease	42,536
Accidents	105,695	Septicemia (blood poisoning)	34,243

Source: Centers for Disease Control and Prevention.

EXAMPLE 3 Interpreting Vital Statistics

Assuming an approximate U.S. population of 300 million, find and compare risks per person and per 100,000 people for pneumonia (and influenza) and cancer.

SOLUTION   We find the risk per person by dividing the number of deaths by the total population of 300 million:

$$\text{Pneumonia/influenza:} \quad \frac{65{,}681 \text{ deaths}}{300{,}000{,}000 \text{ people}} \approx 0.00022 \text{ risk of death per person}$$

$$\text{Cancer:} \quad \frac{554{,}643 \text{ deaths}}{300{,}000{,}000 \text{ people}} \approx 0.0018 \text{ risk of death per person}$$

To put these values in terms of deaths per 100,000 people, we simply multiply the per person rates by 100,000. We find that the pneumonia/influenza death rate is about 22 deaths per 100,000 people, which is much lower than the cancer death rate of 180 deaths per 100,000 people.   ··●

TIME 🕐UT TO THINK

Table 6.7 suggests that the probability of death by stroke is about 50% higher than the probability of death by accident, but these data include all age groups. How do you think the risks of stroke and accident would differ between young people and older people? Explain.

Life Expectancy

One of the most commonly cited vital statistics is *life expectancy*, which is often used to compare overall health at different times or in different countries. The idea will be clearer if we start by looking at death rates. Figure 6.10a shows the overall U.S. death rate (or mortality rate), in deaths per 1,000 people, for different age groups. Note that there is an elevated risk of death near birth, after which the death rate drops to very low levels. At about 15 years of age, the death rate begins a gradual rise.

Figure 6.10b shows the **life expectancy** of Americans of different ages, defined as the number of *additional* years a person of a given age can expect to live on average. As we would expect, life expectancy is higher for younger people because, on average, they have longer left to live. At birth, the life expectancy of Americans today is about 78 years (75 years for men and 80 years for women).

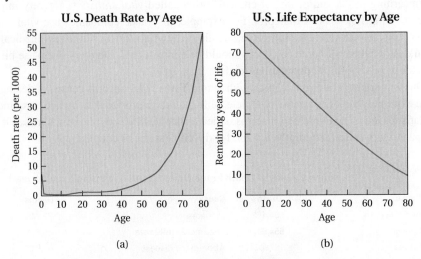

Figure 6.10   (a) The overall U.S. death rate (deaths per 1,000 people) for different ages. (b) Life expectancy (the number of additional years of life that can be expected on average) for different ages.
Source: U.S. National Center for Health Statistics.

> **Definition**
>
> **Life expectancy** is the number of additional years a person with a given age today can expect to live on average.

The subtlety in interpreting life expectancy comes from changes in medical science and public health. Life expectancies are calculated by studying *current* death rates. For example, when we say that the life expectancy of infants born today is 78 years, we mean that the average baby will live to age 78 *if there are no future changes* in medical science or public health. That is, while life expectancy provides a useful measure of current overall health, it should not be considered a *prediction* of future life spans.

In fact, because of advances in both medical science and public health, life expectancies increased dramatically during the 20th century, rising by about 60% (Figure 6.11). If this trend continues, today's infants are likely to live much longer than 78 years on average.

The reports of my death are greatly exaggerated.

—Mark Twain from London, in a cable to the Associated Press

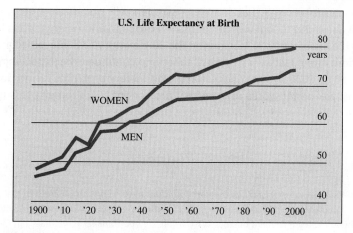

U.S. Life Expectancy at Birth

Figure 6.11   Changes in U.S. life expectancy during the 20th century.
Source: New York Times and National Center for Health Science Statistics.

> **TIME OUT TO THINK**
>
> Using Figure 6.11, compare the life expectancies of men and women. Briefly discuss these differences. Do they have any implications for social policy? For insurance rates? Explain.

EXAMPLE 4 Life Expectancies

Using Figure 6.10b, find the life expectancy of a 20-year-old person and of a 60-year-old person. Are the numbers consistent? Explain.

SOLUTION   The graph shows that the life expectancy at age 20 is about 59 years and at age 60 is about 23 years. This means that an average 20-year-old can expect to live about 59 more years, to age 79. An average 60-year-old can expect to live about 23 more years, to age 83.

It might at first seem strange to calculate a longer average life span for 60-year-olds than 20-year-olds (83 years versus 79 years). But remember that life expectancies are based on *current* data. If there were no changes in medicine or public health, a 60-year-old would have a greater probability of reaching age 83 than a 20-year-old simply because he or she has already made it to age 60. However, if medicine and public health continue to improve, today's 20-year-olds may live to older ages than today's 60-year-olds.   ···●

CASE STUDY Life Expectancy and Social Security

Because of the changing age makeup of the U.S. population, the number of retirees qualifying for Social Security benefits is expected to become much larger in the future, while the number of wage earners paying Social Security taxes is expected to grow much more slowly. As a result, one of the biggest challenges to the future of Social Security is finding a way to make sure there is enough money to pay benefits for future retirees.

Current projections show that, without significant changes, the Social Security program will be bankrupt and unable to pay full benefits after about 2037. Social Security officials have proposed several different ways to solve this problem, including changes in the amounts of benefits paid, increases in the Social Security tax rate, changes in the retirement age, and partially or fully privatizing the Social Security program. Each of these proposals faces political obstacles. But, in addition, all of these proposals are based on assumptions about future life expectancy. Without corresponding changes in retirement age, longer lives mean more years of Social Security benefits.

More specifically, recent proposals by Social Security officials have assumed that life expectancy at birth will rise only slightly during this century. For example, the Social Security projections assume that American women will not reach a life expectancy of 82 years until about 2030—but women in many European and Asian nations have *already* achieved this life expectancy. In fact, during the 20th century, U.S. life expectancies rose an average of about three years per decade. If that trend continues, life expectancy for women will be nearly 86 by 2030, which means that the budgetary problems facing Social Security (and Medicare) will be far worse than current projections might otherwise suggest.

TIME OUT TO THINK

Suppose that life expectancy increases the same amount in this century as it did in the 20th century. What will life expectancy be by 2100? How would that affect programs like Social Security? What other effects would you expect it to have on society? Overall, do you think large increases in life expectancy would be good or bad for society? Defend your opinion.

Section 6.4 Exercises

Statistical Literacy and Critical Thinking

1. **Birth Rate.** The current U.S. birth rate is given as 13.5 per 1,000 population. When comparing population growth in different countries, why is it better to use birth *rates* instead of the actual numbers of births?

2. **Vital Statistics.** What are vital statistics?

3. **Life Expectancy.** What is life expectancy? Does a 30-year-old person have the same life expectancy as a 20-year-old person? Why or why not?

4. **Life Expectancy.** Based on recent data, a 20-year-old person in the United States has a life expectancy of 58.8 years. What does that mean?

Does It Make Sense? For Exercises 5–8, decide whether the statement makes sense (or is clearly true) or does not make sense (or is clearly false). Explain clearly. Not all of these have definitive answers, so your explanation is more important than your chosen answer.

5. **Life Expectancy.** Because people are not expected to live longer than 90 years, a 100-year-old person has a negative life expectancy.

6. **Life Expectancy.** Your life expectancy increases as your age increases.

7. **Expected Age at Death.** As you become older, your expected age at death increases.

8. **Risk of Death.** In a recent year, the total numbers of deaths in the United States due to either accidents or pneumonia were approximately

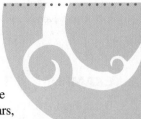

equal. Therefore, the risks of death by accident and pneumonia per 100,000 people are approximately equal.

Commercial Aviation Fatality Rates. For Exercises 9–12, use the following table, which summarizes data on commercial aviation flights in the United States for three separate years.

Year	Departures (thousands)	Fatalities	Passenger miles (billions)	Passengers (millions)
2000	9035	92	692.8	666.2
2004	11,182	14	731.9	697.8
2008	10,437	3	722.8	690.2

9. For each of the three years, find the fatality rate in deaths per 1,000 departures. On the basis of those rates, which year was the safest? Why?

10. For each of the three years, find the fatality rate in deaths per billion passenger miles. On the basis of those rates, which year was the safest? Why?

11. For each of the three years, find the fatality rate in deaths per million passengers. On the basis of those rates, which year was the safest? Why?

12. For the year 2008, find the fatality rate in deaths per passenger mile. Why don't we report the fatality rate in units of deaths per passenger mile?

Life Table. For Exercises 13–16, use the data in the following table for people in the United States between the ages of 16 and 21 years.

Age interval	Probability of dying during the interval	Number surviving to the beginning of the interval	Number of deaths during the interval	Expected remaining life-time (from the beginning of the interval)
16–17	0.000607	98,943	60	61.7
17–18	0.000706	98,883	70	60.7
18–19	0.000780	98,814	77	59.7
19–20	0.000833	98,736	82	58.8
20–21	0.000888	98,654	88	57.8

13. **Expected Lifetime.** How many years is a randomly selected 19-year-old expected to live beyond his or her 19th birthday?

14. **Expected Lifetime.** How many years is a randomly selected 17-year-old expected to live beyond his or her 17th birthday?

15. **Death Rate.** Life insurance companies must carefully monitor death rates. Before issuing a life insurance policy for a 19-year-old, the company needs to know the death rate for that age group. Find the death rate per 10,000 for people during their 20th year (age 19–20).

16. **Death Rate.** Find the death rate per 10,000 for people during their 17th year (age 16–17).

17. **High/Low U.S. Birth Rates.** The highest and lowest birth *rates* in the United States in 2008 were in Utah and Maine, respectively. Utah reported 55,633 births with a population of 2,736,424 people. Maine reported 13,610 births with a population of 1,316,456 people. Use these data to find the birth rate in births per 1,000 people for Utah and Maine.

18. **High/Low U.S. Death Numbers.** In a recent year, there were 235,000 deaths in California, the highest number in the United States. The state with the lowest number of deaths was Alaska, with 3,000 deaths. The populations of California and Alaska were approximately 35,463,000 and 648,000, respectively.

   a. Compute the death *rates* for California and Alaska in deaths per 1,000.

   b. Based on the fact that California and Alaska had the highest and lowest death numbers, respectively, does it follow that California and Alaska had the highest and lowest death *rates*? Why or why not?

19. **U.S. Birth and Death Rates.** In 2011, the estimated U.S. population reached 313 million. The overall birth rate was estimated to be 13.8 births per 1,000, and the overall death rate was estimated to be 8.4 deaths per 1,000.

   a. Approximately how many births were there in the United States?

   b. About how many deaths were there in the United States?

   c. Based on births and deaths alone (i.e., not counting immigration and emigration), about how much did the U.S. population rise during 2011?

   d. Ignoring immigration and emigration, what is the 2011 rate of population growth of the United States? What is the population growth rate expressed as a percentage?

20. **China Birth and Death Rates.** These estimated 2011 values are for China: population = 1,336,718,015; birth rate = 12.3 per 1,000; death rate = 7.0 per 1,000.

   a. Approximately how many births were there in China?

   b. About how many deaths were there in China?

   c. Based on births and deaths alone (i.e., not counting immigration and emigration), about how much did the population of China rise during 2011?

   d. Ignoring immigration and emigration, what is the 2011 rate of population growth of China? What is the population growth rate expressed as a percentage?

PROJECTS FOR THE INTERNET & BEYOND

21. **U.S. vs. World Life Expectancy.** You can find a great deal of data on the Web about life expectancies around the world. How does U.S. life expectancy compare to life expectancy in other developed countries? What might explain the differences you see? Based on your findings, discuss potential implications for social or government policy in the United States.

22. Male and Female Life Expectancies. Find data about how and why male and female life expectancies are changing with time. Why do women have longer life expectancies than men? Should we expect male life expectancies to catch up with female life expectancies in the future? Summarize your findings with a short report.

23. Uganda Case Study. Find data regarding life expectancies in Uganda over the past few decades. You will see that Ugandan life expectancy has risen dramatically in the past decade. What explains this rise, and what lessons might it have for improving life expectancy in other African nations?

24. Life Expectancy Calculations. You will find many life expectancy calculators available on the Internet; try a few of them. Do they seem to give accurate or realistic results? Explore the statistical techniques that are used to make life expectancy tables.

25. Richter Scale for Risk. The Royal Statistical Society has proposed a system of risk magnitudes and risk factors analogous to the Richter scale for measuring earthquakes. Go to the Internet to learn how these measures of risk are defined and computed. Using these measures, discuss the risks of various activities and events.

26. Understanding Risk. The book *Against the Gods: The Remarkable Story of Risk* by Peter Bernstein (John Wiley, 1996) is an award-winning account of the history of probability and risk assessment. Find the book in a library or bookstore (it's a worthwhile purchase) and identify a particular event that changed our understanding of risk. Write a two-page essay on both the history and the consequences of this particular event.

IN THE NEWS

27. Travel Safety. Find a recent news article discussing some aspect of travel safety (such as risk of accidents in automobiles or airplanes, the efficacy of child car seats, or the effects of driving while talking on a cell phone). Summarize any given statistics about risks, and give your overall opinion regarding the safety issue under discussion.

28. Vital Statistics. Find a recent news report that gives current data about vital statistics or life expectancy. Summarize the report and the statistics, and discuss any personal or social implications of the new data.

6.5 COMBINING PROBABILITIES (SUPPLEMENTARY SECTION)

The ideas of probability that we have discussed to this point in the chapter will be sufficient for most of the work we will do in this book. However, probability has many more applications, both in statistics and in other areas of life. In this section, we investigate a few more ideas of probability and a few of their many applications.

And Probabilities

Chance favors only the prepared mind.

—Louis Pasteur,
19th-century scientist

Suppose you toss two fair dice and want to know the probability that *both* will come up 4. One way to find the probability is to consider the two tossed dice as a *single* toss of two dice. Then we can find the probability using the *theoretical* method (see Section 6.2). Because "double 4's" is 1 of 36 possible outcomes, its probability is 1/36.

Alternatively, we can consider the two dice individually. For each die, the probability of a 4 is 1/6. We find the probability that both dice show a 4 by multiplying the individual probabilities:

$$P(\text{double 4's}) = P(4) \times P(4) = \frac{1}{6} \times \frac{1}{6} = \frac{1}{36}$$

By either method, the probability of rolling double 4's is 1/36. In general, we call the probability of event *A and* event *B* occurring an **and probability** (or *joint probability*).

The advantage of the multiplication technique is that it can easily be extended to situations involving more than two events. For example, we might want to find the probability of getting 10 heads on 10 coin tosses or of having a baby *and* getting a pay raise in the same year. However, there is an important distinction that must be made when working with *and* probabilities. We must distinguish between events that are *independent* and events that are *dependent*. Let's investigate each case.

Independent Events

The repeated roll of a single die produces **independent events** because the outcome of one roll does not affect the probabilities of the other rolls. Similarly, coin tosses are independent because one coin toss does not affect others. For independent events, we calculate the *and* probability by multiplying.

> *And* **Probability for Independent Events**
>
> Two events are **independent** if the outcome of one event does not affect the probability of the other event. Consider two independent events A and B with probabilities $P(A)$ and $P(B)$. The probability that A and B occur together is
>
> $$P(A \text{ and } B) = P(A) \times P(B)$$
>
> This principle can be extended to any number of independent events. For example, the probability of A, B, and a third independent event C is
>
> $$P(A \text{ and } B \text{ and } C) = P(A) \times P(B) \times P(C)$$

EXAMPLE 1   Three Coins

Suppose you toss three fair coins. What is the probability of getting three tails?

SOLUTION   Because coin tosses are independent, we multiply the probability of tails on each individual coin:

$$P(3 \text{ tails}) = \underbrace{P(\text{tails})}_{\text{coin 1}} \times \underbrace{P(\text{tails})}_{\text{coin 2}} \times \underbrace{P(\text{tails})}_{\text{coin 3}} = \frac{1}{2} \times \frac{1}{2} \times \frac{1}{2} = \frac{1}{8}$$

The probability that three tossed coins all land on tails is 1/8 (which we also determined in Example 7 of Section 6.2 with much more work). $\cdots \bullet$

Dependent Events

A batch of 15 memory chips contains 5 defective chips. If you select a chip at random from the batch, the probability of getting a defect is 5/15. Now, suppose that you select a defect on the first selection and put it in your pocket. What is the probability of getting a defect on the second selection?

Because you've removed one defective chip from the batch, the batch now contains only 14 chips, of which 4 are defective. Thus, the probability of getting a defective chip on the second draw is 4/14. This probability is less than the 5/15 probability on the first selection because the first selection changed the contents of the batch. Because the outcome of the first event affects the probability of the second event, these are **dependent events**.

Calculating the probability for dependent events still involves multiplying the individual probabilities, but we must take into account how prior events affect subsequent events. In the case of the batch of memory chips, we find the probability of getting two defective chips in a row by multiplying the 5/15 probability for the first selection by the 4/14 probability for the second selection.

$$P(2 \text{ defectives}) = \underbrace{P(\text{defective})}_{\substack{\text{first selection}}} \times \underbrace{P(\text{defective})}_{\substack{\text{second selection} \\ \text{if first selection} \\ \text{defective}}} = \frac{5}{15} \times \frac{4}{14} = 0.0952$$

The probability of drawing two defective chips in a row is 0.0952, which is slightly less than $(5/15) \times (5/15) = 0.111$, the probability we get if we replace the first chip before the second selection.

TECHNICAL NOTE

$P(B$ given $A)$ is called a *conditional probability*. In some texts, it is denoted $P(B|A)$.

And **Probability for Dependent Events**

Two events are **dependent** if the outcome of one event affects the probability of the other event. The probability that dependent events A and B occur together is

$$P(A \text{ and } B) = P(A) \times P(B \text{ given } A)$$

where $P(B$ given $A)$ means the probability of event B given the occurrence of event A.

This principle can be extended to any number of individual events. For example, the probability of dependent events A, B, and C is

$$P(A \text{ and } B \text{ and } C) = P(A) \times P(B \text{ given } A) \times P(C \text{ given } A \text{ and } B)$$

EXAMPLE 2   Bingo

The game of bingo involves drawing labeled buttons from a bin at random, without replacement. There are 75 buttons, 15 for each of the letters B, I, N, G, and O. What is the probability of drawing two B buttons in the first two selections?

SOLUTION   Bingo involves dependent events because selected buttons are not replaced, so removing a button changes the contents of the bin. The probability of drawing a B on the first draw is 15/75. If this occurs, 74 buttons remain in the bin, of which 14 are Bs. Therefore, the probability of drawing a B button on the second draw is 14/74. The probability of drawing two B buttons in the first two selections is

$$P(\text{B and B}) = \underbrace{P(\text{B})}_{\substack{\text{first draw}}} \times \underbrace{P(\text{B})}_{\substack{\text{second draw} \\ \text{given B on} \\ \text{first draw}}} = \frac{15}{75} \times \frac{14}{74} = 0.0378$$

··•

TIME OUT TO THINK

Suppose you have a standard deck of cards (which has 13 hearts among the 52 cards). For your first experiment, you draw three cards from the deck without replacing any of the selected cards. For your second experiment, you draw a card, look at it, and then replace it and reshuffle the deck before drawing again. Without doing any calculations, is the probability of drawing 3 hearts in a row larger in the first experiment or the second? Explain.

EXAMPLE 3   Polling Probability

A polling organization has a list of 1,000 people for a telephone survey. The pollsters know that 433 people out of the 1,000 are members of the Democratic Party. Assuming that a person cannot be called more than once, what is the probability that the first two people called will be members of the Democratic Party?

The importance of probability can only be derived from the judgment that it is rational to be guided by it in action.

—John Maynard Keynes

SOLUTION   This problem involves an *and* probability for dependent events: Once a person is called, that person cannot be called again. The probability of calling a member of the Democratic Party on the first call is 433/1,000. With that person removed from the calling pool, the probability of calling a member of the Democratic Party on the second call is 432/999. Therefore, the probability of calling two members of the Democratic Party on the first two calls is

$$\frac{433}{1,000} \times \frac{432}{999} = 0.1872$$

In this case, it's worth noting what would happen if we were to treat the two calls as independent events (because a selected person is not removed from the list). In that case, the probability of calling a member of the Democratic Party would be 433/1,000 on both calls, and the probability of calling two members of the Democratic Party would be

$$\frac{433}{1,000} \times \frac{433}{1,000} = 0.1875$$

Notice that this result is nearly identical to the result we found when correctly doing the calculation with dependent events. In general, if relatively few items or people are selected from a large pool (in this case, 2 people out of 1,000), then dependent events can be treated as independent events with very little error. A common guideline is that we can treat the events as being independent when the sample size is less than 5% of the population size. This practice is commonly used by polling organizations. $\cdots\bullet$

Either/Or Probabilities

Suppose we want to know the probability that *either* of two events occurs, rather than the probability that both of two events occurs. In that case, we are looking for an **either/or probability**, such as the probability of having either a blue-eyed *or* a green-eyed baby or the probability of losing your home either to a fire *or* to a hurricane. As with *and* probabilities, there are two cases to consider; we call them *overlapping* and *non-overlapping* events.

Non-Overlapping Events

A coin can land on either heads *or* tails, but it can't land on both heads *and* tails at the same time. When two events cannot possibly occur at the same time, they are said to be **non-overlapping** (or *mutually exclusive*). We can represent non-overlapping events with a Venn diagram in which each circle represents an event. If the circles do not overlap, it means that the corresponding events cannot occur together. For example, we show the possibilities of heads and tails in a coin toss as two non-overlapping circles because a coin cannot land on both heads and tails at the same time (Figure 6.12). We can calculate either/or probabilities for non-overlapping events with the following rule.

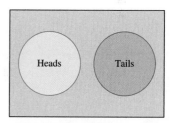

Figure 6.12 Venn diagram for non-overlapping events.

> ### *Either/Or* Probability for Non-Overlapping Events
>
> Two events are **non-overlapping** if they cannot occur at the same time. If *A* and *B* are non-overlapping events, the probability that either *A* or *B* occurs is
>
> $$P(A \text{ or } B) = P(A) + P(B)$$
>
> This principle can be extended to any number of non-overlapping events. For example, the probability that either event *A*, event *B*, or event *C* occurs is
>
> $$P(A \text{ or } B \text{ or } C) = P(A) + P(B) + P(C)$$
>
> provided that *A*, *B*, and *C* are all non-overlapping events.

EXAMPLE ④ *Either/Or* Dice

Suppose you roll a single die. What is the probability of rolling either a 2 or a 3?

SOLUTION The outcomes of 2 and 3 are non-overlapping because a single die can yield only one result. Each probability is 1/6 (because there are 6 ways for the die to land), so the combined probability is

$$P(2 \text{ or } 3) = P(2) + P(3) = \frac{1}{6} + \frac{1}{6} = \frac{2}{6} = \frac{1}{3}$$

The probability of rolling a 2 or a 3 is 1/3. $\cdots\bullet$

TABLE 6.8	Tourism Committee	
	Men	Women
American	2	6
French	4	8

Overlapping Events

To improve tourism between France and the United States, the two governments form a committee consisting of 20 people: 2 American men, 4 French men, 6 American women, and 8 French women (Table 6.8). If you meet one of these people at random, what is the probability that the person will be *either* a woman *or* a French person?

Twelve of the 20 people are French, so the probability of meeting a French person is 12/20. Similarly, 14 of the 20 people are women, so the probability of meeting a woman is 14/20. The sum of these two probabilities is

$$\frac{12}{20} + \frac{14}{20} = \frac{26}{20}$$

This cannot be the correct probability of meeting either a woman or a French person, because probabilities cannot be greater than 1. The Venn diagram in Figure 6.13 shows why simple addition was wrong in this situation. The left circle contains the 12 French people, the right circle contains the 14 women, and the American men are in neither circle. We see that there are 18 people who are either French or women (or both). Because the total number of people in the room is 20, the probability of meeting a person who is *either* French *or* a woman is 18/20 = 9/10.

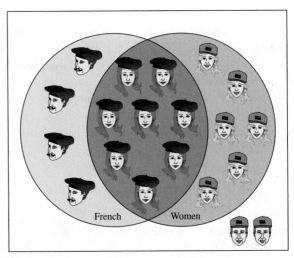

Figure 6.13 Venn diagram for overlapping events.

As the Venn diagram shows, simple addition was incorrect because the region in which the circles overlap contains 8 people who are *both* French *and* women. If we add the two individual probabilities, these 8 people get counted twice: once as women and once as French people. The probability of meeting one of these French women is 8/20. We can correct the double counting error by subtracting out this probability. Thus, the probability of meeting a person who is either French or a woman is

$$P(\text{woman or French}) = \underbrace{\frac{14}{20}}_{\substack{\text{probability} \\ \text{of a woman}}} + \underbrace{\frac{12}{20}}_{\substack{\text{probability of} \\ \text{a French person}}} - \underbrace{\frac{8}{20}}_{\substack{\text{probability of a} \\ \text{French woman}}} = \frac{18}{20} = \frac{9}{10}$$

which agrees with the result found by counting.

We say that meeting a woman and meeting a French person are **overlapping** (or *non–mutually exclusive*) events because both can occur at the same time. Generalizing the procedure we used in this example, we find the following rule.

> ### *Either/Or* Probability for Overlapping Events
>
> Two events *A* and *B* are **overlapping** if they can occur together. For overlapping events, the probability that either *A* or *B* occurs is
>
> $$P(A \text{ or } B) = P(A) + P(B) - P(A \text{ and } B)$$

The last term, $P(A \text{ and } B)$, corrects for the double counting of events in which A and B both occur together. Note that it is not necessary to use this formula. The correct probability can always be found by counting carefully and avoiding double counting.

> ### TIME ◐UT TO THINK
>
> Are the events of being born on a Wednesday or being born in Las Vegas overlapping? Are the events of being born on a Wednesday or being born in March overlapping? Are the events of being born on a Wednesday or being born on a Friday overlapping? Explain.

EXAMPLE 5 Minorities and Poverty

Suppose that the town of Pine Creek has 2,350 citizens, of which 1,950 are white and 400 are nonwhite. Further suppose that 11% of the white citizens, or 215 people, live below the poverty level, while 28% of the nonwhite citizens, or 112 people, live below the poverty level. If you visit Pine Creek, what is the probability of meeting (at random) a person who is *either* nonwhite *or* living below the poverty level?

SOLUTION Meeting a nonwhite citizen and meeting a person living in poverty are over-lapping events. It's useful to make a small table such as Table 6.9, showing how many citizens are in each of the four categories.

TABLE 6.9	Citizens in Pine Creek	
	In poverty	Not in poverty
White	215	1,735
Nonwhite	112	288

You should check that the figures in the table are consistent with the given data and that the total in all four categories is 2,350. Because there are 400 nonwhite citizens, the probability of (randomly) meeting a nonwhite citizen is $400/2{,}350 = 0.170$. Because there are $215 + 112 = 327$ people living in poverty, the probability of meeting a citizen in poverty is $327/2{,}350 = 0.139$. The probability of meeting a person who is *both* nonwhite and living in poverty is $112/2{,}350 = 0.0477$. According to the rule for overlapping events, the probability of meeting *either* a nonwhite citizen *or* a person living in poverty is

$$P(\text{non white or poverty}) = 0.170 + 0.139 - 0.0477 = 0.261$$

The probability of meeting a citizen who is *either* nonwhite or living below the poverty level is about 1 in 4. Notice the importance of subtracting out the term that corresponds to meeting a person who is *both* nonwhite and living in poverty.   ··●

Summary

Table 6.10 provides a summary of the formulas we've used in combining probabilities.

TABLE 6.10	Summary of Combining Probabilities		
***And* probability: independent events**	***And* probability: dependent events**	***Either/or* probability: non-overlapping events**	***Either/or* probability: overlapping events**
$P(A \text{ and } B) =$ $P(A) \times P(B)$	$P(A \text{ and } B) =$ $P(A) \times P(B \text{ given } A)$	$P(A \text{ or } B) =$ $P(A) + P(B)$	$P(A \text{ or } B) =$ $P(A) \times P(B) - P(A \text{ and } B)$

Section 6.5 Exercises

Statistical Literacy and Critical Thinking

1. **Independence.** Let A denote the event of turning on your cell phone and finding that it works, and let B denote the event of turning on your car radio and finding that it works. Are events A and B independent or are they dependent?

2. **Non-overlapping Events.** In your own words, state what it means for two events to be non-overlapping.

3. **Sampling with Replacement?** The professor in a class of 25 students randomly selects a student and then randomly selects a second student. If all 25 students are available for the second selection, is this sampling with replacement or sampling without replacement? Is the second outcome independent of the first?

4. **Complementary Events.** Let A denote some event. Are events A and $\overline{A}$ non-overlapping? Why or why not?

Does It Make Sense? For Exercises 5–8, decide whether the statement makes sense (or is clearly true) or does not make sense (or is clearly false). Explain clearly; not all of these have definitive answers, so your explanation is more important than your chosen answer.

5. **Lottery.** The numbers 5, 17, 18, 27, 36, and 41 were drawn in the last lottery; they should not be bet on in the next lottery because they are now less likely to occur.

6. **Combining Probabilities.** The probability of flipping a coin and getting heads is 0.5. The probability of selecting a red card when one card is drawn from a shuffled deck is also 0.5. When flipping a coin and drawing a card, the probability of getting heads or a red card is $0.5 + 0.5 = 1$.

7. *Either/Or* **Probability.** $P(A) = 0.5$ and $P(A \text{ or } B) = 0.8$.

8. **Lottery.** The probability of your winning the state lottery this week is not affected by whether you won that same lottery last week.

Concepts and Applications

9. **Births.** Assume that boys and girls are equally likely and that the gender of a child is independent of the gender of any brothers or sisters. If a couple already has three girls, find the probability of getting a girl when their fourth baby is born.

10. **Births.** A couple plans to have four children. Find the probability that the first two children are girls and the last two children are boys.

11. **Password.** A new computer owner creates a password consisting of five characters. She randomly selects a letter of the alphabet for the first character and a digit (0, 1, 2, 3, 4, 5, 6, 7, 8, 9) for each of the other four characters; the digits may be reused, so there are 10 possibilities for each of the four characters. What is the probability that her password is A1234?

12. **Wearing Hunter Orange.** A study of hunting injuries and the wearing of hunter orange clothing showed that among 123 hunters injured when mistaken for game, 6 were wearing orange (based on data from the Centers for Disease Control and Prevention). If a follow-up study begins with the random selection of hunters from this sample of 123, find the probability that the first two hunters selected were both wearing orange.

    a. Assume that the first hunter is replaced before the next one is selected.

    b. Assume that the first hunter is not replaced before the second one is selected.

    c. Which makes more sense in this situation: selecting with replacement or selecting without replacement? Why?

13. **Radio Tunes.** An MP3 player is loaded with 60 musical selections: 30 rock selections, 15 jazz selections, and 15 blues selections. The player is set on "random play," so selections are played randomly and can be repeated. What is the probability of each of the following events?

    a. The first four selections are all jazz.

    b. The first five selections are all blues.

    c. The first selection is jazz and the second is rock.

    d. Among the first four selections, none is rock.

    e. The second selection is the same song as the first.

14. **Polling Calls.** A telephone pollster has names and telephone numbers for 45 voters, 20 of whom are registered Democrats and 25 of whom are registered Republicans. Calls are made in random order. Suppose you want to find the probability that the first two calls are to Republicans.

    a. Are these independent or dependent events? Explain.

    b. If you treat them as *dependent* events, what is the probability that the first two calls are to Republicans?

    c. If you treat them as *independent* events, what is the probability that the first two calls are to Republicans?

    d. Compare the results of parts b and c.

Probability and Court Decisions. The data in the following table show the outcomes of guilty and not-guilty pleas in 1,028 criminal court cases. Use the data to answer Exercises 15–20.

	Guilty plea	Not-guilty plea
Sent to prison	392	58
Not sent to prison	564	14

Source: Brereton and Casper, "Does It Pay to Plead Guilty? Differential Sentencing and the Functioning of the Criminal Courts," *Law and Society Review,* Vol. 16, No. 1.

15. What is the probability that a randomly selected defendant either pled guilty or was sent to prison?

16. What is the probability that a randomly selected defendant either pled not guilty or was not sent to prison?

17. If two different defendants are randomly selected, what is the probability that they both entered guilty pleas?

18. If two different defendants are randomly selected, what is the probability that they both were sentenced to prison?

19. If a defendant is randomly selected, what is the probability that the defendant entered a guilty plea and was sent to prison?

20. If a defendant is randomly selected, what is the probability that the defendant entered a guilty plea and was not sent to prison?

Pedestrian Deaths. For Exercises 21–26, use the following table, which summarizes data on 985 pedestrian deaths that were caused by accidents (based on data from the National Highway Traffic Safety Administration).

		Pedestrian intoxicated?	
		Yes	No
Driver intoxicated?	Yes	59	79
	No	266	581

21. If one of the pedestrian deaths is randomly selected, find the probability that the pedestrian was intoxicated or the driver was intoxicated.

22. If one of the pedestrian deaths is randomly selected, find the probability that the pedestrian was not intoxicated or the driver was not intoxicated.

23. If one of the pedestrian deaths is randomly selected, find the probability that the pedestrian was intoxicated or the driver was not intoxicated.

24. If one of the pedestrian deaths is randomly selected, find the probability that the driver was intoxicated or the pedestrian was not intoxicated.

25. If two different pedestrian deaths are randomly selected, find the probability that they both involved intoxicated drivers.

26. If two different pedestrian deaths are randomly selected, find the probability that in both cases the pedestrians were intoxicated.

27. **Drug Tests.** An allergy drug is tested by giving 120 people the drug and 100 people a placebo. A control group consists of 80 people who were given no treatment. The number of people in each group who showed improvement appears in the table below.

	Allergy drug	Placebo	Control	Total
Improvement	65	42	31	138
No improvement	55	58	49	162
Total	120	100	80	300

a. What is the probability that a randomly selected person in the study was given either the drug or the placebo?

b. What is the probability that a randomly selected person either improved or did not improve?

c. What is the probability that a randomly selected person either was given the drug or improved?

d. What is the probability that a randomly selected person was given the drug and improved?

28. **Survey Refusals.** Refer to the following table summarizing results from a study of people who refused to answer survey questions (based on data from "I Hear You Knocking but You Can't Come In," by Fitzgerald and Fuller, Sociological Methods and Research, Vol. 11, No. 1). In each case, assume that one of the subjects is randomly selected.

	Age					
	18–21	22–29	30–39	40–49	50–59	60 and over
Responded	73	255	245	136	138	202
Refused	11	20	33	16	27	49

a. What is the probability that the selected person refused to answer? Does that probability value suggest that refusals are a problem for pollsters? Why or why not?

b. A pharmaceutical company is interested in opinions of the elderly, because they are either receiving Medicare or will receive it soon. What is the probability that the selected subject is someone 60 and over who responded?

c. What is the probability that the selected person responded or is in the 18–21 age bracket?

d. What is the probability that the selected person refused to respond or is over 59 years of age?

29. **Probability Distributions and Genetics.** Many traits are controlled by a dominant gene, denoted by **A**, and a recessive gene, denoted by **a**. Suppose that two parents carry these genes in the proportion 3:1; that is, the probability of either parent giving the **A** gene is 0.75, and the probability of either parent giving the **a** gene is 0.25. Assume that the genes are selected from each parent randomly. To answer the following questions, imagine 100 trial "births."

a. What is the probability that a child receives an **A** gene from both parents?

b. What is the probability that a child receives an **A** gene from one parent and an **a** gene from the other parent? Note that this can occur in two ways.

c. What is the probability that a child receives an **a** gene from both parents?

d. Make a table showing the probability distribution for all events.

e. If the combinations **AA** and **Aa** both result in the same dominant trait (say, brown hair) and **aa** results in the recessive trait (say, blond hair), what is the probability that a child will have the dominant trait?

30. BINGO. The game of BINGO involves drawing numbered and lettered buttons at random from a barrel. The B numbers are 1–15, the I numbers are 16–30, the N numbers are 31–45, the G numbers are 46–60, and the O numbers are 61–75. Buttons are not replaced after they have been selected. What is the probability of each of the following events on the initial selections?

a. Drawing a B button

b. Drawing two B buttons in a row

c. Drawing a B or an O

d. Drawing a B, then a G, then an N, in that order

e. Drawing anything but a B on each of the first five draws

At Least Once Problems. A common problem asks for the probability that an event occurs at least once in a given number of trials. Suppose the probability of a particular event is p (for example, the probability of drawing a heart from a deck of cards is 0.25). Then the probability that the event occurs at least once in N trials is

$$1 - (1 - p)^N$$

For example, the probability of drawing at least one heart in 10 draws (with replacement) is

$$1 - (1 - 0.25)^{10} = 0.944$$

Use this rule to solve Exercises 31 and 32.

31. The Bets of the Chevalier de Mère. It is said that probability theory was invented in the 17th century to explain the gambling of a nobleman named the Chevalier de Mère.

a. In his first game, the Chevalier bet on rolling at least one 6 with four rolls of a fair die. If played repeatedly, is this a game he should expect to win?

b. In his second game, the Chevalier bet on rolling at least one double-6 with 24 rolls of two fair dice. If played repeatedly, is this a game he should expect to win?

32. HIV among College Students. Suppose that 3% of the students at a particular college are known to carry HIV.

a. If a student has 6 sexual partners during the course of a year, what is the probability that at least one of them carries HIV?

b. If a student has 12 sexual partners during the course of a year, what is the probability that at least one of them carries HIV?

c. How many partners would a student need to have before the probability of an HIV encounter exceeded 50%?

PROJECTS FOR THE INTERNET & BEYOND

33. Simulation. A classic probability problem involves a king who wants to increase the proportion of women in his kingdom. He decrees that after a mother gives birth to a son, she is prohibited from having any more children. The king reasons that some families will have just one boy whereas other families will have a few girls and one boy, so the proportion of girls will be increased. Use coin tossing to simulate a kingdom that abides by this decree: "After a mother gives birth to a son, she will not have any other children." If this decree is followed, does the proportion of girls increase?

IN THE NEWS

34. A columnist for the *New York Daily News* (Stephen Allensworth) provided tips for selecting numbers in New York State's lottery. He advocated a system based on the use of "cold digits," which are digits that hit once or not at all in a seven-day period. He made this statement: "That [system] produces the combos $5-8-9$, $7-8-9$, $6-8-9$, $0-8-9$, and $3-8-9$. These five combos have an excellent chance of being drawn this week. Good luck to all." Can this system work? Why or why not?

CHAPTER REVIEW EXERCISES

For Exercises 1–7, use the data in the accompanying table (based on data from "Helmet Use and Risk of Head Injuries in Alpine Skiers and Snowboarders" by Sullheim et al., *Journal of the American Medical Association*, Vol. 295, No. 8).

	Head injured	Head not injured
Wore helmet	96	656
No helmet	480	2330

1. If one of the subjects is randomly selected, find the probability of selecting someone with a head injury.

2. If one of the subjects is randomly selected, find the probability of selecting someone who had a head injury or wore a helmet.

3. If one of the subjects is randomly selected, find the probability of selecting someone who did not wear a helmet or was not injured.

4. If one of the subjects is randomly selected, find the probability of selecting someone who wore a helmet and was injured.

5. If one of the subjects is randomly selected, find the probability of selecting someone who did not wear a helmet and was not injured.

6. If two different study subjects are randomly selected, find the probability that they both wore helmets.

7. If one of the subjects is randomly selected, find the probability of selecting someone who did not wear a helmet, given that the subject had head injuries.

8. Use subjective probability to estimate the probability of randomly selecting a car and selecting one that is black.

9. The Binary Computer Company manufactures computer chips used in DVD players. Those chips are made with a 27% yield, meaning that 27% of them are good and the others are defective.

   a. If one chip is randomly selected, find the probability that it is *not* good.

   b. If two chips are randomly selected, find the probability that they are both good.

   c. If five chips are randomly selected, what is the *expected number* of good chips?

   d. If five chips are randomly selected, find the probability that they are all good. If you did get five good chips among the five selected, would you continue to believe that the yield was 27%? Why or why not?

10. For a recent year, the fatality rate from motor vehicle crashes was reported as 15.2 per 100,000 population.

   a. What is the probability that a randomly selected person will die this year as a result of a motor vehicle crash?

   b. If two people are randomly selected, find the probability that they both die this year as the result of motor vehicle crashes, and express the result using three significant digits.

   c. If two people are randomly selected, find the probability that neither of them dies this year as the result of motor vehicle crashes, and express the result using six decimal places.

1. A Las Vegas handicapper can correctly predict the winning professional football team 70% of the time. What is the probability that she is wrong in her next prediction?

2. For the same handicapper described in Exercise 1, find the probability that she is correct in each of her next two predictions.

3. Estimate the probability that a randomly selected prime-time television show will be interrupted with a news bulletin.

4. When conducting a clinical trial of the effectiveness of a gender selection method, it is found that there is a 0.342 probability that the results could have occurred by chance. Does the method appear to be effective?

5. If $P(A) = 0.4$, what is the value of $P(\overline{A})$?

In Exercises 6–10, use the following results:

In the judicial case of *United States v. City of Chicago*, discrimination was charged in a qualifying exam for the position of fire captain. In the table below, Group A is a minority group and Group B is a majority group.

	Passed	Failed
Group A	10	14
Group B	417	145

6. If one of the test subjects is randomly selected, find the probability of getting someone who passed the exam.

7. Find the probability of randomly selecting one of the test subjects and getting someone who is in Group B or passed.

8. Find the probability of randomly selecting two different test subjects and finding that they are both in Group A.

9. Find the probability of randomly selecting one of the test subjects and getting someone who is in Group A and passed the exam.

10. Find the probability of getting someone who passed, given that the selected person is in Group A.

Focus On

Social Science

Do Lotteries Harm the Poor?

State-sponsored lotteries are a big business in the United States, generating more than $50 billion in annual sales, of which about one-third ($17 billion) ends up as state revenue. (The rest goes to prizes and expenses.) But are lotteries good social policy?

Lottery proponents point to several positive aspects. For example, lottery revenue helps states to fund education and recreation, while also allowing states to keep tax rates lower than they would be otherwise. Proponents also point out that lottery participation is voluntary and that polls show a large majority of Americans to be in favor of state-sponsored lotteries.

This favorable picture is part of the marketing and public relations of state lotteries. For example, Colorado state lottery officials offer statistics on the age, income, and education of lottery players compared to the general population (Figure 6.14). Within a few percentage points, the age of lottery players parallels that of the population as a whole. Similarly, the histogram for the income of lottery players gives the impression that lottery players as a whole are typical citizens—with the exception of the bars for incomes of $15,000−$25,000 and $25,000−$35,000, which show that the poor tend to play more than we would expect for their proportion of the population.

Despite the apparent benefits of lotteries, critics have long argued that lotteries are merely an unfair form of taxation. To investigate the reality, the *New York Times* conducted a study

of data from 48,875 people who had won at least $600 in New Jersey lottery games. (In an ingenious bit of sampling, these winners were taken to be a random sample of all lottery players; after all, lottery winners are determined randomly. However, the sample is not really representative of all lottery players because winners tend to buy more than an average number of tickets.) By identifying the home zip codes of the lottery players, researchers were able to determine whether players came from areas with high or low income, high or low average education, and various other demographic characteristics. The overwhelming conclusion of the *New York Times* study was that lottery spending has a much greater impact *in relative terms* on those players with lower incomes and lower educational backgrounds. For example, the following were among the specific findings:

- People in the state's lowest income areas spend five times as much of their income on lotteries as those in the state's highest income areas (more than $25 per $10,000 of annual

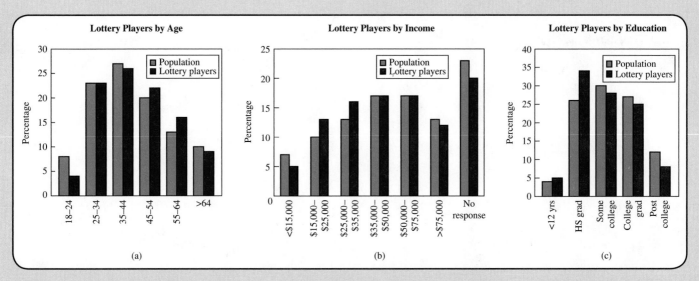

Figure 6.14 Three figures showing (a) age, (b) income, and (c) education of Colorado lottery players compared to population.

income in the lowest income areas, compared to less than $5 per $10,000 of annual income in the highest income areas).

- The number of lottery sales outlets (where lottery tickets can be purchased) is nearly twice as high per 10,000 people in low-income areas as in high-income areas.
- People in areas with the lowest percentage of college education spent over five times as much per $10,000 of annual income as those in areas with the highest percentage of college education.
- Advertising and promotion of lotteries is focused in low-income areas.

Some of the results of the *New York Times* study are summarized in Figure 6.15. It suggests that while New Jersey has a progressive tax system (higher-income people pay a greater percentage of their income in taxes), the "lottery tax" is regressive. Moreover, the study found that the areas that

generate the largest percentage of lottery revenues do *not* receive a proportional share of state funding.

Other studies have found similar patterns in other states. The overall conclusions are inescapable: While lotteries provide many benefits to state governments, the revenue they produce comes disproportionately from poorer and less educated individuals.

QUESTIONS FOR DISCUSSION

1. Study Figure 6.14. Do lottery players appear to be a typical cross-section of American society based on age? Based on income? Based on level of education? Explain. How does the "no response" category affect these conclusions?

2. Based on the more than $50 billion in total annual lottery spending and the current U.S. population, about how much does the average person spend on the lottery each year? Now, using the statistics in Figure 6.14 to estimate the percentage of the population that plays the lottery, about how much does the average lottery player spend each year?

3. Find and study a particular lottery advertisement, and determine whether it is misleading in any way.

4. Considering all factors presented in this section and other facts that you can find, do you think lotteries are fair to poor or uneducated people? Should they remain legal? Should they be restricted in any way?

5. An anonymous quote circulated on the Internet read "Lotteries are a tax on people who are bad at math." Comment on the meaning and accuracy of this quote.

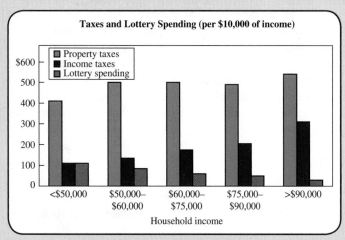

Figure 6.15 Taxes and lottery spending for New Jersey lottery winners (all figures are per $10,000 of income).

Source: New York Times.

• • • • • • • • • • • • • •

FOCUS ON
LAW

Is DNA Fingerprinting Reliable?

DNA fingerprinting (also called DNA profiling or DNA identification) is a major tool of law enforcement, used in criminal cases, in paternity cases, and even in the identification of human remains.

The scientific foundation for DNA identification has been in place for many decades. However, these ideas were not used in the courtroom until 1986. The case involved a 17-year-old boy accused of the rape and murder of two schoolgirls in Narborough, in the Midlands of England. During his interrogation, the suspect asked for a blood test, which was sent to the laboratory of a noted geneticist, Alec Jeffreys, at the nearby University of Leicester. Using methods that he had developed for paternity testing, Jeffreys compared the suspect's DNA to that found in samples from the victims. The tests showed that although both rapes and murders were committed by the same person, the person was *not* the suspect in custody. The following year, after more than 4,500 blood samples were collected, researchers made a positive identification of the murderer using Jeffreys's methods. Word of the British case and Jeffreys's methods spread rapidly. The techniques were swiftly tested, commercialized, and promoted.

To explore the essential roles that probability and statistics play in DNA identification, consider a simple eyewitness analogy. Suppose you are looking for a person who helped you out during a moment of need and you remember only three things about this person:

- The person was female.
- She had green eyes.
- She had long red hair.

If you find someone who matches this profile, can you conclude that this person is the woman who helped you? To answer, you need data telling you the probabilities that randomly selected individuals in the population have these characteristics. The probability that a person is female is about 1/2. Let's say that the probability of green eyes is about 0.06 (6% of the population has green eyes) and the probability of long red hair is 0.0075. If we assume that these characteristics occur independently of one another, then the probability that a randomly selected person has all three characteristics is

$$0.5 \times 0.06 \times 0.0075 = 0.000225$$

or about 2 in 10,000. This may seem relatively low, but it probably is not low enough to draw a definitive conclusion. For example, a profile matched by only 2 in 10,000 people will still be matched by some 200 people in a city with a population of 1 million.

DNA identification is based on a similar idea, but it is designed so that the probability of a profile match is much lower. The DNA of every individual is unique and is the same throughout the individual. A single physical trait is determined by a small piece of DNA called a *gene* at a specific *locus* (location) on a *chromosome* (Figure 6.16); humans have 23 chromosomes and roughly 30,000 genes. A gene can take two or more (often hundreds) of different forms, called *alleles* (pronounced a-leels). Different alleles give rise to variations of a trait (for example, different hair colors or different blood types). Not only can different alleles appear at a locus, but the corresponding piece of DNA can have different lengths in different people (called *variable number of tandem repeats* or VNTRs). The genetic evidence that is collected and analyzed in the lab consists of the allele lengths or allele types at five to eight different loci.

Collecting genetic evidence (from samples of blood, tissues, hair, semen, or even saliva on a postage stamp) and analyzing it are straightforward, at least in theory. Nevertheless, the process is subject to both controversy and sources of error. Suppose, in our analogy, a person is found with "reddish brown" hair instead of "red" hair. Because many characteristics are continuous (not discrete) variables, should you rule this

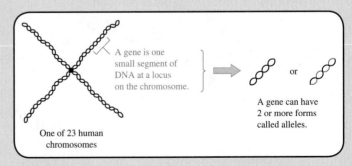

Figure 6.16 Diagram showing the relationship among chromosomes, alleles, genes, and loci.

person out or assume that reddish brown is close enough? For this reason, the issue of *binning* becomes extremely important (see Section 3.1). You might choose to include all people with hair that is any shade of red, or you could choose a narrower bin—say, bright red hair only—which would give a more discriminating test.

The same issue arises in genetic tests. When allele types or lengths are measured in the lab, there is enough variability or error in the measurements that these variables are continuous. Bin widths need to be chosen, and the choice is the source of debate. Bins with a small width give a more refined test, exclude more suspects, and ultimately provide stronger evidence against a defendant.

Other sources of scientific and statistical controversy come from assumptions about the independence of different genetic traits and in the populations chosen for measuring the frequency of different traits. For the former, DNA fingerprinting usually assumes that each genetic trait is independent of other genetic traits—which means that probabilities can be calculated with the multiplication rule for independent events—but some scientists suspect that the presence of a particular trait may affect the probability of other traits. For the latter, consider Figure 6.17, which shows allele data for four different Asian subpopulations. The horizontal axis shows 30 bins for the different allele measurements, and the vertical axis shows the frequency for each bin. Because of the significant variation in the curves for the different subpopulations, calculations will yield different probabilities for the guilt of a suspect depending on whether the suspect is compared to his own Asian subpopulation, to the population of all Asian Americans, or to the population of all Americans. Fortunately, the calculations today yield small enough probabilities that DNA fingerprinting is considered very reliable, as long as no errors are made in examining the evidence.

QUESTIONS FOR DISCUSSION

1. The result of a DNA test is considered physical (as opposed to circumstantial) evidence. Yet it is much more sophisticated and difficult to understand than a typical piece of physical evidence, such as a weapon or a piece of clothing. Some people have therefore argued that DNA evidence should not be used in a criminal trial in which jury members may not fully understand how the evidence is collected and analyzed. What do you think of this argument? Defend your opinion.

2. Suppose that an allele has a *greater* frequency in a suspect's subpopulation than in the full population, but that the suspect is compared to the full population. How would this affect the probability of a match between the suspect and a DNA sample from the crime scene?

3. Evidence from blood tests can identify a suspect with a probability of about 1 in 200. Evidence from DNA tests often provides probabilities claimed to be on the order of 1 in 10 million. If you were a juror, would you accept such a probability as positive identification of a suspect?

4. The Innocence Project uses DNA to try to clear suspects wrongfully convicted of crimes. Is innocence easier to establish than guilt through DNA testing? Explain.

5. Discuss a few other ways DNA tests can be useful, such as in settling issues of paternity. Overall, how much do you think DNA evidence affects our society?

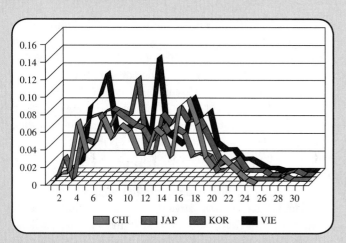

Figure 6.17   Binned frequency data for one allele, showing variability among four ethnic groups.

Source: Kathryn Roeder, "DNA Fingerprinting: A Review of the Controversy," *Statistical Science,* Vol. 9, No. 2, pp. 222–247.

Correlation and Causality

Does smoking cause lung cancer? Are drivers more dangerous when on their cell phones? Is human activity causing global warming? A major goal of many statistical studies is to search for relationships among different variables so that researchers can then determine whether one factor *causes* another. Once a relationship is discovered, we can try to determine whether there is an underlying cause. In this chapter, we will study relationships known as correlations and explore how they are important to the more difficult task of searching for causality.

The person who knows "how"
will always have a job.
The person who knows "why"
will always be his boss.

—Diane Ravitch

7.1   SEEKING CORRELATION

What does it mean when we say that smoking *causes* lung cancer? It certainly does *not* mean that you'll get lung cancer if you smoke a single cigarette. It does not even mean that you'll definitely get lung cancer if you smoke heavily for many years, as some heavy smokers do not get lung cancer. Rather, it is a *statistical* statement meaning that you are *much more likely* to get lung cancer if you smoke than if you don't smoke.

Smoking is one of the leading causes of statistics.

—Fletcher Knebel

How did researchers learn that smoking causes lung cancer? The process began with informal observations, as doctors noticed that a surprisingly high proportion of their patients with lung cancer were smokers. These observations led to carefully conducted studies in which researchers compared lung cancer rates among smokers and nonsmokers. These studies showed clearly that heavier smokers were more likely to get lung cancer. In more formal terms, we say that there is a **correlation** between the variables *amount of smoking* and *likelihood of lung cancer*. A correlation is a special type of relationship between variables, in which a rise or fall in one goes along with a corresponding rise or fall in the other.

> ### Definition
> A **correlation** exists between two variables when higher values of one variable consistently go with higher values of another variable or when higher values of one variable consistently go with lower values of another variable.

Here are a few other examples of correlations:

- There is a correlation between the variables *height* and *weight* for people; that is, taller people tend to weigh more than shorter people.
- There is a correlation between the variables *demand for apples* and *price of apples;* that is, demand tends to decrease as price increases.
- There is a correlation between *practice time* and *skill* among piano players; that is, those who practice more tend to be more skilled.

BY THE WAY
Smoking is linked to many serious diseases besides lung cancer, including heart disease and emphysema. Smoking is also linked with many less lethal health conditions, such as premature skin wrinkling and sexual impotence.

It's important to realize that establishing a correlation between two variables does *not* mean that a change in one variable *causes* a change in the other. The correlation between smoking and lung cancer did not by itself prove that smoking causes lung cancer. We could imagine, for example, that some gene predisposes a person both to smoking and to lung cancer. Nevertheless, identifying the correlation was the crucial first step in learning that smoking causes lung cancer. We will discuss the difficult task of establishing causality later in this chapter. For now, we concentrate on how we look for, identify, and interpret correlations.

> ### TIME ⏲ OUT TO THINK
> Suppose there really were a gene that made people prone to both smoking and lung cancer. Explain why we would still find a strong correlation between smoking and lung cancer in that case, but would not be able to say that smoking causes lung cancer.

Scatterplots

Table 7.1 lists data for a sample of gem-store diamonds—their prices and several common measures that help determine their value. Because advertisements for diamonds often quote only their weights (in carats), we might suspect a correlation between the weights and the prices. We can look for such a correlation by making a **scatterplot** (or *scatter diagram*) showing the relationship between the variables *weight* and *price*.

Diamond	Price	Weight (carats)	Depth	Table	Color	Clarity
1	$6,958	1.00	60.5	65	3	4
2	$5,885	1.00	59.2	65	5	4
3	$6,333	1.01	62.3	55	4	4
4	$4,299	1.01	64.4	62	5	5
5	$9,589	1.02	63.9	58	2	3
6	$6,921	1.04	60.0	61	4	4
7	$4,426	1.04	62.0	62	5	5
8	$6,885	1.07	63.6	61	4	3
9	$5,826	1.07	61.6	62	5	5
10	$3,670	1.11	60.4	60	9	4
11	$7,176	1.12	60.2	65	2	3
12	$7,497	1.16	59.5	60	5	3
13	$5,170	1.20	62.6	61	6	4
14	$5,547	1.23	59.2	65	7	4
15	$7,521	1.29	59.6	59	6	2
16	$7,260	1.50	61.1	65	6	4
17	$8,139	1.51	63.0	60	6	4
18	$12,196	1.67	58.7	64	3	5
19	$14,998	1.72	58.5	61	4	3
20	$9,736	1.76	57.9	62	8	2
21	$9,859	1.80	59.6	63	5	5
22	$12,398	1.88	62.9	62	6	2
23	$11,008	2.03	62.0	63	8	3

TABLE 7.1   Prices and Characteristics of a Sample of 23 Diamonds from Gem Dealers

Notes: Weight is measured in carats (1 carat = 0.2 gram). Depth is defined as 100 times the ratio of height to diameter. Table is the size of the upper flat surface. (Depth and table determine "cut.") Color and clarity are each measured on standard scales, where 1 is best. For color, 1 = colorless, and increasing numbers indicate more yellow. For clarity, 1 = flawless, and 6 indicates that defects can be seen by eye.

Definition

A **scatterplot** (or *scatter diagram*) is a graph in which each point represents the values of two variables.

Figure 7.1 shows the scatterplot, which can be constructed with the following procedure.

1. We assign one variable to each axis and label the axis with values that comfortably fit all the data. Sometimes the axis selection is arbitrary, but if we suspect that one variable depends on the other then we plot the *explanatory variable* on the horizontal axis and the *response variable* on the vertical axis. In this case, we expect the diamond price to depend at least in part on its weight; we therefore say that *weight* is the explanatory variable (because it helps

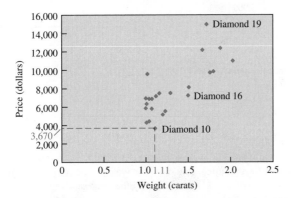

Figure 7.1 Scatterplot showing the relationship between the variables *price* and *weight* for the diamonds in Table 7.1. The dashed lines show how we find the position of the point for Diamond 10.

explain the price) and *price* is the response variable (because it *responds* to changes in the explanatory variable). We choose a range of 0 to 2.5 carats for the *weight* axis and $0 to $16,000 for the *price* axis.

2. For each diamond in Table 7.1, we plot a *single point* at the horizontal position corresponding to its weight and the vertical position corresponding to its price. For example, the point for Diamond 10 goes at a position of 1.11 carats on the horizontal axis and $3,670 on the vertical axis. The dashed lines on Figure 7.1 show how we locate this point.

3. (Optional) We can label some (or all) of the data points, as is done for Diamonds 10, 16, and 19 in Figure 7.1.

Scatterplots get their name because the way in which the points are scattered may reveal a relationship between the variables. In Figure 7.1, we see a general upward trend indicating that diamonds with greater weight tend to be more expensive. The correlation is not perfect. For example, the heaviest diamond is not the most expensive. But the overall trend seems fairly clear.

TIME ◷UT TO THINK
Identify the points in Figure 7.1 that represent Diamonds 3, 7, and 23.

EXAMPLE ① Color and Price

Using the data in Table 7.1, create a scatterplot to look for a correlation between a diamond's *color* and *price*. Comment on the correlation.

SOLUTION We expect price to depend on color, so we plot the explanatory variable *color* on the horizontal axis and the response variable *price* on the vertical axis in Figure 7.2. (You should check a few of the points against the data in Table 7.1.) The points appear much more scattered than in Figure 7.1. Nevertheless, you may notice a weak trend diagonally downward from the upper left toward the lower right. This trend represents a weak correlation in which diamonds with more yellow color (higher numbers for color) are less expensive. This trend is consistent with what we would expect, because colorless diamonds appear to sparkle more and are generally considered more desirable.

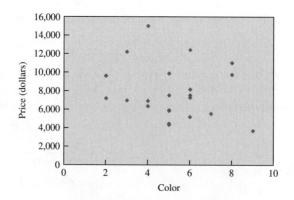

Figure 7.2 Scatterplot for the color and price data in Table 7.1.

TIME ◷UT TO THINK
Thanks to a large bonus at work, you have a budget of $6,000 for a diamond ring. A dealer offers you the following two choices for that price. One diamond weighs 1.20 carats and has color = 4. The other weighs 1.18 carats and has color = 3. Assuming all other characteristics of the diamonds are equal, which would you choose? Why?

Types of Correlation

We have seen two examples of correlation. Figure 7.1 shows a fairly strong correlation between weight and price, while Figure 7.2 shows a weak correlation between color and price. We are now ready to generalize about types of correlation. Figure 7.3 shows eight scatterplots for variables called x and y. Note the following key features of these diagrams:

- Parts a to c show **positive correlations**: The values of y tend to increase with increasing values of x. The correlation becomes stronger as we proceed from a to c. In fact, c shows a perfect positive correlation, in which all the points fall along a straight line.
- Parts d to f show **negative correlations**: The values of y tend to decrease with increasing values of x. The negative correlation becomes stronger as we proceed from d to f. In fact, f shows a perfect negative correlation, in which all the points fall along a straight line.
- Part g shows **no** correlation between x and y: Values of x do not appear to be linked to values of y in any way.
- Part h shows a **nonlinear relationship**: x and y appear to be related but the relationship does not correspond to a straight line. (*Linear* means along a straight line, and *nonlinear* means *not* along a straight line.)

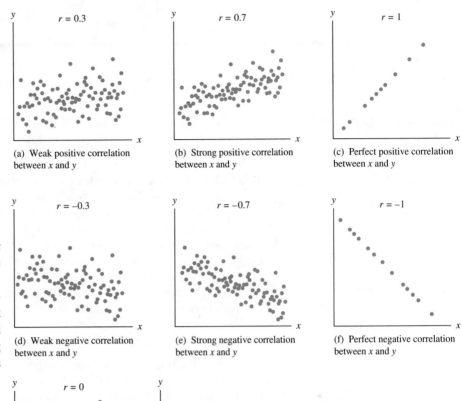

(a) Weak positive correlation between x and y

(b) Strong positive correlation between x and y

(c) Perfect positive correlation between x and y

(d) Weak negative correlation between x and y

(e) Strong negative correlation between x and y

(f) Perfect negative correlation between x and y

(g) No correlation between x and y

(h) Nonlinear relationship between x and y

Figure 7.3 Types of correlation seen on scatterplots.

Types of Correlation

Positive correlation: Both variables tend to increase (or decrease) together.

Negative correlation: The two variables tend to change in opposite directions, with one increasing while the other decreases.

No correlation: There is no apparent (linear) relationship between the two variables.

Nonlinear relationship: The two variables are related, but the relationship results in a scatterplot that does not follow a straight-line pattern.

TECHNICAL NOTE

In this text we use the term *correlation* only for *linear* relationships. Some statisticians refer to nonlinear relationships as "nonlinear correlations." There are techniques for working with nonlinear relationships that are similar to those described in this text for linear relationships.

EXAMPLE 2   Life Expectancy and Infant Mortality

Figure 7.4 shows a scatterplot for the variables *life expectancy* and *infant mortality* in 16 countries. What type of correlation does it show? Does this correlation make sense? Does it imply causality? Explain.

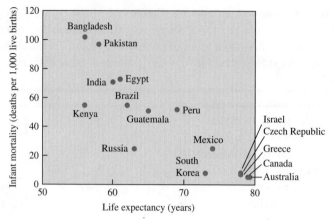

Figure 7.4 Scatterplot for life expectancy and infant mortality data.
Source: United Nations.

SOLUTION   The diagram shows a moderate negative correlation in which countries with *lower* infant mortality tend to have *higher* life expectancy. It is a *negative* correlation because the two variables vary in opposite directions. The correlation makes sense because we would expect that countries with better health care would have both lower infant mortality and higher life expectancy. However, it does *not* imply causality between infant mortality and life expectancy: We would not expect that a concerted effort to reduce infant mortality would increase life expectancy significantly unless it was part of an overall effort to improve health care. (Reducing infant mortality will *slightly* increase life expectancy because having fewer infant deaths tends to raise the mean age of death for the population.)   · · •

Measuring the Strength of a Correlation

For most purposes, it is enough to state whether a correlation is strong, weak, or nonexistent. However, sometimes it is useful to describe the strength of a correlation in more precise terms. Statisticians measure the strength of a correlation with a number called the **correlation coefficient**, represented by the letter r. The correlation coefficient is easy to calculate in principle (see the optional section on p. 243), but the actual work is tedious unless you use a calculator or computer.

We can explore the interpretation of correlation coefficients by studying Figure 7.3, which shows the value of the correlation coefficient r for each scatterplot. Notice that the correlation coefficient is always between −1 and 1. When points in a scatterplot lie close to an ascending straight line, the correlation coefficient is positive and close to 1. When all the points lie close to a descending straight line, the correlation coefficient is negative with a value close to −1. Points that do not fit any type of straight-line pattern or that lie close to a *horizontal* straight line (indicating that the *y* values have no dependence on the *x* values) result in a correlation coefficient close to 0.

TECHNICAL NOTE

For the methods of this section, there is a requirement that the two variables result in data having a "bivariate normal distribution." This basically means that for any fixed value of one variable, the corresponding values of the other variable have a normal distribution. This requirement is usually very difficult to check, so the check is often reduced to verifying that both variables result in data that are normally distributed.

Properties of the Correlation Coefficient, *r*

- The correlation coefficient, *r*, is a measure of the strength of a correlation. Its value can range only from −1 to 1.

- If there is no correlation, the points do not follow any ascending or descending straight-line pattern, and the value of *r* is close to 0.

- If there is a positive correlation, the correlation coefficient is positive ($0 < r \le 1$): Both variables increase together. A perfect positive correlation (in which all the points on a scatterplot lie on an ascending straight line) has a correlation coefficient $r = 1$. Values of *r* close to 1 indicate a strong positive correlation and positive values closer to 0 indicate a weak positive correlation.

- If there is a negative correlation, the correlation coefficient is negative ($-1 \le r < 0$): When one variable increases, the other decreases. A perfect negative correlation (in which all the points lie on a descending straight line) has a correlation coefficient $r = -1$. Values of *r* close to −1 indicate a strong negative correlation and negative values closer to 0 indicate a weak negative correlation.

EXAMPLE ③ U.S. Farm Size

Figure 7.5 shows a scatterplot for the variables *number of farms* and *mean farm size* in the United States. Each dot represents data from a single year between 1950 and 2000; on this diagram, the earlier years generally are on the right and the later years on the left. Estimate the correlation coefficient by comparing this diagram to those in Figure 7.3 and discuss the underlying reasons for the correlation.

BY THE WAY

In 1900, more than 40% of the U.S. population worked on farms; by 2000, less than 2% of the population worked on farms.

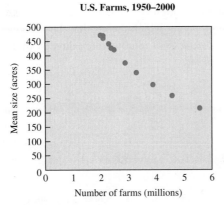

U.S. Farms, 1950–2000

Figure 7.5 Scatterplot for farm size data.
Source: U.S. Department of Agriculture.

SOLUTION   The scatterplot shows a strong negative correlation that most closely resembles the scatterplot in Figure 7.3f, suggesting a correlation coefficient around $r = -0.9$. The correlation shows that when there were fewer farms, they tended to have a larger mean size, and when there were more farms, then tended to have a smaller mean size. This trend reflects a basic change in the nature of farming: Prior to 1950, most farms were small family farms. Over time, these small farms were replaced by large farms owned by agribusiness corporations. ・・●

EXAMPLE ④ Accuracy of Weather Forecasts

The scatterplots in Figure 7.6 show two weeks of data comparing the actual high temperature for the day with the same-day forecast (part a) and the three-day forecast (part b). Estimate the correlation coefficient for each data set and discuss what these coefficients imply about weather forecasts.

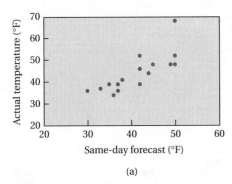

(a)

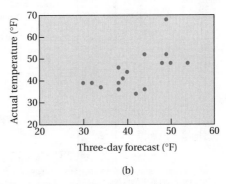

(b)

Figure 7.6 Comparison of actual high temperatures with (a) same-day and (b) three-day forecasts.

SOLUTION   If every forecast were perfect, each actual temperature would equal the corresponding forecasted temperature. This would result in all points lying on a straight line and a correlation coefficient of $r = 1$. In Figure 7.6a, in which the forecasts were made at the beginning of the same day, the points lie fairly close to a straight line, meaning that same-day forecasts are closely related to actual temperatures. By comparing this scatterplot to the diagrams in Figure 7.3, we can reasonably estimate this correlation coefficient to be about $r = 0.8$. The correlation is weaker in Figure 7.6b, indicating that forecasts made three days in advance aren't as close to actual temperatures as same-day forecasts. This correlation coefficient is about $r = 0.6$. These results are unsurprising because we expect longer-term forecasts to be less accurate. ・・●

TIME OUT TO THINK

For further practice, visually estimate the correlation coefficients for the data for diamond weight and price (Figure 7.1) and diamond color and price (Figure 7.2).

Calculating the Correlation Coefficient (Optional Section)

The formula for the (linear) correlation coefficient r can be expressed in several different ways that are all algebraically equivalent, which means that they produce the same value. The following expression has the advantage of relating more directly to the underlying rationale for r:

$$r = \frac{\sum \left[\frac{(x - \bar{x})(y - \bar{y})}{s_x \quad s_y} \right]}{n - 1}$$

⏻ USING TECHNOLOGY—SCATTERPLOTS AND CORRELATION COEFFICIENTS

Excel The screen shot below shows the process for making a scatterplot like that in Figure 7.1:

1. Enter the data, which are shown in Columns B (weight) and C (price).

2. Select the columns for the two variables on the scatterplot; in this case, Columns B and C.

3. Choose "XY Scatter" as the chart type, with no connecting lines. You can then use the "chart options" (which comes up with a right-click in the graph) to customize the design, axis range, labels, and more.

4. To calculate the correlation coefficient, shown in row 26, use the built-in function CORREL.

5. [Optional] The straight line on the graph, called a best-fit line, is added by choosing the option to "Add Trendline"; be sure to choose the "linear" option for the trendline. You'll also find options that add the two items shown in the upper left of the graph: the equation of the line and the value R^2, which is the square of the correlation coefficient. Best-fit lines and R^2 are discussed in Section 7.3.

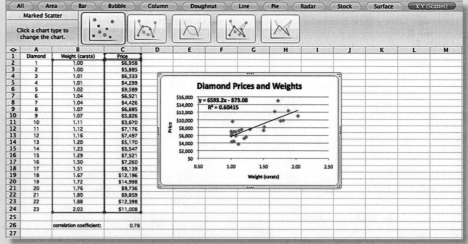

Microsoft Excel 2008 for Mac.

STATDISK Enter the paired data in columns of the STATDISK Data Window. Select **Analysis** from the main menu bar, then select the option **Correlation and Regression**. Select the columns of data to be used, then click on the **Evaluate** button. The STATDISK display will include the value of the linear correlation coefficient r and other. A scatterplot can also be obtained by clicking on the **PLOT** button.

TI-83/84 Plus Enter the paired data in lists L1 and L2, then press **STAT** and select **TESTS**. Using the option of **LinRegTTest** will result in several displayed values, including the value of the linear correlation coefficient r.

To obtain a scatterplot, press **2ND**, then **Y=** (for STAT PLOT). Press **ENTER ENTER** to turn Plot 1 on, then select the first graph type, which resembles a scatterplot. Set the X list and Y list labels to L1 and L2 and press **ZOOM**, then select **ZoomStat** and press **ENTER**.

In the above expression, division by $n - 1$ (where n is the number of pairs of data) shows that r is a type of average, so it does not increase simply because more pairs of data values are included. The symbol s_x denotes the standard deviation of the x values (or the values of the first variable), and s_y denotes the standard deviation of the y values. The expression $(x - \bar{x})/s_x$ is in the same format as the *standard score* introduced in Section 5.2. By using the standard scores for x and y, we ensure that the value of r does not change simply because a different scale of values is used. The key to understanding the rationale for r is to focus on the product of the standard scores for x and the standard scores for y. Those products tend to be positive when there is a positive correlation, and they tend to be negative when there is a negative correlation. For data with no correlation, some of the products are positive and some are negative, with the net effect that the sum is relatively close to 0.

The following alternative formula for r has the advantage of simplifying calculations, so it is often used whenever manual calculations are necessary. The following formula is also easy to program into statistical software or calculators:

$$r = \frac{n \times \Sigma(x \times y) - (\Sigma x) \times (\Sigma y)}{\sqrt{n \times (\Sigma x^2) - (\Sigma x)^2} \times \sqrt{n \times (\Sigma y^2) - (\Sigma y)^2}}$$

This formula is straightforward to use, at least in principle: First calculate each of the required sums, then substitute the values into the formula. Be sure to note that (Σx^2) and $(\Sigma x)^2$ are *not* equal: (Σx^2) tells you to first square all the values of the variable x and then add them; $(\Sigma x)^2$ tells you to add the x values first and then square this sum. In other words, perform the operation within the parentheses first. Similarly, (Σy^2) and $(\Sigma y)^2$ are not the same.

Section 7.1 Exercises

Statistical Literacy and Critical Thinking

1. **Correlation.** In the context of correlation, what does r measure, and what is it called?

2. **Scatterplot.** What is a scatterplot, and how does it help us investigate correlation?

3. **Correlation.** After computing the correlation coefficient r from 50 pairs of data, you find that $r = 0$. Does it follow that there is no relationship between the two variables? Why or why not?

4. **Scatterplot.** One set of paired data results in $r = 1$ and a second set of paired data results in $r = -1$. How do the corresponding scatterplots differ?

Does It Make Sense? For Exercises 5–8, decide whether the statement makes sense (or is clearly true) or does not make sense (or is clearly false). Explain clearly; not all of these statements have definitive answers, so your explanation is more important than your chosen answer.

5. **Births.** A study showed that for one town, as the stork population increased, the number of births in the town also increased. It therefore follows that the increase in the stork population caused the number of births to increase.

6. **Positive Effect.** An engineer for a car company finds that by reducing the weights of various cars, mileage (mi/gal) increases. Because this is a positive result, we say that there is a positive correlation.

7. **Correlation.** Two studies both found a correlation between low birth weight and weakened immune systems. The second study had a much larger sample size, so the correlation it found must be stronger.

8. **Interpreting r.** In investigating correlations between many different pairs of variables, in each case the correlation coefficient r must fall between -1 and 1.

Concepts and Applications

Types of Correlation. Exercises 9–16, list pairs of variables. For each pair, state whether you believe the two variables are correlated. If you believe they are correlated, state whether the correlation is positive or negative. Explain your reasoning.

9. **Weight/Cost.** The weights and costs of 50 different bags of apples

10. **IQ/Hat Size.** The IQ scores and hat sizes of randomly selected adults

11. **Weight/Fuel Efficiency.** The total weights of airliners flying from New York to San Francisco and the fuel efficiency as measured in miles per gallon

12. Weight/Fuel Consumption. The total weights of airliners flying from New York to San Francisco and the total amounts of fuel that they consume

13. Points and DJIA. The total number of points scored in Super Bowl football games and the changes in the Dow Jones Industrial stock index in the years following those games

14. Altitude/Temperature. The outside air temperature and the altitude of aircraft

15. Height/SAT Score. The heights and SAT scores of randomly selected subjects who take the SAT

16. Golf Score/Prize Money. Golf scores and prize money won by professional golfers

17. Crickets and Temperature. One classic application of correlation involves the association between the temperature and the number of times a cricket chirps in a minute. The scatterplot in Figure 7.7 shows the relationship for eight different pairs of temperature/chirps data. Estimate the correlation coefficient and determine whether there appears to be a correlation between the temperature and the number of times a cricket chirps in a minute.

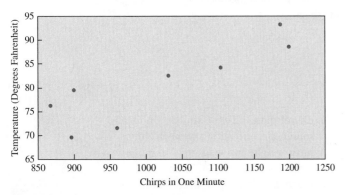

Figure 7.7 Scatterplot for cricket chirps and temperature.

Source: Based on data from *The Song of Insects* by George W. Pierce, Harvard University Press.

18. Two-Day Forecast. Figure 7.8 shows a scatterplot in which the actual high temperature for the day is compared with a forecast made two days in advance. Estimate the correlation coefficient and discuss what these data imply about weather forecasts. Do you think you would get similar results if you made similar diagrams for other two-week periods? Why or why not?

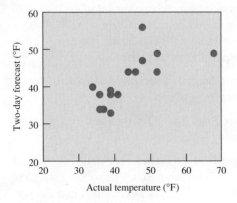

Figure 7.8

19. Safe Speeds? Consider the following table showing speed limits and death rates from automobile accidents in selected countries.

Country	Death rate (per 100 million vehicle-miles)	Speed limit (miles per hour)
Norway	3.0	55
United States	3.3	55
Finland	3.4	55
Britain	3.5	70
Denmark	4.1	55
Canada	4.3	60
Japan	4.7	55
Australia	4.9	65
Netherlands	5.1	60
Italy	6.1	75

Source: D. J. Rivkin, *New York Times.*

a. Construct a scatterplot of the data.

b. Briefly characterize the correlation in words (for example, strong positive correlation, weak negative correlation) and estimate the correlation coefficient of the data. (Or calculate the correlation coefficient exactly with the aid of a calculator or software.)

c. In the newspaper, these data were presented in an article titled "Fifty-five mph speed limit is no safety guarantee." Based on the data, do you agree with this claim? Explain.

20. Population Growth. Consider the following table showing percentage change in population and birth rate (per 1,000 of population) for 10 states over a period of 10 years.

State	Percentage change in population	Birth rate
Nevada	50.1%	16.3
California	25.7%	16.9
New Hampshire	20.5%	12.5
Utah	17.9%	21.0
Colorado	14.0%	14.6
Minnesota	7.3%	13.7
Montana	1.6%	12.3
Illinois	0%	15.5
Iowa	−4.7%	13.0
West Virginia	−8.0%	11.4

Source: U.S. Census Bureau and Department of Health and Human Services.

a. Construct a scatterplot for the data.

b. Briefly characterize the correlation in words and estimate the correlation coefficient.

c. Overall, does birth rate appear to be a good predictor of a state's population growth rate? If not, what other factor(s) may be affecting the growth rate?

21. Brain Size and Intelligence. The table below lists brain sizes (in cm^3) and Wechsler IQ scores of subjects (based on data from "Brain Size, Head Size, and Intelligence Quotient in Monozygatic Twins," by Tramo et al, *Neurology*, Vol. 50, No. 5). Is there sufficient evidence to conclude that there is a linear correlation between brain size and IQ score? Does it appear that people with larger brains are more intelligent?

Brain Size	IQ	Brain Size	IQ
965	90	1,077	97
1,029	85	1,037	124
1,030	86	1,068	125
1,285	102	1,176	102
1,049	103	1,105	114

a. Construct a scatterplot for the data.

b. Briefly characterize the correlation in words and estimate the correlation coefficient.

c. Do these data suggest that people with larger brains are more intelligent? Explain.

22. Movie Data. Consider the following table showing total box office receipts and total attendance for all American films.

Year	Total Gross Receipts (billions of dollars)	Tickets Sold (billions)
2001	8.4	1.49
2002	9.2	1.58
2003	9.2	1.53
2004	9.4	1.51
2005	8.8	1.38
2006	9.2	1.41
2007	9.7	1.40
2008	9.6	1.34
2009	10.6	1.41
2010	10.6	1.34

Source: Motion Picture Association of America.

a. Construct a scatterplot of the data.

b. Briefly characterize the correlation in words and estimate the correlation coefficient.

23. TV Time. Consider the following table showing the average hours of television watched in households in five categories of annual income.

Household income	Weekly TV hours
Less than $30,000	56.3
$30,000–$40,000	51.0
$40,000–$50,000	50.5
$50,000–$60,000	49.7
More than $60,000	48.7

Source: Nielsen Media Research.

a. Construct a scatterplot for the data. To locate the dots, use the midpoint of each income category. Use a value of $25,000 for the category "less than $30,000," and use $70,000 for "more than $60,000."

b. Briefly characterize the correlation in words and estimate the correlation coefficient.

c. Suggest a reason why families with higher incomes watch less TV. Do you think these data imply that you can increase your income simply by watching less TV? Explain.

24. January Weather. Consider the following table showing January mean monthly precipitation and mean daily high temperature for ten Northern Hemisphere cities (National Oceanic and Atmospheric Administration).

City	Mean daily high temperature for January (°F)	Mean January precipitation (inches)
Athens	54	2.2
Bombay	88	0.1
Copenhagen	36	1.6
Jerusalem	55	5.1
London	44	2.0
Montreal	21	3.8
Oslo	30	1.7
Rome	54	3.3
Tokyo	47	1.9
Vienna	34	1.5

Source: The New York Times Almanac.

a. Construct a scatterplot for the data.

b. Briefly characterize the correlation in words and estimate the correlation coefficient.

c. Can you draw any general conclusions about January temperatures and precipitation from these data? Explain.

25. Retail Sales. Consider the following table showing one year's total sales (revenue) and profits for eight large retailers in the United States.

Company	Total sales (billions of dollars)	Profits (billions of dollars)
Wal-Mart	315.6	11.2
Kroger	60.6	0.98
Home Depot	81.5	5.8
Costco	60.1	1.1
Target	52.6	2.4
Starbuck's	7.8	0.6
The Gap	16.0	1.1
Best Buy	30.8	1.1

Source: Fortune.com.

a. Construct a scatterplot for the data.

b. Briefly characterize the correlation in words and estimate the correlation coefficient.

c. Discuss your observations. Does higher sales volume necessarily translate into greater earnings? Why or why not?

26. Calories and Infant Mortality. Consider the following table showing mean daily caloric intake (all residents) and infant mortality rate (per 1,000 births) for 10 countries.

Country	Mean daily calories	Infant mortality rate (per 1,000 births)
Afghanistan	1,523	154
Austria	3,495	6
Burundi	1,941	114
Colombia	2,678	24
Ethiopia	1,610	107
Germany	3,443	6
Liberia	1,640	153
New Zealand	3,362	7
Turkey	3,429	44
United States	3,671	7

a. Construct a scatterplot for the data.

b. Briefly characterize the correlation in words and estimate the correlation coefficient.

c. Discuss any patterns you observe and any general conclusions that you can reach.

Properties of the Correlation Coefficient. For Exercises 27 and 28, determine whether the given property is true, and explain your answer.

27. Interchanging Variables. The correlation coefficient remains unchanged if we interchange the variables *x* and *y*.

28. Changing Units of Measurement. The correlation coefficient remains unchanged if we change the units used to measure *x, y,* or both.

PROJECTS FOR THE INTERNET & BEYOND

29. Unemployment and Inflation. Use the Bureau of Labor Statistics Web page to find monthly unemployment rates and inflation rates over the past year. Construct a scatterplot for the data. Do you see any trends?

30. Success in the NFL. Find last season's NFL team statistics. Construct a table showing the following for each team: number of wins, average yards gained on offense per game, and average yards allowed on defense per game. Make scatterplots to explore the correlations between offense and wins and between defense and wins. Discuss your findings. Do you think that there are other team statistics that would yield stronger correlations with the number of wins?

31. Statistical Abstract. Explore the "frequently requested tables" at the Web site for the *Statistical Abstract of the United States*. Choose data that are of interest to you and explore at least two correlations. Briefly discuss what you learn from the correlations.

32. Height and Arm Span. Select a sample of at least eight people and measure each person's height and arm span. (When you measure arm span, the person should stand with arms extended like the wings on an airplane.) Using the paired sample data, construct a scatterplot and estimate or calculate the value of the correlation coefficient. What do you conclude?

33. Height and Pulse Rate. Select a sample of at least eight people and record each person's pulse rate by counting the number of heartbeats in 1 minute. Also record each person's height. Using the paired sample data, construct a scatterplot and estimate or calculate the value of the correlation coefficient. What do you conclude?

IN THE NEWS

34. Correlations in the News. Find a recent news report that discusses some type of correlation. Describe the correlation. Does the article give any sense of the strength of the correlation? Does it suggest that the correlation reflects any underlying causality? Briefly discuss whether you believe the implications the article makes with respect to the correlation.

35. Your Own Positive Correlations. Give examples of two variables that you expect to be positively correlated. Explain why the variables are correlated and why the correlation is (or is not) important.

36. Your Own Negative Correlations. Give examples of two variables that you expect to be negatively correlated. Explain why the variables are correlated and why the correlation is (or is not) important.

7.2 INTERPRETING CORRELATIONS

Statistics show that of those who contract the habit of eating, very few survive.

—Wallace Irwin

Researchers sifting through statistical data are constantly looking for meaningful correlations, and the discovery of a new and surprising correlation often leads to a flood of news reports. You may recall hearing about some of these discovered correlations: dark chocolate consumption correlated with reduced risk of heart disease; musical talent correlated with good grades in mathematics; or eating less correlated with increased longevity. Unfortunately, the task of *interpreting* such correlations is far more difficult than discovering them in the first place. Long after the news reports have faded, we may still be unsure of whether the correlations are significant and, if so, whether they tell us anything of practical importance. In this section, we discuss some of the common difficulties associated with interpreting correlations.

Beware of Outliers

Examine the scatterplot in Figure 7.9. Your eye probably tells you that there is a positive correlation in which larger values of *x* tend to mean larger values of *y*. Indeed, if you calculate the correlation coefficient for these data, you'll find that it is a relatively high $r = 0.880$, suggesting a very strong correlation.

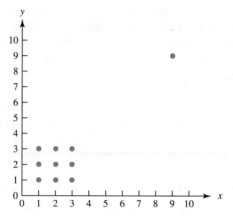

Figure 7.9 How does the outlier affect the correlation?

However, if you place your thumb over the data point in the upper right corner of Figure 7.9, the apparent correlation disappears. In fact, without this data point, the correlation coefficient is zero! In other words, removing this one data point changes the correlation coefficient from $r = 0.880$ to $r = 0$.

This example shows that correlations can be very sensitive to outliers. Recall that an *outlier* is a data value that is extreme compared to most other values in a data set (see Section 4.1). We must therefore examine outliers and their effects carefully before interpreting a correlation. On the one hand, if the outliers are mistakes in the data set, they can produce apparent correlations that are not real or mask the presence of real correlations. On the other hand, if the outliers represent real and correct data points, they may be telling us about relationships that would otherwise be difficult to see.

Note that while we should examine outliers carefully, we should *not* remove them unless we have strong reason to believe that they do not belong in the data set. Even in that case, good research principles demand that we report the outliers along with an explanation of why we thought it legitimate to remove them.

EXAMPLE **1** **Masked Correlation**

You've conducted a study to determine how the number of calories a person consumes in a day correlates with time spent in vigorous bicycling. Your sample consisted of ten women cyclists, all of approximately the same height and weight. Over a period of two weeks, you asked each woman to record the amount of time she spent cycling each day and what she ate on each of those days. You used the eating records to calculate the calories consumed each day. Figure 7.10 shows a scatterplot with each woman's mean time spent cycling on the horizontal axis and mean caloric intake on the vertical axis. Do higher cycling times correspond to higher intake of calories?

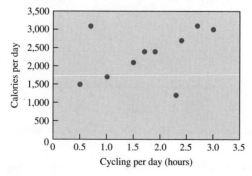

Figure 7.10 Data from the cycling study.

SOLUTION If you look at the data as a whole, your eye will probably tell you that there is a positive correlation in which greater cycling time tends to go with higher caloric intake. But the correlation is very weak, with a correlation coefficient of $r = 0.374$. However, notice that two points are outliers: one representing a cyclist who cycled about a half-hour per day and consumed more than 3,000 calories, and the other representing a cyclist who cycled more than 2 hours per day on only 1,200 calories. It's difficult to explain the two outliers, given that all the women in the sample have similar heights and weights. We might therefore suspect that these two women either recorded their data incorrectly or were not following their usual habits during the two-week study. If we can confirm this suspicion, then we would have reason to delete the two data points as invalid. Figure 7.11 shows that the correlation is quite strong without those two outlier points, and suggests that the number of calories consumed rises by a little more than 500 calories for each hour of cycling. Of course, we should *not* remove the outliers without confirming our suspicion that they were invalid data points, and we should report our reasons for leaving them out.

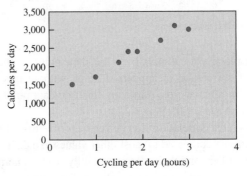

Figure 7.11 The data from Figure 7.10 without the two outliers.

Beware of Inappropriate Grouping

Correlations can also be misinterpreted when data are grouped inappropriately. In some cases, grouping data hides correlations. Consider a (hypothetical) study in which researchers seek a correlation between hours of TV watched per week and high school grade point average (GPA). They collect the 21 data pairs in Table 7.2.

The scatterplot (Figure 7.12) shows virtually no correlation; the correlation coefficient for the data is about $r = -0.063$. The lack of correlation seems to suggest that TV viewing habits are unrelated to academic achievement. However, one astute researcher realizes that some of the students watched mostly educational programs, while others tended to watch comedies, dramas, and movies. She therefore divides the data set into two groups, one for the students who watched mostly educational television and one for the other students. Table 7.3 shows her results with the students divided into these two groups.

TABLE 7.2	Hours of TV and High School GPA (hypothetical data)
Hours per week of TV	**GPA**
2	3.2
4	3.0
4	3.1
5	2.5
5	2.9
5	3.0
6	2.5
7	2.7
7	2.8
8	2.7
9	2.5
9	2.9
10	3.4
12	3.6
12	2.5
14	3.5
14	2.3
15	3.7
16	2.0
20	3.6
20	1.9

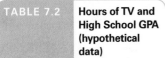

Figure 7.12 The full set of data concerning hours of TV and GPA shows virtually no correlation.

Now we find two very strong correlations (Figure 7.13): a strong positive correlation for the students who watched educational programs ($r = 0.855$) and a strong negative correlation for the other students ($r = -0.951$). The moral of this story is that the original data set hid an important (hypothetical) correlation between TV and GPA: Watching educational TV correlated positively with GPA and watching non-educational TV correlated negatively with GPA. Only when the data were grouped appropriately could this discovery be made.

TABLE 7.3	Hours of TV and High School GPA—Grouped Data (hypothetical data)		
Group 1: watched educational programs		**Group 2: watched regular TV**	
Hours per week of TV	GPA	Hours per week of TV	GPA
5	2.5	2	3.2
7	2.8	4	3.0
8	2.7	4	3.1
9	2.9	5	2.9
10	3.4	5	3.0
12	3.6	6	2.5
14	3.5	7	2.7
15	3.7	9	2.5
20	3.6	12	2.5
		14	2.3
		16	2.0
		20	1.9

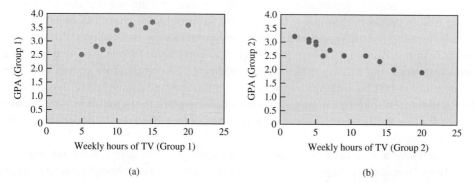

(a)   (b)

Figure 7.13   These scatterplots show the same data as Figure 7.12, separated into the two groups identified in Table 7.3.

In other cases, a data set may show a stronger correlation than actually exists among subgroups. Consider the (hypothetical) data in Table 7.4, showing the relationship between the weights and prices of selected cars. Figure 7.14 shows the scatterplot.

The data set as a whole shows a strong correlation; the correlation coefficient is $r = 0.949$. However, on closer examination, we see that the data fall into two rather distinct categories corresponding to light and heavy cars. If we analyze these subgroups separately, neither shows any correlation: The light cars alone (top six in Table 7.4) have a correlation coefficient $r = 0.019$ and the heavy cars alone (bottom six in Table 7.4) have a correlation coefficient $r = -0.022$. You can see the problem by looking at Figure 7.14. The apparent correlation of the full data set occurs because of the separation between the two clusters of points; there's no correlation within either cluster.

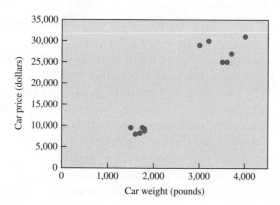

Figure 7.14   Scatterplot for the car weight and price data in Table 7.4.

TABLE 7.4	Car Weights and Prices (hypothetical data)
Weight (pounds)	**Price (dollars)**
1,500	9,500
1,600	8,000
1,700	8,200
1,750	9,500
1,800	9,200
1,800	8,700
3,000	29,000
3,500	25,000
3,700	27,000
4,000	31,000
3,600	25,000
3,200	30,000

TIME (🕐)UT TO THINK

Suppose you were shopping for a compact car. If you looked at only the overall data and correlation coefficient from Figure 7.14, would it be reasonable to consider weight as an important factor in price? What if you looked at the data for light and heavy cars separately? Explain.

(◯)ASE STUDY Fishing for Correlations

Oxford physician Richard Peto submitted a paper to the British medical journal *Lancet* showing that heart-attack victims had a better chance of survival if they were given aspirin within a few hours after their heart attacks. The editors of *Lancet* asked Peto to break down the data into subsets, to see whether the benefits of the aspirin were different for different groups of patients. For example, was aspirin more effective for patients of a certain age or for patients with certain dietary habits?

Breaking the data into subsets can reveal important facts, such as whether men and women respond to the treatment differently. However, Peto felt that the editors were asking him to divide his sample into too many subgroups. He therefore objected to the request, arguing that it would result in purely coincidental correlations. Writing about this story in the *Washington Post*, journalist Rick Weiss said, "When the editors insisted, Peto capitulated, but among other things he divided his patients by zodiac birth signs and demanded that his findings be included in the published paper. Today, like a warning sign to the statistically uninitiated, the wacky numbers are there for all to see: Aspirin is useless for Gemini and Libra heart-attack victims but is a lifesaver for people born under any other sign."

The moral of this story is that a "fishing expedition" for correlations can often produce them. That doesn't make the correlations meaningful, even though they may appear significant by standard statistical measures.

Correlation Does *Not* Imply Causality

Perhaps the most important caution about interpreting correlations is one we've already mentioned: ***Correlation does not necessarily imply causality.*** In general, correlations can appear for any of the following three reasons.

Possible Explanations for a Correlation

1. The correlation may be a *coincidence*.

2. Both correlation variables might be directly influenced by some *common underlying cause*.

3. One of the correlated variables may actually be a *cause* of the other. But note that, even in this case, it may be just one of several causes.

For example, the correlation between infant mortality and life expectancy in Figure 7.4 is a case of common underlying cause: Both variables respond to the underlying variable *quality of health care*. The correlation between smoking and lung cancer reflects the fact that smoking causes lung cancer (see the discussion in Section 7.4). Coincidental correlations are also quite common; Example 2 below discusses one such case.

Caution about causality is particularly important in light of the fact that many statistical studies are designed to look for causes. Because these studies generally begin with the search for correlations, it's tempting to think that the work is over as soon as a correlation is found. However, as we will discuss in Section 7.4, establishing causality can be very difficult.

EXAMPLE 2 How to Get Rich in the Stock Market (Maybe)

Every financial advisor has a strategy for predicting the direction of the stock market. Most focus on fundamental economic data, such as interest rates and corporate profits. But an alternative strategy might rely on a famous correlation between the Super Bowl winner in January and the direction of the stock market for the rest of the year: The stock market tends to rise when a team from the old, pre-1970 NFL wins the Super Bowl and tends to fall when the winner is not from the old NFL. This correlation successfully matched 28 of the first 32 Super Bowls to the stock market, which made the "Super Bowl Indicator" a far more reliable predictor of the stock market than any professional stock broker during the same period. In fact, detailed calculations show that the probability of such success by pure chance is less than 1 in 100,000. Should you therefore make a decision about whether to invest in the stock market based on the NFL origins of the most recent Super Bowl winner?

SOLUTION   The extremely strong correlation might make it seem like a good idea to base your investments on the Super Bowl Indicator, but sometimes you need to apply a bit of common sense. No matter how strong the correlation might be, it seems inconceivable to imagine that the origin of the winning team actually *causes* the stock market to move in a particular direction. The correlation is undoubtedly a coincidence, and the fact that its probability of occurring by pure chance is less than 1 in 100,000 is just another illustration of the fact that you can turn up surprising correlations if you go fishing for them. This fact was borne out in more recent Super Bowls: Following Super Bowl 32, the indicator successfully predicted the stock market direction in only 5 of the next 10 years—exactly the fraction that would be expected by pure chance.   ·· •

CASE STUDY Oat Bran and Heart Disease

If you buy a product that contains oat bran, there's a good chance that the label will tout the healthful effects of eating oats. Indeed, several studies have found correlations in which people who eat more oat bran tend to have lower rates of heart disease. But does this mean that everyone should eat more oats?

Not necessarily. Just because oat bran consumption is correlated with reduced risk of heart disease does not mean that it *causes* reduced risk of heart disease. In fact, the question of causality is quite controversial in this case. Other studies suggest that people who eat a lot of oat bran tend to have generally healthful diets. Thus, the correlation between oat bran consumption and reduced risk of heart disease may be a case of a common underlying cause: Having a healthy diet leads people both to consume more oat bran and to have a lower risk of heart disease. In that case, for some people, adding oat bran to their diets might be a *bad* idea because it could cause them to gain weight, and weight gain is associated with *increased* risk of heart disease.

This example shows the importance of using caution when considering issues of correlation and causality. It may be a long time before medical researchers know for sure whether adding oat bran to your diet actually causes a reduced risk of heart disease.

Useful Interpretations of Correlation

In discussing uses of correlation that might lead to wrong interpretations, we have described the effects of outliers, inappropriate groupings, fishing for correlations, and incorrectly concluding that correlation implies causality. But there are many correct and useful interpretations of correlation, some of which we have already studied. So while you should be cautious in interpreting correlations, they remain a valuable tool in any field in which statistical research plays a role.

Section 7.2 Exercises

Statistical Literacy and Critical Thinking

1. **Correlation and Causality.** In clinical trials of the drug Lisinopril, it is found that increased dosages of the drug correlated with lower blood pressure levels. Based on the correlation, can we conclude that Lisinopril treatments cause lower blood pressure? Why or why not?

2. **SIDS.** An article in the *New York Times* on infant deaths included a statement that, based on the study results, putting infants to sleep in the supine position decreased deaths due to SIDS (sudden infant death syndrome). What is wrong with that statement?

3. **Outliers.** When studying salaries paid to CEOs of large companies, it is found that almost all of them range from a few hundred thousand dollars to several million dollars, but one CEO is paid a salary of $1. Is that salary of $1 an outlier? In general, how might outliers affect conclusions about correlation?

4. **Scatterplot.** Does a scatterplot reveal anything about a cause and effect relationship between two variables?

Does It Make Sense? For Exercises 5–8, decide whether the statement makes sense (or is clearly true) or does not make sense (or is clearly false). Explain clearly; not all of these statements have definitive answers, so your explanation is more important than your chosen answer.

5. **Scatterplot.** A set of paired sample data results in a correlation coefficient of $r = 0$, so the scatterplot will show that there is no pattern of the plotted points.

6. **Causation.** If we have 20 pairs of sample data with a correlation coefficient of 1, then we know that one of the two variables is definitely the cause of the other.

7. **Causation.** If we conduct a study showing that there is a strong negative correlation between resting pulse rate and amounts of time spent in rigorous exercise, we can conclude decreases in resting pulse rates are somehow associated with increases in exercise.

8. **Causation.** If we have two variables with one being the direct cause of the other, then there may or may not be a correlation between those two variables.

Concepts and Applications

Correlation and Causality. Exercises 9–16 make statements about a correlation. In each case, state the correlation clearly. (For example, we might state that "there is a positive correlation between variable A and variable B.") Then state whether the correlation is most likely due to coincidence, a common underlying cause, or a direct cause. Explain your answer.

9. **Guns and Crime Rate.** In one state, the number of unregistered handguns steadily increased over the past several years, and the crime rate increased as well.

10. **Running and Weight.** It has been found that people who exercise regularly by running tend to weigh less than those who do not run, and those who run longer distances tend to weigh less than those who run shorter distances.

11. **Study Time.** Statistics students find that as they spend more time studying, their test scores are higher.

12. **Vehicles and Waiting Time.** It has been found that as the number of registered vehicles increases, the time drivers spend sitting in traffic also increases.

13. **Traffic Lights and Car Crashes.** It has been found that as the number of traffic lights increases, the number of car crashes also increases.

14. **Galaxies.** Astronomers have discovered that, with the exception of a few nearby galaxies, all galaxies in the universe are moving away from us. Moreover, the farther the galaxy, the faster it is moving away. That is, the more distant a galaxy, the greater the speed at which it is moving away from us.

15. **Gas and Driving.** It has been found that as gas prices increase, the distances vehicles are driven tend to get shorter.

16. **Melanoma and Latitude.** Some studies have shown that, for certain ethnic groups, the incidence of melanoma (the most dangerous form of skin cancer) increases as latitude decreases.

17. **Outlier Effects.** Consider the scatterplot in Figure 7.15.

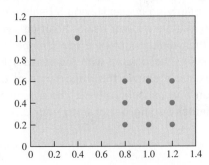

Figure 7.15

a. Which point is an outlier? Ignoring the outlier, estimate or compute the correlation coefficient for the remaining points.

b. Now include the outlier. How does the outlier affect the correlation coefficient? Estimate or compute the correlation coefficient for the complete data set.

18. Outlier Effects. Consider the scatterplot in Figure 7.16.

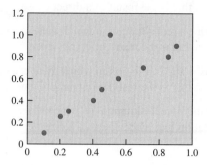

Figure 7.16

a. Which point is an outlier? Ignoring the outlier, estimate or compute the correlation coefficient for the remaining points.

b. Now include the outlier. How does the outlier affect the correlation coefficient? Estimate or compute the correlation coefficient for the complete data set.

19. Grouped Shoe Data. The following table gives measurements of weight and shoe size for 10 people (including both men and women).

a. Construct a scatterplot for the data. Estimate or compute the correlation coefficient. Based on this correlation coefficient, would you conclude that shoe size and weight are correlated? Explain.

Weight (pounds)	Shoe size
105	6
112	4.5
115	6
123	5
135	6
155	10
165	11
170	9
180	10
190	12

b. You later learn that the first five data values in the table are for women and the next five are for men. How does this change your view of the correlation? Is it still reasonable to conclude that shoe size and weight are correlated?

20. Grouped Temperature Data. The following table shows the average January high temperature and the average July high temperature for 10 major cities around the world.

City	January high	July high
Berlin	35	74
Geneva	39	77
Kabul	36	92
Montreal	21	78
Prague	34	74
Auckland	73	56
Buenos Aires	85	57
Sydney	78	60
Santiago	85	59
Melbourne	78	56

a. Construct a scatterplot for the data. Estimate or compute the correlation coefficient. Based on this correlation coefficient, would you conclude that January and July temperatures are correlated for these cities? Explain.

b. Notice that the first five cities in the table are in the Northern Hemisphere and the next five are in the Southern Hemisphere. How does this change your view of the correlation? Would you now conclude that January and July temperatures are correlated for these cities? Explain.

21. Birth and Death Rates. Figure 7.17 shows the birth and death rates for different countries, measured in births and deaths per 1,000 population.

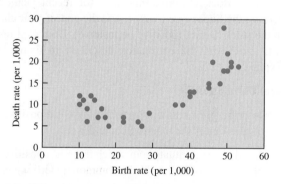

Figure 7.17 Birth and death rates for different countries.
Source: United Nations.

a. Estimate the correlation coefficient and discuss whether there is a strong correlation between the variables.

b. Notice that there appear to be two groups of data points within the full data set. Make a reasonable guess as to the makeup of these groups. In which group might you find a relatively wealthy country like Sweden? In which group might you find a relatively poor country like Uganda?

c. Assuming that your guess about groups in part b is correct, do there appear to be correlations within the groups? Explain. How could you confirm your guess about the groups?

22. Reading and Test Scores. The following (hypothetical) data set gives the number of hours 10 sixth-graders read per week and their performance on a standardized verbal test (maximum of 100).

Reading time per week	Verbal test score
1	50
1	65
2	56
3	62
3	65
4	60
5	75
6	50
10	88
12	38

a. Construct a scatterplot for these data. Estimate or compute the correlation coefficient. Based on this correlation coefficient, would you conclude that reading time and test scores are correlated? Explain.

b. Suppose you learn that five of the children read only comic books while the other five read regular books. Make a guess as to which data points fall in which group. How could you confirm your guess about the groups?

c. Assuming that your guess in part b is correct, how does it change your view of the correlation between reading time and test scores? Explain.

PROJECTS FOR THE INTERNET & BEYOND

23. Football-Stock Update. Find data for recent years concerning the Super Bowl winner and the end-of-year change in the stock market (positive or negative). Do recent results still agree with the correlation described in Example 2? Explain.

24. Real Correlations.

a. Describe a real situation in which there is a positive correlation that is the result of coincidence.

b. Describe a real situation in which there is a positive correlation that is the result of a common underlying cause.

c. Describe a real situation in which there is a positive correlation that is the result of a direct cause.

d. Describe a real situation in which there is a negative correlation that is the result of coincidence.

e. Describe a real situation in which there is a negative correlation that is the result of a common underlying cause.

f. Describe a real situation in which there is a negative correlation that is the result of a direct cause.

IN THE NEWS

25. Misinterpreted Correlations. Find a recent news report in which you believe that a correlation may have been misinterpreted. Describe the correlation, the reported interpretation, and the problems you see in the interpretation.

26. Well-Interpreted Correlations. Find a recent news report in which you believe that a correlation has been presented with a reasonable interpretation. Describe the correlation and the reported interpretation, and explain why you think the interpretation is valid.

7.3   BEST-FIT LINES AND PREDICTION

Suppose you are lucky enough to win a 1.5-carat diamond in a contest. Based on the correlation between weight and price in Figure 7.1, it should be possible to predict the approximate value of the diamond. We need only study the graph carefully and decide where a point corresponding to 1.5 carats is most likely to fall. To do this, it is helpful to draw a **best-fit line** (also called a *regression line*) through the data, as shown in Figure 7.18. This line is a "best fit" in the sense that, according to a standard statistical measure (which we discuss shortly), the data points lie closer to this line than to any other straight line that we could draw through the data.

BY THE WAY

The term *regression* comes from an 1877 study by Sir Francis Galton. He found that the heights of boys with short or tall fathers were closer to the mean than were the heights of their fathers. He therefore said that the heights of the children *regress* toward the mean, from which we get the term *regression.* The term is now used even for data that have nothing to do with a tendency to regress toward a mean.

Figure 7.18   Best-fit line for the data from Figure 7.1.

Definition

The **best-fit line** (or *regression line*) on a scatterplot is a line that lies closer to the data points than any other possible line (according to a standard statistical measure of closeness).

Of all the possible straight lines that can be drawn on a diagram, how do you know which one is the best-fit line? In many cases, you can make a good estimate of the best-fit line simply by looking at the data and drawing the line that visually appears to pass closest to all the data points. This method involves drawing the best-fit line "by eye." As you might guess, there are methods for calculating the precise equation of a best-fit line (see the optional topic at the end of this section), and many computer programs and calculators can do these calculations automatically. For our purposes in this text, a fit by eye will generally be sufficient.

Predictions with Best-Fit Lines

We can use the best-fit line in Figure 7.18 to predict the price of a 1.5-carat diamond. As indicated by the dashed lines in the figure, the best-fit line predicts that the diamond will cost about $9,000. Notice, however, that two actual data points in the figure correspond to 1.5-carat diamonds, and both of these diamonds cost less than $9,000. That is, although the predicted price of $9,000 sounds reasonable, it is certainly not guaranteed. In fact, the degree of scatter among the data points in this case tells us that we should *not* trust the best-fit line to predict accurately the price for any individual diamond. Instead, the prediction is meaningful only in a statistical sense: It tells us that if we examined many 1.5-carat diamonds, their mean price would be about $9,000.

This is only the first of several important cautions about interpreting predictions with best-fit lines. A second caution is to beware of using best-fit lines to make predictions that go beyond the bounds of the available data. Figure 7.19 shows a best-fit line for the correlation between infant mortality and longevity from Figure 7.4. According to this line, a country with a life expectancy of more than about 80 years would have a *negative* infant mortality rate, which is impossible.

> *It is a capital mistake to theorize before one has data.*
>
> —Arthur Conan Doyle

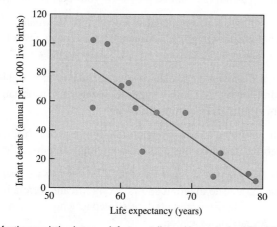

Life Expectancy and Infant Deaths

Figure 7.19 A best-fit line for the correlation between infant mortality and longevity from Figure 7.4.
Source: United Nations.

A third caution is to avoid using best-fit lines from old data sets to make predictions about current or future results. For example, economists studying historical data found a strong negative correlation between unemployment and the rate of inflation. According to this correlation, inflation should have risen dramatically in the mid-2000s when the unemployment rate fell below 6%. But inflation remained low, showing that the correlation from old data did not continue to hold.

Fourth, a correlation discovered with a sample drawn from a particular population cannot generally be used to make predictions about other populations. For example, we can't expect that the correlation between aspirin consumption and heart attacks in an experiment involving only men will also apply to women.

> *It's tough to make predictions, especially about the future.*
>
> —attributed to Niels Bohr, Yogi Berra, and others

Fifth, remember that we can draw a best-fit line through any data set, but that line is meaningless when the correlation is not significant or when the relationship is nonlinear. For example, there is no correlation between shoe size and IQ, so we could not use shoe size to predict IQ.

Cautions in Making Predictions from Best-Fit Lines

1. Don't expect a best-fit line to give a good prediction unless the correlation is strong and there are many data points. If the sample points lie very close to the best-fit line, the correlation is very strong and the prediction is more likely to be accurate. If the sample points lie away from the best-fit line by substantial amounts, the correlation is weak and predictions tend to be much less accurate.

2. Don't use a best-fit line to make predictions beyond the bounds of the data points to which the line was fit.

3. A best-fit line based on past data is not necessarily valid now and might not result in valid predictions of the future.

4. Don't make predictions about a population that is different from the population from which the sample data were drawn.

5. Remember that a best-fit line is meaningless when there is no significant correlation or when the relationship is nonlinear.

EXAMPLE 1 Valid Predictions?

State whether the prediction (or implied prediction) should be trusted in each of the following cases, and explain why or why not.

a. You've found a best-fit line for a correlation between the number of hours per day that people exercise and the number of calories they consume each day. You've used this correlation to predict that a person who exercises 18 hours per day would consume 15,000 calories per day.

b. There is a well-known but weak correlation between SAT scores and college grades. You use this correlation to predict the college grades of your best friend from her SAT scores.

c. Historical data have shown a strong negative correlation between national birth rates and affluence. That is, countries with greater affluence tend to have lower birth rates. These data predict a high birth rate in Russia.

d. A study in China has discovered correlations that are useful in designing museum exhibits that Chinese children enjoy. A curator suggests using this information to design a new museum exhibit for Atlanta-area school children.

e. Scientific studies have shown a very strong correlation between children's ingesting of lead and mental retardation. Based on this correlation, paints containing lead were banned.

f. Based on a large data set, you've made a scatterplot for salsa consumption (per person) versus years of education. The diagram shows no significant correlation, but you've drawn a best-fit line anyway. The line predicts that someone who consumes a pint of salsa per week has at least 13 years of education.

SOLUTION

a. No one exercises 18 hours per day on an ongoing basis, so this much exercise must be beyond the bounds of any data collected. Therefore, a prediction about someone who exercises 18 hours per day should not be trusted.

b. The fact that the correlation between SAT scores and college grades is weak means there is much scatter in the data. As a result, we should not expect great accuracy if we use this weak correlation to make a prediction about a single individual.

c. We cannot automatically assume that the historical data still apply today. In fact, Russia currently has a very low birth rate, despite also having a low level of affluence.

d. The suggestion to use information from the Chinese study for an Atlanta exhibit assumes that predictions made from correlations in China also apply to Atlanta. However, given the cultural differences between China and Atlanta, the curator's suggestion should not be considered without more information to back it up.

e. Given the strength of the correlation and the severity of the consequences, this prediction and the ban that followed seem quite reasonable. In fact, later studies established lead as an actual *cause* of mental retardation, making the rationale behind the ban even stronger.

f. Because there is no significant correlation, the best-fit line and any predictions made from it are meaningless.   ··•

BY THE WAY

In the United States, lead was banned from house paint in 1978 and from food cans in 1991, and a 25-year phaseout of lead in gasoline was completed in 1995. Nevertheless, many young children—especially children living in poor areas—still have enough lead in their blood to damage their health. Major sources of ongoing lead hazards include paint in older housing and soil near major roads, which has high lead content from past use of leaded gasoline.

EXAMPLE 2 Will Women Be Faster Than Men?

Figure 7.20 shows data and best-fit lines for both men's and women's world record times in the 1-mile race. Based on these data, predict when the women's world record will be faster than the men's world record. Comment on the prediction.

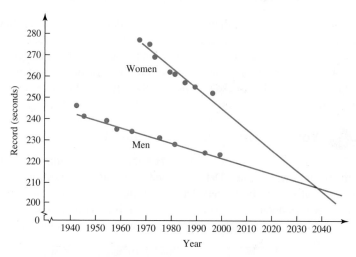

Figure 7.20 World record times in the mile (men and women).

SOLUTION If we accept the best-fit lines as drawn, the women's world record will equal the men's world record by about 2040. However, this is *not* a valid prediction because it is based on extending the best-fit lines beyond the range of the actual data. In fact, notice that the most recent world records (as of 2011) date all the way back to 1999 for men and 1996 for women, while the best-fit lines predict that the records should have fallen by several more seconds since those dates.   ··•

The Correlation Coefficient and Best-Fit Lines

Earlier, we discussed the correlation coefficient as one way of measuring the strength of a correlation. We can also use the correlation coefficient to say something about the validity of predictions with best-fit lines.

For mathematical reasons (not discussed in this text), the *square* of the correlation coefficient, or r^2, is the proportion of the variation in a variable that is accounted for by the best-fit line (or, more technically, by the linear relationship that the best-fit line expresses). For example, the correlation coefficient for the diamond weight and price data (see Figure 7.18) turns out to be $r = 0.777$. If we square this value, we get $r^2 = 0.604$ which we can interpret as follows: About 0.6, or 60%, of the variation in the diamond prices is accounted for by the best-fit line relating weight and price. That leaves 40% of the variation in price that must be due to other factors, presumably such things as depth, table, color, and clarity—which is why predictions made with the best-fit line in Figure 7.18 are not very precise.

A best-fit line can give precise predictions only in the case of a perfect correlation ($r = 1$ or $r = -1$); we then find $r^2 = 1$, which means that 100% of the variation in a variable can be accounted for by the best-fit line. In this special case of $r^2 = 1$, predictions should be exactly correct, except for the fact that the sample data might not be a true representation of the population data.

Best-Fit Lines and r^2

The *square* of the correlation coefficient, or r^2, is the proportion of the variation in a variable that is accounted for by the best-fit line.

EXAMPLE 3   Retail Hiring

You are the manager of a large department store. Over the years, you've found a strong correlation between your September sales and the number of employees you'll need to hire for peak efficiency during the holiday season; the correlation coefficient is 0.950. This year your September sales are fairly strong. Should you start advertising for help based on the best-fit line?

SOLUTION   In this case, we find that $r^2 = 0.950^2 = 0.903$, which means that 90% of the variation in the number of peak employees can be accounted for by a linear relationship with September sales. That leaves only 10% of the variation in the number of peak employees unaccounted for. Because 90% is so high, we conclude that the best-fit line accounts for the data quite well, so it seems reasonable to use it to predict the number of employees you'll need for this year's holiday season.   ·· ●

EXAMPLE 4   Voter Turnout and Unemployment

Political scientists are interested in knowing what factors affect voter turnout in elections. One such factor is the unemployment rate. Data collected in presidential election years since 1964 show a very weak negative correlation between voter turnout and the unemployment rate, with a correlation coefficient of about $r = -0.1$ (Figure 7.21). Based on this correlation, should we use the unemployment rate to predict voter turnout in the next presidential election?

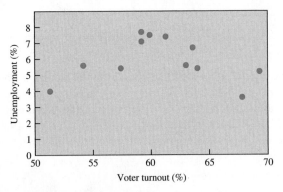

Figure 7.21   Data on voter turnout and unemployment, 1964–2008.
Source: U.S. Bureau of Labor Statistics.

SOLUTION   The square of the correlation coefficient is $r^2 = (-0.1)^2 = 0.01$, which means that only about 1% of the variation in the data is accounted for by the best-fit line. Nearly all of the variation in the data must therefore be explained by other factors. We conclude that unemployment is *not* a reliable predictor of voter turnout.   ·· ●

Multiple Regression

All who drink his remedy recover in a short time, except those whom it does not help, who all die. Therefore, it is obvious that it fails only in incurable cases.

—Galen, Roman "doctor"

If you've ever purchased a diamond, you might have been surprised that we found such a weak correlation between color and price in Figure 7.2. Surely a diamond cannot be very valuable if it has poor color quality. Perhaps color helps to explain why the correlation between weight and price is not perfect. For example, maybe differences in color explain why two diamonds with the same weight can have different prices. To check this idea, it would be nice to look for a correlation between the price and some combination of *weight and color together*.

TIME UT TO THINK

Check this idea in Table 7.1. Notice, for example, that Diamonds 4 and 5 have nearly identical weights, but Diamond 4 costs only $4,299 while Diamond 5 costs $9,589. Can differences in their color explain the different prices? Study other examples in Table 7.1 in which two diamonds have similar weights but different prices. Overall, do you think that the correlation with price would be stronger if we used weight and color together instead of either one alone? Explain.

There is a method for investigating a correlation between one variable (such as price) and a *combination* of two or more other variables (such as weight and color). The technique is called **multiple regression**, and it essentially allows us to find a *best-fit equation* that relates three or more variables (instead of just two). Because it involves more than two variables, we cannot make simple diagrams to show best-fit equations for multiple regression. However, it is still possible to calculate a measure of how well the data fit a linear equation. The most common measure in multiple regression is the *coefficient of determination*, denoted R^2. It tells us how much of the scatter in the data is accounted for by the best-fit equation. If R^2 is close to 1, the best-fit equation should be very useful for making predictions within the range of the data values. If R^2 is close to zero, then predictions with the best-fit equation are essentially useless.

> **Definition**
>
> The use of **multiple regression** allows the calculation of a best-fit equation that represents the best fit between one variable (such as price) and a *combination* of two or more other variables (such as weight and color). The coefficient of determination, R^2, tells us the proportion of the scatter in the data accounted for by the best-fit equation.

In this text, we will not describe methods for finding best-fit equations by multiple regression. However, you can use the value of R^2 to interpret results from multiple regression. For example, the correlation between price and *weight and color together* results in a value of $R^2 = 0.79$. This is somewhat higher than the $r^2 = 0.61$ that we found for the correlation between price and weight alone. Statisticians who study diamond pricing know that they can get stronger correlations by including additional variables in the multiple regression (such as depth, table, and clarity). Given the billions of dollars spent annually on diamonds, you can be sure that statisticians play prominent roles in helping diamond dealers realize the largest possible profits.

EXAMPLE 5 Alumni Contributions

You've been hired by your college's alumni association to research how past contributions were associated with alumni income and years that have passed since graduation. It is found that $R^2 = 0.36$. What does that result tell us?

SOLUTION With $R^2 = 0.36$, we conclude that 36% of the variation in past contributions can be explained by the variation in alumni income and years since graduation. It follows that 64% of the variation in past contributions can be explained by factors other than alumni income level and years since graduation. Because such a large proportion of the variation can be explained by other factors, it would make sense to try to identify any other factors that might have a strong effect on past contributions.   ∙∙●

Finding Equations for Best-Fit Lines (Optional Section)

The mathematical technique for finding the equation of a best-fit line is based on the following basic ideas. If we draw *any* line on a scatterplot, we can measure the *vertical* distance between each data point and that line. One measure of how well the line fits the

BY THE WAY

One study of alumni donations found that, in developing a multiple regression equation, one should include these variables: income, age, marital status, whether the donor belonged to a fraternity or sorority, whether the donor is active in alumni affairs, the donor's distance from the college, and the nation's unemployment rate, used as a measure of the economy (Bruggink and Siddiqui, "An Econometric Model of Alumni Giving: A Case Study for a Liberal Arts College," *The American Economist,* Vol. 39, No. 2).

data is the *sum of the squares* of these vertical distances. A large sum means that the vertical distances of data points from the line are fairly large and hence the line is not a very good fit. A small sum means the data points lie close to the line and the fit is good. Of all possible lines, the best-fit line is the line that minimizes the sum of the squares of the vertical distances. Because of this property, the best-fit line is sometimes called the *least squares line*.

You may recall that the equation of any straight line can be written in the general form

$$y = mx + b$$

where *m* is the *slope* of the line and *b* is the *y-intercept* of the line. The formulas for the slope and *y*-intercept of the best-fit line are as follows:

$$\text{slope} = m = r \times \frac{s_y}{s_x}$$

$$y - \text{intercept} = b = \bar{y} - (m \times \bar{x})$$

In the above expressions, *r* is the correlation coefficient, s_x denotes the standard deviation of the *x* values (or the values of the first variable), s_y denotes the standard deviation of the *y* values, $\bar{x}$ represents the mean of the values of the variable *x*, and $\bar{y}$ represents the mean of the values of the variable *y*. Because these formulas are tedious with manual calculations, we usually use a calculator or computer to find the slope and *y*-intercept of best-fit lines. Statistical software packages and some calculators, such as the TI-83/84 Plus family of calculators, are designed to automatically generate the equation of a best-fit line.

When software or a calculator is used to find the slope and intercept of the best-fit line, results are commonly expressed in the format $y = b_0 + b_1 x$, where b_0 is the intercept and b_1 is the slope, so be careful to correctly identify those two values.

Section 7.3 Exercises

Statistical Literacy and Critical Thinking

1. **Best-Fit Line.** What is a best-fit line (also called a regression line)? How is a best-fit line useful?

2. **r^2.** For a study involving paired sample data, it is found that $r = -0.4$. What is the value of r^2? In general, what is r^2 called, what does it measure, and how can it be interpreted? That is, what does its value tell us about the variables?

3. **Regression.** An investigator has data consisting of heights of daughters and the heights of the corresponding mothers and fathers. She wants to analyze the data to see the effect that the height of the mother and the height of the father has on the height of the daughter. Should she use a (linear) regression or multiple regression? What is the basic difference between (linear) regression and multiple regression?

4. **R^2.** Using data described in Exercise 3, it is found that $R^2 = 0.68$. Interpret that value. That is, what does that value tell us about the data?

Does It Make Sense? For Exercises 5–8, decide whether the statement makes sense (or is clearly true) or does not make sense (or is clearly false). Explain clearly; not all of these statements have definitive answers, so your explanation is more important than your chosen answer.

5. **r^2 Value.** A value of $r^2 = 1$ is obtained from a sample of paired data with one variable representing the amount of gas (gallons) purchased and the total cost of the gas.

6. **r^2 Value.** A value of $r^2 = -0.040$ is obtained from a sample of men, with each pair of data consisting of the height in inches and the SAT score for one man.

7. **Height and Weight.** Using data from the National Health Survey, the equation of the best-fit line for women's heights and weights is obtained, and it shows that a woman 120 inches tall is predicted to weigh 430 pounds.

8. **Old Faithful.** Using paired sample data consisting of the duration time (in seconds) of eruptions of Old Faithful geyser and the time interval (in minutes) after the eruption, a value

of $r^2 = 0.926$ is calculated, indicating that about 93% of the variation in the interval after eruption can be explained by the relationship between those two variables as described by the best-fit line.

Concepts and Applications

Best-Fit Lines on Scatterplots. For Exercises 9–12, do the following.

a. Insert a best-fit line in the given scatterplot.

b. Estimate or compute r and r^2. Based on your value for r^2, determine how much of the variation in the variable can be accounted for by the best-fit line.

c. Briefly discuss whether you could make valid predictions from this best-fit line.

 9. Use the scatterplot for color and price in Figure 7.2.

10. Use the scatterplot for life expectancy and infant mortality in Figure 7.4.

11. Use the scatterplot for number of farms and size of farms in Figure 7.5.

12. Use both scatterplots for actual and predicted temperature in Figure 7.6.

Best-Fit Lines. Exercises 13–20 refer to the tables in the Section 7.1 Exercises. In each case, do the following.

a. Construct a scatterplot and, based on visual inspection, draw the best-fit line by eye.

b. Briefly discuss the strength of the correlation. Estimate or compute r and r^2. Based on your value for r^2, identify how much of the variation in the variable can be accounted for by the best-fit line.

c. Identify any outliers on the scatterplot and discuss their effects on the strength of the correlation and on the best-fit line.

d. For this case, do you believe that the best-fit line gives reliable predictions outside the range of the data on the scatterplot? Explain.

13. Use the data in Exercise 19 of Section 7.1.

14. Use the data in Exercise 20 of Section 7.1.

15. Use the data in Exercise 21 of Section 7.1.

16. Use the data in Exercise 22 of Section 7.1.

17. Use the data in Exercise 23 of Section 7.1. To locate the points, use the midpoint of each income category; use a value of $25,000 for the category "less than $30,000," and use a value of $70,000 for the category "more than $60,000."

18. Use the data in Exercise 24 of Section 7.1.

19. Use the data in Exercise 25 of Section 7.1.

20. Use the data in Exercise 26 of Section 7.1.

PROJECTS FOR THE INTERNET & BEYOND

21. Lead Poisoning. Research lead poisoning, its sources, and its effects. Discuss the correlations that have helped researchers understand lead poisoning. Discuss efforts to prevent it.

22. Asbestos. Research asbestos, its sources, and its effects. Discuss the correlations that have helped researchers understand adverse health effects from asbestos exposure. Discuss efforts to prevent those adverse health effects.

23. Worldwide Population Indicators. The following table gives five population indicators for eleven selected countries. Study these data and try to identify possible correlations. Doing additional research if necessary, discuss the possible correlations you have found, speculate on the reasons for the correlations, and discuss whether they suggest a causal relationship. Birth and death rates are per 1,000 population; fertility rate is per woman.

Country	Birth rate	Death rate	Life expectancy	Percent urban	Fertility rate
Afghanistan	50	22	43	20	6.9
Argentina	21	8	72	88	2.6
Australia	15	7	78	85	1.9
Canada	14	7	78	77	1.6
Egypt	29	8	64	45	3.4
El Salvador	30	6	68	45	3.1
France	13	9	78	73	1.6
Israel	21	7	77	91	2.8
Japan	10	7	79	78	1.5
Laos	45	15	51	22	6.7
United States	16	9	76	76	2.0

Source: The New York Times Almanac.

IN THE NEWS

24. Predictions in the News. Find a recent news report in which a correlation is used to make a prediction. Evaluate the validity of the prediction, considering all of the cautions described in this section. Overall, do you think the prediction is valid? Why or why not?

25. Best-Fit Line in the News. Although scatterplots are rare in the news, they are not unheard of. Find a scatterplot of any kind in a news article (recent or not). Draw a best-fit line by eye. Discuss what predictions, if any, can be made from your best-fit line.

26. Your Own Multiple Regression. Come up with an example from your own life or work in which a multiple regression analysis might reveal important trends. Without actually doing any analysis, describe in words what you would look for through the multiple regression and how the answers might be useful.

7.4   THE SEARCH FOR CAUSALITY

A correlation may *suggest* causality, but by itself a correlation never *establishes* causality. Much more evidence is required to establish that one factor *causes* another. Earlier, we found that a correlation between two variables may be the result of either (1) coincidence, (2) a common underlying cause, or (3) one variable actually having a direct influence on the other. The process of establishing causality is essentially a process of ruling out the first two explanations.

In principle, we can rule out the first two explanations by conducting experiments:

- We can rule out coincidence by repeating the experiment many times (or by using a large number of subjects in the experiment). Because coincidences occur randomly, the same coincidence is unlikely to occur in repeated trials of an experiment.
- We can rule out a common underlying cause by controlling and randomizing the experiment to eliminate the effects of confounding variables (see Section 1.3). If the controls rule out confounding variables, any remaining effects must be caused by the variables of interest.

Unfortunately, these ideas are often difficult to put into practice. In the case of ruling out coincidence, it may be too time-consuming or expensive to repeat an experiment a sufficient number of times. To rule out a common underlying cause, the experiment must control for *everything* except the variables of interest, and this is often impossible. Moreover, there are many cases in which experiments are impractical or unethical, so we can gather only observational data. Because observational studies cannot definitively establish causality, we must find other ways of trying to establish causality.

Establishing Causality

Suppose you have discovered a correlation and suspect causality. How can you test your suspicion? Let's return to the issue of smoking and lung cancer. The strong correlation between smoking and lung cancer did not by itself prove that smoking causes lung cancer. In principle, we could have looked for proof with a controlled experiment. But such an experiment would be unethical because it would require forcing a group of randomly selected people to smoke cigarettes. So how was smoking established as a cause of lung cancer?

The answer involves several lines of evidence. First, researchers found correlations between smoking and lung cancer among many groups of people: women, men, and people of different races and cultures. Second, among groups of people that seemed otherwise identical, lung cancer was found to be more rare in nonsmokers. Third, people who smoked more and for longer periods of time were found to have higher rates of lung cancer. Fourth, when researchers accounted for other potential causes of lung cancer (such as exposure to radon gas or asbestos), they found that almost all the remaining lung cancer cases occurred among smokers (or people exposed to second-hand smoke).

These four lines of evidence made a strong case, but still did not rule out the possibility that some other factor, such as genetics, predisposes people both to smoking and to lung cancer. However, two additional lines of evidence made this possibility highly unlikely. One line of evidence came from animal experiments. In controlled experiments, animals were divided into randomly chosen treatment and control groups. The experiments still found a correlation between inhalation of cigarette smoke and lung cancer, which seems to rule out a genetic factor, at least in the animals. The final line of evidence came from biologists studying small samples of human lung tissue. The biologists discovered the basic process by which ingredients in cigarette smoke create cancer-causing mutations. This process does not appear to depend in any way on specific genetic factors, making it

BY THE WAY

Statistical methods cannot prove that smoking causes cancer, but statistical methods can be used to identify an association, and physical proof of causation can then be sought by researchers. Dr. David Sidransky of Johns Hopkins University and other researchers found a direct physical link that involves mutations of a specific gene among smokers. Molecular analysis of genetic changes allows researchers to determine whether cigarette smoking is the cause of a cancer. (See "Association Between Cigarette Smoking and Mutation of the p53 Gene in Squamous-Cell Carcinoma of the Head and Neck," by Brennan, Boyle et al., *New England Journal of Medicine*, Vol 332, No. 11.)

all but certain that lung cancer is caused by smoking and not by any preexisting genetic factor. The fact that second-hand smoke exposure is also associated with some cases of lung cancer further argues against a genetic factor (since second-hand smoke affects non-smokers) but is consistent with the idea that ingredients in cigarette smoke create cancer-causing mutations.

The following box summarizes these ideas about establishing causality. Generally speaking, the case for causality is stronger when more of these guidelines are met.

Guidelines for Establishing Causality

If you suspect that a particular variable (the suspected cause) is causing some effect:

1. Look for situations in which the effect is correlated with the suspected cause even while other factors vary.

2. Among groups that differ only in the presence or absence of the suspected cause, check that the effect is similarly present or absent.

3. Look for evidence that larger amounts of the suspected cause produce larger amounts of the effect.

4. If the effect might be produced by other potential causes (besides your suspected cause), make sure that the effect still remains after accounting for these other potential causes.

5. If possible, test the suspected cause with an experiment. If the experiment cannot be performed with humans for ethical reasons, consider doing the experiment with animals, cell cultures, or computer models.

6. Try to determine the physical mechanism by which the suspected cause produces the effect.

BY THE WAY

The first four guidelines to the left are called *Mill's methods* after John Stuart Mill (1806–1873). Mill was a leading scholar of his time and an early advocate of women's right to vote. In philosophy, the four methods are called, respectively, the methods of agreement, difference, concomitant variation, and residues.

TIME OUT TO THINK

There's a great deal of controversy concerning whether animal experiments are ethical. What is your opinion of animal experiments? Defend your opinion.

CASE STUDY Air Bags and Children

By the mid-1990s, passenger-side air bags had become commonplace in cars. Statistical studies showed that the air bags saved many lives in moderate- to high-speed collisions. But a disturbing pattern also appeared. In at least some cases, young children, especially infants and toddlers in child car seats, were killed by air bags in low-speed collisions.

At first, many safety advocates found it difficult to believe that air bags could be the cause of the deaths. But the observational evidence became stronger, meeting the first four guidelines for establishing causality. For example, the greater risk to infants in child car seats fit Guideline 3, because it indicated that being closer to the air bags increased the risk of death. (A child car seat sits on top of the built-in seat, thereby putting a child closer to the air bags than the child would be otherwise.)

To seal the case, safety experts undertook experiments using dummies. They found that children, because of their small size, often sit where they could be easily hurt by the explosive opening of an air bag. The experiments also showed that an air bag could impact a child car seat hard enough to cause death, thereby revealing the physical mechanism by which the deaths occurred.

BY THE WAY

Based on these studies, the government now recommends that child car seats *never* be used on the front seat and that children under age 12 (or under 4 feet, 9 inches tall) sit in the back seat whenever possible.

CASE STUDY Cardiac Bypass Surgery

Cardiac bypass surgery is performed on people who have severe blockage of arteries that supply the heart with blood (the coronary arteries). If blood flow stops in these arteries, a patient may suffer a heart attack and die. Bypass surgery essentially involves grafting new blood vessels onto the blocked arteries so that blood can flow around the blocked areas. By the mid-1980s, many doctors were convinced that the surgery was prolonging the lives of their patients.

However, a few early retrospective studies turned up a disconcerting result: Statistically, the surgery appeared to be making little difference. In other words, patients who had the surgery seemed to be faring no better on average than similar patients who did not have it. If this were true, it meant that the surgery was not worth the pain, risk, and expense involved.

Because these results flew in the face of what many doctors thought they had observed in their own patients, researchers began to dig more deeply. Soon, they found confounding variables that had not been accounted for in the early studies. For example, they found that patients getting the surgery tended to have more severe blockage of their arteries, apparently because doctors recommended the surgery more strongly to these patients. Because these patients were in worse shape to begin with, a comparison of longevity between them and other patients was not really valid.

More important, the research soon turned up substantial differences in the results among patients who had the surgery in different hospitals. In particular, a few hospitals were achieving remarkable success with bypass surgery and their patients fared far better than patients who did not have the surgery or had it at other hospitals. Clearly, the surgical techniques used by doctors at the successful hospitals were somehow different and superior. Doctors studied the differences to ensure that all doctors could be trained in the superior techniques.

In summary, the confounding variables of *amount of blockage* and *surgical technique* had prevented the early studies from finding a real correlation between cardiac bypass surgery and prolonged life. Today, cardiac bypass surgery is accepted as a *cause* of prolonged life in patients with blocked coronary arteries. It is now among the most common types of surgery, and it typically adds *decades* to the lives of the patients who undergo it.

Hidden Causality

So far we have discussed how to establish causality after first discovering a correlation. However, sometimes a correlation—or the lack of a correlation—can hide an underlying causality. As the next case study shows, such hidden causality often occurs because of confounding variables.

Confidence in Causality

The six guidelines offer us a way to examine the strength of a case for causality, but we often must make decisions before a case of causality is fully established. Consider, for example, the well-known case of global warming. It may never be possible to prove beyond all doubt that the burning of fossil fuels is causing global warming (see the Focus on Environment at the end of this chapter), so we must decide whether to act while we still face some uncertainty about causation. How much must we know before we decide to act?

In other areas of statistics, accepted techniques help us deal with this type of uncertainty by allowing us to calculate a numerical level of confidence or significance. But there are no accepted ways to assign such numbers to the uncertainty that comes with questions of causality. Fortunately, another area of study has dealt with practical problems of causality for hundreds of years: our legal system. You may be familiar with the three broad ways of expressing a legal level of confidence shown on the top of the next page.

BY THE WAY

As you might guess, it is also difficult to define *reasonable doubt*. For criminal trials, the Supreme Court endorsed this guidance from Justice Ruth Bader Ginsburg: "Proof beyond a reasonable doubt is proof that leaves you firmly convinced of the defendant's guilt. There are very few things in this world that we know with absolute certainty, and in criminal cases the law does not require proof that overcomes every possible doubt. If, based on your consideration of the evidence, you are firmly convinced that the defendant is guilty of the crime charged, you must find him guilty. If on the other hand, you think there is a real possibility that he is not guilty, you must give him the benefit of the doubt and find him not guilty."

Broad Levels of Confidence in Causality

Possible cause: We have discovered a correlation, but cannot yet determine whether the correlation implies causality. In the legal system, possible cause (such as thinking that a particular suspect possibly committed a particular crime) is often the reason for starting an investigation.

Probable cause: We have good reason to suspect that the correlation involves cause, perhaps because some of the guidelines for establishing causality are satisfied. In the legal system, probable cause is the general standard for getting a judge to grant a warrant for a search or wiretap.

Cause beyond reasonable doubt: We have found a physical model that is so successful in explaining how one thing causes another that it seems unreasonable to doubt the causality. In the legal system, cause beyond reasonable doubt is the usual standard for convictions and generally demands that the prosecution have shown how and why (essentially the physical model) the suspect committed the crime. Note that beyond *reasonable* doubt does *not* mean beyond *all* doubt.

While these broad levels remain fairly vague, they give us at least some common language for discussing confidence in causality. If you study law, you will learn much more about the subtleties of interpreting these terms. However, because statistics has little to say about them, we will not discuss them much further in this text.

Section 7.4 Exercises

Statistical Literacy and Critical Thinking

1. **Correlation.** Identify three different explanations for the presence of a correlation between two variables.

2. **Role of Experiments.** In theory, we can use experiments to rule out two of the three different explanations for the presence of a correlation between two variables. Which of the three explanations do we *not* want to rule out? Why would we not want to rule it out?

3. **Confounding Variable.** What is a confounding variable? How can a confounding variable create a situation in which an underlying causality is hidden?

4. **Correlation and Causality.** What is the difference between finding a correlation between two variables and establishing causality between two variables?

Does It Make Sense? For Exercises 5–8, decide whether the statement makes sense (or is clearly true) or does not make sense (or is clearly false). Explain clearly; not all of these statements have definitive answers, so your explanation is more important than your chosen answer.

5. **Value of *r*.** When analyzing paired sample data, the value of the correlation coefficient r allows us to determine whether one variable has a direct causal effect on the other.

6. **Value of *r*.** A variable can have a direct causal effect on another variable only if the correlation coefficient is given by $r = 1$.

7. **Smoking and Cotinine.** A study showed that there is a correlation between exposure to second-hand smoke and the measured amount of cotinine in the body. We can establish that exposure to second-hand smoke is a cause of cotinine if we can rule out coincidence as a possible explanation of the correlation.

8. **Smoking and Cotinine.** When the body absorbs nicotine, it converts it into cotinine. Experiments have ruled out coincidence as an explanation for a correlation between exposure to second-hand smoke and cotinine in the body. The only possible explanations for that correlation are that the exposure causes cotinine or that there is some other underlying cause.

Concepts and Applications

Physical Models. For Exercises 9–12, determine whether the stated causal connection is valid. If the causal connection appears to be valid, provide an explanation.

9. **Test Grades.** Test grades are affected by the amount of time and effort spent studying and preparing for the test.

10. **Magnet Treatment.** Heart disease can be cured by wearing a magnetic bracelet on your wrist.

11. **Drinking and Reaction Time.** Drinking greater amounts of alcohol decreases a person's reaction time.

12. **IQ and Pulse Rate.** People with higher resting pulse rates (beats per minute) tend to have higher IQ scores.

13. **Identifying Causes: Headaches.** You are trying to identify the cause of late-afternoon headaches that plague you several days each week. For each of the following tests and observations, explain which of the six guidelines for establishing causality you used and what you concluded. Then summarize your overall conclusion based on all the observations.

    a. The headaches occur only on days that you go to work.

    b. If you stop drinking Coke at lunch, the headaches persist.

    c. In the summer, the headaches occur less frequently if you open the windows of your office slightly. They occur even less often if you open the windows of your office fully.

14. Smoking and Lung Cancer. There is a strong correlation between tobacco smoking and incidence of lung cancer, and most physicians believe that tobacco smoking causes lung cancer. Yet, not everyone who smokes gets lung cancer. Briefly describe how smoking could cause cancer when not all smokers get cancer.

15. Other Lung Cancer Causes. Several things besides smoking have been shown to be probabilistic causal factors in lung cancer. For example, exposure to asbestos and exposure to radon gas, both of which are found in many homes, can cause lung cancer. Suppose that you meet a person who lives in a home that has a high radon level and insulation that contains asbestos. The person tells you, "I smoke, too, because I figure I'm doomed to lung cancer anyway." What would you say in response? Explain.

16. Longevity of Orchestra Conductors. A famous study in *Forum on Medicine* concluded that the mean lifetime of conductors of major orchestras was 73.4 years, about 5 years longer than that of all American males at the time. The author claimed that a life of music *causes* a longer life. Evaluate the claim of causality and propose other explanations for the longer life expectancy of conductors.

17. Older Moms. A study reported in *Nature* claims that women who give birth later in life tend to live longer. Of the 78 women who were at least 100 years old at the time of the study, 19% had given birth after their 40th birthday. Of the 54 women who were 73 years old at the time of the study, only 5.5% had given birth after their 40th birthday. A researcher stated that "if your reproductive system is aging slowly enough that you can have a child in your 40s, it probably bodes well for the fact that the rest of you is aging slowly too." Was this an observational study or an experiment? Does the study suggest that later child bearing *causes* longer lifetimes or that later child bearing reflects an underlying cause? Comment on how persuasive you find the conclusions of the report.

18. High-Voltage Power Lines. Suppose that people living near a high-voltage power line have a higher incidence of cancer than people living farther from the power line. Can you conclude that the high-voltage power line is the cause of the elevated cancer rate? If not, what other explanations might there be for it? What other types of research would you like to see before you concluded that high-voltage power lines cause cancer?

19. Gun Control. Those who favor gun control often point to a positive correlation between the availability of handguns and murder rates to support their position that gun control would save lives. Does this correlation, by itself, indicate that handgun availability causes a higher murder rate? Suggest some other factors that might support or weaken this conclusion.

20. Vasectomies and Prostate Cancer. An article titled "Does Vasectomy Cause Prostate Cancer?" (*Chance*, Vol. 10, No. 1) reports on several large studies that found an increased risk of prostate cancer among men with vasectomies. In the absence of a direct cause, several researchers attribute the correlation to *detection bias*, in which men with vasectomies are more likely to visit the doctor and thereby are more likely to have any prostate cancer found by the doctor. Briefly explain how this detection bias could affect the claim that vasectomies cause prostate cancer.

PROJECTS FOR THE INTERNET & BEYOND

21. Air Bags and Children. Starting from the Web site of the National Highway Traffic Safety Administration, research the latest studies on the safety of air bags, especially with regard to children. Write a short report summarizing your findings and offering recommendations for improving child safety in cars.

22. Dietary Fiber and Coronary Heart Disease. In the largest study of how dietary fiber prevents coronary heart disease (CHD) in women (*Journal of the American Medical Association*, Vol. 281, No. 21), researchers detected a reduced risk of CHD among women who have a high-fiber diet. Find the research paper, summarize its findings, and discuss whether a cause for the correlation is proposed.

23. Coffee and Gallstones. Writing in the *Journal of the American Medical Association* (Vol. 281, No. 22), researchers reported finding a negative correlation between incidence of gallstone disease and coffee consumption in men. Find the research paper, summarize its findings, and discuss whether a cause for the correlation is proposed.

24. Alcohol and Stroke. Researchers reported in the *Journal of the American Medical Association* (Vol. 281, No. 1) that moderate alcohol consumption is correlated with a decreased risk of stroke in people 40 years of age and older. (Heavy consumption of alcohol was correlated with deleterious effects.) Find the research paper, summarize its findings, and discuss whether a cause for the correlation is proposed.

25. Tobacco Lawsuits. Tobacco companies have been the subject of many lawsuits related to the dangers of smoking. Research one recent lawsuit. What were the plaintiffs trying to prove? What statistical evidence did they use? How well do you think they established causality? Did they win? Summarize your findings in one to two pages.

IN THE NEWS

26. Causation in the News. Find a recent news report in which a statistical study led to a conclusion of causation. Describe the study and the claimed causation. Do you think the claim of causation is legitimate? Explain.

27. Legal Causation. Find a news report concerning an ongoing legal case, either civil or criminal, in which establishing causality is important to the outcome. Briefly describe the issue of causation in the case and how the ability to establish or refute causality will influence the outcome of the case.

CHAPTER REVIEW EXERCISES

For Exercises 1–3, refer to the combined city–highway fuel economy ratings (mi/gal) for different cars. The old ratings are based on tests used before 2008 and the new ratings are based on tests that went into effect in 2008.

Old	16	27	17	33	28	24	18	22	20	29	21
New	15	24	15	29	25	22	16	20	18	26	19

1. Construct a scatterplot. What does the result suggest?

2. Estimate the value of the correlation coefficient. What does that value suggest?

3. Can we conclude that the old ratings have a direct causal effect on the new ratings? Explain briefly.

4. In a study of casino size (square feet) and revenue, the value of $r = 0.445$ is obtained. Find the value of r^2. What does that value tell us?

5. In a study of global warming, assume that we have found a strong positive correlation between carbon dioxide concentration and temperature. Identify three possible explanations for this correlation.

6. For 10 pairs of sample data, the correlation coefficient is computed to be $r = -1$. What do you know about the scatterplot?

7. In a study of randomly selected subjects, it is found that there is a strong correlation between household income and number of visits to dentists. Is it valid to conclude that higher incomes cause people to visit dentists more often? Is it valid to conclude that more visits to dentists cause people to have higher incomes? How might the correlation be explained?

8. You are considering the most expensive purchase that you are likely to make: the purchase of a home. Identify at least five different variables that are likely to affect the actual value of a home. Among the variables that you have identified, which single variable is likely to have the greatest influence on the value of the home? Identify a variable that is likely to have little or no effect on the value of a home.

9. A researcher collects paired sample data and computes the value of the linear correlation coefficient to be 0. Based on that value, he concludes that there is no relationship between the two variables. What is wrong with this conclusion?

10. Examine the scatterplot in Figure 7.22 and estimate the value of the correlation coefficient.

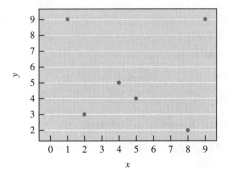

Figure 7.22

1. Fill in the blanks: Every possible correlation coefficient must lie between the values of _____ and _____.

2. Which of the following are likely to have a correlation?

   a. SAT scores and weights of randomly selected subjects

   b. Reaction times and IQ scores of randomly selected subjects

   c. Height and arm span of randomly selected subjects

   d. Proportion of seats filled and amount of airline profit for randomly selected flights

   e. Value of cars owned and annual income of randomly selected car owners

3. For a collection of paired sample data, the correlation coefficient is found to be -0.099. Which of the following statements best describes the relationship between the two variables?

   a. There is no correlation.

   b. There is a weak correlation.

   c. There is a strong correlation.

   d. One of the variables is the direct cause of the other variable.

   e. Neither of the variables is the direct cause of the other variable.

4. Estimate the correlation coefficient for the data in Figure 7.23.

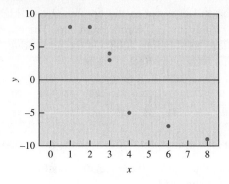

Figure 7.23

5. Refer again to the scatterplot in Figure 7.23. Does there appear to be a significant correlation between the two variables?

In Exercises 6–10, determine whether the given statement is true or false.

6. If $r = 0.200$, then $r^2 = 0.040$ and 4% of the plotted points lie on the line of best fit.

7. If $r = 1$ or $r = -1$, then all points in the scatterplot lie directly on the line of best fit.

8. If the value of the correlation coefficient is negative, the value of r^2 must also be negative.

9. A scatterplot is a graph in which the points are scattered throughout, without any noticeable pattern.

10. If the line of best fit is inserted in a scatterplot, it must pass through every point in the graph.

F OCUS ON
EDUCATION

What Helps Children Learn to Read?

Everyone has an idea about how best to teach reading to children. Some advocate a phonetic approach, teaching students to "sound out" words. Some advocate a "whole language" approach, teaching students to recognize words from their context. Others advocate a combination of these approaches, or something else entirely. These differing ideas would be unimportant if they were merely opinions. But in a nation that spends more than a *trillion dollars* per year on education, differing approaches to teaching reading involve major political confrontations among groups with different special interests.

The huge stakes involved in teaching reading demand statistics to measure the effectiveness of various approaches. Some of the most important educational statistics are those that come from the National Assessment of Educational Progress (NAEP), often known more simply as "the Nation's Report Card." The NAEP is an ongoing survey of student achievement conducted by a government agency, the National Center for Education Statistics, with authorization and funding from the U.S. Congress.

The NAEP uses stratified random sampling (see Chapter 1) to choose representative samples of fourth-, eighth-, and 12th-grade students of varying ethnicity, family income, type of school attended, and so on. Students chosen for the samples are given tests designed to measure their academic achievement in a particular subject area, such as reading, mathematics, or history. Samples are chosen on both state and national levels. Overall, a few thousand students are chosen for each test. Results from NAEP tests inevitably make the news, with articles touting improvements or decrying drops in test scores.

But what really causes improvement in reading performance? Researchers begin by searching for correlations between reading performance and other factors. Sometimes the correlations are clear, but offer no direction for improving reading. For example, parental education is clearly correlated with reading achievement—children with more highly educated parents tend to read more proficiently than those with uneducated parents—but this correlation doesn't offer much guidance for the schools because children do not choose their parents. Other times the correlations may suggest ways to improve reading. For example, students who report reading more pages daily in school and for homework tend to score higher than students who read fewer pages. This suggests that schools should assign more reading.

Of course, the high stakes involved in education make education statistics particularly prone to misinterpretation or misuse. Consider just a few of the problems that make the NAEP reading tests difficult to interpret:

- They are standardized tests that are mostly multiple choice. Some people believe that such tests are inevitably biased and cannot truly measure reading ability.
- Because the tests generally don't affect students' grades, some students may not take the tests seriously, in which case test results may not reflect actual reading ability.
- State-by-state comparisons may not be valid if the makeup of the student population (particularly in its fraction of students for whom English is a second language) varies significantly among states.
- There is some evidence of cheating on the part of the *adults* involved in the NAEP tests by, for example, choosing samples that are not truly representative but instead skewed toward students who read better.

You can probably think of a dozen other problems that make it difficult to interpret NAEP results. So what can you do, as an individual, to help a child to read? Fortunately, the NAEP studies also reveal a few correlations that are uncontroversial and agree with common sense. For example, higher reading performance correlates with each of the following factors:

- more total reading, both for school and for pleasure
- more choice in reading—that is, allowing children to pick their own books to read
- more writing, particularly of extended pieces such as essays or long letters
- more discussion of reading material with friends and family
- less television watching

These correlations give at least some guidance on how to help a child learn to read and should be good starting points for discussions of how to increase literacy.

QUESTIONS FOR DISCUSSION

1. One result of the NAEP reading tests is that students in private schools tend to score significantly higher than students in public schools. Does this imply that private schools are "better" than public schools? Defend your opinion.

2. Do you think that standardized tests like those of the NAEP are valid ways to measure academic achievement? Why or why not?

3. Currently, the NAEP tests are given to only a few thousand of the millions of school children in the United States. Some people advocate giving similar tests to all students, on either a voluntary or a mandatory basis. Do you think such "standardized national testing" is a good idea? Why or why not?

4. Have you ever helped a child learn to read? Compare your experiences with those of other classmates who have worked with young children.

5. Read the latest edition of the *NAEP Reading Report Card* (available online). What are some of the latest results with regard to the teaching of reading in the United States?

FOCUS ON
ENVIRONMENT

What Is Causing Global Warming?

Global warming is one of the most important issues of our time, yet surveys and media reports suggest that many people doubt that it is real or that humans are responsible for it. In this Focus, we will investigate the evidence that has led the vast majority of climate scientists to conclude that human activity is the cause of global warming.

As we discussed in the Focus on Environment for Chapter 3 (page 117), measurements clearly show that the atmospheric carbon dioxide concentration is rising rapidly and is now significantly higher than it has been at any time during at least the past 800,000 years (see Figure 3.46). Chemical analysis shows the added carbon dioxide is coming primarily from human activity, especially the burning of fossil fuels. Moreover, data from ice cores show that the carbon dioxide concentration is strongly *correlated* with the global average temperature. The key question, then, is whether this correlation implies causality. To answer it, we must understand how a gas like carbon dioxide can affect the temperature and then investigate whether the recent increase in the concentration is having the expected effects.

The fact that some atmospheric gases—called **greenhouse gases**—can trap heat has been well-known for more than 150 years, ever since Irish physicist John Tyndall measured the heat-absorbing effects of carbon dioxide and water vapor in his laboratory in 1859. Other scientists, most notably Swedish scientist Svante Arrhenius (1859–1927), later pointed out that the burning of fossil fuels releases carbon dioxide, and that this might therefore cause global warming. Today, the mechanism by which carbon dioxide and other greenhouse gases (the most important others being water vapor and methane) warm a planet is called the **greenhouse effect**. It is well understood and summarized in Figure 7.24.

Scientists can further test this understanding of the greenhouse effect by checking to see whether it successfully accounts for the temperatures of various planets. In the absence of greenhouse gases, a world's average temperature would be determined by only two major factors: its distance from the Sun and the fraction of the incoming sunlight that its surface absorbs (the rest is reflected back into space). There is a simple equation that allows the calculation of the temperature in this case, and it successfully predicts the temperatures of worlds with no atmosphere, such as the Moon and the planet Mercury. For planets with atmospheres, however, scientists can successfully predict their temperatures only by taking the

greenhouse effect into account, and the results clearly show that more greenhouse gases mean more excess heating. Our planetary neighbors vividly demonstrate this fact. Mars has a very thin carbon dioxide atmosphere that gives it a fairly weak greenhouse effect, making the planet about 11°F warmer than it would be otherwise. Venus, which has an extremely dense atmosphere containing nearly 200,000 times as much carbon dioxide as Earth's atmosphere, has a correspondingly extreme greenhouse effect that makes its surface about 850°F hotter than it would be otherwise—giving it a surface hot enough to melt lead.

Earth is the lucky intermediate case. Without the greenhouse effect, Earth's average temperature would be well below freezing, at about −16°C (3°F). But thanks to the carbon

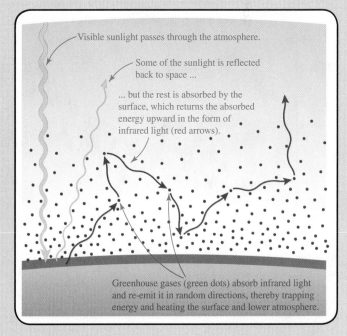

Visible sunlight passes through the atmosphere.

Some of the sunlight is reflected back to space ...

... but the rest is absorbed by the surface, which returns the absorbed energy upward in the form of infrared light (red arrows).

Greenhouse gases (green dots) absorb infrared light and re-emit it in random directions, thereby trapping energy and heating the surface and lower atmosphere.

Figure 7.24 This diagram shows the basic mechanism of the greenhouse effect. The greater the abundance of greenhouse gases, the more the escape of infrared light is slowed and the warmer the planet becomes.

dioxide, methane, and water vapor in our atmosphere, the actual temperature is close to 15°C (60°F). From that standpoint, the greenhouse effect is a very good thing, because our lives would not be possible without it. Just keep in mind that the case of Venus offers proof that it's possible to have too much of this good thing.

The evidence from laboratory measurements and studies of other planets leave no reasonable doubt that carbon dioxide and other greenhouse gases cause a planet's temperature to be hotter than it would be otherwise. Nevertheless, because carbon dioxide is not the only thing that affects our planet's temperature, we might wonder whether any effects due to the recent rise in carbon dioxide might be offset by, say, reductions in the amount of other greenhouse gases or an increase in how much sunlight our planet reflects. Scientists can test this idea in two basic ways. First, they look at data showing changes in Earth's average temperature. As you'll see in Figure 7.25, Earth's average temperature has indeed been rising, and the data give at least some hint that the rise has been accelerating in recent decades.

The second way to test the idea that human burning of fossil fuels is causing global warming is to conduct experiments. We obviously cannot perform controlled experiments with our entire planet, so scientists instead build *computer models* designed to simulate the way Earth's climate works. Earth's climate is incredibly complex, so the models cannot be perfect. Nevertheless, today's models match real climate data quite well, giving scientists confidence that the models have predictive value. Figure 7.26 compares real data to models with and without the human contribution to the greenhouse

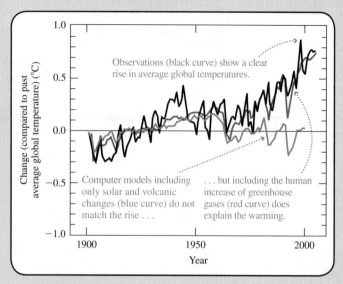

Figure 7.26 This graph compares observed temperature changes (black curve) with the predictions of climate models that include only natural factors such as changes in the brightness of the Sun and effects of volcanoes (blue curve), and models that also include the human-made increase in the greenhouse gas (red curve). Only the red curve matches the observations well. (The red and blue model curves are each averages of many scientists' independent models of global warming, which generally agree with each other to within 0.1°C − 0.2°C.)

gas concentration. We see a good match only for models that include the human contribution.

The conclusion is clear: Laboratory measurements of the greenhouse effect, studies of other planets, data for Earth's rising carbon dioxide concentration and temperature, and computer models of the climate all provide evidence in favor of the claim that human activity is causing global warming. It is the fact that so many lines of evidence are all in agreement that makes scientists so confident that the causality is real.

QUESTIONS FOR DISCUSSION

1. Look back at the six guidelines for establishing causality on page 265. Discuss whether or how each guideline is met by current data and understanding of global warming.

2. Look back at the legal levels of confidence in causality discussed in Section 7.4. Would you say that the case for human activity as the cause of global warming is now at the level of possible cause, probable cause, or cause beyond reasonable doubt? Defend your opinion.

3. Investigate some of the likely consequences of global warming. If current trends continue, what changes can you expect in the world by the year 2050? 2100?

4. Based on what you've learned about the cause of global warming and its potential consequences, what do *you* think we should be doing about it, if anything?

• • • • • • • • • • • • • • • •

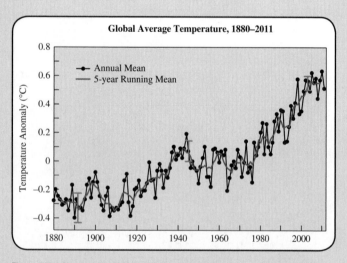

Figure 7.25 Clear evidence of a warming Earth: The black curve shows the mean global average temperature during each year; the red curve shows a running mean computed over 5-year periods. The vertical scale ("temperature anomaly") shows the difference between each year's actual average temperature and the average during the period 1951–1980. The blue bars represent the uncertainty ranges in the data at three different times; the uncertainty is lower for recent times because measurements have become more precise.

From Samples to Populations

Did you ever wonder how the outcome of a national election can be predicted hours before the polls close? Or how a large retailer can make critical marketing decisions based on a survey of only a few hundred people? These examples illustrate perhaps the most powerful aspect of statistics: the capability to use information gathered from a small sample to make conclusions about a much larger population. This process, called *making inferences*, is the subject of the branch of statistics called *inferential statistics*.

Some people hate the very name statistics, but I find them full of beauty and interest. Whenever they are not brutalized, but delicately handled by the higher methods, and are warily interpreted, their power of dealing with complicated phenomena is extraordinary.

—Sir Francis Galton (1822–1912)

LEARNING GOALS

8.1 Sampling Distributions
Understand the fundamental ideas of sampling distributions, enabling you to work with both a distribution of sample *means* and a distribution of sample *proportions*.

8.2 Estimating Population Means
Learn to estimate population means and compute the associated margins of error and confidence intervals.

8.3 Estimating Population Proportions
Learn to estimate population proportions and compute the associated margins of error and confidence intervals.

FOCUS TOPICS

p. 303 Focus on History: Where Did Statistics Begin?

p. 306 Focus on Literature: How Many Words Did Shakespeare Know?

8.1   SAMPLING DISTRIBUTIONS

Consider the following statements taken from recent news articles or research reports.

- The mean daily protein consumption by Americans is 67 grams.
- Nationwide, the mean hospital stay after delivery of a baby decreased from 3.2 days in 1980 to the current mean of 2.0 days.
- Thirty percent of high school girls in this country believe they would be happier being married than not being married.
- About 5% of all American children live with a grandparent.

Each statement makes a claim about a population, based on statistics obtained from a much smaller sample. More specifically, the sample statistics were used to make estimates of the corresponding population parameters (see Section 1.1). In other words, the sample data were used to make *inferences* about the population, which means the methods used are part of the branch of statistics known as *inferential statistics*.

Public opinion in this country is everything.

—Abraham Lincoln

Note also that the four statements are based on two different types of sample statistics. The first two statements give estimates concerning the *mean* of a quantity—mean protein consumption of 67 grams and mean hospital stays of 3.2 days and 2.0 days. The last two statements say something about a *proportion* of the population—30% of high school girls and 5% of American children. The methods for dealing with sample means and sample proportions are slightly different, so we will consider each separately.

Sample Means: The Basic Idea

Consider a family with three children of ages 4, 5 and 9. Imagine that, for some reason, we wanted to study this family alone, so that the three children represented the entire population for our study. Such small populations rarely arise in real statistical applications, but using this small one will make it easier for us to look carefully at the sampling process. Notice that the population mean age for the three children is:

$$\frac{4 + 5 + 9}{3} = 6.0 \text{ years}$$

The smallest possible sample size is $n = 1$, and there are three possible samples that we could draw: the child of age 4, the child of age 5, and the child of age 9. The mean of each sample of size $n = 1$ is simply the age of the child in the sample. Figure 8.1 shows a histogram of the means of all three samples of size $n = 1$.

The distribution of any sample statistic from *all* possible samples of a particular size is called a **sampling distribution**, and the distribution in Figure 8.1 is called a **sampling distribution of sample means**, because it shows the means of all three samples of size $n = 1$. Because the three sample means in this case are simply the ages of the three children, we can say that the "mean of the sample means" is equal to the mean of the entire population, or 6.0 years.

TECHNICAL NOTE

In the large populations that we usually deal with in statistics, it makes no significant difference whether we sample with or without replacement. However, much as we found for *and* probabilities in Section 6.5, independent events are easier to analyze than dependent events, which is why we focus here on sampling with replacement.

> **Definition**
>
> The **sampling distribution of sample means** is the distribution that results when we find the means of *all* possible samples of a given size.

Let's move on to samples of size $n = 2$, in which each sample consists of two children. Although it may sound a little strange for a population of only three children, we will

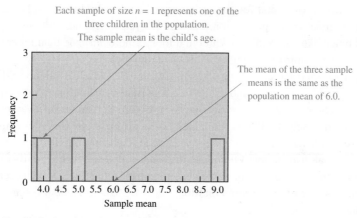

Figure 8.1 Sampling distribution of sample means for sample size $n = 1$, drawn from a population of three children ages 4, 5, and 9.

assume that it's possible for a sample of size $n = 2$ to have the same child twice. This type of sampling is called sampling *with replacement*, because we choose one child at random as the first member of the sample, then we put that child back in the pool of possible choices before choosing the next member of the sample. As a result, there are three children to choose from for each of the two members of the sample, so the total number of possible samples of size $n = 2$ is $3^2 = 9$. Table 8.1 lists the nine different possible samples and the mean for each sample. Be sure to notice that each sample mean is simply the mean of the two values in the sample; for example, the sample consisting of the ages 4 and 9 has a sample mean of $(4 + 9)/2 = 6.5$.

To construct the sampling distribution of the sample means, we must look through Table 8.1 to find the frequency with which each sample mean occurs. For example, a sample mean of 4.0 occurs only once (for the sample {4, 4}), while a sample mean of 4.5 occurs twice (for the samples {4, 5} and {5, 4}). Table 8.2 shows the frequency for each of the sample means with sample size $n = 2$, and Figure 8.2 shows a histogram of this sampling distribution of sample means. You should notice two key facts. First, we can use the data in Table 8.2 to calculate the mean of the nine individual sample means for sample size $n = 2$:

TABLE 8.1	Sample Means for a Three-Child Family of Ages 4, 5, 9; Sample Size $n = 2$
Sample	**Sample Mean**
4, 4	4.0
4, 5	4.5
4, 9	6.5
5, 4	4.5
5, 5	5.0
5, 9	7.0
9, 4	6.5
9, 5	7.0
9, 9	9.0

$$\frac{4.0 + (2 \times 4.5) + 5.0 + (2 \times 6.5) + (2 \times 7.0) + 9.0}{9} = 6.0 \text{ years}$$

TABLE 8.2	Frequency of Sample Means from Table 8.1
Sample Mean	**Frequency**
4.0	1
4.5	2
5.0	1
6.5	2
7.0	2
9.0	1

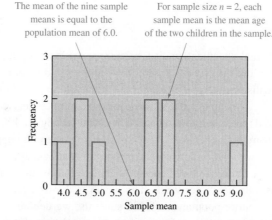

Figure 8.2 Sampling distribution of sample means for sample size $n = 2$, drawn from a population of three children ages 4, 5, and 9.

In other words, just as we found for our earlier sampling distribution ($n = 1$), the mean of the sample means is equal to the population mean of 6.0 years. Second, although Figure 8.2 still does not look much like a normal distribution, it looks "more normal" than Figure 8.1, because we are starting to see some clustering of values near the mean.

In fact, if we used larger sample sizes, we would find that the sampling distribution of sample means would look more and more like a normal distribution (a consequence of the Central Limit Theorem, discussed in Section 5.3). Figure 8.3 shows an example in which a computer was used to select 1,000 random samples of size $n = 10$ from our same population of three children with ages 4, 5, and 9. As you can see, the sampling distribution is beginning to look like a normal distribution. Of course, drawing samples of size $n = 10$ from a population of only three children creates the somewhat unrealistic situation in which each sample will always have at least one of the children multiple times. So to continue our discussion, let's turn to a slightly more realistic situation.

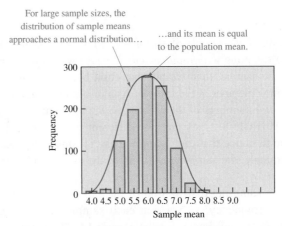

Figure 8.3   The distribution of sample means for 1,000 randomly selected samples of size $n = 10$, drawn from a population of three children ages 4, 5, and 9.

TIME OUT TO THINK

Figure 8.3 shows 1,000 random samples of size $n = 10$ drawn from the population of three children, but this is not all of the possible samples. How many samples are possible in this case? (Hint: Remember that there are three children to choose from for each of the 10 selections that you make for the sample.)

Sample Means with Larger Populations

In real statistical studies, we rarely work with populations as small as three individuals. To explore the process of selecting samples and forming distributions of sample means in a more realistic setting, let's consider a somewhat larger population.

Imagine that you work for the computer services department of a small college. In order to develop networking strategies, you survey all 400 students at the college to determine how many hours per week they spend using a search engine on the Internet. The responses (hours per week) are shown at the top of the next page.

You could, of course, calculate the mean of all 400 responses. You would find that it is 3.88 hours per week. This mean is the true **population mean**, because it is the mean for the entire population of 400 students; we denote the population mean by the Greek letter μ (pronounced "mew"). Similarly, a calculation shows that the population standard deviation is $\sigma = 2.40$ hours.

3.4	6.8	6.7	3.4	0.0	5.0	5.4	1.8	0.7	1.6	2.1	3.5	3.4	6.4	7.2	1.8	7.4	3.0	4.0	5.2
1.2	7.8	7.0	0.4	7.2	4.8	3.6	8.0	5.4	6.4	3.5	5.3	4.7	5.4	5.6	3.8	0.1	2.4	0.5	4.0
4.5	8.0	4.2	1.0	6.2	7.1	3.8	0.7	5.5	1.7	2.6	1.6	0.7	1.3	6.5	2.4	3.0	0.3	2.2	0.4
1.9	5.0	2.0	5.3	7.5	5.0	0.3	7.4	6.0	4.3	1.3	0.8	7.2	6.6	0.2	3.4	1.6	2.2	3.0	4.5
5.5	5.3	6.5	0.1	0.3	4.2	2.2	6.2	7.3	3.1	5.4	1.3	6.3	4.5	7.1	5.8	6.1	0.5	0.4	4.1
7.0	6.0	1.1	0.8	1.4	2.9	7.3	0.8	2.7	0.6	3.0	0.7	2.8	6.5	1.9	3.6	1.6	2.6	2.6	6.6
6.8	6.1	3.6	1.4	7.7	5.2	3.8	6.0	2.2	7.5	6.7	4.4	4.1	7.3	5.2	5.7	6.7	2.4	0.6	6.7
1.0	2.3	0.7	1.2	4.5	3.3	4.2	2.1	5.9	3.0	7.2	7.9	2.5	7.1	8.0	6.7	4.1	4.9	0.0	3.1
6.0	0.5	4.2	2.7	0.1	1.4	2.1	2.5	3.9	5.8	5.9	2.7	2.8	3.7	7.3	0.7	6.9	4.4	0.7	1.6
3.1	2.1	7.4	3.6	6.5	2.9	5.4	3.9	3.0	0.8	0.3	0.8	3.3	0.8	8.0	5.6	7.1	1.3	0.2	5.2
7.8	4.7	7.2	0.9	5.1	0.9	1.7	1.2	0.4	6.9	0.6	3.0	3.6	6.1	1.6	6.0	3.8	0.4	1.1	4.0
3.8	4.0	1.8	0.9	1.1	3.9	1.7	1.7	2.6	0.1	4.0	1.4	1.9	0.9	0.2	4.2	4.7	0.2	5.3	2.2
5.8	7.5	5.8	5.2	3.9	3.4	7.3	4.1	0.5	7.9	7.7	7.7	5.0	2.3	7.8	2.3	5.6	6.5	7.9	5.0
2.0	5.5	5.4	6.6	6.7	4.4	7.2	2.5	4.9	7.0	2.1	7.2	4.1	1.2	6.2	3.3	6.3	2.3	4.9	2.2
6.4	7.2	0.1	5.3	3.0	0.7	1.5	1.2	1.1	7.4	5.1	7.2	7.2	3.0	7.1	4.5	6.7	7.2	7.2	0.9
2.9	4.3	2.5	0.7	7.6	3.9	0.7	5.8	6.6	3.4	0.3	6.5	7.5	0.7	6.1	6.1	4.8	1.9	1.9	5.0
1.1	7.8	6.8	4.9	3.0	6.5	5.2	2.2	5.1	3.4	4.7	7.0	3.8	5.7	6.8	1.2	1.7	6.5	0.1	4.3
6.3	1.2	0.8	0.7	0.6	7.0	4.0	6.6	6.9	0.5	4.3	1.0	0.5	3.1	0.9	2.3	5.7	6.7	7.3	0.5
0.3	0.9	2.4	2.5	7.8	5.6	3.2	0.7	5.4	0.0	5.7	0.3	7.2	5.1	2.5	3.2	3.1	2.8	5.0	5.6
3.1	0.7	0.5	3.9	2.6	7.3	1.4	1.2	7.1	5.5	3.1	5.0	6.8	6.5	1.7	2.1	7.3	4.0	2.2	5.6

In typical statistical applications, populations are huge and it is impractical or too expensive to survey every individual in the population; consequently, we rarely know the true population mean, μ. Therefore, it makes sense to consider using the mean of a *sample* as an estimate of the mean of the entire population. Although a sample is easier to work with, it cannot possibly represent the entire population exactly. Therefore, we should not expect an estimate of the population mean obtained from a sample to be perfect. The error that we introduce by working with a sample is called the **sampling error**. We can explore this idea by considering samples drawn from the 400 responses about Internet use.

> ## Sampling Error
>
> The **sampling error** is the error introduced because a random sample is used to estimate a population parameter. It does not include other sources of error, such as those due to biased sampling, bad survey questions, or recording mistakes.

> ## TIME ꧁UT TO THINK
>
> Would you expect the sampling error to increase or decrease if the sample size were increased? Explain.

Suppose you select a random sample of $n = 32$ responses from the population of 400 responses shown above and calculate their mean. For example, the random sample might be the following values (in hours):

| 1.1 | 7.8 | 6.8 | 4.9 | 3.0 | 6.5 | 5.2 | 2.2 | 5.1 | 3.4 | 4.7 | 7.0 | 3.8 | 5.7 | 6.5 | 2.7 |
| 2.6 | 1.4 | 7.1 | 5.5 | 3.1 | 5.0 | 6.8 | 6.5 | 1.7 | 2.1 | 1.2 | 0.3 | 0.9 | 2.4 | 2.5 | 7.8 |

The mean of this sample, which we call the **sample mean**, is $\bar{x} = 4.17$ hours; we use the notation $\bar{x}$ (read as "x-bar") to denote a sample mean.

Notation for Population and Sample Means

n = sample size

μ = population mean

$\bar{x}$ = sample mean

σ = population standard deviation

s = sample standard deviation (not used until Section 8.2)

Now, suppose you collect a second sample of size $n = 32$ from the same data set, such as this sample:

1.8 0.4 4.0 2.4 0.8 6.2 0.8 6.6 5.7 7.9 2.5 3.6 5.2 5.7 6.5 1.2
5.4 5.7 7.2 5.1 3.2 3.1 5.0 3.1 0.5 3.9 3.1 5.8 2.9 7.2 0.9 4.0

For this sample, the sample mean is $\bar{x} = 3.98$ hours.

Note that the two sample means found so far ($\bar{x} = 4.17$ and $\bar{x} = 3.98$) do not perfectly match each other, and neither one matches the true population mean ($\mu = 3.88$). This shouldn't be surprising, because we don't expect samples to be exact representations of a population. But suppose you continue to select samples of 32 responses, and each time you calculate the sample mean, $\bar{x}$. To get a good picture of all these sample means, you could make a histogram showing the number of samples that have various sample means. Figure 8.4 shows a histogram that results from 100 different samples, each with 32 students. Notice that this histogram is very close to a *normal distribution* and that its mean is very close to the population mean, $\mu = 3.88$ hours.

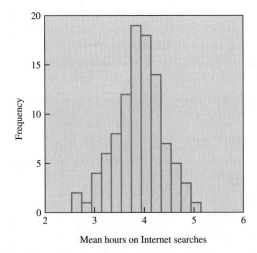

Figure 8.4 A distribution of 100 sample means, with a sample size of $n = 32$, appears close to a normal distribution with a mean of 3.88 hours.

TIME ( ) UT TO THINK

Suppose you choose only one sample of size $n = 32$. According to Figure 8.4, are you more likely to choose a sample with a mean less than 2.5 or a sample with a mean less than 3.5? Explain.

What would happen if we made a histogram of the complete distribution of sample means (rather than only of 100 sample means) by analyzing *all* possible samples of size $n = 32$? We can't actually compute all these sample means, because there are too many: The number of possible samples of size $n = 32$ drawn from a population of 400 is 400^{32} (because there are 400 choices for each of the 32 sample selections), which turns out be a number greater than the total estimated number of atoms in the observable universe!

Nevertheless, if we could compute all of those sample means, we would find that their distribution would be indistinguishable from a perfectly normal distribution, that it would have a mean equal to the population mean (3.88 hours in this case), and that it would have a standard deviation equal to the population standard deviation divided by the square root of the sample size, or $\sigma/\sqrt{n} = 2.40/\sqrt{32} = 0.42$ hours. The following box generalizes these ideas.

Characteristics of the Distribution of Sample Means

For any distribution of sample means:

- The larger the sample size, the more closely the distribution of sample means approximates a normal distribution.

- The mean of the distribution of sample means equals the population mean.

- The standard deviation of the distribution of sample means is given by the expression $\dfrac{\sigma}{\sqrt{n}}$.

TECHNICAL NOTE

A common guideline is to assume that the distribution of sample means is close to normal if the sample size is greater than 30.

We can use these ideas to make a quantitative statement about how well the mean of a particular sample approximates the population mean. Consider one more sample of 32 responses from the data on page 279:

| 5.8 | 7.5 | 5.8 | 5.2 | 3.9 | 3.4 | 7.3 | 4.1 | 0.5 | 7.9 | 7.7 | 7.7 | 5.0 | 2.3 | 7.8 | 2.3 |
| 5.0 | 6.8 | 6.5 | 1.7 | 2.1 | 7.3 | 4.0 | 2.2 | 5.6 | 4.7 | 5.3 | 3.5 | 6.5 | 3.4 | 6.6 | 5.0 |

The mean of this sample is $\bar{x} = 5.01$. Given that the mean of the distribution of sample means is 3.88 and the standard deviation is 0.42, the sample mean of $\bar{x} = 5.01$ has a *standard score* (see Section 5.2) of

$$z = \frac{\text{sample mean} - \text{pop.mean}}{\text{standard deviation}} = \frac{5.01 - 3.88}{0.42} = 2.7$$

In other words, the sample we have selected has a mean that is 2.7 standard deviations above the mean of the sampling distribution. From Table 5.1, this standard score corresponds to the 99.65th percentile. Therefore, the probability of selecting another sample with a mean *less* than 5.01 is about 0.9965, and the probability of selecting another sample with a mean *greater* than 5.01 is about $1 - 0.9965 = 0.0035$. This low probability tells us that the sample we selected is rather extreme within this distribution.

EXAMPLE 1 Sampling Farms

Texas has roughly 225,000 farms, more than any other state in the United States. The actual mean farm size is $\mu = 582$ acres, and the standard deviation is $\sigma = 150$ acres. For random samples of $n = 100$ farms, find the mean and standard deviation of the distribution of sample means. What is the probability of selecting a random sample of 100 farms with a mean greater than 600 acres?

SOLUTION   Because the distribution of sample means is approximately a normal distribution, its mean should be the same as the mean of the entire population, which is 582 acres. The standard deviation of the sampling distribution is $\sigma/\sqrt{n} = 150/\sqrt{100} = 15$. A sample mean of $\bar{x} = 600$ acres therefore has a standard score of

$$z = \frac{\text{sample mean} - \text{pop.mean}}{\text{standard deviation}} = \frac{600 - 582}{15} = 1.2$$

BY THE WAY

The total number of U.S. farms peaked at about 6.8 million in 1935. Today, there are approximately 2 million farms in the United States, but their mean acreage is significantly larger than the mean acreage of the past.

According to Table 5.1, this standard score is in the 88th percentile. Therefore, the probability of selecting a sample with a mean less than 600 acres is about 0.88, and the probability of selecting a sample with a mean greater than 600 acres is about 0.12.   ·· •

Sample Proportions

Much of what we have learned about distributions of sample means carries over to distributions of *sample proportions*. We can see the parallels by considering another survey of 400 students. Suppose your goal is to determine the proportion (or percentage) of all 400 students who own a car. Each Y (for *yes*) or N (for *no*) below is one person's answer to the question "Do you own a car?"

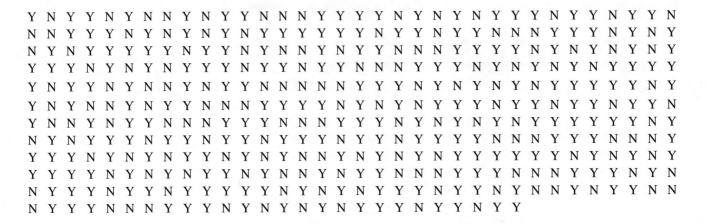

If you counted carefully, you would find that 240 of the 400 responses are Y's, so the *exact* proportion of car owners in the 400-student population is

$$p = \frac{240}{400} = 0.6$$

This *population proportion, p = 0.6*, is another example of a *population parameter*. Of the 400 students in the population, it is the true proportion of car owners.

> **TIME (◯)UT TO THINK**
>
> Give another survey question that would result in a population proportion rather than a population mean.

Once again, in typical statistical problems, it is impractical or prohibitively expensive to survey every individual in the population. Therefore, it's reasonable to consider the idea of using a random sample of, say, *n* = 32 people to estimate the population proportion. Suppose you randomly draw 32 responses from the list of Y's and N's to generate the following sample:

Y N Y Y N Y Y N Y N Y Y N Y Y Y N N Y Y Y N N Y Y N Y Y Y N Y Y

The proportion of Y responses in this list is

$$\hat{p} = \frac{21}{32} = 0.656$$

This proportion is another example of a *sample statistic*. In this case, it is a **sample proportion** because it is the proportion of car owners within a *sample;* we use the symbol $\hat{p}$ (read "*p*-hat") to distinguish this sample proportion from the population proportion, *p*.

Notation for Population and Sample Proportions

$$n = \text{sample size}$$

$$p = \text{population proportion}$$

$$\hat{p} = \text{sample proportion}$$

Proceeding as we did with sample means, suppose that we next selected many more samples of 32 responses, each time calculating the sample proportion, $\hat{p}$. If we make a histogram of the many sample proportions, it will show the number of samples that have particular values of $\hat{p}$ from 0 through 1. Figure 8.5 shows such a histogram, in this case resulting from 100 samples of size $n = 32$. As we found for sample means, this distribution of sample proportions is very close to a normal distribution, and its mean is very close to the population proportion of 0.6.

There are three sorts of opinion: informed, uninformed and inconsequential. Uninformed opinions encompass most of American opinion.

—Paul Talmey, pollster

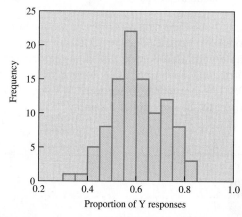

Figure 8.5 The distribution of 100 sample proportions, with a sample size of 32, appears to be close to a normal distribution.

If it were possible to select *all* possible samples of size $n = 32$, the resulting distribution would be called a **sampling distribution of sample proportions**. As the following box indicates, this distribution shares many similarities with a distribution of sample means and can also be used to determine the probability of a particular sample mean.

The Distribution of Sample Proportions

The **sampling distribution of sample proportions** is the distribution that results when we find the proportions ($\hat{p}$) in all possible samples of a given size. Note that:

- The larger the sample size, the more closely the distribution of sample proportions approximates a normal distribution.

- The mean of the distribution of sample proportions equals the population proportion.

- The standard deviation of the distribution of sample proportions is given by the expression $\sqrt{\dfrac{p(1-p)}{n}}$.

EXAMPLE 2 Analyzing a Sample Proportion

Consider the distribution of sample proportions shown in Figure 8.5. Suppose you randomly select the following sample of 32 responses:

Y Y N Y Y Y Y N Y Y Y Y Y Y Y N Y Y N Y Y Y Y N Y Y N Y Y N Y N Y Y

Compute the sample proportion, $\hat{p}$, for this sample. How far does it lie from the population mean? What is the probability of selecting another sample with a proportion greater than the one you selected?

SOLUTION The proportion of Y responses in this sample is

$$\hat{p} = \frac{24}{32} = 0.75$$

Using the population mean of 0.6 (given earlier), the standard deviation for the distribution of sample proportions is

$$\sqrt{\frac{p(1-p)}{n}} = \sqrt{\frac{0.6(1-0.6)}{32}} \approx 0.09$$

The sample statistic, $\hat{p} = 0.75$, therefore has a standard score of

$$z = \frac{\text{sample proportion} - \text{pop. proportion}}{\text{standard deviation}} = \frac{0.75 - 0.6}{0.09} = 1.7$$

The sample proportion is 1.7 standard deviations above the mean of the distribution. Using Table 5.1, a standard score of 1.7 corresponds to about the 95th percentile. The probability of selecting another sample with a proportion less than the one we selected is about 0.95, and the probability of selecting another sample with a proportion greater than the one we selected is about 0.05. In other words, if we were to select 100 random samples of 32 responses, we should expect to see only about 5 samples with a higher proportion than the one we selected. $\quad\cdots\bullet$

Section 8.1 Exercises

Statistical Literacy and Critical Thinking

1. **Sampling Distribution.** Pollsters often use randomly selected digits between 0 and 9 to generate parts of telephone numbers to be called. What is the distribution of such randomly selected digits? If we repeat the process of randomly generating 50 digits and finding the mean, what is the distribution of the resulting sample means?

2. **Polls.** America OnLine published a survey question. Among the 2,300 Internet users who responded, 20% answered "yes" and the others answered "no." For the "yes" responses, what is the value of $\hat{p}$? What is fundamentally wrong with this survey?

3. **Notation.** What does $\bar{x}$ denote, what does μ denote, and what is the difference between them?

4. **Notation.** What does $\hat{p}$ denote, what does p denote, and what is the difference between them?

Does It Make Sense? For Exercises 5–8, decide whether the statement makes sense (or is clearly true) or does not make sense (or is clearly false). Explain clearly. Not all of these statements have definitive answers, so your explanation is more important than your chosen answer.

5. **Larger Sample Size.** Two different surveys will be conducted to estimate the mean salary of employees who have taken a statistics course. If the first survey has a larger sample size than the second survey, the first survey will result in a sample mean that is closer to the population mean (when compared with the second survey).

6. **Larger Sample Size.** When a random sample is used to estimate a population mean, the sample mean tends to become a better estimate of the population mean as the sample size increases.

7. **Large Sample Size.** A researcher for a car dealership wants to estimate the mean age of cars in his county. He goes to the largest shopping mall, he randomly selects 3,000 cars in the parking lot, and he obtains their ages from the registration stickers on the car windows. He then computes the mean age of the sample of 3,000 cars. The sample mean is a good estimate of the population mean, because the large sample size compensates for the fact that he is using a convenience sample.

8. **Survey.** In an Adecco Staffing survey of 1,000 adults in the United States, 140 (or 14%) said that salary was the most important feature of their job. The sample proportion of 0.14 cannot be a good estimate of the population proportion because this survey is based on such a small proportion of the population of adults.

Concepts and Applications

9. Estimating Population Proportions. In a survey of children 5 to 17 years old, 1,050 children were randomly selected from the nine states in the Northeast, and the proportion who spoke a language other than English in their home was 0.19. Is that sample proportion a good estimate for children in the United States? Why or why not?

10. Estimating Population Means. When 40 women were randomly selected and tested for their cholesterol levels, a mean of 240.9 milligrams was obtained. Would you be more confident of your estimate if the sample included measurements from 500 women? Explain.

11. Distribution of Sample Means. Assume that cans of Coke are filled so that the actual amounts have a mean of 12.00 ounces. A random sample of 36 cans has a mean amount of 12.19 ounces. The distribution of sample means of size 36 is normal with an assumed mean of 12.00 ounces, and those sample means have a standard deviation of 0.02 ounce.

a. How many standard deviations is the sample mean from the mean of the distribution of sample means?

b. In general, what is the probability that a random sample of size 36 has a mean of at least 12.19 ounces?

c. Does it appear that consumers are being cheated? Why or why not?

12. Distribution of Sample Means. Assume that the population of heights of men has a normal distribution with a mean of 69.5 in. and random samples of 100 men result in sample mean heights with a mean of 69.5 in. and a standard deviation of 0.24 in.

a. If one sample of 100 men results in a mean height of 69.0 in., how many standard deviations is the sample mean from the population mean?

b. What is the probability that a second sample of 100 men would have a mean less than 69.0 in.?

13. Sample and Population Proportions. Suppose that, in a suburb of 12,345 people, 6,523 people moved there within the past five years. You survey 500 people and find that 245 of the people in your sample moved to the suburb in the past five years.

a. What is the population proportion of people who moved to the suburb in the past five years?

b. What is the sample proportion of people who moved to the suburb in the past five years?

c. Does your sample appear to be representative of the population of the suburb? Discuss.

14. Sample and Population Proportions. The College of Portland has 2,444 students and 269 of them are left-handed. You conduct a survey of 50 students and find that 8 of them are left-handed.

a. What is the population proportion of left-handed students?

b. What is the sample proportion of left-handed students?

c. If the sample mean misses the population mean by a large amount, does that imply that the sampling method is flawed? Explain.

15. Estimating Population Proportions. You select a random sample of 150 people at a medical convention attended by 1,608 people. Within your sample, you find that 73 people have traveled from abroad. Based on this sample statistic, estimate how many people at the convention traveled from abroad. Would you be more confident of your estimate if you sampled 300 people? Explain.

16. Estimating Population Proportions. A random sample of 500 people was selected from the 103,219 people in attendance at the Super Bowl game between the Green Bay Packers and the Pittsburgh Steelers. Within the sample, 290 people supported the Packers. Based on this result, estimate how many people at the game supported the Packers. Would you be more confident of your estimate if the sample size were 2,000 people? Explain.

17. Distribution of Sample Proportions. Suppose you know that the distribution of sample proportions of nonresidents in samples of 200 students is normal with a mean of 0.34 and a standard deviation of 0.03. Suppose you select a random sample of 200 students and find that the proportion of nonresident students in the sample is 0.32.

a. How many standard deviations is the sample proportion from the mean of the distribution of sample proportions?

b. What is the probability that a second sample selected would have a proportion less than 0.32?

18. Distribution of Sample Proportions. Suppose you know that the distribution of sample proportions of women employees is normal with a mean of 0.42 and a standard deviation of 0.21. Suppose you select a random sample of employees and find that the proportion of women in the sample is 0.45.

a. How many standard deviations is the sample proportion from the mean of the distribution of sample proportions?

b. What is the probability that a second sample selected would have a proportion greater than 0.45?

19. Sampling Distribution. A quarterback threw 1 interception in his first game, 2 interceptions in his second game, 5 interceptions in his third game, and then he retired. Consider the values of 1, 2, and 5 to be a population. Assume that samples of size 2 are randomly selected (with replacement) from the population.

a. List the 9 different possible samples and find the mean of each sample.

b. What is the mean of the sample means from part a?

c. Is the mean of the sampling distribution from part b equal to the mean of the population of the three listed values? Are those means *always* equal?

20. Sampling Distribution. Here is the population of all five U.S. presidents who had professions in the military, along with their ages at inauguration: Eisenhower, 62; Grant, 46; Harrison, 68; Taylor, 64; and Washington, 57. Assume that samples of size 2 are randomly selected (with replacement) from the population of five ages.

a. List the 25 different possible samples and find the mean of each sample.

b. What is the mean of the sample means from part a?

c. Is the mean of the sampling distribution (from part b) equal to the mean of the population of the five listed values? Are those means *always* equal?

21. Forming Sampling Distributions. Three states and their areas (in thousands of square miles) are given in the following table. Consider these three states to be the entire population from which samples of size $n = 2$ will be selected (with replacement).

a. Find the mean of each of the 9 different possible samples.

b. What is the mean of the sample means from part a?

c. Is the mean of the sampling distribution from part b equal to the mean of the population of the three listed values? Are those means *always* equal?

State	Area (thousands of square miles)
Connecticut	5
Alabama	52
Georgia	60

22. Forming Sampling Distributions. The ages (years) of the four U.S. presidents when they were assassinated in office are 56 (Lincoln), 49 (Garfield), 58 (McKinley), and 46 (Kennedy). Consider these four ages to be a population.

a. Assuming that two of the ages are randomly selected to form samples of size $n = 2$ with replacement, find the mean of each of the 16 possible samples.

b. What is the mean of the sample means from part a?

c. Is the mean of the sample means equal to the mean of the population of the four ages? Are those means *always* equal?

PROJECTS FOR THE INTERNET & BEYOND

23. Distributions of Sample Means. Consider the large data set of hours students spend using an Internet search engine, listed in this section on page 279. Discuss methods for selecting a random sample from this population. Let each person in the class select a random sample of $n = 10$ individuals from the population and find the mean of his or her sample. Find the mean of the individual sample means and compare it to the population mean. Repeat the process with samples of size $n = 20$. How does the mean of the sample means compare to the population mean?

24. Distributions of Sample Proportions. Consider the data set on page 282 showing 400 *yes* and *no* responses to a survey question ("Do you own a car?"). Discuss methods for selecting a random sample from this population. Let each person in the class select a random sample of $n = 10$ responses from the population and find the proportion of *yes* responses for his or her sample. Find the mean of the individual sample proportions and compare it to the population proportion of 0.6. Repeat the process with samples of size $n = 20$. How does the mean of the sample proportions compare to the population proportion?

IN THE NEWS

25. Sample Means in the News. Find a news or research report in which a sample mean is cited. Discuss how it is used to estimate a population mean.

26. Sample Proportions in the News. Find a news or research report in which a sample proportion is cited. Discuss how it is used to estimate a population proportion.

8.2 ESTIMATING POPULATION MEANS ··

In Section 8.1, we explored distributions of sample means and sample proportions through cases in which we had data for the entire population. This was useful because it allowed us to compare individual sample statistics (sample mean or sample proportion) to actual population parameters (population mean or population proportion). However, it was also somewhat artificial, because in most real statistical studies, we have *only* the sample data, not the complete population data. In the rest of this chapter, we will consider how we deal with more realistic situations, focusing on population means in this section and population proportions in Section 8.3.

Estimating a Population Mean: The Basics

Let's begin with an example. We all need protein in our diets, and while different organizations and government agencies make slightly different recommendations, the recommended daily allowance (RDA) for protein is generally around 55 to 60 grams per day for men and 45 to 50 grams per day for women (except when pregnant or nursing, when it is higher). How does the amount of protein that Americans actually eat compare with these recommended daily allowances?

We obviously cannot measure protein intake for the entire population, so we will need to focus on some sample that we hope will be representative of the population. Figure 8.6 shows the sample we will use; it shows the daily protein intake (in grams) for a sample of $n = 267$ men. The sample data have a mean of $\bar{x} = 77.0$ grams and a standard deviation of $s = 58.6$ grams. Our goal is to use these sample data to make an inference about the average protein intake for the population of all American men. There are two basic steps to this process, summarized in the following box.

BY THE WAY

Why do we need protein? Proteins are a key building block of all living cells and are therefore necessary for all forms of life on Earth. Our bodies build proteins from smaller building blocks called *amino acids* and therefore need a source of amino acids for this biochemistry. The source is food; our bodies obtain amino acids by breaking down the proteins that were built in plants and animals that we eat.

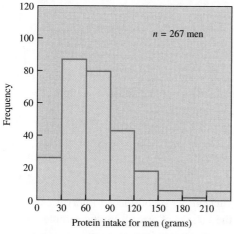

Figure 8.6 Histogram of daily protein intake for men taken from a sample of $n = 267$ men.

Source: National Center for Health Statistics. This sample comes from the Third National Health and Nutrition Examination Survey, or NHANES III, which surveyed approximately 30,000 participants on different aspects of their health and diet.

Using a Sample Mean to Estimate a Population Mean

Suppose that we have computed the sample mean, $\bar{x}$, for a single sample drawn from a large population. To estimate the population mean:

1. Because we have only a single sample mean, we use it as our best (and only) estimate of the population mean.

2. Based on the sample size and sample standard deviation (s), we compute a *margin of error* and use it to construct a *confidence interval* around our estimate. We can then make a statement that gives us some idea of how good our estimate is.

In this case, we choose the sample mean of $\bar{x} = 77.0$ grams per day as our best estimate of the population mean. We know that although this is our best (and only) estimate, it may or may not be a *good* estimate. So our next step is to calculate the margin of error and construct the confidence interval, an idea we first discussed back in Section 1.1.

Finding the Confidence Interval

As we discussed in Chapter 1, a **confidence interval** is a range of values likely to contain the true value of the population mean. In this text, we will work only with 95% confidence levels, although other levels of confidence (such as 90% or 99%) are sometimes used.

The idea of a confidence interval comes directly from the work we did with sampling distributions in Section 8.1. Recall that the sampling distribution of means is a nearly normal distribution with a mean equal to the population mean of μ. Therefore, the 68-95-99.7 rule for normal distributions (see Section 5.2) tells us that approximately 95% of all sample means lie within 2 standard deviations of the population mean.

The difficulty, of course, is that we don't know the true values of either the population mean or population standard deviation. Therefore, we cannot know the true range in which 95% of all sample means would lie. However, we can use the sample standard deviation and the sample size to calculate a **margin of error**, as defined in the following box.

> **The Margin of Error for a Sample Mean**
>
> When we use a sample mean as an estimate of the population mean, the **margin of error** for the 95% confidence interval is
>
> $$\text{margin of error} = E \approx \frac{2s}{\sqrt{n}}$$
>
> where s is the standard deviation of the sample.

Continuing our protein example, we are using a sample of $n = 267$ men with a sample mean of $\bar{x} = 77.0$ grams and a sample standard deviation of $s = 58.6$ grams. The margin of error for 95% confidence is therefore

$$E \approx \frac{2s}{\sqrt{n}} = \frac{2 \times 58.6}{\sqrt{267}} = 7.2 \text{ grams}$$

We can now construct the 95% confidence interval as we discussed in Chapter 1. We simply add and subtract the margin of error from the sample mean of $\bar{x} = 77.0$ grams, which gives a 95% confidence interval extending approximately from $77.0 - 7.2 = 69.8$ grams to $77.0 + 7.2 = 84.2$ grams. We can write this result more formally as 69.8 grams $< \mu <$ 84.2 grams; note that this suggests that American men are consuming a significantly different amount of protein than the recommended daily allowance of 55 to 60 grams.

More generally, the confidence interval for a population mean always has the sample mean $\bar{x}$ at the center and extends a distance equal to the margin of error in either direction (Figure 8.7).

95% confidence interval

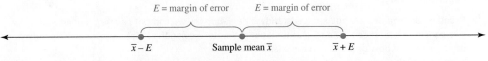

Figure 8.7 The 95% confidence interval extends a distance equal to the margin of error on either side of the sample mean.

The Confidence Interval for a Population Mean

We find the **95% confidence interval** by adding and subtracting the margin of error from the sample mean. That is, the 95% confidence interval ranges

$$\text{from } (\bar{x} - \text{margin of error}) \text{ to } (\bar{x} + \text{margin of error})$$

We can write this confidence interval formally as $\bar{x} - E < \mu < \bar{x} + E$ or in the format of $\bar{x} \pm E$.

Interpreting the Confidence Interval

Confidence intervals are easy to construct and are very useful, but we need to be very careful in how we interpret them. The proper interpretation requires thinking about the sampling distribution of means. Recall that 95% of all possible samples have a sample mean lying within approximately 2 standard deviations of the population mean. Therefore, if we were to repeat the process of obtaining samples and constructing confidence intervals many times, 95% of the confidence intervals would contain the population mean (μ) and 5% would not include it.

Interpreting a 95% Confidence Interval

When we choose a single sample and construct its 95% confidence interval, we have no way to know whether or not it contains the population mean, (μ). However, if we could repeat the sampling process many times, we would find that 95% of the confidence intervals would contain the population mean and 5% would not.

Figure 8.8 shows a visual interpretation of this idea. The true population mean (μ) is represented by the blue vertical line running down the middle of the figure. The region to the left of this line represents values below the population mean and the region to the right represents values above the population mean. The figure shows confidence intervals for 20 different samples; in each case, the sample mean ($\bar{x}$) is the black dot at the center and the red line shows the 95% confidence interval around that sample mean. Notice that 19 of the 20 samples, or 95%, have a confidence interval that actually contains the population mean. One sample (the 6th one from the top), or 5%, has a confidence interval that does not contain the population mean.

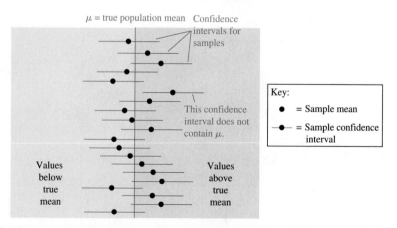

Figure 8.8 This figure illustrates the idea behind confidence intervals. The central vertical line represents the true population mean, μ. Each of the 20 horizontal lines represents the 95% confidence interval for a particular sample, with the sample mean marked by the dot in the center of the confidence interval. With a 95% confidence interval, we expect that 95% of all samples will give a confidence interval that contains the population mean, as is the case in this figure, for 19 of the 20 confidence intervals do indeed contain the population mean. We expect that the population mean will not be within the confidence interval in 5% of the cases; here, 1 of the 20 confidence intervals (the sixth from the top) does not contain the population mean.

The subtlety arises from the fact that "confidence" and "probability" are not quite the same thing. When we construct a confidence interval around a single sample mean, we have no way to know whether our particular confidence interval is one of the 95% that contain the population mean or one of the 5% that do not. In other words, our confidence interval either does or does not contain the population mean, and any other probability statement would be misleading. All we can say is that 95% of all possible confidence intervals for our sample size would contain the population mean, and 5% would not.

EXAMPLE 1 **Protein Intake for Women**

Figure 8.9 shows a histogram for a random sample of $n = 264$ women (drawn from the same NHANES III study used for the sample of men in Figure 8.6). The mean of these data is $\bar{x} = 59.6$ grams and the standard deviation is $s = 30.5$ grams. Estimate the population mean and give a 95% confidence interval. Comment on how these values compare to the recommended daily allowance (RDA) for women of 45–50 grams.

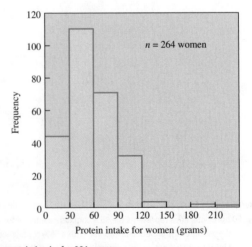

Figure 8.9 Histogram for daily protein intake for 264 women.

Source: National Center for Health Statistics.

SOLUTION We use the sample mean, $\bar{x} = 59.6$ grams, as our best (and only) estimate of the population mean. To find the 95% confidence interval, we must first find the margin of error using the sample size ($n = 264$) and sample standard deviation $s = 30.5$ grams:

$$E \approx \frac{2s}{\sqrt{n}} = \frac{2 \times 30.5 \text{ grams}}{\sqrt{264}} = 3.8 \text{ grams}$$

The 95% confidence interval therefore ranges from $59.6 - 3.8 = 55.8$ grams to $59.6 + 3.8 = 63.4$ grams, or

$$55.8 \text{ grams} < \mu < 63.4 \text{ grams}$$

We have 95% confidence that the interval from 55.8 grams to 63.4 grams contains the true value of the population mean. This means that if we were to repeat this process many times, 95% of the resulting confidence intervals should contain the true value of the population mean. We conclude that protein consumption by women is very likely to be different than the recommended daily allowance of 45–50 grams for women.   ···●

> ### TIME ⏱UT TO THINK
> Recall that the standard deviation of the data on protein intake for the sample of men was $s = 58.6$ grams—almost double the standard deviation for the sample of women ($s = 30.5$ grams). How does this difference affect the margins of error? Speculate on why the standard deviations are different.

EXAMPLE ❷ Garbage Production

A study conducted by the Garbage Project at the University of Arizona analyzed the contents of garbage discarded by $n = 62$ households; the households ranged in size from 2 to 11 members. The histogram in Figure 8.10 shows the total weekly garbage production (in pounds) for the households in the sample. (The complete study gives the breakdown of garbage by various categories.) The mean for the sample is $\bar{x} = 27.4$ pounds and the standard deviation is $s = 12.5$ pounds. Estimate the population mean for weekly garbage production with a 95% confidence interval.

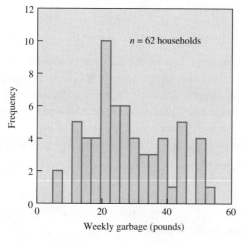

Figure 8.10 Histogram for total garbage production for $n = 62$ households.

BY THE WAY

According to the Garbage Project data, the leading component of most household garbage is paper (comprising a third to a half of the total garbage), followed by food and glass. That is why recycling and composting can drastically reduce the amount of waste going to landfills.

SOLUTION We use the sample mean, $\bar{x} = 27.4$ pounds, as our best (and only) estimate of the population mean. We use the margin of error formula to find that

$$E \approx \frac{2s}{\sqrt{n}} = \frac{2 \times 12.5 \text{ grams}}{\sqrt{62}} = 3.2 \text{ pounds}$$

The 95% confidence interval therefore ranges from $27.4 - 3.2 = 24.2$ pounds to $27.4 + 3.2 = 30.6$ pounds, or

$$24.2 \text{ pounds} < \mu < 30.6 \text{ pounds}$$

We have 95% confidence that the interval from 24.2 pounds to 30.6 pounds contains the mean amount of garbage discarded by American households in a week.   • • ●

EXAMPLE 3   Mean Body Temperature

A study by University of Maryland researchers investigated the body temperatures of $n = 106$ subjects. The sample mean of the data set is $\bar{x} = 98.20°F$ and the standard deviation for the sample is $s = 0.62°F$. Estimate the population mean body temperature with a 95% confidence interval.

SOLUTION   The sample mean, $\bar{x} = 98.20°F$, is our best (and only) estimate of the population mean body temperature. For a 95% confidence interval, the margin of error is

$$E \approx \frac{2s}{\sqrt{n}} = \frac{2 \times 0.62°F}{\sqrt{106}} = 0.12°F$$

The 95% confidence interval therefore ranges from $98.20°F - 0.12°F = 98.08°F$ to $98.20°F + 0.12°F = 98.32°F$, or

$$98.08°F < \mu < 98.32°F$$

We have 95% confidence that the interval from 98.08°F to 98.32°F contains the true population mean. This means that if we were to select many different samples of size $n = 106$ and compute confidence intervals for all of the samples, we expect that 95% of the confidence intervals would contain the true population mean. Notice that the commonly cited mean body temperature for humans (98.6°F) is *not* contained in the confidence interval. Based on this sample, it is likely that the accepted mean body temperature of 98.6°F is wrong.   • • ●

Choosing Sample Size

In planning statistical surveys and experiments, we often know in advance the margin of error we would like to achieve. For example, we might want to estimate the mean cost of a new car to within $200. We can estimate the sample size, n, needed to ensure this margin of error by solving the margin of error formula ($E \approx 2s/\sqrt{n}$) for n. With a little bit of algebra, we find

$$n \approx \left(\frac{2s}{E}\right)^2$$

The exact sample size formula uses the population standard deviation, σ, in place of s. In practice, we rarely know the population standard deviation, because we study only samples. Therefore, to use the exact sample size formula, we usually estimate the population standard deviation based on previous studies, pilot studies, or educated guesses. The size of any actual sample must be a whole number, so we round the result of the sample size formula *up* to the nearest whole number. Any sample larger than this size will give us a margin of error as small as or smaller than the one we seek.

TECHNICAL NOTE

The given formula for sample size assumes we are working with a 95% level of confidence; a more accurate formula would use 1.96 instead of 2, as discussed earlier. For the methods of this section, if the population standard deviation is not known, we can use the sample standard deviation, s, as an estimate of that value. However, the results will be good only if the sample size is large. If a small sample is taken from a non-normal population, the use of s in place of σ may lead to very poor results.

Choosing the Correct Sample Size

In order to estimate the population mean with a specified margin of error of at most E, the size of the sample should be at least

$$n = \left(\frac{2\sigma}{E}\right)^2$$

where σ is the population standard deviation (often estimated by the sample standard deviation s).

EXAMPLE 4 Mean Housing Costs

You want to study housing costs in the country by sampling recent house sales in various (representative) regions. Your goal is to provide a 95% confidence interval estimate of the housing cost. Previous studies suggest that the population standard deviation is about $7,200. What sample size (at a minimum) should be used to ensure that the sample mean is within

a. $500 of the true population mean?

b. $100 of the true population mean?

SOLUTION

a. With $E = \$500$ and σ estimated as $7,200, the minimum sample size that meets the requirements is

$$n = \left(\frac{2\sigma}{E}\right)^2 = \left(\frac{2 \times 7,200}{500}\right)^2 = 28.8^2 = 829.4$$

Because the sample size must be a whole number, we conclude that the sample should include *at least* 830 prices.

b. With $E = \$100$ and $\sigma = \$7,200$, the minimum sample size that meets the requirements is

$$n = \left(\frac{2\sigma}{E}\right)^2 = \left(\frac{2 \times 7,200}{100}\right)^2 = 144^2 = 20,736$$

Notice that to decrease the margin of error by a factor of 5 (from $500 to $100), we must increase the sample size by a factor of 25. That is why achieving greater accuracy generally comes with a high cost.
      • • ●

TIME ◯UT TO THINK

If you decide you want a smaller margin of error for a confidence interval, should you increase or decrease the sample size? Explain.

Section 8.2 Exercises

Statistical Literacy and Critical Thinking

1. **Confidence Interval.** From a random sample of weights of dollar coins, we construct this 95% confidence interval estimate of the mean:

    8.0518 grams $< \mu <$ 8.0902 grams

    Interpret this confidence interval.

2. **Margin of Error.** Based on a random sample of hospital costs for car crash victims, the sample mean is $9,004 and the margin of error for a 95% confidence interval is $266. Identify the confidence interval.

3. **Confidence Intervals in the Media.** Here is a typical statement made by the media: "Based on a recent study, pennies weigh an average of 2.5 grams with a margin of error of 0.006 gram." What important and relevant piece of information is omitted from that statement? Is it OK to use the word "average"?

4. **Sample Size.** The National Health Examination involves measurements from about 25,000 people, and the results are used to estimate values of various population means. Is it valid to criticize this survey because the sample size is only about 0.01% of the population of all Americans? Explain.

Does It Make Sense? For Exercises 5–8, decide whether the statement makes sense (or is clearly true) or does not make sense (or is clearly false). Explain clearly. Not all of these statements have definitive answers, so your explanation is more important than your chosen answer.

5. **Interpreting a Confidence Interval.** For the confidence interval given in Exercise 1, we can interpret that result

by saying that "95% of sample means will fall between 8.0518 grams and 8.0902 grams."

6. **Margin of Error.** The mean income of high school mathematics teachers was estimated to be $48,213 with a margin of error of five percentage points.

7. **Margin of Error.** When sample data are used to estimate the value of a population mean, the margin of error decreases as the sample size increases.

8. **Best Estimate.** When sample data were used to estimate the value of the mean weight of all pennies, this 95% confidence interval was obtained:

$$2.512 \text{ grams} < \mu < 2.512 \text{ grams}$$

Concepts and Applications

Finding Margins of Error and Confidence Intervals. For Exercises 9–12, assume that population means are to be estimated from the samples described. In each case, use the sample results to approximate the margin of error and 95% confidence interval.

9. sample size = 49, sample mean = 25.2 cm, sample standard deviation = 2.2 cm

10. Sample size = 81, sample mean = 4.5 km, sample standard deviation = 3.1 km

11. $n = 100$, $\bar{x} = 8.0$ ft, $s = 2.0$ ft

12. $n = 64$, $\bar{x} = \$550$, $s = \$60$

Sample Sizes. For Exercises 13–16, assume that you want to construct a 95% confidence interval estimate of a population mean. Find an estimate of the sample size needed to obtain the specified margin of error for the 95% confidence interval. The sample standard deviation is given.

13. Margin of error = $5, standard deviation = $20

14. Margin of error = 18.2 cm, standard deviation = 95.2 cm

15. Margin of error = 3.5 ml, standard deviation = 155.2 ml

16. Margin of error = 0.5 g, standard deviation = 8.7 g

17. **Sample Size for TV Survey.** Nielsen Media Research wishes to estimate the mean number of hours that high school students spend watching TV on a weekday. A margin of error of 0.25 hour is desired. Past studies suggest that a population standard deviation of 1.7 hours is reasonable. Estimate the minimum sample size required to estimate the population mean with the stated accuracy.

18. **Sample Size for Housing Prices.** A government survey conducted to estimate the mean price of houses in a large metropolitan area is designed to have a margin of error of $10,000. Pilot studies suggest that the population standard deviation is $65,500. Estimate the minimum sample size needed to estimate the population mean with the stated accuracy.

19. **Sample Size for Mean IQ of Statistics Students.** The Wechsler IQ test is designed so that the mean is 100 and the standard deviation is 15 for the population of normal adults. Find the sample size necessary to estimate the mean IQ score of Delaware residents. We want to be 95% confident that our sample mean is within 3 IQ points of the true mean. Assume that $\sigma = 15$ and determine the required sample size.

20. **Sample Size for Estimating Income.** An economist wants to estimate mean annual income from the first year of work for college graduates who have had the profound wisdom to take a statistics course. How many such incomes must be found if she wants to be 95% confident that the sample mean is within $750 of the true population mean? Assume that a previous study has revealed that for such incomes, $\sigma = \$6,250$.

21. **Weight of Quarters.** You want to estimate the mean weight of quarters in circulation. A sample of 40 quarters has a mean weight of 5.639 grams and a standard deviation of 0.062 gram. Use a single value to estimate the mean weight of all quarters. Also, find the 95% confidence interval.

22. **Weights of Babies.** A sample of 186 newborn babies has a mean weight of 3103 g and a standard deviation of 696 g. Use a single value to estimate the mean weight of a newborn baby. Also, find the 95% confidence interval.

23. **Time to Graduation.** Data from the National Center for Education Statistics on 4,400 college graduates show that the mean time required to graduate with a bachelor's degree is 5.15 years with a standard deviation of 1.68 years. Use a single value to estimate the mean time required to graduate for all college graduates. Also, find the 95% confidence interval.

24. **Garbage Production.** Based on a sample of 62 households, the mean weight of discarded plastic is 1.91 pounds and the standard deviation is 1.07 pounds (data from the Garbage Project at the University of Arizona). Use a single value to estimate the mean weight of discarded plastic for all households. Also, find the 95% confidence interval.

25. **Weight of Bears.** The health of the bear population in Yellowstone National Park is monitored by periodic measurements taken from anesthetized bears. A sample of the weights (pounds) of such bears is given below. Find a 95% confidence interval estimate of the mean of the population of all such bear weights.

80	344	416	348	166	220	262	360	204
144	332	34	140	180	105	166	204	26
120	436	125	132	90	40	220	46	154
116	182	150	65	356	316	94	86	150

26. **Cotinine Levels of Smokers.** When people smoke, the nicotine they absorb is converted to cotinine, which can

be measured. A sample of cotinine levels of 40 smokers is listed below. Find a 95% confidence interval estimate of the mean cotinine level of all smokers.

1	0	131	173	265	210	44	277
32	3	35	112	477	289	227	103
222	149	313	491	130	234	164	198
17	253	87	121	266	290	123	167
250	245	48	86	284	1	208	173

27. Family Size. You select a random sample of $n = 31$ families in your neighborhood and find the following family sizes (number of people in the family):

2	3	6	5	4	2	3	3	1	2	3
2	3	4	5	3	1	3	3	4	7	3
2	3	2	2	3	4	1	5	2		

a. What is the mean family size for the sample?

b. What is the standard deviation for the sample?

c. What is the best estimate for the mean family size for the population of all American families?

d. What is the 95% confidence interval for the estimate?

e. Comment on the reliability of the estimate.

28. TV Sets. A random sample of $n = 31$ households is asked the number of TV sets in the household. The responses are as follows:

1	0	2	3	2	3	4	2	1	1	2
4	3	2	3	3	0	1	0	1	3	2
4	3	2	1	4	0	1	2	3		

a. What is the mean number of TVs for the sample?

b. What is the standard deviation for the sample?

c. What is the best estimate for the mean number of TVs for the population of all American households?

d. What is the 95% confidence interval for the estimate?

e. Comment on the reliability of the estimate.

PROJECTS FOR  THE INTERNET & BEYOND

29. Car Ages. Assume that you want to estimate the mean age of cars driven by students at your college. A previous study shows that the standard deviation of those ages is approximately 3.7 years. How many car ages must you randomly select in order to be 95% confident that your sample mean is within 1 year of the population mean? Using that sample size, collect your own sample data, consisting of the ages of cars driven by students at your college. Then use the methods of this section to construct a 95% confidence interval. Write a statement summarizing your results.

30. Polling Organizations. Three leading public polling organizations are the Gallup Organization, Harris Poll, and Yankelovich Partners. Visit their Web sites. Describe the history of each organization and the polling services it provides. Which organization has the best description on its Web site of its polling methods?

31. Network Polls. All of the major television networks conduct regular polls on a variety of issues. Visit the Web site of at least one major network and gather the results of a particular poll that involves estimation of a population mean. Be sure to include all information that is given about the sample size, margin of error, and confidence intervals.

32. International Corruption. Transparency International uses surveys to determine a Bribe Payer's Index and Corruption Perception Index that measure the degree of corruption in many countries worldwide. Visit the Transparency International Web site and review the extensive documentation describing the methods used by this organization. Discuss the results and their validity.

IN THE NEWS

33. Estimating Population Means. Find a news article or report in which a population mean is estimated from a sample. The article should include a margin of error and/or a confidence interval. Discuss the methods used in the study and how the conclusions were reached.

8.3 ESTIMATING POPULATION PROPORTIONS

We are now ready to turn our attention to estimates of population *proportions*. Many well-known polls and surveys rely on the techniques that we will discuss. For example, the Nielsen ratings estimate the proportion of the population tuned in to certain radio and television shows, the monthly unemployment figures released by the Bureau of Labor Statistics are estimates of the proportion of Americans who are unemployed, and the opinion polls that dominate American politics estimate the proportion of the population that supports a particular candidate.

The Basics of Estimating a Population Proportion

The Bureau of Labor Statistics estimates the unemployment rate from a monthly survey of 60,000 households (see Example 2 in Section 1.1). The unemployment rate for this sample is the *sample proportion* (the proportion of people in the sample who are unemployed), denoted $\hat{p}$. As we did with means, we use the sample proportion as our best and only estimate of *population proportion* (the proportion of people in the population who are unemployed), denoted p.

Just as with population means, an estimate of the population proportion can be better understood if we describe its accuracy using margins of error and confidence intervals. The only change from estimating population means is in the definition of the margin of error, which is given in the following box.

> **The Margin of Error for a Population Proportion**
>
> For a population proportion, the **margin of error** for the 95% confidence interval is
>
> $$E \approx 2\sqrt{\frac{\hat{p}(1 - \hat{p})}{n}}$$
>
> where $\hat{p}$ is the sample proportion.

Again, once we have the margin of error, we construct the 95% confidence interval simply by adding and subtracting it from the sample proportion, as illustrated in Figure 8.11.

95% confidence interval

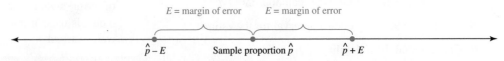

Figure 8.11 The confidence interval extends a distance equal to the margin of error on either side of the sample proportion, $\hat{p}$.

> **The Confidence Interval for a Population Proportion**
>
> We find the **95% confidence interval** by adding and subtracting the margin of error from the sample proportion. That is, the 95% confidence interval ranges
>
> from $\hat{p} -$ margin of error to $\hat{p} +$ margin of error
>
> We can write this confidence interval more formally as
>
> $$\hat{p} - E < p < \hat{p} + E$$

EXAMPLE 1 Unemployment Rate

The Bureau of Labor Statistics finds 5,160 unemployed people in a sample of $n = 60,000$ people. Estimate the population unemployment rate and give a 95% confidence interval.

SOLUTION   The sample proportion is the unemployment rate for the sample:

$$\hat{p} = \frac{5,160}{60,000} = 0.086$$

We use this as the best estimate for the population unemployment rate. The margin of error is

$$E \approx 2\sqrt{\frac{\hat{p}(1 - \hat{p})}{n}} = 2\sqrt{\frac{0.086(1 - 0.086)}{60,000}} = 0.0023$$

(The approximation is valid because of the large sample size.) The 95% confidence interval ranges from $0.0860 - 0.0023 = 0.0837$ to $0.0860 + 0.0023 = 0.0883$ or

$$0.0837 < p < 0.0883$$

We have 95% confidence that the interval from 8.37% to 8.83% contains the true unemployment rate for the population. We interpret this result as follows: If we computed confidence intervals for many samples of size $n = 60,000$, we should expect 95% of the confidence intervals to contain the true population proportion.   ⋯●

BY THE WAY

Here's how the Bureau of Labor Statistics describes uncertainty in its unemployment survey: "A sample is not a total count and the survey may not produce the same results that would be obtained from interviewing the entire population. But the chances are 90 out of 100 that the monthly estimate of un-employment from the sample is within 290,000 of the figure obtainable from a total census. Since monthly unemployment totals have ranged between about 7 and 11 million in re-cent years, the possible error resulting from sampling is not large enough to distort the total unemployment picture."

EXAMPLE 2 TV Nielsen Ratings

The Nielsen ratings for television use a random sample of households. A Nielsen survey results in an estimate that a women's World Cup soccer game had 72.3% of the entire viewing audience. Assuming that the sample consists of $n = 5,000$ randomly selected households, find the margin of error and the 95% confidence interval for this estimate.

SOLUTION   The sample proportion, $\hat{p} = 72.3\% = 0.723$, is the best estimate of the population proportion. The margin of error is

$$E \approx 2\sqrt{\frac{\hat{p}(1 - \hat{p})}{n}} = 2\sqrt{\frac{0.723(1 - 0.723)}{5,000}} = 0.013$$

⏻ **USING TECHNOLOGY—CONFIDENCE INTERVAL ESTIMATE OF A POPULATION PROPORTION**

Excel: Use XLSTAT. Click on **XLSTAT** at the top, click on **Parametric Tests**, then select **Tests for one proportion**. In the screen that appears, enter the sample proportion in the "Proportion" box, enter the sample size in the "Sample size" box, select the "Data Format" of **Proportion**, and be sure that the box next to "z test" is checked. For the "Range" box, enter A1 so that the results will start at cell A1. Click on the **Options** tab, to enter the desired "Significance level (%)." Enter 5 for a 95% confidence interval. For the "Variance (confidence interval)" options, select **Sample** (so that the sample variance is used in the computation of the confidence interval). There are four options for the type of confidence interval; accept the default of **Wald**. Click **OK**. After the results are displayed, look for "confidence interval on the proportion (Wald)."

Statdisk. Select **Analysis** and then **Confidence Intervals**. Then select **Proportion – One Sample**. In the dialog box that appears, first enter the significance level as a decimal number. Enter 0.95 for a 95% confidence level. Proceed to enter the other required items. Then click on **Evaluate** and the confidence interval will be displayed.

TI-83/84 Plus Press STAT, select **TESTS**, then select **1-PropZInt** and enter the required items. Like many technologies, the TI-83/84 Plus calculator requires entry of the number of successes. Also like many technologies, the confidence interval limits are expressed in the format of $(\hat{p} - E, \hat{p} + E)$.

The 95% confidence interval is $0.723 - 0.013 < p < 0.723 + 0.013$, or

$$0.710 < p < 0.736$$

With 95% confidence, we conclude that between 71.0% and 73.6% of the entire viewing audience watched the women's World Cup soccer game. $\cdots\bullet$

EXAMPLE ③ Pete Rose Poll

The Gallup Organization conducted a survey of 1,016 randomly selected adults who were asked:

> *As you may know, former major league player Pete Rose is ineligible for baseball's Hall of Fame due to charges that he had gambled on baseball games. Do you think he should or should not be eligible for admission to the Hall of Fame?*

Among those surveyed, 59% believed that Pete Rose should be eligible. Find the margin of error and confidence interval for this survey. The reported results cited a margin of error of "no more than 5 percentage points." Is this claim consistent with the sample size?

SOLUTION   A sample with $n = 1,016$ respondents and a sample proportion of $\hat{p} = 0.59$ has a margin of error of

$$E \approx 2\sqrt{\frac{\hat{p}(1 - \hat{p})}{n}} = 2\sqrt{\frac{0.59(1 - 0.59)}{1,016}} = 0.031$$

This margin of error is about 3 percentage points, so the confidence interval ranges from approximately $59 - 3 = 56\%$ to $59 + 3 = 62\%$. Note that the cited margin of error of "no more than 5 percentage points" is, in fact, an overestimate. $\cdots\bullet$

Choosing Sample Size

Designers of surveys and polls often specify a certain level of accuracy for the results. For example, it might be desirable to estimate a population proportion with a 95% confidence interval and a margin of error of no more than 1.5 percentage points. In such situations, it's necessary to determine how large the sample must be to guarantee this accuracy. As long as we use a 95% level of confidence, we can work with this simplified, approximate formula for the margin of error:

$$E \approx \frac{1}{\sqrt{n}}$$

TECHNICAL NOTE

You can derive the $E \approx 1/\sqrt{n}$ formula from the more precise formula given earlier for the margin of error by replacing the product $\hat{p}(1 - \hat{p})$ by its maximum possible value of 0.25. This approximation overestimates the actual margin of error and is most accurate when p is near 0.5.

This formula gives a conservative (higher than necessary) estimate for the margin of error. Solving for n yields the sample size needed to achieve a margin of error E:

$$n \approx \frac{1}{E^2}$$

Any sample size equal to or larger than this value will suffice.

> **Choosing the Correct Sample Size**
>
> In order to estimate a population proportion with a 95% degree of confidence and a specified margin of error of E, the size of the sample should be at least
>
> $$n = \frac{1}{E^2}$$

EXAMPLE ④ Minimum Sample Size for Survey

You plan a survey to estimate the proportion of students on your campus who carry an iPad regularly. How many students should be in the sample if you want (with 95% confidence) a margin of error of no more than 4 percentage points?

SOLUTION Note that 4 percentage points means a margin of error of 0.04. From the given formula, the minimum sample size is

$$n = \frac{1}{E^2} = \frac{1}{0.04^2} = 625$$

You should survey at least 625 students. $\cdots\bullet$

EXAMPLE ⑤ Yankelovich Poll

A poll by Yankelovich Partners concluded that 61% of all households have a computer, with a margin of error of 3.5 percentage points. Approximately what sample size must have been used in this poll?

SOLUTION A margin of error of 3.5 percentage points (or 0.035) could be achieved with a sample size of

$$n = \frac{1}{E^2} = \frac{1}{0.035^2} = 816.3$$

Because we must round up to the next larger whole number, we conclude that approximately 817 households were surveyed. $\cdots\bullet$

Section 8.3 Exercises

Statistical Literacy and Critical Thinking

1. **Confidence Interval.** In an Accountemps survey, senior executives were asked to identify the most common error made in interviews of job applicants. The following 95% confidence interval estimates the population proportion p for "little or no knowledge of the company." Interpret that confidence interval.

$$0.393 < p < 0.553$$

2. **Margin of Error.** In a study of 1,228 randomly selected medical malpractice lawsuits, it is found that the proportion that were dropped or dismissed is 0.697. When a 95% confidence interval is constructed for the population proportion of all lawsuits, the margin of error is found to be 0.026. Identify the confidence interval.

3. **Confidence Intervals in the Media.** Here is a typical statement made by the media: "Based on a survey of 1050 likely voters, 44% plan to vote for the Republican candidate, and this survey has a margin of error of 3 percentage points." What important and relevant piece of information is omitted from that statement?

4. **Small Sample.** In a Pew Research Center poll, 73% of 3011 adults surveyed said that they use the Internet. A reporter claims that the results are not very good because they are based on a survey of only 0.001% of the adults in the United States, and that percentage is too small to be meaningful. Is that claim valid? Why or why not?

Does It Make Sense? For Exercises 5–8, decide whether the statement makes sense (or is clearly true) or does not make sense (or is clearly false). Explain clearly. Not all of these statements have definitive answers, so your explanation is more important than your chosen answer.

5. **Interpreting a Confidence Interval.** We can interpret the confidence interval given in Exercise 1 by saying that there is a 95% chance that the true population proportion will fall between 0.393 and 0.553.

6. **Confidence Interval in the Media.** The *Kingston Chronicle* publishes an article stating that, based on survey results, 82% of Orange County residents oppose an increase in the sales tax, with a margin of error of 4 percentage points. A reader says that this can be expressed as the confidence interval $0.78 < p < 0.86$.

7. Sample Size. The 95% confidence interval given in Example 1 is based on a sample size of 150. If the sample size is increased, the confidence interval will become smaller (or narrower).

8. Sample Size. A reporter for the *Kingston Chronicle* claims that any good confidence interval should be based on a sample that is at least 5% of the population size.

Concepts and Applications

Margins of Error and Confidence Intervals. For Exercises 9–12, assume that population proportions are to be estimated from the samples described. In each case, find the approximate margin of error and 95% confidence interval.

9. Sample size = 1,000, sample proportion = 0.4

10. Sample size = 800, sample proportion = 0.25

11. $n = 550$, $\hat{p} = 0.1$

12. $n = 780$, $\hat{p} = 0.160$

Sample Size. For Exercises 13–16, estimate the minimum sample size needed to achieve the given margin of error.

13. $E = 0.04$

14. $E = 0.05$

15. $E = 0.015$

16. $E = 0.035$

17. Nielsen Ratings. Nielsen Media Research uses samples of 5,000 households to rank TV shows. Nielsen reported that *60 Minutes* had 15% of the TV audience. What is the 95% confidence interval for this result?

18. Nielsen Ratings. Repeat Exercise 17 assuming that the sample size is doubled to 10,000. Given that the large cost and effort of conducting the Nielsen survey would be doubled, does this increase in sample size appear to be justified by the increased reliability?

19. Hazing of Athletes. A study done by researchers at Alfred University concluded that 80% of all student athletes in this country have been subjected to some form of hazing. The study is based on responses from 1,400 athletes. What are the margin of error and 95% confidence interval for the study?

20. Student Opinions. An annual survey of first-year college students, conducted by the Higher Education Research Institute at UCLA, asks approximately 276,000 students about their attitudes on a variety of subjects. According to a recent survey, 51% of first-year students believe that abortion should be legal (down from 65% in 1990) and 40% believe that casual sex is acceptable (down from 50% in 1975). What are the margins of error and 95% confidence intervals for these estimates?

21. Gender Selection. The Genetics and IVF Institute conducted clinical trials of the YSORT method designed to increase the probability of conceiving a boy. Among 152 babies born to parents using the YSORT method, 127 were boys. Identify the margin of error and the 95% confidence interval for these clinical trials.

22. Global Warming. A Pew Research Center poll included 1708 randomly selected adults who were asked whether "global warming is a problem that requires immediate government action." Results showed that 939 of those surveyed indicated that immediate government action is required. A news reporter wants to determine whether these survey results constitute strong evidence that the majority (more than 50%) of people believe that immediate government action is required.

  a. What is the best estimate of the proportion of adults who believe that immediate government action is required?

  b. Construct a 95% confidence interval estimate of the proportion of adults believing that immediate government action is required.

  c. Is there strong evidence supporting the claim that the majority is in favor of immediate government action? Why or why not?

23. Drugs in Movies. A study by Stanford University researchers for the Office of National Drug Control Policy and the Department of Health and Human Services concluded that 98% of the top rental films involve drugs, drinking, or smoking. Assume that this study is based on the top 400 rental films.

  a. Use the results of this sample to estimate the proportion of all films that involve drugs, drinking, or smoking.

  b. What is the 95% confidence interval?

  c. Do you believe that the top 400 films represent a random sample? Explain.

24. Teen Pressure. A study commissioned by the U.S. Department of Education concluded that 44% of teenagers cite grades as their greatest source of pressure. The study was based on responses from 1,015 teenagers. What is the 95% confidence interval?

25. Pre-Election Polls. Prior to a statewide election for the U.S. Senate, three polls are conducted. In the first poll, 780 of 1,500 voters favor candidate Martinez. In the second poll, 1,285 of 2,500 voters favor Martinez. In the third poll, 1,802 of 3,500 voters favor Martinez. Find the 95% confidence intervals for all three polls. Discuss Martinez's prospects for victory based on these polls.

26. Unemployment Survey. The Bureau of Labor Statistics estimates the unemployment rate in the United States monthly by surveying 60,000 individuals.

  a. In one month, 3.4% of the 60,000 individuals surveyed are found to be unemployed. Find the margin of error for this estimate. Is the precision (nearest tenth of a percent) reasonable? Explain.

  b. Suppose that the number of individuals surveyed were increased by a factor of four (to 240,000). By how much would the margin of error change?

c. Suppose that the number of individuals surveyed were decreased by a factor of one-fourth (to 15,000). By how much would the margin of error change?

27. Opinion Poll. A poll finds that 54% of the population approves of the job that the President is doing; the poll has a margin of error of 4% (assuming a 95% degree of confidence).

a. What is the 95% confidence interval for the true population percentage that approves of the President's performance?

b. What was the size of the sample for this poll?

28. Concealed Weapons. Two-thirds (or 66.6%) of 626 Colorado residents polled by Talmey-Drake Research & Strategy Inc. said they backed a bill pending in the legislature that would standardize laws on granting concealed-weapon permits to gun owners. The bill would force local law enforcement to grant such permits to anyone who can legally carry a gun. The margin of error in the poll was reported as 4 percentage points.

a. Is the reported margin of error consistent with the sample size for this estimate?

b. What sample size would be needed to give a margin of error of 2 percentage points?

PROJECTS FOR THE INTERNET & BEYOND

29. Who's the Vice President? Assume that you want to estimate the proportion of students at your college who can correctly identify the Vice President of the United States.

How many students must you randomly select in order to be 95% confident that your sample proportion is within 0.1 of the population proportion? Using that sample size, collect your own sample data by randomly selecting and surveying students at your college. Then use the methods of this section to construct a 95% confidence interval. Write a statement summarizing your results.

30. Nielsen Methods. Visit the Nielsen Media Research Web site and report on the actual methods used to estimate population proportions and confidence intervals in Nielsen ratings.

31. Network Polls. All of the major television networks conduct regular polls on a variety of issues. Visit the Web sites of the major networks and gather the results of a particular poll that involves estimation of a population proportion. Be sure to include all information that is given about the sample size, margin of error, and confidence intervals. Include any details about the actual polling procedure.

IN THE NEWS

32. Estimating Population Proportions. Find a news article or report in which a population proportion is estimated from a sample. The article should include a margin of error and/or a confidence interval. Discuss the methods used in the study and how the conclusions were reached.

1. In a Pew Research Center poll of 745 randomly selected adults, 589 said that it is morally wrong to not report all income on tax returns.

   a. What single value is the best estimate of the population proportion of all adults who say that it is morally wrong to not report all income on tax returns?

   b. Use the sample results to construct a 95% confidence interval estimate of the population proportion of adults who say that it is morally wrong to not report all income on tax returns.

   c. What is the margin of error?

   d. Write a statement that correctly interprets the confidence interval found in part b.

   e. The sample proportion of 0.791 was obtained from one specific sample of 745 adults. Suppose many different samples of 745 adults are obtained and the sample proportion is found for each sample. What do we know about the shape of the distribution of the sample proportions?

   f. Assume that we plan to conduct another poll to determine whether the given sample results are accurate. How many adults would we need to poll in order to have 95% confidence that the margin of error is 0.02?

2. We want to estimate the mean IQ score on the Stanford-Binet test for the population of college students. We know that for people randomly selected from the general population, the standard deviation of IQ scores on the Stanford-Binet test is 16.

   a. Using a standard deviation of 16, how many college students must we randomly select for IQ tests if we want to have 95% confidence that the sample mean is within 3 IQ points of the population mean?

   b. How is our estimate of the mean IQ of college students affected if the sample size is larger than necessary? Smaller than necessary?

   c. Is the actual standard deviation of IQ scores for college students likely to be equal to 16, more than 16, or less than 16? Explain. If we used the actual value instead of 16 in part a, how would the answer to part a be affected? Would it be the same, smaller, or larger?

3. A sample of 40 randomly selected women is obtained, and the blood platelet count of each subject is measured. The mean is 279.5 and the standard deviation is 65.2.

   a. Use these sample results to construct a 95% confidence interval estimate of the population mean.

   b. What is the margin of error?

   c. Write a statement that correctly interprets the confidence interval found in part a.

   d. If the standard deviation is 65.2, how many subjects must be included if we want 95% confidence that the sample mean is in error by at most 10?

4. a. You have been hired by Intel to determine the proportion of computer owners who plan to upgrade to a new operating system. Assuming that you want to be 95% confident that your sample proportion is within 0.02 of the true population proportion, how many people must you survey?

   b. Suppose that, in conducting the survey described in part a, you find that half of the people called refuse to answer the survey questions because they believe that you are trying to sell them something. If you proceed by calling twice as many people so that your sample size is large enough, will your results be good? Explain.

CHAPTER QUIZ

1. If many different random samples of size 100 are selected from the population of pulse rates of adult women, what is the shape of the distribution of the sample means?

2. If many different random samples of size 500 are selected from the population of college students, what is the shape of the distribution of the proportions of women?

3. What does the notation $\hat{p}$ represent?

4. A journal article provides a confidence interval for a proportion in the format 0.60 ± 0.08. Express this confidence interval in the format $a < p < b$. (That is, rewrite $a < p < b$ using specific values in place of a and b.)

5. Assume that we want to estimate the mean grip strength of adult males in the United States. If a random sample of grip strengths is obtained, which of the following is the best estimate of the population mean?

   a. Median of the sample

   b. Mean of the sample

   c. Standard deviation of the sample

   d. Range of the sample

   e. The sample proportion

6. Find the margin of error corresponding to this 95% confidence interval: $0.440 < p < 0.500$.

7. Find the margin of error corresponding to this 95% confidence interval: $98.0 < \mu < 98.6$.

8. When a 95% confidence interval is constructed for the population mean, a sample mean is found to be 69.6 and the margin of error is found to be 1.2. Identify the 95% confidence interval.

9. When a 95% confidence interval is constructed for the population proportion, a sample proportion is found to be 0.720 and the margin of error is found to be 0.025. Identify the 95% confidence interval.

10. Identify what is wrong with this 95% confidence interval for the population proportion: $0.950 < p < 1.250$.

Where Did Statistics Begin?

The origins of many disciplines are lost in antiquity, but the roots of statistics can be identified with some certainty. Systematic record keeping began in London in 1532 with weekly data collection on deaths. Later in the same decade, official data collection on baptisms, deaths, and marriages began in France. In 1608, the collection of similar vital statistics began in Sweden. Canada conducted the first official census in 1666.

Of course, statistics is more than the collection of data. If there is a founder of statistics, that person must be someone who worked with the data in clever and systematic ways and who used the data to reach conclusions that were not previously evident. Many experts believe that an Englishman named John Graunt deserves the title of the founder of statistics.*

John Graunt was born in London in 1620. As the eldest child in a large family, he took up his father's business as a draper (a dealer in clothing and dry goods). He spent most of his life as a prominent London citizen, until he lost his house and possessions in the Fire of London in 1666. Eight years later, he died in poverty.

It's not clear how John Graunt became interested in the weekly records of baptisms and burials—known as bills of mortality—that had been kept in London since 1563. In the preface of his book *Natural and Political Observations on the Bills of Mortality*, he noted that others "made little other use of them" and wondered "what benefit the knowledge of the same would bring to the World." He must have worked on his statistical projects for many years before his book was first published in 1662.

Graunt worked primarily with the annual bills, which were year-end summaries of the weekly bills of mortality. Figure 8.12 shows the annual bill for 1665 (the year of the Great Plague). The top third of the bill shows the numbers of burials and baptisms (christenings) in each parish. Total burials and baptisms are noted in the middle of the bill, with deaths due to the plague recorded separately. The lower third of the bill shows deaths due to a variety of other causes, with totals given for males and females.

Graunt was aware of rough estimates of the population of London that were made periodically for taxation purposes, but he must have been skeptical of one estimate that put the population of London at 6 or 7 million in 1661. Using the annual bills, comparing burials and baptisms, and estimating the density of families in London (with an average family size of eight), he arrived at a population estimate of 460,000 by three different methods—quite a change from 6 or 7 million! He also found that the population of London was increasing while the populations of towns in the countryside were decreasing, showing an early trend toward urbanization. He raised awareness of the high rates of infant mortality. He also refuted a popular theory that plagues arrive with new kings.

Graunt's most significant contribution may have been his construction of the first life table. Although detailed data on age at death were not available, Graunt knew that of 100 new babies, "36 of them die before they be six years old, and that perhaps but one surviveth 76." With these two data points, he filled in the intervening years as shown in Table 8.5, using methods that he did not fully explain.

With estimates of deaths for various ages, he was able to make the companion table of survivors shown in Table 8.6. Although some modern statisticians have doubted the methods used to construct these tables, Graunt appears to have appreciated their value and anticipated the actuarial tables now used by life insurance companies. It wasn't until 1693 that Edmund Halley, of comet fame, constructed life tables using age-based mortality rates.

---

*Some historians claim that Graunt's book was actually written by his life-long friend and collaborator William Petty, though most statisticians believe Graunt wrote his own book. Either way, we know that Petty continued Graunt's work, publishing later editions of Graunt's book and creating the field of "political arithmetic," which we now call *demography*.

A generall Bill for this present year, ending the 19 of *December* 1665. according to the Report made to the KINGS most Excellent Majesty.

By the Company of Parish Clerks of *London*, &c.

The Total of all the Christnings	9967
The Total of all the Burials this year	97306
Whereof, of the Plague	68596

The Diseases and Casualties this year.

Abortive and Stilborne	617	Executed	21	Palsie	30
Aged	1545	Flox and Small Pox	655	Plague	68596
Ague and Feaver	5257	Found dead in streets, fields, &c.	20	Planner	6
Appoplex and Suddenly	116	French Pox	86	Plurisie	15
Bedrid	10	Frighted	23	Poysoned	1
Blasted	5	Gout and Sciatica	27	Quinsie	35
Bleeding	16	Grief	46	Rickets	557
Bloody Flux, Scowring & Flux	185	Griping in the Guts	1288	Rising of the Lights	397
Burnt and Scalded	8	Hangd & made away themselves	7	Rupture	34
Calenture	3	Headmouldshot & Mouldfallen	14	Scurvy	105
Cancer, Gangrene and Fistula	56	Jaundies	110	Shingles and Swine pox	2
Canker, and Thrush	111	Impostume	227	Sores, Ulcers, broken and bruised Limbs	82
Childbed	625	Kild by severall accidents	46	Spleen	14
Chrisomes and Infants	1258	Kings Evill	86	Spotted Feaver and Purples	1929
Cold and Cough	68	Leprosie	2	Stopping of the stomack	332
Collick and Winde	134	Lethargy	14	Stone and Strangury	98
Consumption and Tissick	4808	Livergrown	20	Surfet	1251
Convulsion and Mother	2036	Meagrom and Headach	12	Teeth and Worms	2614
Distracted	5	Measles	7	Vomiting	51
Dropsie and Timpany	1478	Murthered and Shot	9	VVenn	1
Drowned	50	Overlaid & Starved	45		

Christned: Males 5114, Females 4853, In all 9967
Buried: Males 48569, Females 48737, In all 97306. Of the Plague 68596

Increased in the Burials in the 130 Parishes and at the Pest-house this year 79009
Increased of the Plague in the 130 Parishes and at the Pest-house this year 68590

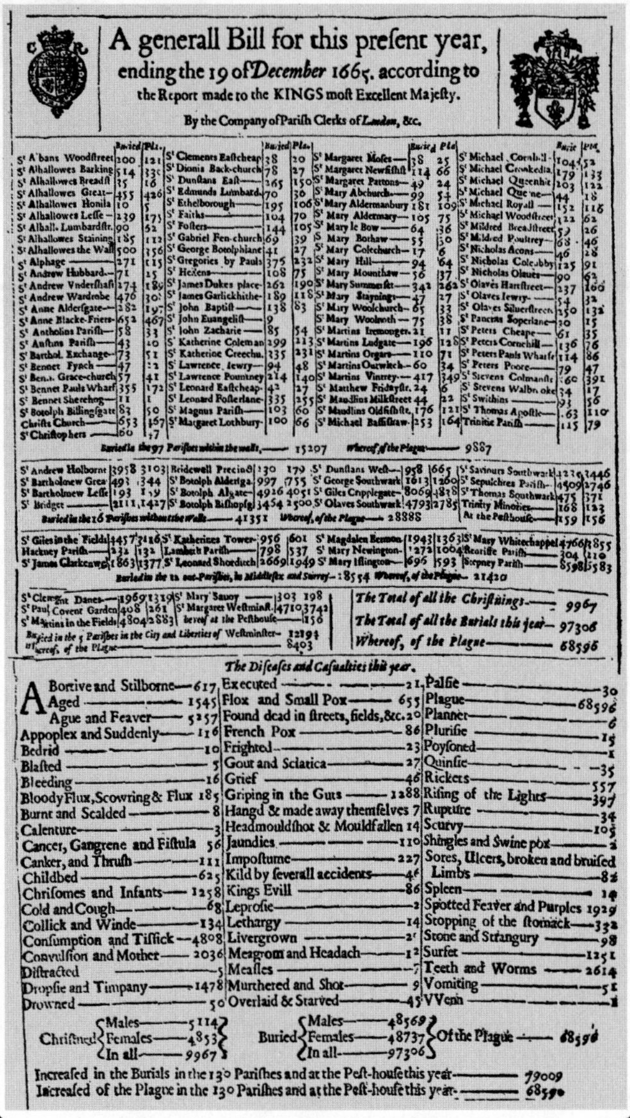

Figure 8.12 Reproduction of 1665 annual bill of mortality.

Source: From the Wellcome Historical Medical Library, reprinted in *Journal of the Royal Statistical Society*, Vol. 126, part 4, 1963, pp. 537–557.

TABLE 8.5 Graunt's Life Table: Number of Deaths by Age for 100 People

Age	Deaths
Within the first six years	36
The next ten years	24
The second decade	15
The third decade	9
The fourth	6
The next	4
The next	3
The next	2
The next	1

TABLE 8.6 Graunt's Table of Survivors at Various Ages

Age	Survivors
At sixteen years end	40
At twenty six	25
At thirty six	16
At forty six	10
At fifty six	6
At sixty six	3
At seventy six	1
At eighty	0

QUESTIONS FOR DISCUSSION

1. Is it surprising that accurate estimates of the population of London were not available in 1660? How would you have suggested making such estimates at that time?

2. Do you think that records of burials and baptisms would have given accurate counts of actual births and deaths? Why or why not?

3. Assuming that the estimates of deaths in Table 8.5 are accurate, are the estimates of survivors in Table 8.6 consistent? Note that the numbers in Table 8.6 do not total to 100; should they? Explain.

4. Based on Graunt's tables, estimate the average life expectancy in 1660. Explain your reasoning.

FOCUS ON LITERATURE

How Many Words Did Shakespeare Know?

Imagine that you go to an orientation party for international students. During the course of the evening, you meet 12 Swedish people, 9 Chinese people, 6 French people, 4 Israelis, 3 Koreans, and 1 Iranian. You know there are people of other nationalities whom you did *not* meet at the party. Based only on the people you met, is it possible to estimate the *total* number of nationalities represented at the party—including those of the people you did not meet? Thanks to ideas of sampling, the answer is yes.

The party problem may be a bit frivolous. However, essentially the same question arose when Oxford University marine biologist Charles Paxton wondered how many "sea monsters" (creatures more than 2 meters in length) remain to be discovered. In this case, the nationalities at the party correspond to species of sea monsters. Using statistical methods, Paxton was able to estimate that, in addition to the roughly 220 sea monsters already known (as of 1998), another 47 wait to be discovered.

Similar methods have been used to analyze the works of Shakespeare. Statisticians Bradley Efron and Ronald Thisted wondered about the number of words Shakespeare actually knew, which must have been larger than the number he used in his writings. Here, the nationalities at the party correspond to different words in Shakespeare's plays and poems. The data collected at the party can be regarded as the *first sample*. For the Shakespeare question, the first sample consists of the complete known works of Shakespeare—specifically, the numbers of words that are used in these works once, twice, three times, and so forth. Table 8.7 shows a (small) part of the first sample. For example, the table says that in the works of Shakespeare,

14,376 words were used exactly once, 4,343 words were used exactly twice, and so forth. (The full table is much larger and continues far beyond 10 occurrences.)

Given the full table for the first sample, we can now ask a hypothetical question. Suppose a second, new and different sample of Shakespeare's works of the same size as the first sample were discovered. How many words could we expect to find in the second sample that were *not* used in the first sample? We would expect there to be fewer new words in the second sample, because in the first sample *every* first occurrence of a word is new, even a common word like "the"; in the second sample, those common words are no longer new. Efron and Thisted estimated that 11,430 words would appear in the second sample that did not appear in the first sample.

They repeated this argument with a third sample, fourth sample, fifth sample, and so on. With each new sample, the number of new words decreases, but the total number of words used (among all samples) increases. Efron and Thisted eventually found that the number of new words approached about 35,000. This means that in addition to the 31,534 words that Shakespeare knew and used, there were approximately 35,000 words that he knew but didn't use. Thus, they estimated that Shakespeare knew approximately 66,500 words.

The analysis of Efron and Thisted, which required advanced methods, was done in 1976. More than 10 years later, the ideas were put to practical use when a new sonnet by an unknown author of Shakespeare's time period was discovered. As before, the complete volume of Shakespeare's works was considered the first sample. Now, the second sample was the new sonnet with 429 words. The same statistical method was used to predict that, if Shakespeare was the author, the new sonnet should have seven new words that did not appear in the complete works of Shakespeare. In fact, the new sonnet had nine new words that did not previously appear. Similarly, the method predicted that the new sonnet should have four words that were used exactly once in the complete works. In fact, there were seven words that were used exactly once in the complete works. And the number of words in the sonnet that

TABLE 8.7	Numbers of Words in Complete Works of Shakespeare Used from One to 10 Times
Occurrences	Number of words
1	14,376
2	4,343
3	2,292
4	1,463
5	1,043
6	837
7	638
8	519
9	430
10	364

were used exactly twice in the complete works was predicted to be three, when in fact there were five such words. The authors concluded that the agreement between the predictions and the actual word frequencies in the sonnet was good enough to attribute authorship to Shakespeare.

These statistical methods have been used to distinguish the works of Shakespeare from those of other Elizabethan writers such as Marlowe, Donne, and Jonson. They have been used to strengthen the belief that James Madison wrote certain of the Federalist Papers whose authorship was in doubt. They have even been used to establish the order of Plato's works.

This example illustrates the remarkable applicability of statistics to problems in a seemingly unrelated discipline. Perhaps more important, it demonstrates how two disciplines as distant from each other as literature and marine biology can be related by a common statistical thread.

QUESTIONS FOR DISCUSSION

1. Comment on whether you believe that literature is enhanced by statistical analysis. For example, is your appreciation of Shakespeare improved by knowing how many words Shakespeare knew?

2. Do you believe that statistical analysis is useful for identifying authors of "lost works"?

3. Suggest another discipline, besides biology and literature, in which the ideas described in this Focus could be used to estimate an unknown quantity.

• • • • • • • • • • • • • • • •

Hypothesis Testing

Everyone makes claims. Advertisers make claims about their products. Universities claim their programs are superb. Governments claim their programs are effective. Lawyers make claims about a suspect's guilt or innocence. Medical diagnoses are claims about the presence or absence of disease. Pharmaceutical companies make claims about the effectiveness of their drugs. But how do we know whether any of these claims are true? Statistics offers a way to test many claims, through a powerful set of techniques called *hypothesis testing.*

It is the mark of an educated mind to be able to entertain a thought without accepting it.

—Aristotle

9.1 FUNDAMENTALS OF HYPOTHESIS TESTING ·················

A company called ProCare Industries, Ltd. once claimed that its product, called Gender Choice, could increase a woman's chance of giving birth to a baby girl. The company claimed that the chance of a baby girl could be increased "up to 80%," but let's focus simply on the claim that the chance of having a baby girl is greater than the approximately 50%, or 0.5, expected under ordinary conditions. How could we test whether the Gender Choice claim is true?

One way would be to study a random sample of, say, 100 babies born to women who used the Gender Choice product. If the product does not work, we would expect about half of these babies to be girls. If it does work, we would expect significantly more than half of the babies to be girls. The key question, then, is what constitutes "significantly more." If there were 97 girls among the 100 births in the sample, we would all agree that this constituted significantly more than half and that the product is probably effective. If there were only 52 girls among the 100 births, we'd probably agree that 52 was so close to half that we had no reason to think the product had any effect. But would we consider, say, 64 girls in the sample of 100 babies to be "significantly more" than half, and therefore think that the product might really work?

In statistics, we answer such questions through *hypothesis testing*. Before we discuss the specific terminology and procedures of hypothesis testing, let's look at the procedure we would use to draw a conclusion about Gender Choice based on a sample in which there are 64 girls among 100 babies born to women who used the product.

- We begin by assuming that Gender Choice does *not* work; that is, it does *not* increase the percentage of girls. If this is true, then we should expect about 50% girls among the *population* of all births to women using the product. That is, if Gender Choice does not work, then the proportion of girls should be 0.50.
- We now use our *sample* (with 64 girls among 100 births) to test the above assumption. We conduct this test by calculating the likelihood of drawing a random sample of 100 births in which 64% (or more) of the babies are girls from a population in which the overall proportion of girls is only 50%.
- If we find that a random sample of births is fairly likely to have 64% (or more) girls, then we do *not* have evidence that Gender Choice works. However, if we find that a random sample of births is *unlikely* to have at least 64% girls, then we conclude that the result in the Gender Choice sample is probably due to something other than chance—meaning that the product may be effective. Note, however, that even in this case we will not have *proved* that the product is effective, as there could still be other explanations for the result (such as that we happened to choose an unusual sample or that there were confounding variables we did not account for).

TIME OUT TO THINK

Suppose Gender Choice is effective and you select a new random sample of 1,000 babies born to women who used the product. Based on the first sample (with 64 girls among 100 births), how many baby girls would you expect in the new sample? Next, suppose Gender Choice is *not* effective; how many baby girls would you expect in a random sample of 1,000 babies born to women who used the product?

Formulating the Hypotheses

The key remaining issues in our test of the effectiveness of Gender Choice are that we have not yet described how to calculate the likelihood of drawing a random sample like the one with 64% girls, and we have not yet defined exactly what we mean when we ask whether such a sample is "fairly likely" or "unlikely" to be drawn at random. Most of this chapter is devoted to the formal procedures used to resolve these issues. The first step in resolving them is to define exactly what we are testing.

BY THE WAY

The Gender Choice product was based on the principle that the gender of a baby can be determined through careful timing of conception. There was no evidence that the product actually worked, and it is no longer available. However, newer technologies have proven successful in allowing gender selection (with artificial insemination), raising ethical questions about whether or when gender selection should be an option for parents.

In the context of statistics, a **hypothesis** is a claim about a population parameter or some other characteristic of a population. For example, in the Gender Choice case, the population parameter is the proportion of girls born to all women who use the product. A **hypothesis test**, then, is a test of whether a particular claim about a population is supported or unsupported by the available sample evidence.

> ### Definitions
>
> A **hypothesis** is a claim about a population parameter (such as a population proportion p or population mean μ) or some other characteristic of a population.
>
> A **hypothesis test** is a standard procedure for testing a claim about a population parameter or some characteristic of a population.

There are always at least two hypotheses in any hypothesis test. In the Gender Choice case, the two hypotheses are essentially: (1) that Gender Choice works in raising the population proportion of girls from the 50% that we would normally expect; and (2) that Gender Choice does *not* work in raising the population proportion of girls. The starting point for the hypothesis test is the second of those hypotheses—that it does *not* raise the proportion of baby girls. We call this starting point the **null hypothesis**, denoted by H_0 (read "*H*-naught"). The other hypothesis (1) is called the **alternative hypothesis**, denoted by H_a (read "*H*-a"). To summarize, for the Gender Choice example:

BY THE WAY

The word *null* comes from the Latin *nullus*, meaning "nothing." A null hypothesis usually states that there is no special effect or difference.

- The *null hypothesis* is the claim that Gender Choice does *not* work, in which case the population proportion of girls born to women who use the product should be 50%, or 0.50. Using p to represent the population proportion, we write this null hypothesis as

$$H_0 \text{ (null hypothesis): } p = 0.50$$

- The *alternative hypothesis* is the claim that Gender Choice *does* work, in which case the population proportion of girls born to women who use the product should be *greater than* 0.50. We write this alternative hypothesis as

$$H_a \text{ (alternative hypothesis): } p > 0.50$$

In this chapter, the null hypothesis will always include the condition of equality, just as the null hypothesis for the Gender Choice example is the equality $p = 0.50$. (We will see a few examples of other types of null hypotheses in Chapter 10.) However, the alternative hypothesis for the Gender Choice case ($p > 0.50$) is only one of three general forms of alternative hypotheses that we will encounter in this chapter:

$$\text{population parameter} < \text{claimed value}$$
$$\text{population parameter} > \text{claimed value}$$
$$\text{population parameter} \neq \text{claimed value}$$

As we will see in Section 9.2, these three different types of alternative hypotheses require slightly different calculations in the hypothesis test; for that reason, it is useful to give them names. The first form ("less than") leads to a **left-tailed** hypothesis test, because it requires testing whether the population parameter lies to the *left* (lower values) of the claimed value. Similarly, the second form ("greater than") leads to a **right-tailed** hypothesis test, because it requires testing whether the population parameter lies to the *right* (higher values) of the claimed value. The third form ("not equal to") leads to a **two-tailed** hypothesis test, because it requires testing whether the population parameter lies significantly far to *either* side of the claimed value.

> ### Null and Alternative Hypotheses
>
> The **null hypothesis**, or H_0, is the starting assumption for a hypothesis test. For the types of hypothesis tests in this chapter, the null hypothesis always claims a specific value for a population parameter and therefore takes the form of an equality:
>
> H_0 (null hypothesis): population parameter = claimed value
>
> The **alternative hypothesis**, or H_a, is a claim that the population parameter has a value that differs from the value claimed in the null hypothesis. It may take one of the following forms:
>
> (left-tailed) H_a: population parameter < claimed value
>
> (right-tailed) H_a: population parameter > claimed value
>
> (two-tailed) H_a: population parameter $\neq$ claimed value

EXAMPLE 1 Identifying Hypotheses

In each case, identify the population parameter about which a claim is made, state the null and alternative hypotheses for a hypothesis test, and indicate whether the hypothesis test will be left-tailed, right-tailed, or two-tailed.

a. Nissan claims that the Leaf, an electric car, has a mean range of 110 miles between charges. A consumer group claims that the mean range is less than 110 miles.

b. The Ohio Department of Health claims that the average stay in Ohio hospitals after childbirth is greater than the national mean of 2.0 days.

c. A wildlife biologist working in the African savanna claims that the actual proportion of female zebras in the region is different from the accepted proportion of 50%.

SOLUTION

a. The population parameter about which a claim is made is a population *mean* (μ)—the mean range of the car between charges. The null hypothesis must state that this population mean is *equal* to some specific value. We therefore recognize the null hypothesis as the advertised claim that the mean range for the cars is 110 miles. The alternative hypothesis is the consumer group's claim that the true range is *less* than advertised. To summarize:

$$H_0 \text{ (null hypothesis): } \mu = 110 \text{ miles}$$
$$H_a \text{ (alternative hypothesis): } \mu < 110 \text{ miles}$$

Because the alternative hypothesis has a "less than" form, the hypothesis test will be left-tailed.

b. The population is all women in Ohio who have recently given birth, and the population parameter is their mean (μ) stay after childbirth in Ohio hospitals. The null hypothesis must be an equality, so in this case it is the claim that the mean hospital stay equals the national average of 2.0 days. The alternative hypothesis is the Health Department's claim that the mean stay in Ohio is *greater than* this national average. To summarize:

$$H_0 \text{ (null hypothesis): } \mu = 2.0 \text{ days}$$
$$H_a \text{ (alternative hypothesis): } \mu > 2.0 \text{ days}$$

Because the alternative hypothesis has a "greater than" form, the hypothesis test will be right-tailed.

c. In this case, the claim is about a population *proportion* (p)—the proportion of female zebras among the zebra population of the region. The accepted population proportion is $p = 0.5$, which becomes the null hypothesis. The wildlife biologist claims that the actual population proportion is different from the value in the null hypothesis. Because "different" can be either greater or less than, the alternative hypothesis has a "not equal to" form:

$$H_0 \text{ (null hypothesis): } p = 0.5$$
$$H_a \text{ (alternative hypothesis): } p \neq 0.5$$

The "not equal to" form of this alternative hypothesis leads to a two-tailed test. $\cdots\bullet$

Possible Outcomes of a Hypothesis Test

A hypothesis test always begins with the assumption that the null hypothesis is true. We then test to see whether the data give us reason to think otherwise. As a result, there are generally only two possible outcomes to a hypothesis test, summarized in the following box.

Two Possible Outcomes of a Hypothesis Test

There are two possible outcomes to a hypothesis test:

1. *Reject* the null hypothesis, H_0, in which case we have evidence in support of the alternative hypothesis.

2. *Not reject* the null hypothesis, H_0, in which case we do not have enough evidence to support the alternative hypothesis.

Notice that "accepting the null hypothesis" is *not* a possible outcome, because the null hypothesis is always the starting assumption. The hypothesis test may not give us reason to reject this starting assumption, but it cannot by itself give us reason to conclude that the starting assumption is true.

The fact that there are only two possible outcomes makes it extremely important that the null and alternative hypotheses be chosen in an unbiased way. In particular, both hypotheses must always be formulated *before* a sample is drawn from the population for testing. Otherwise, data from the sample might inappropriately bias the selection of the hypotheses to test.

TIME ⏰UT TO THINK

The idea that a hypothesis test cannot lead us to accept the null hypothesis is an example of the old dictum that "absence of evidence is not evidence of absence." As an illustration of this idea, explain why it would be easy in principle to prove that some legendary animal (such as Bigfoot or the Loch Ness Monster) exists, but it is nearly impossible to prove that it does *not* exist.

EXAMPLE ❷ Hypothesis Test Outcomes

For each of the three cases from Example 1, describe the possible outcomes of a hypothesis test and how we would interpret these outcomes.

BY THE WAY

The first recorded example of hypothesis testing is attributed to the Scotsman John Arbuthnot (1667–1735). Using 82 years of data, he noticed that the annual number of male baptisms was consistently higher than the annual number of female baptisms. Knowing that there was no bias in baptisms based on gender, he argued that such a regular pattern could not be explained by chance and had to be due to "divine providence," meaning boys must represent slightly more than 50% of all births. In modern terminology, he rejected the null hypothesis that the data could be explained by chance alone.

SOLUTION

a. Recall that the null hypothesis is the advertised claim that the mean range for the cars is $\mu = 110$ miles. The alternative hypothesis is the consumer group's claim that the true range is *less* than advertised, or $\mu < 110$ miles. The possible outcomes are
   - Reject the null hypothesis of $\mu = 110$ miles, in which case we have evidence in support of the consumer group's claim that the range is less than advertised.
   - Do not reject the null hypothesis, in which case we lack evidence to support the consumer group's claim. Note, however, that this option does not imply that the advertised claim is true.

b. The null hypothesis is that the mean hospital stay is the national average of 2.0 days. The alternative hypothesis is the Health Department's claim that the mean stay in Ohio is *greater than* the national average. The possible outcomes are
   - Reject the null hypothesis of $\mu = 2.0$ days, in which case we have evidence in support of the Health Department's claim that the mean stay in Ohio is greater than the national average.
   - Do not reject the null hypothesis, in which case we lack evidence to support the Health Department's claim. Note, however, that this option does not imply that the Ohio average stay is actually equal to the national average stay of 2.0 days.

c. The null hypothesis is that the proportion of female zebras is the accepted population proportion of 50% ($p = 0.5$). The alternative hypothesis is the biologist's claim that the accepted value is wrong, meaning that the actual proportion of female zebras is not 50% (it may be either more or less than 50%). The possible outcomes are

- Reject the null hypothesis of $p = 0.5$, in which case we have evidence in support of the biologist's claim that the accepted value is wrong.

- Do not reject the null hypothesis, in which case we lack evidence to support the biologist's claim. Note, however, that this does not imply that the accepted value is correct.

$\cdots\bullet$

Drawing a Conclusion from a Hypothesis Test

Let's return to the Gender Choice example in which we imagine drawing a random sample of 100 babies (born to women using the Gender Choice product) and find that 64 of these babies are girls. How do we decide whether this sample result should lead us to reject or not reject the null hypothesis? The answer comes down to deciding whether the sample result was *likely* or *unlikely* to have occurred by chance if the null hypothesis is true.

Remember that the null hypothesis for this case is that the true proportion of baby girls among the population of Gender Choice users is 50%, or $p = 0.50$. Using the notation introduced in Chapter 8, the sample we are studying has a sample size of $n = 100$ and a sample proportion of $\hat{p} = 0.64$. The precise question, then, is this: If the true population proportion is $p = 0.50$ (as the null hypothesis claims), what is the probability that by chance alone a sample of size $n = 100$ would have a sample proportion of at least $\hat{p} = 0.64$? If the probability is low, then it is highly unlikely that we would have found such a sample by chance; we therefore have reason to reject the null hypothesis. If the probability of observing the sample result is moderate or high, then it is fairly likely that we could have found such a result by chance, so we cannot reject the null hypothesis.

There are multiple ways to make the decision about rejecting or not rejecting the null hypothesis. Here, we'll look at two closely related options: making the decision based on the statistical significance of the result and making the decision based on the actual probability, or "*P*-value," of the test result.

Statistical Significance

We introduced the idea of statistical significance in Section 6.1. Recall that if the probability of a particular result is 0.05 or less, we say that the result is statistically significant at the 0.05 level; if the probability is 0.01 or less, the result is statistically significant at the 0.01 level. The 0.01 level therefore means stronger significance than the 0.05 level. The following box summarizes how we can apply these ideas directly to hypothesis tests.

Probable truth relies on statistical arguments to weigh which of several possibilities is more likely to be true. When something must be proven beyond a reasonable doubt, just what does that mean? What level of doubt is acceptable? One in twenty? One in a trillion?

—K. C. Cole

> **Hypothesis Test Decisions Based on Levels of Statistical Significance**
>
> We decide the outcome of a hypothesis test by comparing the actual sample statistic (mean or proportion) to the result expected if the null hypothesis is true. We must choose a significance level for the decision.
>
> - If the chance of a sample statistic at least as extreme as the observed statistic is less than 1 in 100 (or 0.01), then the test is statistically significant at the 0.01 level and offers strong evidence for rejecting the null hypothesis.
>
> - If the chance of a sample statistic at least as extreme as the observed statistic is less than 1 in 20 (or 0.05), then the test is statistically significant at the 0.05 level and offers moderate evidence for rejecting the null hypothesis.
>
> - If the chance of a sample statistic at least as extreme as the observed statistic is greater than the chosen level of significance (0.05 or 0.01), then we do not reject the null hypothesis.

EXAMPLE ③ Statistical Significance in Hypothesis Testing

Consider the Gender Choice example in which the sample size is $n = 100$ and the sample proportion is $\hat{p} = 0.64$. Using techniques that we will discuss later in this chapter, it is possible to calculate the probability of randomly choosing such a sample proportion (or a more extreme proportion with $\hat{p} > 0.64$) under the assumption that the null hypothesis ($p = 0.50$) is true; the result is that the probability is 0.0026. (The result is 0.0033 if technology is used to obtain very accurate results.) Based on this result, should you reject or not reject the null hypothesis?

SOLUTION The probability of 0.0026 means that if the null hypothesis is true (meaning that the true population proportion is 50%), the probability of drawing a random sample with a sample proportion of at least $\hat{p} = 0.64$ is less than 1 in 100, so this result is statistically significant at the 0.01 level. It therefore gives us good reason to reject the null hypothesis, which means it provides support for the alternative hypothesis that Gender Choice raises the proportion of girls to more than 50%.   ·· ●

P-Values

In Example 3 above, we concluded that the sample result gave us reason to reject the null hypothesis, because the sample proportion was statistically significant at the 0.01 level. In fact, the result was even better than that: The computed probability of 0.0026 is called a **P-value** (short for *probability value*), and it gives us more information than simply stating a level of statistical significance; notice the capital P, used to avoid confusion with the lowercase p that stands for a population proportion. In other words, for Example 3 we say that the P-value for the hypothesis test is 0.0026. We will discuss the actual calculation of P-values in Sections 9.2 and 9.3; here, we focus only on their interpretation.

Hypothesis Test Decisions Based on *P*-Values

The **P-value** (probability value) for a hypothesis test of a claim about a population parameter is the probability of selecting a sample with a statistic that is at least as extreme as the observed statistic, assuming that the null hypothesis is true:

• A small P-value (such as less than or equal to 0.05) indicates that the sample result is unlikely and therefore provides reason to reject the null hypothesis.

• A large P-value (such as greater than 0.05) indicates that the sample result could easily occur by chance, so we cannot reject the null hypothesis.

EXAMPLE ④ Fair Coin?

You suspect that a coin may have a bias toward landing tails more often than heads, and decide to test this suspicion by tossing the coin 100 times. The result is that you get 40 heads (and 60 tails). A calculation (not shown here) indicates that the probability of getting 40 or fewer heads in 100 tosses with a fair coin is 0.0284. Find the P-value and level of statistical significance for your result. Should you conclude that the coin is biased against heads?

SOLUTION The null hypothesis is that the coin is fair, in which case the proportion of heads should be about 50% ($H_0: p = 0.50$). The alternative hypothesis is your suspicion that the coin is biased against heads, in which case the proportion of heads would be less than 50% ($H_a: p < 0.50$). Your 100 coin tosses represent a random sample of size $n = 100$, and the result of 40 heads is the sample proportion ($\hat{p} = 0.40$) for the hypothesis test. The P-value for the test is the probability of getting a sample proportion at least as extreme as the one you found ($\hat{p} \leq 0.40$), *assuming* that the coin is fair and the population proportion of heads is 0.5. The given probability for this occurrence is 0.0284, which is the P-value for the test. Because this P-value is smaller than 0.05, the result is statistically significant at the 0.05 level. Because it is not smaller than 0.01, the result is not significant at the 0.01 level. Statistical significance at the 0.05 level gives you moderate reason to reject the null hypothesis and conclude that the coin is biased against heads.   ·· ●

Putting It All Together

We have now covered all the basic ideas of hypothesis testing, except for the actual calculations that are required. We will discuss these calculations for hypothesis tests of population means in Section 9.2 and of population proportions in Section 9.3. The following box summarizes the steps that go into a hypothesis test.

The Hypothesis Test Process

Step 1. Formulate the null and alternative hypotheses, each of which must make a claim about a *population* parameter, such as a population mean (μ) or a population proportion (p), be sure this is done before drawing a sample or collecting data. Based on the form of the alternative hypothesis, decide whether you will need a left-, right-, or two-tailed hypothesis test.

Step 2. Draw a sample from the population and measure the sample statistics, including the sample size (n) and the relevant sample statistic, such as the sample mean ($\bar{x}$) or sample proportion ($\hat{p}$).

Step 3. Determine the likelihood of observing a sample statistic (mean or proportion) at least as extreme as the one you found *under the assumption that the null hypothesis is true*. The precise probability of such an observation is the *P*-value (probability value) for your sample result.

Step 4. Decide whether to reject or not reject the null hypothesis, based on your chosen level of significance (usually 0.05 or 0.01, but other significance levels are sometimes used).

Again, be sure to avoid confusion among the three different uses of the letter p:

- A lowercase p represents a *population proportion*—that is, the true proportion among a complete population.
- A lowercase $\hat{p}$ ("*p*-hat") represents a *sample proportion*—that is, the proportion found in a sample drawn from a population.
- An uppercase P stands for probability in a *P*-value.

EXAMPLE 5   Mean Rental Car Mileage

In the United States, the average car is driven about 12,000 miles each year. The owner of a large rental car company suspects that for his fleet, the mean distance is greater than 12,000 miles each year. He selects a random sample of $n = 225$ cars from his fleet and finds that the mean annual mileage for this sample is $\bar{x} = 12{,}375$ miles. A calculation shows that if you assume the fleet mean is the national mean of 12,000 miles, then the probability of selecting a 225-car sample with a mean annual mileage of at least 12,375 miles is 0.01. Based on these data, describe the process of conducting a hypothesis test and drawing a conclusion.

SOLUTION   We follow the four steps listed in the box above:

Step 1. The population parameter of interest is a *population mean* (μ)—the mean annual mileage for the population of all cars in the rental car fleet. The null hypothesis is that the population mean is the national average of 12,000 annual miles per car. The alternative hypothesis is the owner's claim that the population mean for his fleet is greater than this national average. That is,

$$H_0: \mu = 12{,}000 \text{ miles}$$
$$H_a: \mu > 12{,}000 \text{ miles}$$

Because the alternative hypothesis has a "greater than" form, the hypothesis test will be right-tailed.

Step 2. This step asks that we select the sample and measure the sample statistics; here, we are given the sample size of $n = 225$ and the sample mean of $\bar{x} = 12{,}375$ miles.

Step 3. This step asks us to determine the likelihood that, under the assumption that the fleet average is really 12,000 miles (the null hypothesis), we would by chance select a sample with a mean of at least 12,375 miles. Here, we don't need to do the calculation, because we are told that the probability is 0.01. (The calculation method is given in Section 9.2.) This probability is the *P*-value; it tells us that if the null hypothesis were true, there would be only a 0.01 probability of selecting a sample as extreme as the one observed.

Step 4. The *P*-value of 0.01 tells us that the result is significant at the 0.01 level. We therefore reject the null hypothesis and conclude that the test provides strong support for the alternative hypothesis, implying that the mean annual mileage for the rental car fleet is greater than the national average of 12,000 miles. $\cdots\bullet$

A Legal Analogy of Hypothesis Testing

A legal analogy might help clarify the idea of hypothesis testing. In American courts of law, the fundamental principle is that a defendant is presumed innocent until proven guilty. Because the starting assumption is innocence, this represents the null hypothesis:

H_0: The defendant is innocent.

H_a: The defendant is guilty.

The job of the prosecutor is to present evidence so compelling that the jury is persuaded to reject the null hypothesis and find the defendant guilty. If the prosecutor does not make a sufficiently compelling case, then the jury will not reject H_0 and the defendant will be found "not guilty." Note that finding a person innocent (accepting H_0) is not an option: A verdict of not guilty means the evidence is not sufficient to establish guilt, but it does not prove innocence.

> **TIME ◖OUT TO THINK**
>
> Consider two situations. In one, you are a juror in a case in which the defendant could be fined a maximum of $2,000. In the other, you are a juror in a case in which the defendant could receive the death penalty. Compare the significance levels you would use in the two situations. In each situation, what are the consequences of wrongly rejecting the null hypothesis?

Section 9.1 Exercises

Statistical Literacy and Critical Thinking

1. **Hypothesis Test.** What is a hypothesis test?

2. **Hypotheses.** What is a null hypothesis? What notation is used for a null hypothesis? What is an alternative hypothesis? What notation is used for an alternative hypothesis?

3. **Alternative Hypotheses.** A researcher wants to test the claim that the mean weight of male airline passengers, including carry-on baggage, is 195 lb. Identify the three different possible expressions that could be used for the alternative hypothesis.

4. **P-Value.** What is a *P*-value for a hypothesis test?

Does It Make Sense? For Exercises 5–12, decide whether the statement makes sense (or is clearly true) or does not make sense (or is clearly false). Explain clearly. Not all of these statements have definitive answers, so your explanation is more important than your chosen answer.

5. **Interpreting a Hypothesis Test.** To test the claim that a particular drug lowers blood pressure, sample data are collected and a hypothesis test is used, and we form this conclusion: "The study proves that the drug lowers blood pressure in everyone."

6. **P-Value.** A researcher is convinced that she can show that a new drug is effective in lowering LDL cholesterol. She claims that the *P*-value of 0.20 supports her claim of a lower mean level of LDL cholesterol.

7. **Null Hypothesis.** In testing a claim that the mean LDL cholesterol level is less than 130 mg/dL, the researcher states the alternative hypothesis as $\mu < 130$ mg/dL.

8. **Null Hypothesis.** After conducting a hypothesis test, a researcher concludes that there is sufficient sample evidence to support the null hypothesis that $\mu = 75$.

9. Drug Treatment. In a test of the claim that, among patients treated with Ziac, the proportion who experience dizziness is less than 0.06, the null hypothesis is $p = 0.06$.

10. Sample Mean. A study is designed to determine the proportion of men who weigh more than 195 pounds, so the null hypothesis is $\mu = 195$ pounds and the alternative hypothesis is $\mu > 195$ pounds.

11. P-Value. In interpreting a *P*-value of 0.45, a researcher states that the results are statistically significant because the *P*-value is less than 0.5, indicating that the results are not likely to occur by chance.

12. P-Value. In interpreting a *P*-value of 0.001, a researcher states that the results are statistically significant because the *P*-value is very small, indicating that the results are not likely to occur by chance.

Concepts and Applications

13. What Is Significant? In testing a method of gender selection, 40 couples are given a treatment designed to increase the likelihood of a girl, and each couple has one baby.

   a. If the 40 babies include exactly 22 girls, would you consider this result statistically significant or would you attribute it to random fluctuations?

   b. If the 40 babies include exactly 35 girls, would you consider this result statistically significant or would you attribute it to random fluctuations?

   c. The *P*-value of 0.0000007 is obtained for the results of either part a or b. Which is it? Why?

14. What Is Significant? In testing a method of gender selection, 200 couples are given a treatment designed to increase the likelihood of a girl, and each couple has one baby.

   a. If the 200 babies include exactly 130 girls, would you consider this result statistically significant or would you attribute it to random fluctuations?

   b. If the 200 babies include exactly 105 girls, would you consider this result statistically significant or would you attribute it to random fluctuations?

   c. The *P*-value of 0.262 is obtained for the results of either part a or b. Which is it? Why?

Formulating Hypotheses. In Exercises 15–22, formulate the null and alternative hypotheses for a hypothesis test. State clearly the two possible conclusions that address the given claim.

15. Calories. Package labeling indicates that Twix candy bars have an amount of caloric content equal to 250 calories.

16. SAT Scores. A high school principal claims that the mean SAT score of seniors at his school is less than the national average of 1518.

17. Gender Selection. The Chief Operations Officer of a medical facility claims that treatments can increase the

probability that a baby will be a girl so that the proportion of girls is greater than 0.7.

18. Quality Control. The quality control manager at a manufacturing company claims that the proportion of defective carbon monoxide detectors is less than 0.04.

19. Vending Machines. A sales representative claims that her vending machines dispense coffee so that the mean amount supplied is equal to 10 ounces.

20. Aspirin. The Food and Drug Administration claims that a pharmaceutical company is producing aspirin tablets with a mean amount of aspirin that is less than 350 milligrams.

21. Holocaust. A high school teacher claims that the majority of her students do not know what the term *Holocaust* refers to.

22. Smoking. An educator claims that less than 20% of college graduates smoke.

P-Values and Births. Assume that male births and female births are equally likely. The following table shows the probabilities of various numbers of male babies in a random sample of 100 births. Use this information for Exercises 23–28, and assume that we are testing for a bias against males.

Number of males among 100	Probability
35 or fewer	0.002
40 or fewer	0.028
45 or fewer	0.184
48 or fewer	0.382

23. A random sample of 100 births has 48 male babies. Is this result significant at the 0.05 level? What is the *P*-value for this result?

24. A random sample of 100 births has 45 male babies. Is this result significant at the 0.01 level? What is the *P*-value for this result?

25. A random sample of 100 births has 40 male babies. Is this result significant at the 0.01 level? What is the *P*-value for this result?

26. A random sample of 100 births has 40 male babies. Is this result significant at the 0.05 level? What is the *P*-value for this result?

27. A random sample of 100 births has 35 male babies. Is this result significant at the 0.01 level? What is the *P*-value for this result?

28. A random sample of 100 births has 32 male babies. Is this result significant at the 0.01 level? What is the *P*-value for this result?

PROJECTS FOR THE INTERNET & BEYOND

29. Professional Journals. Many professional journals, such as *Journal of the American Medical Association*, contain articles that include information about formal tests of hypotheses. Find such an article and identify the null hypothesis and alternative hypothesis. In simple terms, state the objective of the hypothesis test and the conclusion that was reached.

30. **Coin Activity.** Select a particular quarter and test the claim that it favors heads when flipped. State the null and alternative hypotheses; then flip the quarter 100 times. Applying only common sense, what do you conclude about the claim that the quarter favors heads?

9.2 HYPOTHESIS TESTS FOR POPULATION MEANS

In Section 9.1, we sketched the basic outline of the hypothesis testing process. Remember that in all cases, we need to make a decision about whether to reject or not reject the null hypothesis, which is the starting assumption for the test. In this section, we describe the calculations used to make that decision for hypothesis tests with population means. We first investigate the procedure for one-tailed (left-tailed or right-tailed) tests and then discuss the small differences in procedure necessary for two-tailed tests.

One-Tailed Hypothesis Tests

Consider the following hypothetical situation. Columbia College advertises that the mean starting salary of its graduates is $39,000. The Committee for Truth in Advertising, an independent organization, suspects that this claim is exaggerated and decides to conduct a hypothesis test to seek evidence to support its suspicion.

The parameter of interest is the mean starting salary of the population of all Columbia College graduates, so the hypothesis test will concern a *population mean* (μ). The null hypothesis is the college's claim that the mean starting salary is $39,000. The alternative hypothesis is the committee's claim that the college has exaggerated the mean starting salary, in which case the mean starting salary is *less than* $39,000. The null and alternative hypotheses are

$$H_0: \mu = \$39,000$$
$$H_a: \mu < \$39,000$$

Because the alternative hypothesis has a "less than" form, we are dealing with a left-tailed hypothesis test. The general procedure for left- and right-tailed tests is the same, so we will consider both together as *one-tailed* tests.

Having formed the hypotheses, the Committee for Truth in Advertising selects a random sample of 100 recent graduates from the college. The mean salary of the graduates in the sample turns out to be $37,000. The sample size is $n = 100$ and the sample mean is $\bar{x} = \$37,000$.

If you look back at the four-step hypothesis test process on page 315, you'll see that the first two steps are already complete: The mean starting salary has been identified as the population parameter of interest, the null and alternative hypotheses have been stated, and the sample has been drawn and measured to determine its sample size and sample mean. We are therefore ready for Steps 3 and 4, in which we assume the null hypothesis is true and then determine whether the sample statistics give us reason to reject this assumption.

The Sampling Distribution

Step 3 of the hypothesis test process is finding the likelihood of observing a sample mean as extreme as the one found, under the assumption that the null hypothesis is true. For the Columbia College example, the question becomes this: How likely are we to select a sample (of size $n = 100$) with a mean of $37,000 or less when the mean for the whole population is $39,000? To answer this question, we need a crucial observation based on our work with sampling distributions in Chapter 8: *The observed sample mean ($\bar{x} = \$37,000$) is just one point in a distribution of sample means.*

To understand the importance of this fact, imagine that instead of drawing just one sample of size $n = 100$, the Committee for Truth in Advertising drew many samples of that size. Each sample would have a unique sample mean, $\bar{x}$, so we could construct a graph of the distribution of all these sample means. As discussed in Section 8.1, for a reasonably large sample size such as $n = 100$, the resulting sampling distribution will be approximately normal with a mean that is equal to the population mean. In this case, the null hypothesis claims that the population mean is $\mu = \$39,000$, so the sampling distribution will peak at this value if the null hypothesis is correct.

Figure 9.1 shows the idea graphically. The red curve represents the sampling distribution under the assumption that the null hypothesis is true. Remember that this curve is what we would expect to find if we plotted the distribution of means from *many* samples. When we have only a single sample mean (such as the sample mean $\bar{x} = \$37,000$ for the Columbia College example), it represents just one point along this curve. If this point lies close to the peak of the curve, it tells us that the sample mean is near the population mean expected with the null hypothesis. In that case, the probability of finding such a sample mean is not small and there is no reason to reject the null hypothesis. In contrast, if the sample mean lies far from the population mean claimed by the null hypothesis, then the probability of finding such a sample mean is small if the null hypothesis is true. We would therefore conclude that the true population mean probably is *not* what the null hypothesis claims, in which case we have reason to reject the null hypothesis.

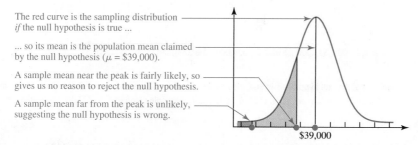

The red curve is the sampling distribution *if* the null hypothesis is true ...

... so its mean is the population mean claimed by the null hypothesis ($\mu = \$39,000$).

A sample mean near the peak is fairly likely, so gives us no reason to reject the null hypothesis.

A sample mean far from the peak is unlikely, suggesting the null hypothesis is wrong.

$39,000

Figure 9.1 Graphical interpretation of the hypothesis test for the Columbia College example. If the null hypothesis is true, then the population mean is $\mu = \$39,000$. In that case, if we took many samples from the population, the distribution of sample means would be approximately normal with a mean of $39,000. Under this assumption, the distance of a single sample mean from the population mean allows us to decide whether to reject or not reject the null hypothesis.

Finding the Standard Score

As illustrated in Figure 9.1, the decision of whether to reject the null hypothesis depends on whether the sample mean is "near" to or "far" from the population mean claimed by the null hypothesis. Because the sampling distribution is a (nearly) normal distribution, the *standard score* of the sample mean represents a quantitative measure of the distance between the sample mean and the claimed population mean. Recall from Section 5.2 that the standard score (or z-score) of a data value in a normal distribution is the number of standard deviations that it lies above or below the mean of the distribution. From the Central Limit Theorem (Section 5.3), the standard deviation of a distribution of sample means is $\sigma/\sqrt{n}$, where σ is the population standard deviation and n is the sample size. Putting these ideas together, we find the following formula for the standard score of a sample mean $\bar{x}$:

$$z = \frac{\text{sample mean} - \text{population mean}}{\text{standard deviation of sampling distribution}} = \frac{\bar{x} - \mu}{\sigma/\sqrt{n}}$$

The only remaining problem is that we generally do not know the population standard deviation, σ. For now, let's assume that we can approximate the population standard deviation with the sample standard deviation, s, and that $s = \$6,150$ for the 100 salaries in the sample. (In Section 10.1, we'll discuss a better way to proceed when σ is not known.) In that case, we set $\sigma = \$6,150$ and the standard deviation for the distribution of sample means is

$$\frac{\sigma}{\sqrt{n}} = \frac{\$6,150}{\sqrt{100}} = \$615$$

Using this value in the previous equation tells us that the standard score for the sample mean of $\bar{x} = \$37{,}000$ in a sampling distribution with a population mean of $\mu = \$39{,}000$ is

$$z = \frac{\bar{x} - \mu}{\sigma/\sqrt{n}} = \frac{\$37{,}000 - \$39{,}000}{\$615} = -3.25$$

In other words, the sample mean of $\bar{x} = \$37{,}000$ lies 3.25 standard deviations below the mean of the sampling distribution.

Figure 9.2 shows that the standard score of -3.25 places the sample result far out on the left tail of the distribution, indicating that it would be a rare sample result *if* the population mean really were the \$39,000 claimed by the null hypothesis. This suggests that the null hypothesis is incorrect and should be rejected. However, rather than rejecting the null hypothesis for this visual reason alone, let's look at a formal procedure for analyzing the standard score and making the hypothesis test decision.

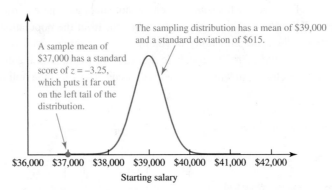

Figure 9.2 Assuming that the null hypothesis is true, the distribution of sample means for the Columbia College example has a mean of \$39,000 and a standard deviation of \$615. In that case, a sample mean of $\bar{x} = \$37{,}000$ has a standard score of -3.25.

> ### Computing the Standard Score for the Sample Mean in a Hypothesis Test
>
> When we draw a random sample for a hypothesis test, we can consider it to be one of many possible samples in the sampling distribution. Given the sample size (n), the sample mean ($\bar{x}$), the population standard deviation (σ), and the claimed population mean (μ), we make the following computations:
>
> $$\text{standard deviation for the distribution of sample means} = \frac{\sigma}{\sqrt{n}}$$
>
> $$\text{standard score for the sample mean, } z = \frac{\bar{x} - \mu}{\sigma/\sqrt{n}}$$
>
> Note: In reality, it is rare that we know the population standard deviation σ, so we often approximate it with the sample standard deviation s (see Section 10.1).

TECHNICAL NOTE

The standard score for the sample mean is often called the *test statistic*.

Critical Values for Statistical Significance

Recall that the hypothesis test is significant at the 0.05 level if the probability of finding a result as extreme as the one actually observed is 0.05 or less (assuming the null hypothesis is true). For a left-tailed test, then, we are looking for a standard score that is at or below the 5th percentile of the sampling distribution. From Table 5.1 (page 176), the 5th percentile has a standard score between $z = -1.6$ and $z = -1.7$; the more precise standard score tables in Appendix A show that the 5th percentile has a standard score of $z = -1.645$. Therefore, a left-tailed hypothesis test is significant at the 0.05 level if the standard score of the sample mean is less than or equal to $z = -1.645$. This standard score represents the **critical value** for significance at the 0.05 level in a left-tailed hypothesis test.

A similar argument applies to right-tailed tests with alternative hypotheses of the form $H_a: \mu >$ claimed value. In such cases, significance at the 0.05 level requires that the sample mean lie at or above the 95th percentile—which requires a standard score greater than or equal to $z = 1.645$. Figure 9.3 illustrates these critical values for both left- and right-tailed tests. To find critical values for significance at the 0.01 level, we look for the standard scores of the 1st and 99th percentiles (rather than the 5th and 95th); Appendix A shows that these are -2.33 and 2.33, respectively.

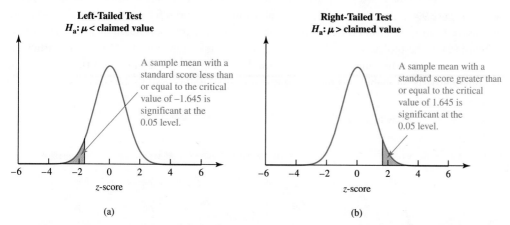

Figure 9.3 These graphs illustrate the meaning of critical values for one-tailed hypothesis tests when testing for significance at the 0.05 level. The locations of the critical values correspond to the 5th percentile for left-tailed tests and the 95th percentile for right-tailed tests.

Decisions Based on Statistical Significance for One-Tailed Hypothesis Tests

We decide whether to reject or not reject the null hypothesis by comparing the standard score (z) for a sample mean to critical values for significance at a given level. Table 9.1 summarizes the decisions for one-tailed hypothesis tests at the 0.05 and 0.01 levels of significance.

TABLE 9.1   **Testing for Significance at the 0.05 and 0.01 Levels**

Type of test	Form of H_a	For 0.05 level: Reject H_0 if standard score is	For 0.01 level: Reject H_0 if standard score is
Left-tailed test	$H_a: \mu <$ claimed value	$z \leq -1.645$	$z \leq -2.33$
Right-tailed test	$H_a: \mu >$ claimed value	$z \geq 1.645$	$z \geq 2.33$

TIME ⏲ OUT TO THINK

Suppose that, in right-tailed tests, one study finds a sample mean with $z = 3$ and another study finds a sample mean with $z = 10$. Both are significant at the 0.01 level, but which result provides stronger evidence for rejecting the null hypothesis? Explain.

EXAMPLE ① Columbia College Test Significance

Assuming that the null hypothesis is true and the mean starting salary for Columbia College graduates is $39,000, is it statistically significant to find a sample in which the mean is only $37,000? Based on your answer, should we reject or not reject the null hypothesis?

SOLUTION   The hypothesis test is left-tailed (because the alternative hypothesis has the "less than" form of $\mu <$ $39,000) and we already found that a sample mean of $\bar{x} =$ $37,000 has a standard score of -3.25. This result is significant at the 0.05 level because the standard score is less than the critical value of $z = -1.645$. In fact, it is also significant at the 0.01 level, because the standard score is less than -2.33. We therefore have strong reason to reject the null hypothesis and conclude that Columbia College officials did indeed exaggerate the mean starting salary of its graduates.

⋯●

Finding the *P*-Value

We can use *P*-values to be more precise about the significance of a result. Recall that the *P*-value is the probability of finding a sample mean as extreme (or more extreme) as the one found, under the assumption that the null hypothesis is true. For the hypothesis tests we discuss in this chapter, we generally find the *P*-value from the standard score of the sample mean. Figure 9.4 shows the idea. The total area under the curve for the distribution of sample means is 1, so we can interpret areas under the curve as probabilities. For a left-tailed test (Figure 9.4a), the probability of finding a sample mean less than or equal to some particular value is simply the area under the curve to the *left* of the sample mean. For a right-tailed test (Figure 9.4b), the probability is the area under the curve to the *right* of the sample mean.

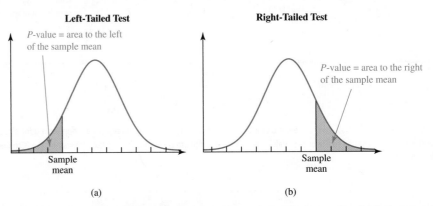

Figure 9.4 The *P*-value for one-tailed hypothesis tests corresponds to an area under the sampling distribution curve. Once we compute the standard score for a particular sample mean, we can find the corresponding area from standard score tables like those in Appendix A. (Note: Appendix A lists areas to the left of each standard score; therefore, for right-tailed tests, finding the area to the right of the sample mean requires subtracting the value given in the table from 1.)

The P-value ... is often used as a gauge of the degree of contempt in which the null hypothesis deserves to be held.

—Robert Abelson, *Statistics as a Principled Argument*

As an example, let's find the *P*-value for the Columbia College hypothesis test. We have already determined that the sample mean ($\bar{x} = 37{,}000$) has a standard score of $z = -3.25$. If you look in Appendix A, you'll see that this standard score corresponds to a "cumulative area from the left" of 0.0006—so this is the *P*-value we are seeking. This small *P*-value provides strong evidence against the null hypothesis. Be sure to notice that this *P*-value is less than 0.01, which is why the result is significant at the 0.01 level; the fact that it is much less than 0.01 tells us that we have very strong evidence for rejecting the null hypothesis.

Summary of One-Tailed Tests for Population Means

We began this section with the goal of learning how to carry out Steps 3 and 4 of the four-step hypothesis test process on page 315. We have now succeeded in doing this for one-tailed hypothesis tests for population means. To summarize:

- Because we are dealing with population means, the null hypothesis has the form μ = claimed value. To decide whether to reject or not reject the null hypothesis, we must determine whether a sample as extreme as the one found in the hypothesis test is likely or unlikely to occur if the null hypothesis is true.
- We determine this likelihood from the standard score (z) of the sample mean, which we compute from the formula

$$z = \frac{\bar{x} - \mu}{\sigma/\sqrt{n}}$$

where n is the sample size, $\bar{x}$ is the sample mean, μ is the population mean claimed by the null hypothesis, and σ is the population standard deviation.
- We can then assess the standard score in two ways:
  1. We can assess its level of statistical significance by comparing it to the critical values given in Table 9.1 (page 321).
  2. We can determine its *P*-value with standard score tables like those in Appendix A. For a left-tailed test, the *P*-value is the area under the normal curve to the left of the

standard score; for a right-tailed test, it is the area under the normal curve to the right of the standard score.

- If the result is statistically significant at the chosen level (usually either the 0.05 or the 0.01 significance level), we reject the null hypothesis. If it is not statistically significant, we do not reject the null hypothesis.

EXAMPLE 2 Mean Rental Car Mileage (Revisited)

Recall the case of the rental car fleet owner (Example 5 of Section 9.1) who suspects that the mean annual mileage of his cars is greater than the national mean of 12,000 miles. He selects a random sample of $n = 225$ cars and calculates the sample mean to be $\bar{x} = 12,375$ miles and the sample standard deviation to be s $= 2,415$ miles. Determine the level of statistical significance and P-value for this hypothesis test, and interpret your findings.

SOLUTION   To determine the significance and P-value we must first calculate the standard score for the sample mean of $\bar{x} = 12,375$ miles. We are given the claimed population mean ($\mu = 12,000$ miles) and the sample size ($n = 225$), we do not know the population standard deviation, σ, but we will assume it is the same as the sample standard deviation and set $\sigma = 2,415$ miles. We find

$$z = \frac{\bar{x} - \mu}{\sigma/\sqrt{n}} = \frac{12,375 - 12,000}{2,415/\sqrt{225}} = 2.33$$

This standard score is greater than the critical value of $z = 1.645$ for significance at the 0.05 level and equal to the value of $z = 2.33$ for significance at the 0.01 level. We can find the P-value from Appendix A, which shows that the area to the *left* of a standard score of $z = 2.33$ is 0.9901; because this is a right-tailed test, the probability is the area to the *right* (see Figure 9.4b), which is $1 - 0.9901 = 0.0099$. Therefore, the P-value is 0.0099, which is very close to 0.01. We therefore have strong reason to reject the null hypothesis and conclude that the rental car fleet mean really is greater than the national mean. $\cdot\cdot\bullet$

Two-Tailed Tests

The same basic ideas apply to two-tailed hypothesis tests in which the alternative hypothesis has the "not equal to" form of H_a: $\mu \neq$ claimed value. However, the critical values and the calculations for P-values are slightly different for two-tailed tests.

As always, the hypothesis test is significant at the 0.05 level if the probability of finding a result as extreme as the one actually found is 0.05 or less. For one-tailed tests, a probability of 0.05 corresponds to standard scores at the 5th percentile for left-tailed tests and 95th percentile for right-tailed tests (see Figure 9.3). For two-tailed tests, however, a value "at least as extreme as the one actually found" can lie *either* on the left or on the right of the sampling distribution (Figure 9.5). A probability of 0.05, or 5%, therefore corresponds to standard scores either in the first 2.5% of the sampling distribution on the left or in the last 2.5% on the right. From Appendix A, the 2.5th percentile corresponds to a standard score of -1.96 and the 97.5th

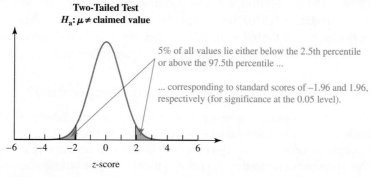

Figure 9.5   For two-tailed tests, the critical values for significance at the 0.05 level correspond to the 2.5th and 97.5th percentiles (as opposed to the 5th or 95th percentiles for one-tailed tests).

percentile corresponds to a standard score of 1.96. These standard scores become the critical values for two-tailed tests to be significant at the 0.05 level. Similarly, the two-tailed critical values for significance at the 0.01 level correspond to standard scores for the 0.5th and 99.5th percentiles; from Appendix A, these values are -2.575 and 2.575, respectively.

Similar considerations apply to P-values for two-tailed tests. Recall that for a one-tailed test, the P-value is the area under the curve to the left of the sample mean for a left-tailed test or to the right of the sample mean for a right-tailed test. Because a two-tailed test asks us to consider extremes on both sides of the claimed mean, the P-value for a two-tailed test must be twice what it would be if the test were one-tailed.

> **Two-Tailed Test (H_a: $\mu \neq$ claimed value)**
>
> **Statistical significance:** A two-tailed test is significant at the 0.05 level if the standard score of the sample mean is at or below a critical value of -1.96 *or* at or above a critical value of 1.96. For significance at the 0.01 level, the critical values are -2.575 and 2.575.
>
> ***P-value:*** To find the P-value for a sample mean in a two-tailed test, first use the standard score of the sample mean to find the P-value assuming the test is one-tailed; then *double* this value to find the P-value for the two-tailed test.

An example should help clarify these ideas. Consider a drug company that seeks to be sure that its "500-milligram" aspirin tablets really contain 500 milligrams of aspirin. If the tablets contain less than 500 milligrams, consumers are not getting the advertised dose. If the tablets contain more than 500 milligrams, consumers are getting too much of the drug. The null hypothesis says that the population mean of the aspirin content is 500 milligrams:

$$H_0: \mu = 500 \text{ milligrams}$$

The drug company is interested in the possibility that the mean weight is *either* less than *or* greater than 500 milligrams. Because the company is interested in possibilities on both sides of the claimed mean of 500 milligrams, we have a two-tailed test in which the alternative hypothesis is

$$H_a: \mu \neq 500 \text{ milligrams}$$

Suppose the company selects a random sample of $n = 100$ tablets and finds that they have a mean weight of $\bar{x} = 501.5$ milligrams; further suppose that the population standard deviation is $\sigma = 7.0$ milligrams. Then the standard score (z) for this sample mean is

$$z = \frac{\bar{x} - \mu}{\sigma/\sqrt{n}} = \frac{501.5 - 500}{7/\sqrt{100}} = 2.14$$

This standard score is above the critical value of 1.96 for a two-tailed test, so it is significant at the 0.05 level. (It is not significant at the 0.01 level, because it is not above the critical value of 2.575.) This gives us good reason to reject the null hypothesis and conclude that the mean weight of the "500-milligram" aspirin tablets is different from 500 milligrams.

We find the P-value by using Appendix A, which shows that the area to the *left* of a standard score of $z = 2.14$ is 0.9838. Therefore, if this were a one-tailed test, the P-value would be the area to the *right*, or $1 - 0.9838 = 0.0162$. But because this is a two-tailed test, we *double* this number to find that the P-value is $2 \times 0.0162 = 0.0324$. In other words, *if* the null hypothesis is true, the probability of drawing a sample with a mean at least as extreme as the one found is 0.0324.

EXAMPLE 3   What Is Human Mean Body Temperature?

Consider again the study in which University of Maryland researchers measured body temperatures in a sample of $n = 106$ healthy adults (see Example 5 in Section 8.2), finding a sample mean body temperature of $\bar{x} = 98.20°\text{F}$ with a sample standard deviation of $s = 0.62°\text{F}$. We will assume that the population standard deviation is the same as the sample standard deviation. Determine whether this sample provides enough evidence for rejecting the common belief that mean human body temperature is $\mu = 98.6°\text{F}$.

SOLUTION The null hypothesis is the claim that mean human body temperature is 98.6°F, or H_0: $\mu = 98.6$°F. We are asked to test the alternative hypothesis that mean body temperature is *not* 98.6°F, or H_a: $\mu \neq 98.6$°F. The "not equal to" form of the alternative hypothesis tells us that we need a two-tailed test. We are given the sample size $n = 106$, the sample mean $\bar{x} = 98.2$°F, and the assumed population standard deviation $\sigma = 0.62$°F. The standard score of the sample mean is

$$z = \frac{\bar{x} - \mu}{\sigma/\sqrt{n}} = \frac{98.20 - 98.60}{0.62/\sqrt{106}} = -6.64$$

This standard score is much less than the critical values of -1.96 for significance at the 0.05 level and -2.575 for significance at the 0.01 level (Figure 9.6). Because the result is significant at the 0.01 level, it provides strong evidence against the null hypothesis. We therefore reject the null hypothesis and conclude that human mean body temperature is *not* equal to 98.6°F. (In this case, the standard score of -6.64 is so extreme that we cannot find the precise P-value from Appendix A; a calculation shows that the P-value is about 3×10^{-11}.)

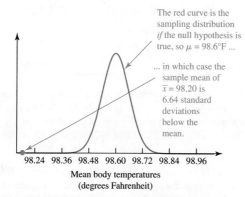

The red curve is the sampling distribution *if* the null hypothesis is true, so $\mu = 98.6$°F …

… in which case the sample mean of $\bar{x} = 98.20$ is 6.64 standard deviations below the mean.

Mean body temperatures
(degrees Fahrenheit)

Figure 9.6 The null hypothesis states that human mean body temperature is 98.6°F. If this is true, then the distribution of sample means has a mean of 98.6°F. Based on the sample size and standard deviation, the sample mean of 98.20°F. is more than 6 standard deviations below the claimed mean temperature—which makes it seem likely that the claim is wrong and human mean body temperature is *not* equal to 98.6°F.

$\bullet \, \bullet \, \bullet$

Common Errors in Hypothesis Testing

We have now covered all the fundamentals of hypothesis testing, as well as specific methods used for hypothesis tests for population means. However, even if a hypothesis test is carried out correctly, two common types of error may affect the conclusions. To understand these errors, recall the legal analogy at the end of Section 9.1, in which the null hypothesis is H_0: The defendant is innocent. One type of error occurs if we conclude that the defendant is guilty when, in reality, he or she is innocent. In this case, the null hypothesis (innocence) has been wrongly rejected. The other type of error occurs if we find the defendant *not guilty* when he or she actually is guilty. In this case, we have wrongly failed to reject the null hypothesis.

For a statistical example, consider again the pharmaceutical company testing the claim that the mean amount of aspirin in its tablets is 500 milligrams (H_0: $\mu = 500$ milligrams). In drawing a conclusion from the test, the company could make the following two types of errors:

- The company might reject the null hypothesis and conclude that the mean amount is not 500 milligrams when it really is. This error can lead to wasting time and money trying to fix a process that isn't broken. An error of this type, in which H_0 is *wrongly rejected*, is called a **type I error**.
- The company might fail to reject the null hypothesis when, in fact, the mean amount of aspirin is not 500 milligrams. In this case, the company will distribute tablets that have too much or too little aspirin; consumers could suffer or sue. An error of this type, in which we *wrongly fail to reject H_0*, is called a **type II error**.

USING TECHNOLOGY

HYPOTHESIS TEST FOR A POPULATION MEAN

Excel Use **XLSTAT**. First enter the list of original sample values in a list. Click on **XLSTAT** at the top, click on **Parametric tests**, then select **One sample t test and z test**. In the screen that appears, for the "Data" box enter the range of data, such as A1:A11 for 11 data values in column A. For "Data Format" select **One sample**. Click on the the "z test" box. Click on the **Options** tab to select the type of test; select the option including $\neq$ for a two-tailed test, select the option including $<$ for a left-tailed test, or select the option including $>$ for a right-tailed test. For the "Theoretical mean" box, enter the claimed value of the population mean, which is the same value used in the state-ment of the null hypothesis. Enter the desired "Significance level (%)." For example, enter 5 for a 0.05 significance level. Click **OK**. After the results are displayed, look for the test statistic identified as "z (Observed value)." The P-value will also be displayed. See the accompanying display that results from Example 3. See that the P-value in this display is so small that it is given as less than 0.0001.

XLSTAT

Difference	-0.4000
z (Observed value)	-6.6423
|z| (Critical value)	1.9600
p-value (Two-tailed)	< 0.0001
alpha	0.05

STATDISK If working with a list of the original sample values, first find the sample size, sample mean, and sample standard deviation by selecting **Analysis**, then **Descriptive Statistics**. After find-ing the values of n, $\bar{x}$, and s, select **Analysis** from the main menu, then select **Hypothesis Testing**, followed by **Mean-One Sample**. In the dialog box that appears, select the format of the claim being tested and proceed to enter the other required items, then click on **Evaluate**.

TI-83/84 PLUS If using a TI-83/84 Plus calculator, press **STAT**, then select **TESTS** and choose the menu item of **Z-Test**. You can use the original data **(Data)** or the summary statistics **(Stats)** by providing the entries indicated in the window display. The first three items of the TI-83/84 Plus calculator results will include the alternative hypothesis, the test statistic, and the P-value.

Table 9.2 summarizes the four possible cases.

TABLE 9.2	Decision Table for H_0 and H_a		
		Reality	
		H_0 true	**H_a true**
Decision	Reject H_0	Type I error	Correct decision
	Do not reject H_0	Correct decision	Type II error

If you think about it, you'll realize that there is an important connection between the sig-nificance level of a hypothesis test and type I errors (wrongly rejecting H_0): *The significance level is the probability of making a type I error.* For example, if we conduct a hypothesis test using a 0.05 significance level, there is a 0.05 probability of making the mistake of rejecting the null hypothesis when it is actually true. If we conduct the test at a 0.01 significance level, there is a 0.01 probability of wrongly rejecting the null hypothesis. (The probability of a type II error can also be quantified but is beyond the scope of this text.)

EXAMPLE 4 Errors in the Body Temperature Test

Consider the null hypothesis from Example 3—that mean body temperature equals $98.6°F$ (H_0: $\mu = 98.6°F$).

a. What correct decisions are possible with this null hypothesis?

b. Explain the meaning of type I and type II errors in this case.

c. In Example 3, we rejected the null hypothesis. What is the probability that we made a type I error in doing so?

SOLUTION

a. Any hypothesis test has two possible correct decisions. In this case, one correct decision occurs if the mean body temperature really is 98.6°F and we do *not* reject H_0. The other correct decision occurs if the mean body temperature is *not* 98.6°F and we reject H_0.

b. A type I error occurs if we reject H_0 when it is actually true. In this case, a type I error occurs if mean body temperature really is 98.6°F but we conclude that it is not. A type II error occurs if we do not reject H_0 when it is actually false. In this case, a type II error occurs if the mean body temperature is not 98.6°F but we fail to reach this conclusion.

c. We rejected the null hypothesis at a 0.01 significance level. Therefore, there is a 0.01 probability that we made a type I error of rejecting a null hypothesis that is actually true. $\cdots\bullet$

Bias in Choosing Hypotheses

As we have seen, a well-conducted hypothesis test follows a fairly strict process that tends to minimize the possibility of tainting the test with bias. Nevertheless, bias can be introduced in many ways, and in particular it can occur in the initial selection of the hypotheses. Consider a situation in which a factory is investigated for releasing pollutants into a nearby stream at a level that may (or may not) exceed the maximum level allowed by the government. The logical choice for the null hypothesis is that the actual mean level of pollutants equals the maximum allowed level:

$$H_0: \text{mean level of pollutants} = \text{maximum level allowed}$$

However, this leaves us with two reasonable choices for the alternative hypothesis, each of which introduces some bias into the ultimate conclusions.

If *factory officials* conducted the test, they would be inclined to claim that the mean level of pollutants is *less than* the maximum allowed level. That is, they would choose

$$H_a: \text{mean level of pollutants} < \text{maximum level allowed}$$

With this alternative hypothesis, the two possible outcomes are

- Reject H_0, in which case we conclude that the mean level of pollutants is less than the allowed maximum. This outcome would please the factory officials.
- Not reject H_0, in which case the test is inconclusive.

In other words, by choosing the left-tailed ("less than") hypothesis test, the factory officials ensure that at worst we end up with no evidence that they have exceeded the allowed maximum, while it's possible that the conclusion will be in their favor.

Now suppose an *environmental group* conducts the test. Because the group suspects that the factory is violating the government standards, its choice for the alternative hypothesis would be that the mean level of pollutants is *greater than* the maximum allowed level, or

$$H_a: \text{mean level of pollutants} > \text{maximum level allowed}$$

With this choice of alternative hypothesis, rejection of H_0 implies that the factory has violated the standards, while not rejecting H_0 is again inconclusive. In other words, choosing the right-tailed ("greater than") test creates a situation in which the factory may be found in violation but cannot be proved to be in compliance.

EXAMPLE 5 Mining Gold

The success of precious metal mines depends on the purity (or grade) of ore removed and the market price for the metal. Suppose the purity of gold ore must be at least 0.5 ounce of gold per ton of ore in order to keep a particular mine open. Samples of gold ore are used to estimate the purity of the ore for the entire mine. Discuss the impact of type I and type II errors on two of the possible alternative hypotheses:

$$H_a: \text{purity} < 0.5 \text{ ounce per ton}$$
$$H_a: \text{purity} > 0.5 \text{ ounce per ton}$$

SOLUTION   For the left-tailed case, the null and alternative hypotheses are

$$H_0: \text{purity} = 0.5 \text{ ounce per ton}$$
$$H_a: \text{purity} < 0.5 \text{ ounce per ton}$$

The two possible decisions in this case are

- Reject H_0, which means concluding that the purity of the ore is less than that needed to keep the mine open—so the mine is closed.
- Not reject H_0, which means we have insufficient evidence to conclude that the purity of the ore is less than that needed—so the mine stays open.

A type I error (wrongly rejecting a true H_0) means the mine is closed when, in fact, the purity of the gold ore is sufficient to operate the mine. For the mine's operators, this means the loss of potential profits from the mine; for the employees, it means unnecessary loss of jobs. A type II error (failing to reject a false H_0) means that the mine continues to operate but is actually unprofitable.

Now suppose that the mine will be kept open only if the purity is greater than 0.5 ounce per ton. For this right-tailed case, the null and alternative hypotheses are

$$H_0: \text{purity} = 0.5 \text{ ounce per ton}$$
$$H_a: \text{purity} > 0.5 \text{ ounce per ton}$$

The two possible decisions are

- Reject H_0, which means concluding that the purity of the ore is *more* than that needed to keep the mine open—so the mine stays open.
- Not reject H_0, which means there is insufficient evidence to conclude that the purity of the ore is high enough to keep the mine open—so the mine is closed.

A type I error (wrongly rejecting a true H_0) means the mine is left open when it is actually unprofitable. A type II error (failing to reject a false H_0) means closing the mine when it is actually profitable—putting the employees out of work for no reason. $\cdot \cdot \bullet$

Section 9.2 Exercises

Statistical Literacy and Critical Thinking

1. **Notation.** Briefly describe what each of the variables $n, \bar{x}, s, \sigma,$ and μ represent in hypothesis tests of a claim made about a population mean.

2. **Waiting Times.** A student in a statistics class conducts a project designed to test the claim that wait times at McDonald's drive-up windows have a mean that is less than 5 minutes. She collects sample data from wait times at five randomly selected McDonald's locations that are within 30 miles of her home. What is wrong with this project?

3. **Types of Errors.** What is a type I error? What is a type II error?

4. **Statistical Significance.** Consider a test of the claim that with a new device installed, the mean fuel consumption of cars is reduced by 0.08 miles per gallon. The resulting P-value is 0.009. Is the mean fuel consumption reduction statistically significant? Does the mean fuel consumption reduction of 0.08 miles per gallon have practical significance? Why or why not?

Does It Make Sense? For Exercises 5–12, decide whether the statement makes sense (or is clearly true) or does not make sense

(or is clearly false). Explain clearly. Not all of these statements have definitive answers, so your explanation is more important than your chosen answer.

5. **Hypothesis Test.** In hypothesis tests, the P-value is always the same as the significance level.

6. **P-Value.** In testing a claim about a population mean, a larger z test statistic always results in a larger P-value.

7. **P-Value.** In a hypothesis test, a P-value of 0.001 indicates that you should support the alternative hypothesis.

8. **Alternative Hypothesis.** In testing the claim that the mean IQ score of statistics students is greater than 100, the alternative hypothesis is expressed as $\mu > 100$.

9. **Significance.** A significance level of 0.05 indicates that the probability of making a type I error is 0.05.

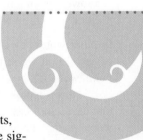

10. **Type I Error.** When a consumer group is testing the claim that the mean amount of aspirin in tablets is 350 milligrams, it is extremely important not to reject a true null hypothesis wrongly. Thus, it is better to choose 0.01 than 0.05 for the significance level.

11. **Significance Level.** Because the significance level is the probability of making a type I error, it is wise to select a significance level of zero so that there is no probability of making that error.

12. **Mnemonic.** A handy mnemonic for interpreting the *P*-value in a hypothesis test is this: "If the *P* (value) is low, then the null must go."

Concepts and Applications

Alternative Hypothesis Supported? In Exercises 13–20, find the value of the standard score, z, and determine whether the alternative hypothesis is supported at a 0.05 significance level.

13. H_a: $\mu < 75$, $n = 100$, $\bar{x} = 70$, $\sigma = 15$.

14. H_a: $\mu < 75$, $n = 36$, $\bar{x} = 72$, $\sigma = 15$.

15. H_a: $\mu > 12$, $n = 64$, $\bar{x} = 14$, $\sigma = 2$.

16. H_a: $\mu > 1007$, $n = 225$, $\bar{x} = 1021$, $\sigma = 35$.

17. H_a: $\mu \neq 2.55$, $n = 100$, $\bar{x} = 2.58$, $\sigma = 0.29$.

18. H_a: $\mu \neq 156.2$, $n = 225$, $\bar{x} = 155.5$, $\sigma = 29$.

19. H_a: $\mu \neq 0.88$, $n = 50$, $\bar{x} = 0.75$, $\sigma = 0.18$.

20. H_a: $\mu \neq 877$, $n = 90$, $\bar{x} = 921$, $\sigma = 52$.

Finding *P*-Values. In Exercises 21–34, use Table 5.1 to find the *P*-value that corresponds to the standard *z*-score, and determine whether the alternative hypothesis is supported at the 0.05 significance level.

21. $z = -0.4$ for H_a: $\mu < 25$

22. $z = -2.4$ for H_a: $\mu < 727$

23. $z = 1.9$ for H_a: $\mu > 36.35$

24. $z = 1.5$ for H_a: $\mu > 0.227$

25. $z = -1.6$ for H_a: $\mu \neq 0.389$

26. $z = -2.0$ for H_a: $\mu \neq 172$

27. $z = 1.7$ for H_a: $\mu \neq 75$

28. $z = 1.9$ for H_a: $\mu \neq 25.7$

29. $z = 2.7$ for H_a: $\mu > 19.4$

30. $z = 3.5$ for H_a: $\mu > 75$

31. $z = -2.1$ for H_a: $\mu < 1{,}007$

32. $z = -1.1$ for H_a: $\mu < 149.6$

33. $z = 0.15$ for H_a: $\mu \neq 90.3$

34. $z = 3.5$ for H_a: $\mu \neq 1{,}022$

35. **Interpreting *P*-Values.** Assume that you are testing an alternative hypothesis of the form H_a: $\mu > $ claimed value. If the sample mean has a standard score of $z = -1.0$, what

do you conclude? Why is it not necessary to actually conduct the formal hypothesis test?

36. **Interpreting *P*-Values.** Assume that you are testing an alternative hypothesis of the form H_a: $\mu < $ claimed value. If the sample mean has a standard score of $z = 0.5$, what do you conclude? Why is it not necessary to actually conduct the formal hypothesis test?

Hypothesis Tests for Means. For Exercises 37–50, use a 0.05 significance level and conduct a full hypothesis test using the four-step process described in the text. Be sure to state your conclusion.

37. **Roper Poll.** A Roper poll used a sample of 100 randomly selected car owners. Within the sample, the mean time of ownership for a single car was 7.01 years with a standard deviation of 3.74 years. Test the claim by the owner of a large dealership that the mean time of ownership for all cars is less than 7.5 years.

38. **Hospital Time.** According to a study by the Centers for Disease Control, the national mean hospital stay after childbirth is 2.0 days. Reviewing records at her own hospital, a hospital administrator calculates that the mean hospital stay for a sample of 81 women after childbirth is 2.2 days with a standard deviation of 1.2 days. Assuming that the patients represent a random sample of the population, test the claim that this hospital keeps new mothers longer than the national average.

39. **Cola.** A random sample of 36 cans of regular Coke is obtained and the contents are measured. The sample mean is 12.19 oz and the standard deviation is 0.11 oz. Test the claim that the contents of all such cans have a mean different from 12.00 oz, as indicated by the label.

40. **Motorcycle Manufacturing.** The manufacturers of motorcycles must produce axles that meet specified dimensions. In particular, the diameters of the axles must be 8.50 centimeters. The axles in a sample of $n = 64$ axles have a mean diameter of 8.56 centimeters with a standard deviation of 0.24 centimeter. Test the claim that the axles actually have the specified mean diameter.

41. **Drug Amounts.** The cold medicine Dozenol lists 600 milligrams of acetaminophen per fluid ounce as an active ingredient. The Food and Drug Administration tests 65 one-ounce samples of the medicine and finds that the mean amount of acetaminophen for the sample is 589 milligrams with a standard deviation of 21 milligrams. Test the claim of the FDA that the medicine does not contain the required amount of acetaminophen.

42. **Red Blood Cell Count.** A simple random sample of 50 adults is obtained, and each person's red blood cell count (in cells per microliter) is measured. The sample mean is 5.23. The population standard deviation for red blood cell counts is 0.54. Test the claim that the sample is from a population with a mean less than 5.4, which is a value often used for the upper limit of the range of normal values. What do the results suggest about the sample group?

43. Fuel Consumption. According to the Energy Information Administration (Federal Highway Administration data), the average gas mileage of all automobiles is 21.4 miles per gallon. For a random sample of 40 sport utility vehicles (SUVs), the mean gas mileage is 19.8 miles per gallon with a standard deviation of 3.5 miles per gallon. Test the claim that the mean mileage of all SUVs is less than 21.4 miles per gallon.

44. Baseballs. A random sample of 40 new baseballs is obtained. Each ball is dropped onto a concrete surface, and the bounce heights have a mean of 92.67 inches and a standard deviation of 1.79 inches (based on data from *USA Today*). Test the claim that the new baseballs have a mean bounce height that is less than the mean bounce height of 92.84 inches found for older baseballs.

45. Compulsive Buyers. Researchers developed a questionnaire to identify compulsive buyers. A random sample of 32 subjects who identified themselves as compulsive buyers was obtained, and they had a mean questionnaire score of 0.83 with a standard deviation of 0.24 (based on data from "A Clinical Screener for Compulsive Buying," by Faber and Guinn, *Journal of Consumer Research*, Vol. 19). Test the claim that the population of self-identified compulsive buyers has a mean greater than the mean of 0.21 for the general population.

46. Coin Weight. According to the U.S. Department of the Treasury, the mean weight of a quarter is 5.670 grams. A random sample of 50 quarters has a mean weight of 5.622 grams with a standard deviation of 0.068 gram. Test the claim that the mean weight of quarters in circulation is 5.670 grams.

47. Birth Weight. The mean birth weight of male babies born to 121 mothers taking a vitamin supplement is 3.67 kilograms with a standard deviation of 0.66 kilogram (based on data from the New York State Department of Health). Test the claim that the mean birth weight of all babies born to mothers taking the vitamin supplement is equal to 3.39 kilograms, which is the mean for the population of all male babies.

48. Weights of Bears. The health of the bear population in Yellowstone National Park is monitored by periodic measurements taken from anesthetized bears. A sample of 54 bears has a mean weight of 182.9 lb. Assuming that σ is known to be 121.8 lb, test the claim that the population mean of all such bear weights is greater than 150 lb.

49. Convicted Embezzlers. When 70 convicted embezzlers were randomly selected, the mean length of prison terms was found to be 22.1 months and the standard deviation was 8.6 months (based on data from the U.S. Department of Justice). Jane Fleming is running for political office on a platform of tougher treatment of convicted criminals. Test her claim that prison terms for convicted embezzlers have a mean of less than 24 months.

50. NCAA Football Coach Salaries. A simple random sample of 40 salaries of NCAA football coaches in the NCAA has a mean of $415,953. The standard deviation of all salaries of NCAA football coaches is $463,364. Test the claim that the mean salary of a football coach in the NCAA is less than $500,000.

Type I and Type II Errors. In Exercises 51–54, a null and alternative hypothesis are given. Without using the terms "null hypothesis" and "alternative hypothesis," identify the type I error and identify the type II error.

51. H_0: The patient is free of a particular disease.
H_a: The patient has the disease.

52. H_0: The defendant is not guilty.
H_a: The defendant is guilty.

53. H_0: The lottery is fair.
H_a: The lottery is biased.

54. H_0: The mean length of a bolt in the suspension system of new Audi cars is 3.456 centimeters.
H_a: The mean length of a bolt in the suspension system of new Audi cars is not equal to 3.456 centimeters.

PROJECTS FOR THE INTERNET & BEYOND

55. Comparisons with National Averages. Choose several variables that are relatively easy to measure in a class or sample of students. The variables should involve a quantity that can be averaged (for example, height, weight, family size, blood pressure, heart rate, reaction time). Use the Internet or other references to determine national averages (means) for these variables (by age categories, if appropriate). Collect data on the variables, using a random sample of at least 50 individuals. Carry out the relevant hypothesis test to determine whether the sample mean differs significantly from the population mean.

56. County Data. The *Statistical Abstract of the United States* and the Current Population Survey provide an extensive supply of social, economic, and vital statistics at the county, state, and local levels. Use their Web sites to compare state data to national data in the following way.

   a. Choose a variable of interest that involves a mean for a particular state (for example, the mean household size in Illinois).

   b. Find the current national value for that variable (for example, the national mean household size).

   c. Choose a particular county within the state and obtain the corresponding data for the county, as well as the sample size (for example, the mean household size in Cook County, Illinois).

   d. Assuming that the county is a random sample of the state, test the claim that the state is above or below the national level in terms of that variable.

   e. Discuss and interpret your results. The hypothesis test depends on the sample (the county) being a random sample of the population (the state). Be sure to discuss this factor in your conclusions.

57. Hypothesis Test Applet. Using a search engine such as Google, search for "hypothesis testing" and "applet." Find an applet and run it. Describe how the applet works and what it illustrates.

58. Power of a Test. Using a search engine such as Google, search for "power" of a hypothesis test. Describe what the power of a hypothesis test is.

9.3 HYPOTHESIS TESTS FOR POPULATION PROPORTIONS

We now consider hypothesis testing with *proportions*. All the ideas from previous sections apply, except we need a different method for calculating the standard deviation of the sampling distribution. An example will illustrate the process.

Suppose a political candidate commissions a poll in advance of a close election. Using a random sample of $n = 400$ likely voters, the poll finds that 204 people support the candidate. Should the candidate be confident of winning? In Chapter 8, we discussed how to determine the margin of error and confidence interval for this type of poll. We now cast the question as a hypothesis test.

For a hypothesis test, we ask whether the poll results (which are the sample statistics) support the hypothesis that the candidate has more than 50% of the vote. As usual, we let p represent the proportion of people in the voting *population* who favor the candidate, and we let $\hat{p}$ denote the proportion of people in the *sample* who favor the candidate. Because 204 of the 400 people in the sample support the candidate, the sample proportion is

$$\hat{p} = \frac{204}{400} = 0.510$$

We can now formulate the null and alternative hypotheses. As usual, we set up our null hypothesis as an equality:

$$H_0: p = 0.5 \text{ (50\% of voters favor the candidate)}$$

The candidate wants to know if she has majority support, so the alternative hypothesis is right-tailed:

$$H_a: p > 0.5 \text{ (more than 50\% of voters favor the candidate)}$$

Calculations for Hypothesis Tests with Proportions

How do we determine whether there is enough evidence in the sample to reject the null hypothesis? Following the hypothesis test process on page 315, we see that we have completed the first two steps (formulating the hypotheses and collecting the sample data). Now, just as we did for sample means in Section 9.2, we must determine the likelihood that the sample result could have arisen by chance assuming that the null hypothesis is true.

Proceeding as we did with sample means, we imagine selecting many samples of size $n = 400$. For each sample, we compute the proportion of people who favor the candidate. Because we have a reasonably large sample size, the distribution of sample proportions should be very close to a normal distribution. Under the starting assumption that the null hypothesis is true (that the proportion of people in the population who favor the candidate is 0.5), the peak of this distribution will be the population proportion claimed by the null hypothesis, $p = 0.5$. The standard deviation of the sampling distribution is given by the following formula (the derivation is beyond the scope of this text):

$$\text{standard deviation of distribution of sample proportions} = \sqrt{\frac{p(1-p)}{n}}$$

For this case, the standard deviation is

$$\sqrt{\frac{p(1-p)}{n}} = \sqrt{\frac{0.5(1-0.5)}{400}} = 0.025$$

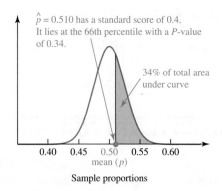

$\hat{p} = 0.510$ has a standard score of 0.4.
It lies at the 66th percentile with a P-value
of 0.34.

34% of total area
under curve

0.40   0.45   0.50   0.55   0.60
mean (p)

Sample proportions

Figure 9.7 The distribution of sample proportions for an election poll with a mean of $p = 0.5$ and a standard deviation of 0.025. The sample proportion $\hat{p} = 0.510$ has a standard score of 0.4 and a P-value of 0.34.

Figure 9.7 shows the distribution of sample proportions. We consider the sample proportion from the poll ($\hat{p} = 0.510$) as a single point in this sampling distribution. Much as in hypothesis tests with means (see Figure 9.1), we observe the following:

- If the sample result is close to the peak of the sampling distribution, then we have no reason to think the null hypothesis is wrong and we do not reject the null hypothesis.
- If the sample result is far from the peak of the sampling distribution, then the more likely explanation is that the sampling distribution does not really peak where the null hypothesis claims, in which case we reject the null hypothesis.

Again as we did for sample means, we decide whether the sample proportion is "near" to or "far" from the peak of the sampling distribution (assuming the null hypothesis is true) by quantifying the distance with a standard score. The formula for the standard score (z) always has the same general form (see page 173); for the distribution of sample proportions, the formula becomes

$$z = \frac{\text{sample proportion} - \text{population proportion}}{\text{standard deviation of sampling distribution}} = \frac{\hat{p} - p}{\sqrt{p(1 - p)/n}}$$

We can now compute and interpret the standard score for our current example. The sample size is the $n = 400$ people interviewed, the sample proportion is the 51% of the sample that supports the candidate ($\hat{p} = 0.510$), and the population proportion is the $p = 0.5$ claimed by the null hypothesis. Therefore, the standard score for the sample proportion is

$$z = \frac{\hat{p} - p}{\sqrt{p(1 - p)/n}} = \frac{0.510 - 0.5}{0.025} = 0.4$$

In other words, the sample used in the poll has a proportion that is 0.4 standard deviation away from the peak of the distribution of sample proportions. As you can see in Figure 9.7, this sample result is *not* very extreme, indicating that we should not reject the null hypothesis. That is, the candidate cannot be confident of majority support.

Standard Score for the Sample Proportion in a Hypothesis Test

Given the sample size (n), the sample proportion ($\hat{p}$), and the claimed population proportion (p), the standard score for the sample proportion is

$$z = \frac{\hat{p} - p}{\sqrt{p(1 - p)/n}}$$

Significance Levels and *P*-Values

We can be more quantitative by using a significance level or *P*-value. Let's start with significance. Because the election poll example uses a right-tailed test, we find the critical values for significance in Table 9.1 (page 321). The standard score of $z = 0.4$ is *not* greater than the critical value of $z = 1.645$ for significance at the 0.05 level, confirming that the result should not cause us to reject the null hypothesis. To find the *P*-value, we use the tables in Appendix A. The tables show that the area to the *left* of a standard score of $z = 0.4$ is 0.6554. Because this is a right-tailed test, we subtract this value from 1 to find the area to the right (shown in Figure 9.7), which is the *P*-value for this hypothesis test. That is, the *P*-value is $1 - 0.6554 = 0.3446$, telling us that if the null hypothesis is true, there is a more than 0.34 chance of randomly selecting a sample as extreme as the one found in this poll. With such a high probability of drawing such a sample by chance, we have no reason to reject the null hypothesis, so the candidate cannot assume that she has the support of more than 50% of voters.

Summary of Hypothesis Tests with Proportions

We can now summarize the procedure for doing a hypothesis test with a population proportion. Be sure to notice that we are still following the general four-step hypothesis test process on page 315, and here focus only on the specifics of dealing with proportions.

- Because we are dealing with population proportions, the null hypothesis has the form p = claimed value. To decide whether to reject or not reject the null hypothesis, we must determine whether a sample as extreme as the one found in the hypothesis test is likely or unlikely to occur if the null hypothesis is true.
- We determine this likelihood from the standard score (*z*) of the sample proportion, which we compute from the formula

$$z = \frac{\hat{p} - p}{\sqrt{p(1 - p)/n}}$$

  where *n* is the sample size, $\hat{p}$ is the sample proportion, and *p* is the population proportion claimed by the null hypothesis.
- We then use the standard score in the same two ways it was used for hypothesis tests with means:
  1. We can assess its level of statistical significance by comparing the standard score to the critical values given in Table 9.1 (page 321) for one-tailed tests and in the box on page 324 for two-tailed tests.
  2. We can determine its *P*-value with standard score tables like those in Appendix A. For a left-tailed test, the *P*-value is the area under the normal curve to the left of the standard score; for a right-tailed test, it is the area under the normal curve to the right of the standard score; and for a two-tailed test, it is double the value we would find if we were calculating the *P*-value for a one-tailed test.
- If the result is statistically significant at the chosen level (usually either the 0.05 or the 0.01 significance level), we reject the null hypothesis. If it is not statistically significant, we do not reject the null hypothesis.

EXAMPLE ❶ Local Unemployment Rate

Suppose the national unemployment rate is 9.5%. In a survey of $n = 450$ people in a rural Wisconsin county, 54 people are found to be unemployed. County officials apply for state aid based on the claim that the local unemployment rate is higher than the national average. Test this claim at a 0.05 significance level.

⏻ USING TECHNOLOGY

HYPOTHESIS TEST FOR A POPULATION PROPORTION

EXCEL Use XLSTAT. Click on **XLSTAT** at the top, click on **Parametric Tests**, then select **Tests for one proportion**. In the screen that appears, enter the sample proportion in the "Proportion" box, enter the sample size in the "Sample size" box, and enter the claimed value of the population proportion in the "Test proportion" box. (This is the same proportion used in the null hypothesis.) Select the "Data Format" of **Proportion,** and be sure that the box next to "z test" is checked. For the "Range" box, enter A1 so that the results will start at cell A1. Click on the **Options** tab to select the type of test; select the option including ≠ for a two-tailed test, select the option including < for a left-tailed test, or select the option including > for a right-tailed test. Enter the desired "Significance level (%)." For example, enter 5 for a 0.05 significance level. Click **OK**. After the results are displayed, look for the test statistic identified as "z (Observed value)" and the *P*-value. Critical values will also be displayed.

STATDISK Select **Analysis**, then select **Hypothesis Testing**. For the methods discussed in this section, select **Proportion–One Sample**. A dialog box will appear. Click on the box in the upper left corner and select the item that matches the claim being tested. Proceed to enter the other items in the dialog box, then click on **Evaluate**. The results will include the test statistic and *P*-value.

TI-83/84 PLUS Press **STAT**, select **TESTS**, and then select **1-PropZTest**. Enter the claimed value of the population proportion for p0, then enter the values for *x* and *n*, and then select the type of test. Highlight **Calculate,** then press **ENTER**.

SOLUTION   Let's follow the original four-step process from page 315.

Step 1.   The unemployment rate is a population proportion (as opposed to a population mean). The null hypothesis is the assumption that the local unemployment rate is equal to the national rate, or H_0: $p = 0.095$. The alternative hypothesis is the county's claim that the local unemployment rate is higher than the national average, or H_a: $p > 0.095$. This is a right-tailed test.

Step 2.   The sample statistics are the sample size, $n = 450$, and sample proportion, $\hat{p}$. From the given data, the sample proportion is

$$\hat{p} = \frac{54}{450} = 0.12$$

Step 3.   We now determine the likelihood that, *under the assumption that the null hypothesis is true,* chance alone would yield a sample proportion at least as extreme as the $\hat{p} = 0.12$ found for this particular sample. To find this likelihood, we start by calculating the standard score for this sample proportion:

$$z = \frac{\hat{p} - p}{\sqrt{p(1 - p)/n}} = \frac{0.12 - 0.095}{\sqrt{0.095(1 - 0.095)/450}} \approx 1.81$$

This standard score is above the critical value of $z = 1.645$ for a right-tailed test, so the result is significant at the 0.05 level. From Appendix A, the standard score of 1.81 has a *P*-value of $1 - 0.9649 = 0.0351$, confirming that there is a smaller than 0.05 probability that this sample would arise by chance if the null hypothesis is true. Figure 9.8 shows these ideas on the sampling distribution.

Step 4.   Because the test meets the criterion for significance at the 0.05 level, we reject the null hypothesis. In other words, this sample provides some evidence to support the alternative hypothesis that the county unemployment rate is above the national average.

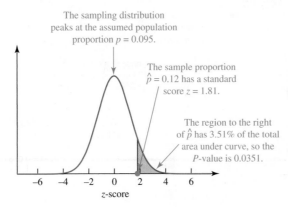

The sampling distribution peaks at the assumed population proportion $p = 0.095$.

The sample proportion $\hat{p} = 0.12$ has a standard score $z = 1.81$.

The region to the right of $\hat{p}$ has 3.51% of the total area under curve, so the P-value is 0.0351.

z-score

Figure 9.8 The distribution of sample proportions for Example 1. The sample proportion has a standard score of 1.81, which is above the critical value $z = 1.645$ for significance at the 0.05 level.

EXAMPLE 2   Left-Handed Population

A random sample of $n = 750$ people is selected, of whom 92 are left-handed. Use these sample data to test the claim that 10% of the population is left-handed.

SOLUTION   We again follow the four-step process.

Step 1.   The claim concerns the *proportion* of the population that is left-handed, so this is a test with a population proportion. The null hypothesis is the claim that 10% of the population is left-handed, or $H_0: p = 0.1$. To test this claim, we need to account for the possibility that the actual population proportion is either less than *or* greater than 10%. Therefore, the alternative hypothesis is $H_a: p \neq 0.1$, which calls for a two-tailed test.

Step 2.   The sample statistics are the sample size, $n = 750$, and the proportion of left-handed people in the sample:

$$\hat{p} = \frac{92}{750} = 0.123$$

Step 3.   The standard score for this sample proportion ($\hat{p} = 0.123$) is

$$z = \frac{\hat{p} - p}{\sqrt{p(1 - p)/n}} = \frac{0.123 - 0.1}{\sqrt{0.1(1 - 0.1)/750}} = 2.1$$

From the box on page 324, the critical values for significance at the 0.05 level in a two-tailed test are standard scores less than -1.96 or greater than 1.96. The standard score of 2.1 for this test is greater than 1.96, so we conclude that the test is significant at the 0.05 level. From Appendix A, the area to the right of a standard score of 2.1 is $1 - 0.9821 = 0.0179$. This would be the P-value if we were conducting a one-tailed test, but because we have a two-tailed test, we *double* it to find that the P-value is $2 \times 0.0179 = 0.0358$. Figure 9.9 shows the meaning of this P-value on the sampling distribution.

Step 4.   Because the test is significant at the 0.05 level, we reject the null hypothesis and conclude that the proportion of the population that is left-handed is *not* equal to 10%. Remembering that the significance level is the probability of a type I error, we recognize that there is a 0.05 probability that we made a type I error of rejecting a hypothesis that is actually true.

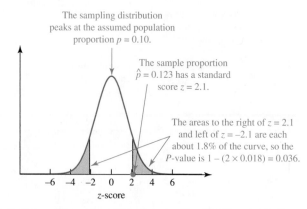

The sampling distribution peaks at the assumed population proportion $p = 0.10$.

The sample proportion $\hat{p} = 0.123$ has a standard score $z = 2.1$.

The areas to the right of $z = 2.1$ and left of $z = -2.1$ are each about 1.8% of the curve, so the P-value is $1 - (2 \times 0.018) = 0.036$.

z-score

Figure 9.9 This graph shows the position of the sample proportion ($\hat{p} = 0.123$) on the distribution of sample proportions for Example 2. Because this is a two-tailed test, the P-value corresponds to the area under the curve more than 2.1 standard deviations from the peak in *either* direction.

Section 9.3 Exercises

Statistical Literacy and Critical Thinking

1. **Notation.** What do p, $\hat{p}$, and P-value represent?

2. **Distribution.** In conducting a hypothesis test as described in this section, which term best describes the particular distribution that is used: uniform, normal, probability, weighted?

3. **Scream.** A survey of 61,647 people included several questions about office relationships. Of the respondents, 26% reported that bosses scream at employees. The survey is an *Elle*/MSNBC.COM survey in which Internet users chose whether to respond. Is it valid to use the sample results for testing the claim that less than 50% of employees have their bosses scream at them? Why or why not?

4. **P-Value.** A P-value of 0.00001 is obtained when sample data are used to test the claim that the majority of car crashes occur within 5 miles of home. What does this P-value tell us?

Does It Make Sense? For Exercises 5–8, decide whether the statement makes sense (or is clearly true) or does not make sense (or is clearly false). Explain clearly. Not all of these statements have definitive answers, so your explanation is more important than your chosen answer.

5. **Null Hypothesis.** In a test of the claim that a majority of Americans favor registration of all handguns, the null hypothesis is $p = 0.5$ and the alternative hypothesis is $p > 0.5$.

6. **Null and Alternative Hypotheses.** In a hypothesis test involving a claim made about a population proportion, if there is sufficient evidence to reject the null hypothesis, then there is sufficient evidence to support the alternative hypothesis.

7. **Alternative Hypothesis.** In a two-tailed hypothesis test of a claim about a proportion, the P-value is the area to the right of the standard score, z.

8. **Alternative Hypothesis.** The claim of $p < 0.25$ can never be supported if the sample proportion $\hat{p}$ is greater than 0.25.

Concepts and Applications

Hypothesis Tests. For Exercises 9–18, use a 0.05 significance level to conduct a hypothesis test using the four-step procedure described in the text. Be sure to state your conclusion.

9. **Voter Poll.** In a pre-election poll, a candidate for district attorney receives 205 of 400 votes. Assuming that the people polled represent a random sample of the voting population, test the claim that a majority of voters support the candidate.

10. **Smoking and College Education.** A survey showed that among 785 randomly selected subjects who completed four years of college, 144 smoke and the others do not smoke (based on data from the American Medical Association). Test the claim that the rate of smoking among those with four years of college is less than the 27% rate for the general population.

11. **Grade Pressure.** A study commissioned by the U.S. Department of Education, based on responses from 1,015 randomly selected teenagers, concluded that 44% of teenagers cite grades as their greatest source of pressure. Test the claim that fewer than half of all teenagers in the population feel that grades are their greatest source of pressure.

12. Married Adults. In a recent year, 125.8 million adults, or 58.6% of the adult American population, were married. In a New England town, a simple random sample of 1,445 adults includes 56.0% who are married. Test the claim that this sample comes from a population with a married percentage of less than 58.6%.

13. Drug Use. A Department of Health and Human Services study of illegal drug use among 12- to 17-year-olds reported a decrease in use (from 11.4% in 1997) to 9.9% now. Suppose a survey in a large high school reveals that, in a random sample of 1,050 students, 98 report using illegal drugs. Test the principal's claim that illegal drug use in her school is below the current national average.

14. Poverty. According to recent estimates, 12.1% of the 4,342 people in Custer County, Idaho, live in poverty. Assume that the people in this county represent a random sample of all people in Idaho. Based on this sample, test the claim that the poverty rate in Idaho is less than the national rate of 13.3%.

15. Abortion Survey. An annual survey of first-year college students, conducted by the Higher Education Research Institute at UCLA, asks approximately 276,000 students about their attitudes on a variety of subjects. According to a recent survey, 51% of first-year students believe that abortion should be legal (down from 65% in 1990). Test the claim that more than half of all first-year students believe that abortion should be legal.

16. Internet Use. When 3011 adults were surveyed in a Pew Research Center poll, 2198 said that they use the Internet. Is it OK for a newspaper reporter to write that "3/4 of all adults use the Internet"? Why or why not?

17. Natural Gas Use. According to the Energy Information Administration, 53.0% of households nationwide used natural gas for heating in 1997. A recent survey of 3,600 randomly selected households showed that 54.0% used natural gas. Use a 0.05 significance level to test the claim that the 53.0% national rate has changed.

18. Clinical Test. In clinical tests of the drug Lipitor, 863 patients were treated with the drug and 19 of them experienced flu symptoms (based on data from Parke-Davis). Test the claim that the percentage of treated patients with flu symptoms is greater than the 1.9% rate for patients not given treatments.

PROJECTS FOR THE INTERNET & BEYOND

19. Left-Handedness. Given the claim that 10% of Americans are left-handed, randomly select at least 50 students at your college and determine whether they are left-handed. Test the claim with a formal hypothesis test.

20. Smoking. Use the Internet or library references to determine the proportion of Americans who smoke. Test the claim that the proportion of students at your college who smoke is different from the proportion of all Americans. Collect sample data from at least 50 randomly selected students.

21. Women College Students. Use the Internet or library references to find the proportion of college students in the United States who are women. Test the claim that the proportion of women students at your college is different from the proportion of all U.S. college students. Collect sample data from at least 100 randomly selected students.

22. County Data. The *Statistical Abstract of the United States* and the Current Population Survey provide an extensive supply of social, economic, and vital statistics at the county, state, and local levels. Use their Web sites to compare state data to national data in the following way.

  a. Choose a variable of interest that involves a proportion for a particular state (for example, the percentage of people living in poverty in Arizona).

  b. Find the current national value for that variable (for example, the national poverty rate).

  c. Choose a particular county within the state and obtain the corresponding data for the county (for example, the poverty rate in Pima County, Arizona).

  d. Assuming that the county is a random sample of the state, test the claim that the state is above or below the national level in terms of that variable.

  e. Discuss and interpret your results. The hypothesis test depends on the sample (the county) being a random sample of the population (the state). Be sure to discuss this factor in your conclusions.

IN THE NEWS

23. Hypothesis Testing in the News. Find a news article or research report that describes (perhaps not explicitly) a hypothesis test for a population proportion. Attach the article and summarize the method used.

CHAPTER REVIEW EXERCISES

1. Randomly selected cans of Coke are measured for the amount of cola, in ounces. The sample values listed below have a mean of 12.19 ounces and a standard deviation of 0.11 ounce. Assume that we want to use a 0.05 significance level to test the claim that cans of Coke have a mean amount of cola greater than 12 ounces. Assume that the population has a standard deviation of $\sigma = 0.115$ ounce.

```
12.3   12.1   12.2   12.3   12.2   12.3   12.0   12.1   12.2
12.1   12.3   12.3   11.8   12.3   12.1   12.1   12.0   12.2
12.2   12.2   12.2   12.2   12.2   12.4   12.2   12.2   12.3
12.2   12.2   12.3   12.2   12.2   12.1   12.4   12.2   12.2
```

a. What is the null hypothesis?

b. What is the alternative hypothesis?

c. What is the value of the standard score for the sample mean of 12.19 ounces?

d. What is the critical value?

e. What is the P-value?

f. What do you conclude? (Be sure to address the original claim that the mean is greater than 12 ounces.)

g. Describe a type I error for this test.

h. Describe a type II error for this test.

i. Find the P-value if the test is modified to test the claim that the mean is *different from* 12 ounces (instead of being greater than 12 ounces).

2. In a study of smokers who tried to quit smoking with nicotine patch therapy, 39 were smoking one year after the treatment, and 32 were not smoking one year after the treatment (based on data from "High Dose Nicotine Patch Therapy," by Dale et al., *Journal of the American Medical Association*, Vol. 274, No. 17). We want to use a 0.05 significance level to test the claim that among smokers who try to quit with nicotine patch therapy, the majority are smoking a year after the treatment.

a. What is the null hypothesis?

b. What is the alternative hypothesis?

c. What is the value of the standard score for the sample proportion?

d. What is the critical value?

e. What is the P-value?

f. What do you conclude? (Be sure to address the original claim that among smokers who try to quit with nicotine patch therapy, the majority are smoking a year the treatment.)

g. Describe a type I error for this test.

h. Describe a type II error for this test.

i. What is the P-value if the claim is modified to state that the proportion is *equal* to 0.5?

3. **Finding a Job Through Networking.** In a survey of 703 randomly selected workers, 429 got their jobs through networking (based on data from Taylor Nelson Sofres Research).

a. Use the sample data with a 0.05 significance level to test the claim that most (more than 50% of) workers get their jobs through networking.

b. If given a claim that less than 50% of workers get their jobs through networking, why is it not necessary to go through the steps of conducting a formal hypothesis test? What should we conclude?

4. A medical student wants to test the claim that males who smoke have pulse rates with a mean greater than 70. She collects sample data by surveying her fellow medical students. What is the fundamental flaw in his procedure?

1. What is the alternative hypothesis that results from the claim that the mean brain volume of adults is larger than $1{,}126 \text{ cm}^3$?

2. What is the null hypothesis that is used for testing the claim that the proportion of college graduates who voted in the last election is greater than 0.5?

3. What is the alternative hypothesis that results from the claim that the proportion of convicted felons who serve time in prison is equal to 0.6?

4. Is a test of the claim that $p \neq 0.75$ left-tailed, right-tailed, or two-tailed?

In Exercises 5–10, assume that we want to use a 0.05 significance level to test the claim that the mean IQ score of professional comedians is greater than 110.

5. What is the null hypothesis?

6. What is the alternative hypothesis?

7. If the test results in a P-value of 0.007, what do you conclude about the given claim?

8. What are the two possible conclusions that can be reached about the null hypothesis?

9. What are the two possible conclusions that can be reached about the claim being tested?

10. If you incorrectly conclude that professional comedians have a mean IQ score greater than 110 when their actual mean IQ score is 100, have you made a type I error or a type II error?

HEALTH & EDUCATION

Will Your Education Help You Live Longer?

In this chapter, we focused on hypothesis testing in its simplest form, in which we test a single claim about a population and determine whether it is supported by evidence collected from a single sample. Not surprisingly, many of the most interesting statistical problems require much more complex analysis. Consider, for example, the question of what you can do to help you live longer.

Although this question is much more difficult to answer than anything we encountered in this chapter, it can be addressed through the same basic principles. Suppose, for example, that a researcher suspects that eating oats can make you live longer. To test this suspicion, the researcher conducts a hypothesis test. He or she starts with the null hypothesis that oat consumption has no effect on life span, then examines data to look for evidence that would support rejecting this null hypothesis and concluding that oat consumption really does increase life span. The difficult parts of this research are collecting the data—for example, finding people who consume oats to be

compared to people who don't—and then finding a way to separate the effects of oat consumption from those of the huge number of other variables that may also affect life span.

Over the past few decades, researchers have identified many factors that appear to contribute to longer lives. For example, greater wealth is correlated with longer life. Race plays a role in life span. Diet has numerous health effects. Exercise is generally a positive factor for longer life. Somewhat surprisingly, however, one factor appears to be more important than all the others: years of education. The longer you stay in school, the longer you'll live, at least on average. Figure 9.10 shows some of the data that support this conclusion.

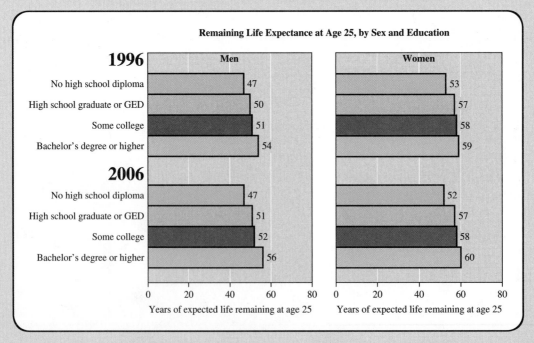

Figure 9.10 These bar charts show how the remaining life expectancy for 25-year-olds in the United States varies with education level for both men and women, and compares the results from 1996 and 2006. Notice the significant increase in life expectancy with education in all cases. Also notice that from 1996 to 2006, life expectancy remained stagnant or declined for the least educated people, while it increased for the most educated. *Source: US Department of Health and Human Services, "Health, United States, 2011."*

Why might more education lead to longer life? Researchers have suggested many possible reasons. For example, people with more education tend to be less likely to engage in life-shortening behaviors such as smoking or excessive drinking, while also undertaking exercise and other activities that can increase life span. One particularly interesting hypothesis is that getting an education involves some measure of short-term sacrifice for longer-term gain (such as paying for your education now in hopes of getting a higher-paying job later), and this willingness to accept delayed gratification helps people make all kinds of decisions that contribute to longer life. Of course, no one yet knows for sure—so you can be sure there is much more hypothesis testing to be done.

Nevertheless, based on the current data, we can draw at least one very interesting conclusion directly relevant to your reading of this text: You may be taking a statistics class because it is required, but it may also help you live longer.

QUESTIONS FOR DISCUSSION

1. Figure 9.10 presents a wealth of data. Interpret it carefully. How does it support the hypothesis that more education leads to longer life? Explain.

2. Why do *you* think education contributes to longer life? For example, do you think the mere fact of staying in school makes people live longer, or is it the extra learning associated with more years of schooling that makes the difference? How would you test your hypothesis?

3. Smoking is well known to be one of the biggest risk factors in disease and early death. As a result, the government and health care groups have often undertaken expensive advertising campaigns in hopes of convincing people to stop smoking (or never start). But the smoking rate drops dramatically with years of education. Therefore, if the goal is to reduce smoking, one might argue that it would make more sense to spend money on programs to help people stay in school than to spend money on anti-smoking campaigns. Do you think this is a good idea? Defend your opinion.

4. Education correlates with healthier lives as well as longer lives, which means that more educated people tend to have lower health care costs. Can you think of any ways in which this fact could be used to help get health care costs under control? Discuss with friends or classmates.

Focus on

AGRICULTURE

Are Genetically Modified Foods Safe?

British newspapers call them "Frankenfoods" (a word play on *Frankenstein*). Europeans, by and large, don't eat them at all. But today, many foods sold in the United States already fall in this category, including an estimated 65% of American corn and 94% of American soybeans. We are speaking of "genetically modified foods," or GM foods. GM foods are a fairly recent agricultural invention, dating only from about the mid-1990s.* Nevertheless, they are at the forefront of one of the biggest debates in agricultural history.

   A genetically modified organism is an organism into which scientists have inserted a gene that does not naturally exist in the organism or its close relatives. Genetically modified organisms are being tested and used for many purposes. For example, scientists have developed bacteria containing genes that produce drugs like insulin.

   In agriculture, genetic modification allows scientists to create in crops traits that would be difficult or impossible to achieve by traditional breeding techniques. One of the first widely used genetically modified crops, often called "Bt corn," illustrates the idea behind genetic modification. Corn is usually susceptible to destruction by a variety of insect pests. As a result, farmers usually spray crops with pesticides. Unfortunately, many pesticides are toxic to animals besides the insect pests and hence can cause environmental damage (and may not be ideal for humans to consume).

   Here's where Bt, a bacterium known as *bacillus thuringiensis*, comes in. Bt lives naturally in soil. As early as 1911, scientists discovered that Bt produces a toxin that kills certain types of insects. Different strains of Bt kill different insects, but are generally harmless to other animals and humans. By the 1960s, these traits had led to the use of Bt as an "environmentally friendly" pesticide that could be sprayed on crops (in the form of killed bacteria). Unfortunately, Bt pesticide proved to be expensive and often ineffective, mainly because it works only if insects eat it (more effective pesticides kill on contact). And, because it breaks down rapidly in sunlight and washes off plants in rain, it kills pests only if its application is timed just right.

Genetic modification solves the problems inherent in Bt pesticide. The pesticide action of Bt bacteria arises from particular proteins that the bacteria produce. Scientists identified the genes responsible for these proteins, then transferred these genes into plants such as corn. Once the corn contains the necessary genes, it produces the same pest-killing proteins as the Bt bacteria. The application of a sprayed pesticide is no longer necessary because the corn itself is now toxic to the pests. Moreover, because the corn continually produces the pest-killing proteins, there are no more concerns about the pesticide breaking down or washing away.

   The advantages of the Bt corn over traditional strains of corn are clear, but are there any disadvantages? This is where the great debate over GM foods begins. On one side, many scientists argue that GM foods are completely safe and that their benefits will help improve nutrition for people throughout the world. On the other side stand the people, including some scientists, who label GM foods as "Frankenfoods" and argue that they are one of the most dangerous technologies ever invented.

   Broadly speaking, the issue of GM food safety can be broken down into three major questions:

1. Do GM foods have any toxic effects in humans?

2. Do the new proteins contained in GM foods cause allergic reactions in some people?

3. Can the GM crops cause any unforeseen environmental damage, such as transferring their genes into weeds (thereby making "superweeds") or killing animals besides the insect pests?

---

*More than 50 new GM crops have been approved for sale in the United States, including corn and soybeans that produce their own pesticide and tomatoes engineered for longer shelf life. GM crops under development include potatoes that resist bruising, better-tasting soybeans, and grains containing vitamins and other nutrients that could improve nutrition in impoverished countries.

These questions can be addressed through hypothesis testing. In each case, we begin with a null hypothesis that states that there is no safety difference between traditional foods and GM foods. For example, the null hypothesis for Question 1 says that GM foods are no more toxic than traditional foods (which often contain low levels of toxic chemicals). The alternative hypothesis states that there is a difference in toxicity. Scientists then develop experiments to test the hypotheses. If the evidence provided by the experiments reveals a significant difference in toxicity between the food groups (beyond what would be expected by chance alone), there is reason to reject the null hypothesis.

Unfortunately, experiments to date have been unable to resolve the controversy. For example, some genetically modified crops *do* contain products toxic to humans—and therefore never receive approval for sale. To proponents of GM foods, this fact provides an argument in their favor, because it appears to show that current regulatory procedures (for example, requiring approval of GM foods by the U.S. Food and Drug Administration) adequately ensure that only safe foods enter the marketplace. Similarly, while some of the proteins in GM foods undoubtedly can cause allergic reactions, many scientists think they understand these reactions well enough to ensure that only safe foods are approved.

Opponents of GM foods use the very same experimental results to support their case. They argue that even if scientists have identified toxicity that is obvious in experiments, they may still be missing long-term effects that might not show up in the population for many years. Similarly, they claim that we cannot be sure we understand *all* allergic reactions and therefore might inadvertently approve GM foods that could have severe allergic consequences in at least some people.

The environmental issues are even more difficult to study. For example, one study has shown that Bt corn is toxic to the Monarch butterfly, a species that is not a pest and that no one wants to kill. But the study was conducted in a laboratory and may not accurately represent what occurs in the real environment. Similarly, the possibility of "gene jumping," in which the Bt genes might spread from corn to other plants, is not well understood and therefore is very difficult to study.

The debate over GM foods is likely to continue, and even to intensify, for many years to come. After all, when it comes to food, everyone has an interest.

QUESTIONS FOR DISCUSSION

1. Propose an experiment that could be used to test the safety of a GM food product, such as corn. Describe your experiment in detail, and discuss any practical difficulties that might be involved in carrying it out or interpreting its results.

2. Ignoring any safety issues of GM foods, make a list of as many ways as you can think of in which GM foods might be beneficial to humanity. Then, ignoring any benefits of GM foods, make a list of as many ways as you can think of in which GM foods might be dangerous. Overall, do you think that further use of GM foods should be encouraged or discouraged? Defend your opinion.

3. Some people advocate giving the choice about GM foods to consumers by requiring labeling on all products that contain genetically modified ingredients. Do you think this is a good idea? Why or why not?

4. Investigate recent developments in the debate over GM foods. Does the new information alter any of the major arguments in the debate? Explain.

10

t Tests, Two-Way Tables, and ANOVA

We have explored many core ideas and applications of statistics in Chapters 1–9, but you will encounter many more if you continue your study of statistics. In this final chapter, we explore three particularly common applications of statistics that you may see in future course work; they will also further your understanding of the role of statistics in your daily life. All three of these applications build upon the important technique of hypothesis testing introduced in Chapter 9. We begin with the *t* distribution, which applies to confidence intervals as well as hypothesis tests, and then investigate hypothesis tests with two variables and with the method known as analysis of variance, or ANOVA.

> The web of this world is woven of necessity and chance. Woe to him who has accustomed himself to find something capricious in what is necessary, and who would ascribe something like reason to chance.
>
> —Johann Goethe

10.1 *t* DISTRIBUTION FOR INFERENCES ABOUT A MEAN

We discussed confidence interval estimates of a population mean in Section 8.2. Using the assumption that the distribution of sample means is a normal distribution, we estimated the margin of error for the 95% confidence interval, *E*, to be

$$E \approx \frac{2s}{\sqrt{n}}$$

As was stated in a Technical Note, the precise formula for the margin of error uses 1.96 rather than 2, because that is the standard score *z* with an area of 0.025 (half of 5%) to its right (see Appendix A).

We discussed hypothesis tests for claims about a population mean in Section 9.2, again based on the assumption that the sampling distribution is normal. We used the following formula for the standard score of the sample mean:

$$z = \frac{\bar{x} - \mu}{\sigma/\sqrt{n}}$$

We then presented the following criteria for rejecting the null hypothesis (at the 0.05 level of significance): $z \leq -1.645$ for a left-tailed test, $z \geq 1.645$ for a right-tailed test, and $z \leq -1.96$ or $z \geq 1.96$ for a two-tailed test. The values of -1.645, 1.645, -1.96, and 1.96 are all derived from the standard normal distribution.

Notice that the above formula requires that we know the value of the population standard deviation σ. It is very rare that we test a claim about an unknown population mean while we somehow know the value of the population standard deviation. For this and other reasons, statisticians generally prefer an approach that does not require knowing the value of σ. One common approach is to use what is known as the **Student *t* distribution**, or ***t* distribution** for short. This approach works well when we do not know the population standard deviation *and* either of these two conditions is satisfied: (1) The population has a normal distribution or (2) the sample size is greater than 30.

> ### BY THE WAY
>
> The Student *t* distribution was developed by William Gosset (1876–1937), a Guinness Brewery employee who needed a distribution that could be used with relatively small samples. The Irish brewery where he worked did not allow publication of research results, so Gosset published under the pseudonym *Student*.

> ### Inferences about a Population Mean: Choosing between *t* and Normal Distributions
>
> **t distribution:** Population standard deviation is not known and the population is normally distributed.
>
> or   Population standard deviation is not known and the sample size is greater than 30.
>
> ---
>
> **Normal distribution:** Population standard deviation is known and the population is normally distributed.
>
> or   Population standard deviation is known and the sample size is greater than 30.

We won't go into detail about the nature of the *t* distribution in this text, but the basic idea is easy to understand: The *t* distribution is very similar in shape and symmetry to the normal distribution, but it accounts for the greater variability that is expected with small samples. Figure 10.1 contrasts the *t* distribution with the standard normal distribution for sample sizes of $n = 3$ and $n = 12$. Notice that for the larger sample size, the *t* distribution is closer to the normal distribution. In fact, the larger the sample, the more closely the *t* distribution matches the normal distribution.

An important advantage of the *t* distribution is that it allows us to extend ideas of confidence intervals or hypothesis tests to many cases in which we cannot use the normal distribution because we do not know the population standard deviation. Keep in mind, however, that it still does not work for all cases. For example, if we have a small sample of size 30 or less and the sample data suggest that the population has a distribution that is radically different from a normal distribution, then neither the *t* distribution nor the normal distribution applies. Such cases require other methods (such as bootstrapping methods or nonparametric methods) not discussed in this text.

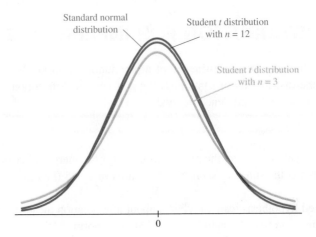

TABLE 10.1	Critical Values of *t*	
	Area in one tail	
Degrees	0.025	0.05
of freedom	**Area in two tails**	
(*n* − 1)	0.05	0.10
1	12.706	6.314
2	4.303	2.920
3	3.182	2.353
4	2.776	2.132
5	2.571	2.015
6	2.447	1.943
7	2.365	1.895
8	2.306	1.860
9	2.262	1.833
10	2.228	1.812
11	2.201	1.796
12	2.179	1.782
13	2.160	1.771
14	2.145	1.761
15	2.131	1.753
16	2.120	1.746
17	2.110	1.740
18	2.101	1.734
19	2.093	1.729
20	2.086	1.725
21	2.080	1.721
22	2.074	1.717
23	2.069	1.714
24	2.064	1.711
25	2.060	1.708
26	2.056	1.706
27	2.052	1.703
28	2.048	1.701
29	2.045	1.699
30	2.042	1.697
31	2.040	1.696
32	2.037	1.694
34	2.032	1.691
36	2.028	1.688
38	2.024	1.686
40	2.021	1.684
50	2.009	1.676
100	1.984	1.660
Large	1.960	1.645

Figure 10.1   This figure compares the standard normal distribution to the *t* distribution for two different sample sizes. Notice that as the sample size gets larger, the *t* distribution more closely approximates the normal distribution.

Confidence Intervals Using the *t* Distribution

Much as we did in Sections 8.2 and 9.2 under the assumption of a normal distribution, we can use the *t* distribution to construct a confidence interval for a population mean or to conduct a hypothesis test for a population mean. However, instead of determining significance based on standard *z*-scores such as −1.645, 1.645, −1.96, and 1.96, we use values of *t*, such as those shown in Table 10.1.

To specify a confidence interval, we first calculate the margin of error, *E*. With a *t* distribution, the formula is

$$E = t \times \frac{s}{\sqrt{n}}$$

where *n* is the sample size, *s* is the sample standard deviation, and *t* is a value found in Table 10.1. The only "trick" is in finding the correct value of *t* from the table, which we do as follows:

- First, determine the number of **degrees of freedom** (column 1 in Table 10.1) for the sample data. For the purposes of this section, the number of degrees of freedom is the sample size minus 1:

  degrees of freedom for *t* distribution = *n* − 1

- Based on the number of degrees of freedom, find the appropriate *t* value. For confidence interval estimates of population means, the *t* values correspond to 95% confidence in the middle column and 90% confidence in the right column. (We use the table values for the "area in two tails" because the margin of error can be either below the mean or above it; for example, 95% confidence means we are looking for a total area of 0.05 both to the far left and to the far right of a *t* distribution like those shown in Figure 10.1.)

Once you find the *t* value for your data and confidence level, you can determine the confidence interval just as we did in Section 8.2, except we use a new formula for the margin of error, *E*.

Confidence Interval for a Population Mean (*μ*) with the *t* Distribution

If conditions require use of the *t* distribution (*σ* not known and population normally distributed or *n* > 30), the confidence interval estimate of the true value of the population mean (*μ*) extends from the sample mean minus the margin of error ($\bar{x} - E$) to the sample mean plus the margin of error ($\bar{x} + E$). That is, the confidence interval for the population mean is

$$\bar{x} - E < \mu < \bar{x} + E \ \text{(or, equivalently, } \bar{x} \pm E\text{)}$$

where the margin of error is

$$E = t \times \frac{s}{\sqrt{n}}$$

and we find *t* from Table 10.1 using degrees of freedom = *n* − 1.

EXAMPLE 1   Confidence Interval for Diastolic Blood Pressure

Here are five measurements of diastolic blood pressure from randomly selected adult men: 78, 54, 81, 68, 66. These five values result in these sample statistics: $n = 5, \bar{x} = 69.4, s = 10.7$. Using this sample, construct the 95% confidence interval estimate of the mean diastolic blood pressure level for the population of all adult men.

SOLUTION   Because the population standard deviation is not known and because it is reasonable to assume that blood pressure levels of adult men are normally distributed, we use the *t* distribution instead of the normal distribution. With a sample of size $n = 5$, the number of degrees of freedom is

$$\text{degrees of freedom for } t \text{ distribution} = n - 1 = 5 - 1 = 4$$

For 95% confidence, we use the middle column in Table 10.1 to find that $t = 2.776$. We now use this value along with the given sample size ($n = 5$) and sample standard deviation ($s = 10.7$) to calculate the margin of error, E:

$$E = t \times \frac{s}{\sqrt{n}} = 2.776 \times \frac{10.7}{\sqrt{5}} = 13.3$$

Finally, we use the margin of error and the sample mean to find the 95% confidence interval:

$$\bar{x} - E < \mu < \bar{x} + E$$
$$69.4 - 13.3 < \mu < 69.4 + 13.3$$
$$56.1 < \mu < 82.7$$

Based on the five sample measurements, we have 95% confidence that the limits of 56.1 and 82.7 contain the mean diastolic blood pressure level for the population of all adult men.   •• •

Hypothesis Tests Using the *t* Distribution

When the *t* distribution is used for a hypothesis test of a claim about a population mean ($H_0: \mu =$ claimed value), the *t* value plays the role that the standard score *z* played when we studied these hypothesis tests in Section 9.2. With the *t* distribution, instead of calculating the standard score *z*, we use the following formula to calculate *t*:

$$t = \frac{\bar{x} - \mu}{s/\sqrt{n}}$$

where *n* is the sample size, $\bar{x}$ is the sample mean, *s* is the sample standard deviation, and μ is the population mean claimed by the null hypothesis. We then determine statistical significance by comparing the *t* value to critical values or by finding its *P*-value. Finding *P*-values is difficult with Table 10.1, but technology can be used to find *P*-values. If using critical values instead of *P*-values, the critical values depend on the type of test as follows.

Right-tailed test: Reject the null hypothesis if the computed test statistic *t* is greater than or equal to the value of *t* found in the column of Table 10.1 labeled "Area in one tail." Notice that for the one-tailed test, the middle column gives critical values for significance at the 0.025 level and the right column gives critical values for significance at the 0.05 level.

Left-tailed test: Reject the null hypothesis if the computed test statistic *t* is less than or equal to the negative of the value of *t* found in the column of Table 10.1 labeled "Area in one tail." Again, because this is a one-tailed test, the middle column gives critical values for significance at the 0.025 level and the right column gives critical values for significance at the 0.05 level.

Two-tailed test: Reject the null hypothesis if the absolute value of the computed test statistic *t* is greater than or equal to the value of *t* found in the column of Table 10.1 labeled "Area in two tails." For this case, the middle column gives critical values for significance at the 0.05 level and the right column gives critical values for significance at the 0.10 level.

EXAMPLE ② Right-Tailed Hypothesis Test for a Mean

Listed below are 10 randomly selected IQ scores of statistics students:

$$111 \quad 115 \quad 118 \quad 100 \quad 106 \quad 108 \quad 110 \quad 105 \quad 113 \quad 109$$

Using methods from Chapter 4, you can confirm that these data have the following sample statistics: $n = 10$, $\bar{x} = 109.5$, $s = 5.2$. Using a 0.05 significance level, test the claim that statistics students have a mean IQ score greater than 100, which is the mean IQ score of the general population.

SOLUTION Based on the claim that the mean IQ of statistics students is greater than 100, we use the null hypothesis H_0: $\mu = 100$ and the alternative hypothesis H_a: $\mu > 100$. Because the standard deviation of IQ scores for the population of all statistics students is not known and because it is reasonable to assume that IQ scores of statistics students are normally distributed, we use the t distribution instead of the normal distribution. The value of the t test statistic is computed as follows:

$$t = \frac{\bar{x} - \mu}{s/\sqrt{n}} = \frac{109.5 - 100}{5.2/\sqrt{10}} = 5.777$$

We now compare this value to the appropriate critical value from Table 10.1:

- We find the correct row by recognizing that this data set has $n - 1 = 10 - 1 = 9$ degrees of freedom.
- Because it is a one-tailed test and we are asked to test for significance at the 0.05 level, we use the values from the right column.
- Looking in the row for 9 degrees of freedom and column 3, we find that the critical value for significance at the 0.05 level is $t = 1.833$.

Because the sample test statistic $t = 5.777$ is greater than the critical value $t = 1.833$, we reject the null hypothesis. We conclude that there is sufficient evidence to support the claim that the mean IQ score is greater than 100.

We can be more precise by using software to compute the *P*-value for this hypothesis test, which turns out to be 0.000135. Notice that this *P*-value is much less than 0.05, so we can be quite confident in the decision to reject the null hypothesis and support the claim that the mean IQ score is greater than 100. $\cdot\cdot\bullet$

EXAMPLE ③ Two-Tailed Hypothesis Test for a Mean

Using the sample data in Example 2 and the same significance level of 0.05, test the claim that the mean IQ score of statistics students is *equal to* 100.

SOLUTION Based on the claim that the mean IQ score of statistics students is equal to 100, we use the null hypothesis H_0: $\mu = 100$ and the alternative hypothesis H_a: $\mu \neq 100$. The "not equal to" form of the alternative hypothesis means that this is a two-tailed test.

The value of the sample test statistic is still the same as in Example 2 ($t = 5.777$), but the critical values are different for the two-tailed test. Because we are using the same data set, the number of degrees of freedom is still 9, telling us which row to look at in Table 10.1. For a two-tailed test, we find the critical value for significance at the 0.05 level in the middle column rather than in column 3; the value is $t = 2.262$.

Notice that the absolute value of the test statistic $t = 5.777$ is greater than the critical value $t = 2.262$, so again we reject the null hypothesis. We conclude that there is sufficient evidence to reject the claim that the mean IQ score of statistics students is equal to 100. (With software, we find that the test statistic has a *P*-value of 0.000269.) $\cdot\cdot\bullet$

⏻ **USING TECHNOLOGY**—**THE t DISTRIBUTION**

Confidence Intervals Using the t Distribution

EXCEL Use XLSTAT. Click on **XLSTAT** at the top, click on **Parametric tests**, then select **One sample t test and z test.** In the screen that appears, for the "Data" box enter the range of data, such as A1:A12 for 12 data values in column A. For "Data Format" select **One sample.** Click on the "Student's t test" box (or click on the "z test" box if σ is known). Click on the **Options** tab, to enter the desired "Significance level (%)." Enter 5 for a 95% confidence interval. Click **OK.** After the results are displayed, look for "confidence interval on the mean." (The use of Excel's **CONFIDENCE** tool is not recommended, for a variety of reasons.)

STATDISK: Select **Analysis,** then **Confidence Intervals,** then **Mean - One Sample.** In the dialog box that appears, first enter the significance level as a decimal number. Enter 0.95 for a 95% confidence level. Proceed to enter the other required items and then click on **Evaluate;** the confidence interval will be displayed.

TI-83/84 PLUS The TI-83/84 Plus calculator can be used to generate confidence intervals for original sample values stored in a list, or you can use the summary statistics n, $\bar{x}$, and s. Either enter the data in list L1 or have the summary statistics available, then press **STAT**. Now select **TESTS** and choose **TInterval** if σ is not known. (Choose **ZInterval** if σ is known.) After making the required entries, the calculator display will include the confidence interval in the format of $(\bar{x} - E, \bar{x} + E)$.

Hypothesis Tests with the t Distribution

EXCEL Use XLSTAT. First enter the list of original sample values in a list. Click on **XLSTAT** at the top, click on **Parametric tests,** then select **One sample t test and z test.** In the screen that appears, for the "Data" box enter the range of data, such as A1: A11 for 11 data values in column A. For "Data Format" select **One sample.** Click on the "Student's t test" box (or click on the "z test" box if σ is known). Click on the **Options** tab to select the type of test; select the option including $\neq$ for a two-tailed test, select the option including $<$ for a left-tailed test, or select the option including $>$ for a right-tailed test. For the "Theoretical mean" box, enter the claimed value of the population mean, which is the same value used in the statement of the null hypothesis. Enter the desired "Significance level (%)." For example, enter 5 for a 0.05 significance level. Click **OK.** After the results are displayed, look for the test statistic identified as "t (Observed value)" or "z (Observed value)." The P-value and critical value(s) will also be displayed.

STATDISK: Select **Analysis,** then select **Hypothesis Testing.** For the methods discussed in Section 10.1, select **Mean - One Sample.** A dialog box will appear. Click on the box in the upper left corner and select the item that matches the claim being tested. Proceed to enter the other items in the dialog box, and then click on **Evaluate.** The results will include the test statistic and P-value.

TI-83/84 PLUS If using a TI-83/84 Plus calculator, press **STAT**, then select **TESTS** and choose the menu item of **T-Test.** (For σ known, select Z-Test.) You can use the original data **(Data)** or the summary statistics **(Stats)** by providing the entries indicated in the window display. The first three items of the TI-83/84 Plus calculator results will include the alternative hypothesis, the test statistic, and the P-value.

EXAMPLE ④   Left-Tailed Hypothesis Test for a Mean

Because of the expense involved, car crash tests often use small samples. In one study, five BMW cars are crashed under standard conditions, and the repair costs (in dollars) are used to test the claim that the mean repair cost for all BMW cars is less than $3,000. The five sample repair costs have a mean of $2,835 with a standard deviation of $883. Use a 0.05 significance level to test the claim that the population of BMW repair costs has a mean less than $3,000.

SOLUTION   Based on the claim that the mean repair cost is less than $3,000, the null hypothesis is H_0: $\mu = \$3,000$ and the alternative hypothesis is H_a: $\mu < \$3,000$. The population standard deviation is not known. We will assume that repair costs are normally distributed. (The hypothesis test for t is not very sensitive to small departures from normal distributions, so this is a reasonable assumption.) The value of the t test statistic is

$$t = \frac{\bar{x} - \mu}{s/\sqrt{n}} = \frac{2835 - 3000}{883/\sqrt{5}} = -0.418$$

For this example, the sample size is $n = 5$, so the number of degrees of freedom is $n - 1 = 5 - 1 = 4$. Because this is a one-tailed test, we find the critical values for significance at the 0.05 level in column 3. If you look at row 4 and column 3, you will find the critical value $t = 2.132$. For a *left*-tailed test, we are looking for values less than or equal to the negative of this value, or $t = -2.132$.

Because the sample test statistic $t = -0.418$ is *not* less than or equal to the critical value $t = -2.132$, we do not reject the null hypothesis. There is not sufficient evidence to support the claim that the mean repair cost is less than $3,000. With software, you would find that the *P*-value for the test statistic is 0.3488. In other words, if the null hypothesis is true, the probability of selecting a sample at least as extreme as the one found here is about 35%. This probability is not small enough to suggest that the null hypothesis should be rejected. $\cdots\bullet$

Section 10.1 Exercises

Statistical Literacy and Critical Thinking

1. **Sampling Method.** You obtain IQ scores from 15 of your fellow statistics classmates. Can you use the *t* distribution to test the claim that the mean IQ score of all college students is greater than 100? Explain.

2. **t Distribution.** In constructing a confidence interval estimate of a population mean or testing a hypothesis about a population mean, why is the *t* distribution used so much more often than the normal distribution?

3. **Terminology.** What is the difference between a Student *t* distribution and a *t* distribution?

4. **Sample Size.** You want to estimate the mean white blood cell count for the population of approximately 310 million people in the United States. Can you obtain a reasonable estimate using only 25 randomly selected people in the United States?

Does It Make Sense? For Exercises 5–8, decide whether the statement makes sense (or is clearly true) or does not make sense (or is clearly false). Explain clearly. Not all of these statements have definitive answers, so your explanation is more important than your chosen answer.

5. **t Test.** In testing a hypothesis about a population mean with a sample that has fewer than 30 values, the *t* distribution is always used.

6. **t Test.** In testing a claim about a population mean, the *t* distribution is always used when the population standard deviation σ is used.

7. **t Test.** Sample data consist of five annual incomes of movie stars, and we plan to use those data to test the claim that all movie stars have a mean annual income greater than $1 million. Because the sample is small and the population has a distribution that is far from normal, the *t* distribution should not be used.

8. **t vs. Normal Distribution.** Because the *t* test does not require a known value of the population standard deviation and because the value of the population standard deviation is rarely known, the *t* test is used in real situations much more often than the hypothesis test using the normal distribution.

Concepts and Applications

Confidence Intervals. In Exercises 9–16, use the *t* distribution to construct the confidence interval estimate of the population mean.

9. **IQ Scores.** A simple random sample of IQ scores is selected from a normally distributed population of statistics professors. The sample statistics are $n = 16$, $\bar{x} = 130$, $s = 10$. Construct the 95% confidence interval estimate of the population mean.

10. **Heights of NBA Players.** A simple random sample of heights of basketball players in the NBA is obtained, and the population has a distribution that is approximately normal. The sample statistics are $n = 16$, $\bar{x} = 77.9$ inches, $s = 3.50$ inches. Construct the 95% confidence interval estimate of the population mean.

11. **Elbow to Fingertip Length of Men.** A simple random sample of men is obtained, and the elbow to fingertip length of each man is measured. The population of those lengths has a distribution that is normal. The sample statistics are $n = 35$, $\bar{x} = 14.5$ inches, $s = 0.7$ inch. Construct the 95% confidence interval estimate of the population mean.

12. **SAT Scores.** A simple random sample of SAT scores is obtained, and the population has a distribution that is approximately normal. The sample statistics are $n = 41$, $\bar{x} = 1503$, $s = 352$. Construct the 95% confidence interval estimate of the population mean.

13. **Crash Hospital Costs.** A study was conducted to estimate hospital costs for accident victims who wore seat belts. Twenty randomly selected cases have a distribution that appears to be approximately bell-shaped with a mean of $9,004 and a standard deviation of $5,629 (based on data from the U.S. Department of Transportation). *(continued)*

a. Construct the 95% confidence interval for the mean of all such costs.

b. If you are a manager for an insurance company that provides lower rates for drivers who wear seat belts and you want a conservative estimate for a worst-case scenario, what amount should you use as the possible hospital cost for an accident victim who wears a seat belt?

14. **Forecast and Actual Temperatures.** One of the authors compiled a list of actual high temperatures and the corresponding list of three-day-forecast high temperatures. The difference for each day was then found by subtracting the three-day-forecast high temperature from the actual high temperature; the result was a list of 31 values with a mean of $-0.419°$ and a standard deviation of $3.704°$.

a. Construct a 95% confidence interval estimate of the mean difference between all actual high temperatures and three-day-forecast high temperatures.

b. Does the confidence interval include $0°$? If the confidence interval does include $0°$, can we conclude that the three-day-forecast temperatures are inaccurate?

15. **Estimating Car Pollution.** Each car in a sample of seven cars was tested for nitrogen-oxide emissions (in grams per mile), and the following results were obtained: 0.06, 0.11, 0.16, 0.15, 0.14, 0.08, 0.15 (based on data from the Environmental Protection Agency).

a. Assuming that this sample is representative of cars in use, construct a 95% confidence interval estimate of the mean amount of nitrogen-oxide emissions for all cars.

b. The Environmental Protection Agency requires that nitrogen-oxide emissions be less than 0.165 gram/mile. Someone claims that nitrogen-oxide emissions have a mean equal to 0.165 gram/mile. Does the confidence interval suggest that this claim is not valid? Why or why not?

16. **Movie Lengths.** Listed below are lengths (in minutes) of randomly selected movies.

110 96 125 94 132 120 136 154 149 94 119 132

a. Construct a 95% confidence interval estimate of the mean length of all movies.

b. Is it reasonable for a manager of a movie theater to plan for cleaning between movies by assuming that all movies run less than 130 minutes?

Hypothesis Tests. In Exercises 17–24, test the given claim.

17. **Sugar in Cereal.** A simple random sample of 16 different cereals is obtained, and the sugar content (in grams of sugar per gram of cereal) is measured for each cereal selected. Those amounts have a mean of 0.295 gram and a standard deviation of 0.168 gram. Use a 0.05 significance level to test the claim of a cereal lobbyist that the mean for all cereals is less than 0.3 gram.

18. **Testing Wristwatch Accuracy.** Students randomly selected 30 people and measured the accuracy of their wristwatches, with positive errors representing watches that were ahead of the correct time and negative errors representing watches that were behind the correct time. The 30 values have a mean of 117.3 seconds and a standard deviation of 185.0 seconds. Use a 0.05 significance level to test the claim that the population of all watches has a mean equal to 0 seconds. What can be concluded about the accuracy of people's wristwatches?

19. **Weights of Pennies.** The U.S. Mint has a specification that pennies have a mean weight of 2.5 g. Thirty randomly selected pennies have a mean weight of 2.49910 g and a standard deviation of 0.01648 g. Use a 0.05 significance level to test the claim that this sample is from a population with a mean weight equal to 2.5 g. Do pennies appear to conform to the specifications of the U.S. Mint?

20. **Reliability of Aircraft Radios.** The mean time between failures for a Telektronic Company radio used in light aircraft is 420 hours. After 15 new radios were modified in an attempt to improve reliability, tests were conducted to measure the times between failures. The 15 radios had a mean time between failures of 442 hours with a standard deviation of 44.0 hours. Use a 0.05 significance level to test the claim that modified radios have a mean time between failures that is greater than 420 hours. Does it appear that the modifications improved reliability?

21. **Effect of Vitamin Supplement on Birth Weight.** When birth weights were recorded for a simple random sample of 16 male babies born to mothers taking a special vitamin supplement, the sample had a mean of 3.675 kilograms and a standard deviation of 0.657 kilogram (based on data from the New York State Department of Health). Use a 0.05 significance level to test the claim that the mean birth weight for all male babies of mothers given vitamins is different from 3.39 kilograms, which is the mean for the population of all males. Based on these results, does the vitamin supplement appear to have an effect on birth weight?

22. **Pulse Rates.** One of the authors claimed that his pulse rate was lower than the mean pulse rate of statistics students. The author's pulse rate was measured and found to be 60 beats per minute, and the 20 students in his class measured their pulse rates. The 20 students had a mean pulse rate of 74.5 beats per minute, and their standard deviation was 10.0 beats per minute. Is there sufficient evidence to support the claim that the mean pulse rate of statistics students is greater than 60 beats per minute? Use a 0.05 significance level.

23. **Tests of Child Booster Seats.** The National Highway Traffic Safety Administration conducted crash tests of child booster seats for cars. Listed below are results from those tests, with the measurements given in "hic" (standard *head injury condition* units). The safety requirement is that the hic measurement should be less than 1,000 hic. Assume that the sample data are from a normally distributed population

and use a 0.05 signficance level to test the claim that the sample is from a population with a mean less than 1,000 hic. Do the results suggest that all of the child booster seats meet the specified requirement?

774	649	1210	546	431	612

24. **Nicotine in Cigarettes.** The Carolina Tobacco Company advertised that its best-selling nonfiltered cigarettes contain 40 milligrams of nicotine or less, but *Consumer Advocate* magazine ran tests of 10 randomly selected cigarettes and found the amounts (in milligrams) shown below.

47.3	39.3	40.3	38.3	46.3
43.3	42.3	49.3	40.3	46.3

Use a significance level of 0.05 to test the magazine's claim that the mean nicotine content is greater than 40 milligrams.

PROJECTS FOR THE INTERNET & BEYOND

25. **Noted Personalities.** The *World Almanac and Book of Facts* includes a section called "Noted Personalities," with subsections for architects, artists, business leaders, cartoonists, and several other categories. Select a sample from one group and find the mean and standard deviation of the life spans. Test the claim that the group has a mean life span that is different from 78 years, which is the current mean life span for the general population.

26. **Gosset.** The *t* distribution was originally developed by William Sealey Gosset. Use the Internet to search for "William Sealey Gosset" or "William S. Gosset." Write a paragraph describing important information about Gosset and his accomplishments.

10.2 HYPOTHESIS TESTING WITH TWO-WAY TABLES

The hypothesis tests we have considered so far have had null hypotheses in which a population mean (μ) or proportion (p) is claimed to be *equal* to some value. But there are many situations in which the null hypothesis takes a different form. In this section, we examine hypothesis tests designed to look for a relationship between two variables. The basic process is the same as always: Identify the null and alternative hypotheses, test the null hypothesis with sample data, and then decide whether the evidence from the sample supports rejecting or not rejecting the null hypothesis. If we reject the null hypothesis, it means this sample supports accepting the alternative hypothesis.

Identifying the Hypotheses with Two Variables

Suppose that administrators at a college are concerned that there may be gender bias in the way degrees are awarded, so they collect data on the number of degrees awarded to men and women in different departments. These data concern two variables: *major* and *gender*. The variable *major* can take on many values, such as biology, business, mathematics, and music. The variable *gender* can take on only two values: male or female.

To test whether there is bias in the awarding of degrees, the administrators ask the following question: *Do the data suggest a relationship between the two variables?* If there is a relationship, then men and women are choosing majors at different rates, suggesting that a person's gender somehow influences his or her choice of major (either by choice or because of bias within different departments). If there is no relationship, it means there is no evidence that gender influences a person's choice of major.

This idea suggests the following choices for the two hypotheses. The null hypothesis, H_0, states that the two variables are *independent* (there is *no relationship* between them); in our current example, it states that there is no relationship between gender and major. The alternative hypothesis, H_a, states the opposite: There *is* a relationship between the variables, which in this case implies that gender influences a person's choice of major.

Null and Alternative Hypotheses with Two Variables

The **null hypothesis**, H_0, states that the variables are independent (there is *no relationship* between them).

The **alternative hypothesis**, H_a, states that there *is* a relationship between the two variables.

Displaying the Data in Two-Way Tables

With the hypotheses identified, the next step in the hypothesis test is to examine the data set to see if it supports rejecting or not rejecting the null hypothesis. Collecting the data for our current example means finding the numbers of men and women awarded degrees in various majors. Once the data have been collected, we need to find an efficient way to display them. Because we are dealing with two variables, we can display the data efficiently with a **two-way table** (also called a **contingency table**), so named because it displays two variables.

Table 10.2 shows what the two-way table might look like for data on the variables *major* and *gender*. Note that one variable (*major*) is displayed along the columns and the other (*gender*) along the rows. For this example, there are only two rows because *gender* can be only either male or female; there are many columns for the majors, with just the first few shown. Each cell shows a frequency (or count) for one combination of the two variables. For example, the cell in row *Women* and column *Biology* shows that 32 bachelor's degrees were awarded to women in biology. Similarly, the cell in row *Men* and column *Business* shows that 87 bachelor's degrees were awarded to men in business.

variable 1 *major* →

TABLE 10.2	**Two-Way Table for the Variables *Major* and *Gender***				
	Biology	**Business**	**Mathematics**	**Psychology**	...
Women	32	110	18	75	...
Men	21	87	15	70	...

↑
variable 2 *gender*

> **Two-Way Tables**
>
> A **two-way table** shows the relationship between two variables by listing one variable in the rows and the other variable in the columns. The entries in the table's cells are called frequencies (or counts).

If we were looking for *any* relationship between major and gender, we would need a complete set of data for all majors, which means Table 10.2 would have dozens of columns. In addition, as we'll see shortly, carrying out the calculations for the hypothesis test requires that we find totals for all the rows and columns. Here, to simplify the calculations, let's focus on just two majors, biology and business. That is, instead of asking if there is a relationship between major and gender across all majors, we will look only at this simpler question: Does a person's gender influence whether he or she chooses to major in biology or business? Table 10.3 shows the biology and business data extracted from Table 10.2, along with row and column totals.

I cannot do it without counters.

—William Shakespeare,
The Winter's Tale

TABLE 10.3	**Two-Way Table for Biology and Business Degrees**		
	Biology	**Business**	**Total**
Women	32	110	142
Men	21	87	108
Total	53	197	250

BY THE WAY

Across all American colleges and universities, men outnumber women in declared majors of engineering, computer science, and architecture. Women outnumber men in psychology, fine art, accounting, biology, and elementary education.

> **TIME ⏱ OUT TO THINK**
>
> Use Table 10.3 to answer the following questions: (a) How many business degrees were awarded to men? (b) How many business degrees were awarded in total? (c) Compare the total number of degrees awarded to men and women to the total number of degrees awarded in business and biology. Are these totals the same or different? Why?

EXAMPLE 1   A Two-Way Table for a Survey

Table 10.4 shows the results of a pre-election survey on gun control. Use the table to answer the following questions.

TABLE 10.4	Two-Way Table for Gun Control Survey (with totals)			
	Favor stricter laws	Oppose stricter laws	Undecided	Total
Democrat	456	123	43	622
Republican	332	446	21	799
Total	788	569	64	1,421

Source: Adapted from Malcolm W. Browne, "Following Benford's Law, or Looking Out for No. 1," *New York Times*, August 4, 1998.

a. Identify the two variables displayed in the table.

b. What percentage of Democrats favored stricter laws?

c. What percentage of all voters favored stricter laws?

d. What percentage of those who opposed stricter laws are Republicans?

SOLUTION   Note that the total of the row totals and the total of the column totals are equal.

a. The columns show the variable *survey response*, which can be either "favor stricter laws," "oppose stricter laws," or "undecided." The rows show the variable *party affiliation*, which in this table can be either Democrat or Republican.

b. Of the 622 Democrats polled, 456 favored stricter laws. The percentage of Democrats favoring stricter laws is $456/622 = 0.733$, or 73.3%.

c. Of the 1,421 people polled, 788 favored stricter laws. The percentage of all respondents favoring stricter laws is $788/1,421 = 0.555$, or 55.5%.

d. Of the 569 people polled who opposed stricter laws, 446 are Republicans. Since $446/569 = 0.783$, 78.3% of those opposed to stricter laws are Republicans.   ⋯●

Carrying Out the Hypothesis Test

With the hypotheses identified and the data organized in the two-way table, we are now ready to carry out the hypothesis test. The basic idea is the same as always—to decide whether the data provide enough evidence to reject the null hypothesis. For the case of a hypothesis test with a two-way table, the specific steps are as follows:

- As always, we start by assuming that the null hypothesis is true, meaning there is no relationship between the two variables. In that case, we would expect the frequencies (the numbers in the individual cells) in the two-way table to be those that would occur by pure chance. Our first step, then, is to find a way to calculate the frequencies we would expect by chance.
- We next compare the frequencies expected by chance to the observed frequencies from the sample, which are the frequencies displayed in the table. We do this by calculating what is called the *chi-square statistic* (pronounced "ky-square") for the sample data, which here plays a role similar to the role of the standard score *z* in the hypothesis tests we carried out in Chapter 9 or the role of the *t* test statistic in Section 10.1.
- Recall that for the hypothesis tests in Chapter 9, we made the decision about whether to reject or not reject the null hypothesis by comparing the computed value of the standard score for the sample data to critical values given in tables; similarly, in Section 10.1 we compared computed values of the *t* test statistic to values found in a table. Here, we do the same thing, except rather than using critical values for the standard score or *t*, we use critical values for the chi-square statistic.

As an example of the process, let's work through these steps with the data in Table 10.3.

Finding the Frequencies Expected by Chance

Our first step is to find the frequencies we would expect in Table 10.3 *if there were no relationship* between the variables, which is equivalent to the frequency expected by chance alone. Let's start by finding the frequency we would expect by chance for male business majors. To do this, we first calculate the fraction of *all* students in the sample who received business degrees:

$$\frac{\text{total business degree}}{\text{total degrees}} = \frac{197}{250}$$

As discussed in Chapter 6, we can interpret this result as a *relative frequency probability*. That is, if we select a student *at random* from the sample, the probability that he or she earned a business degree is 197/250. Using the notation for probability, we write

$$P(\text{business}) = \frac{197}{250}$$

Similarly, if we select a student at random from the sample, the probability that this student is a man is

$$P(\text{man}) = \frac{\text{total men}}{\text{total men and women}} = \frac{108}{250}$$

We now have all the information needed to find the frequency we would expect by chance for male business majors. Recall from Section 6.5 that if two events A and B are independent (the outcome of one does not affect the probability of the other), then

$$P(A \text{ and } B) = P(A) \times P(B)$$

We can apply this rule to determine the probability that a student is *both* a man and a business major (assuming the null hypothesis that gender is independent of major):

$$P(\text{man and business}) = P(\text{man}) \times P(\text{business}) = \frac{108}{250} \times \frac{197}{250} \approx 0.3404$$

This probability is equivalent to the fraction of the total students whom we expect to be male business majors *if there is no relationship* between gender and major. We therefore multiply this probability by the total number of students in the sample (250) to find the number (or frequency) of male business majors that we expect by chance:

$$\frac{108}{250} \times \frac{197}{250} \times 250 \approx 85.104$$

We call this value the **expected frequency** for the number of male business majors. (Notice the similarity between the idea of expected frequency and that of *expected value* discussed in Section 6.3.)

> **Definition**
>
> The **expected frequencies** in a two-way table are the frequencies we would expect by chance *if there were no relationship* between the row and column variables.

EXAMPLE 2   Expected Frequencies for Table 10.3

Find the frequencies expected by chance for the three remaining cells in Table 10.3. Then construct a table showing both observed frequencies and frequencies expected by chance.

SOLUTION   We follow the procedure used above to find the expected number of male business majors. We already have P(man) and P(business). We'll also need P(woman) and P(biology):

$$P(\text{woman}) = \frac{\text{total women}}{\text{total men and women}} = \frac{142}{250} = 0.5680$$

$$P(\text{biology}) = \frac{\text{total biology degrees}}{\text{total degrees}} = \frac{53}{250} = 0.2120$$

Millions saw the apple fall,
but Newton was the one who
asked why.

—Bernard Baruch

We combine the individual probabilities to find the probability for each of the three remaining cells:

$$P(\text{woman and business}) = P(\text{woman}) \times P(\text{business}) = \frac{142}{250} \times \frac{197}{250} \approx 0.4476$$

$$P(\text{man and biology}) = P(\text{man}) \times P(\text{biology}) = \frac{108}{250} \times \frac{53}{250} \approx 0.0916$$

$$P(\text{woman and biology}) = P(\text{woman}) \times P(\text{biology}) = \frac{142}{250} \times \frac{53}{250} \approx 0.1204$$

Notice that, as we should expect, the total of the probabilities for all four cells is $0.3404 + 0.4476 + 0.09158 + 0.1204 = 1.0000$.

We now find the expected frequencies by multiplying the cell probabilities by the total number of students (250):

$$\text{Expected frequency of women business majors} = 250 \times \frac{142}{250} \times \frac{197}{250} \approx 111.896$$

$$\text{Expected frequency of men biology majors} = 250 \times \frac{108}{250} \times \frac{53}{250} \approx 22.896$$

$$\text{Expected frequency of women biology majors} = 250 \times \frac{142}{250} \times \frac{53}{250} \approx 30.104$$

Table 10.5 repeats the data from Table 10.3, but this time it also shows the expected frequency for each cell (in parentheses). To check that we did our work correctly, we confirm that the total of all four expected frequencies equals the total of 250 students in the sample:

$$85.104 + 111.896 + 22.896 + 30.104 = 250.000$$

Notice also that the values in the "Total" row and "Total" column are the same for both the observed frequencies and the frequencies expected by chance. This should always be the case, providing another good check on your work.

TECHNICAL NOTE

Table 10.5 includes expected frequencies rounded to three decimal places, but it is better to round to as many decimal places as your calculator can carry. Final answers can vary slightly depending on the rounding method used.

TABLE 10.5	Observed Frequencies and Expected Frequencies (in parentheses) for Table 10.3		
	Biology	**Business**	**Total**
Women	32 (30.104)	110 (111.896)	142 (142.000)
Men	21 (22.896)	87 (85.104)	108 (108.000)
Total	53 (53.000)	197 (197.000)	250 (250.000)

Source: Adapted from Malcolm W. Browne, "Following Benford's Law, or Looking Out for No. 1," *New York Times*, August 4, 1998.

Computing the Chi-Square Statistic

Notice that the expected frequencies in Table 10.5 appear to agree fairly well with the observed frequencies. For example, the expected frequency of about 85.1 for male business majors is quite close to the observed frequency of 87. We might therefore already guess that the data do not give us any reason to reject the null hypothesis of no relationship between the variables *gender* and *major*. However, we can be more specific by finding a way to quantify the difference between the observed and expected frequencies.

Let's denote the observed frequencies by O and the expected frequencies by E. With this notation, $O - E$ ("O minus E") tells us the difference between the observed frequency and the expected frequency for each cell. We are looking for a measure of the total difference for the whole table. We cannot get such a measure by simply adding the individual differences, $O - E$, because they always sum to zero. Instead, we consider the square of the difference in each cell, $(O - E)^2$. We then make each value of $(O - E)^2$ a relative difference by dividing it by the corresponding expected frequency; this gives us the quantity $(O - E)^2/E$ for each cell. Summing the individual values of $(O - E)^2/E$ gives us the **chi-square statistic**, denoted χ^2 (χ is the Greek letter chi)

Finding the Chi-Square Statistic

Step 1. For each cell in the two-way table, identify O as the observed frequency and E as the expected frequency if the null hypothesis is true (no relationship between the variables).

Step 2. Compute the value $(O - E)^2/E$ for each cell.

Step 3. Sum the values from step 2 to get the chi-square statistic:

$$\chi^2 = \text{sum of all values } \frac{(O - E)^2}{E}$$

The larger the value of χ^2, the greater the average difference between the observed and expected frequencies in the cells.

To do this calculation in an organized way, it's best to make a table such as Table 10.6, with a row for each of the cells in the original two-way table. As shown in the lower right cell, the result for the gender/major data is $\chi^2 = 0.350$.

TABLE 10.6	Calculation of χ^2 Statistic for Data in Table 10.5				
Outcome	O	E	$O - E$	$(O - E)^2$	$(O - E)^2/E$
Women/business	110	111.896	−1.896	3.595	0.032
Women/biology	32	30.104	1.896	3.595	0.119
Men/business	87	85.104	1.896	3.595	0.042
Men/biology	21	22.896	−1.896	3.595	0.157
Totals	250	250.000	0.000	14.380	$\chi^2 = 0.350$

TIME OUT TO THINK

Why must the numbers in the $O - E$ column always sum to zero?

Making the Decision

The value of χ^2 gives us a way of testing the null hypothesis of no relationship between the variables. If χ^2 is small, then the average difference between the observed and expected frequencies is small and we should *not* reject the null hypothesis. If χ^2 is large, then the average difference between the observed and expected frequencies is large and we have reason to reject the null hypothesis of independence. To quantify what we mean by "small" or "large," we compare the χ^2 value found for the sample data to critical values:

- If the calculated value of χ^2 is *less than* the critical value, the differences between the observed and expected values are small and there is *not* enough evidence to reject the null hypothesis.
- If the calculated value of χ^2 is *greater than or equal to* the critical value, then there is enough evidence in the sample to reject the null hypothesis (at the given level of significance).

Table 10.7 (on the next page) gives the critical values of χ^2 for two significance levels, 0.05 and 0.01. Notice that the critical values differ for different table sizes, so you must make sure you read the critical values for a data set from the appropriate table size row. For the gender/major data we have been studying in Tables 10.3 and 10.5, there are two rows and two columns (do not count the "total" rows or columns), which means a table size of 2×2. Looking in the first row of Table 10.7, we see that the critical value of χ^2 for significance at the 0.05 level is 3.841. The chi-square value that we found for the gender/major data is $\chi^2 = 0.350$; because this is less than the critical value of 3.841, we cannot reject the null hypothesis. Of course, failing to reject the null hypothesis does not *prove* that major and gender are independent. It simply means that we do not have enough evidence to justify rejecting the null hypothesis of independence.

TECHNICAL NOTE

The χ^2 test statistic is technically a discrete variable, whereas the actual χ^2 distribution is continuous. That discrepancy does not cause any substantial problems as long as the expected frequency for every cell is at least 5. We will assume that this condition is met for all the examples in this text.

TABLE 10.7	Critical Values of χ^2: Reject H_0 Only If $\chi^2 >$ Critical Value	
Table size (rows × columns)	Significance level	
	0.05	0.01
2 × 2	3.841	6.635
2 × 3 or 3 × 2	5.991	9.210
3 × 3	9.488	13.277
2 × 4 or 4 × 2	7.815	11.345
2 × 5 or 5 × 2	9.488	13.277

EXAMPLE 3   Vitamin C Test

A (hypothetical) study seeks to determine whether vitamin C has an effect in preventing colds. Among a sample of 220 people, 105 randomly selected people took a vitamin C pill daily for a period of 10 weeks and the remaining 115 people took a placebo daily for 10 weeks. At the end of 10 weeks, the number of people who got colds was recorded. Table 10.8 summarizes the results. Determine whether there is a relationship between taking vitamin C and getting colds. Use a 0.01 significance level.

TABLE 10.8	Two-Way Table for Observed Number in Each Category		
	Cold	No cold	Total
Vitamin C	45	60	105
Placebo	75	40	115
Total	120	100	220

SOLUTION   We begin by stating the null and alternative hypotheses.

H_0 (null hypothesis): There is no relationship between taking vitamin C and getting colds; that is, vitamin C has no more effect on colds than the placebo.

H_a (alternative hypothesis): There is a relationship between taking vitamin C and getting colds; that is, the numbers of colds in the two groups are not what we would expect if vitamin C and the placebo were equally effective (or equally ineffective).

As always, we assume that the null hypothesis is true and calculate the expected frequency for each cell in the table. Noting that the sample size is 220 and proceeding as in Example 2, we find the following expected frequencies:

$$\textit{Vitamin C and cold:} \quad 220 \times \underbrace{\frac{105}{220}}_{P(\text{vit. C})} \times \underbrace{\frac{120}{220}}_{P(\text{cold})} = 57.273$$

$$\textit{Vitamin C and no cold:} \quad 220 \times \underbrace{\frac{105}{220}}_{P(\text{vit. C})} \times \underbrace{\frac{100}{220}}_{P(\text{no cold})} = 47.727$$

$$\textit{Placebo and cold:} \quad 220 \times \underbrace{\frac{115}{220}}_{P(\text{placebo})} \times \underbrace{\frac{120}{220}}_{P(\text{cold})} = 62.727$$

$$\textit{Placebo and no cold:} \quad 220 \times \underbrace{\frac{115}{220}}_{P(\text{placebo})} \times \underbrace{\frac{100}{220}}_{P(\text{no cold})} = 52.273$$

Table 10.9 shows the two-way table with the expected frequencies in parentheses.

TABLE 10.9	Observed and Expected Frequencies for Vitamin C Study		
	Cold	No cold	Total
Vitamin C	45 (57.273)	60 (47.727)	105 (105.000)
Placebo	75 (62.727)	40 (52.273)	115 (115.000)
Total	120 (120.000)	100 (100.000)	220 (220.000)

We now compute the chi-square statistic for the sample data. Table 10.10 shows how we organize the work; you should confirm all the calculations shown.

TABLE 10.10	Table for Computing χ^2 Statistic for Vitamin C Study				
Outcome	O	E	$O - E$	$(O - E)^2$	$(O - E)^2/E$
Vitamin C/cold	45	57.273	−12.273	150.627	2.630
Vitamin C/no cold	60	47.727	12.273	150.627	3.156
Placebo/cold	75	62.727	12.273	150.627	2.401
Placebo/no cold	40	52.273	−12.273	150.627	2.882
Totals	220	220.000	0.000	602.508	$\chi^2 = 11.069$

To make the decision about whether to reject the null hypothesis, we compare the value of chi-square for the sample data, $\chi^2 = 11.069$, to the critical values from Table 10.7. We look in the row for a table size of 2×2, because the original data in Table 10.8 have two rows and two columns (not counting the "total" values). We see that the critical value of χ^2 for significance at the 0.01 level is 6.635. Because our sample value of $\chi^2 = 11.069$ is greater than this critical value, we reject the null hypothesis and conclude that there is a relationship between vitamin C and colds. That is, based on the data from this sample, there is reason to believe that vitamin C *does* have more effect on colds than a placebo. $\cdots\bullet$

EXAMPLE 4   To Plead or Not to Plead

The two-way table in Table 10.11 shows how a plea of guilty or not guilty affected the sentence in 1,028 randomly selected burglary cases in the San Francisco area. Test the claim that the sentence (prison or no prison) is independent of the plea.

TABLE 10.11	Observed Frequencies for Plea and Sentence		
	Prison	No prison	Total
Guilty plea	392	564	956
Not-guilty plea	58	14	72
Total	450	578	1,028

Source: Law and Society Review, Vol. 16, No. 1.

SOLUTION   The null and alternative hypotheses for the problem are

H_0 (null hypothesis): The sentence in burglary cases is independent of the plea.

H_a (alternative hypothesis): The sentence in burglary cases depends on the plea.

We find the following *expected* number of people in each category, assuming that the row variables are independent of the column variables:

$$\text{Guilty and prison:} \quad 1{,}028 \times \frac{956}{1{,}028} \times \frac{450}{1{,}028} = 418.482$$

$$\text{Guilty and no prison:} \quad 1{,}028 \times \frac{956}{1{,}028} \times \frac{578}{1{,}028} = 537.518$$

$$\text{Not guilty and prison:} \quad 1{,}028 \times \frac{72}{1{,}028} \times \frac{450}{1{,}028} = 31.518$$

$$\text{Not guilty and no prison:} \quad 1{,}028 \times \frac{72}{1{,}028} \times \frac{578}{1{,}028} = 40.482$$

BY THE WAY

Dozens of careful studies have been conducted on the question of vitamin C and colds. Some have found high levels of confidence in the effects of vitamin C, but others have not. Because of these often-conflicting results, the issue of whether vitamin C helps to prevent colds remains controversial.

Table 10.12 summarizes the observed frequencies and expected frequencies.

TABLE 10.12	Observed Frequencies and Frequencies Expected by Chance (in parentheses) for Plea and Sentence		
	Prison	**No prison**	**Total**
Guilty	392 (418.482)	564 (537.518)	956
Not guilty	58 (31.518)	14 (40.482)	72
Total	450	578	1,028

Source: *Law and Society Review*, Vol. 16, No. 1.

As usual, we calculate χ^2 by organizing our work as shown in Table 10.13, where O denotes observed frequency and E denotes expected frequency. As shown in the lower right cell, the chi-square statistic for these data is $\chi^2 = 42.556$. This value is much greater than the critical values (for a 2 $\times$ 2 table) for significance at both the 0.05 level ($\chi^2 = 3.841$) and the 0.01 level ($\chi^2 = 6.635$). The data therefore support rejecting the null hypothesis and accepting the alternative hypothesis. Based on these data, there is reason to believe that the sentence given in a burglary case is associated with the plea. Specifically, of the people who pled guilty, fewer actually went to prison than expected (by chance) and more avoided prison than expected. Of the people who pled not guilty, more actually went to prison than expected and fewer avoided prison than expected. Remember that the test does not prove a causal relationship between the plea and the sentence.

TABLE 10.13	Calculation of χ^2				
Outcome	**O**	**E**	**O − E**	**(O − E)²**	**(O − E)²/E**
Guilty/prison	392	418.482	−26.482	701.296	1.676
Guilty/no prison	564	537.518	26.482	701.296	1.305
Not guilty/prison	58	31.518	26.482	701.296	22.251
Not guilty/no prison	14	40.482	−26.482	701.296	17.324
Totals	**1,028**	**1,028.000**	**0.000**	**2805.184**	$\chi^2 = 42.556$

TIME OUT TO THINK

If you were the lawyer for a burglary suspect, how might the results of the previous example affect your strategy in defending your client? Explain.

⏻ USING TECHNOLOGY—HYPOTHESIS TESTING WITH TWO-WAY TABLES

EXCEL Use XLSTAT. First enter the contingency table in rows and columns. Click on **XLSTAT** at the top. Select **Correlation/Association tests**, then select **Test on contingency table**. In the "Contingency table" box, enter the range of cells containing the frequency counts of the contingency table. For example, enter A1:B4 for a contingency table with two columns (A and B) and four rows. For the "Data format" select the **Contingency Table** option. Click on the **Options** tab, leave a check mark next to "Chi-square test," and enter a value for "Significance level (%)." For example, enter 5 for a 0.05 significance level. Click **OK** and results including the chi-square test statistic and *P*-value will be displayed.

STATDISK Enter the observed frequencies in the Data Window as they appear in the contingency table. Select **Analysis** from the main menu, then select **Contingency Tables**. Enter a significance level and proceed to identify the columns containing the frequencies. Click on **Evaluate**. The STATDISK results include the test statistic, critical value, and *P*-value.

TI-83/84 PLUS First enter the contingency table as a matrix by pressing [2ND] [x⁻¹] to get the MATRIX menu (or the MATRIX key on the TI-83). Select EDIT, and press [ENTER]. Enter the dimensions of the matrix (rows by columns) and proceed to enter the individual frequencies. When finished, press [STAT], select TESTS, and then select the option. χ^2-Test. Be sure that the observed matrix is the one you entered, such as matrix A. The expected frequencies will be automatically calculated and stored in the separate matrix identified as "Expected." Scroll down to Calculate and press [ENTER] to get the test statistic, *P*-value, and number of degrees of freedom.

Statistical Literacy and Critical Thinking

In Exercises 1–4, refer to the following data describing a clinical trial of Chantix, a drug used as an aid for those who want to stop smoking. Assume that we want to use a 0.05 significance level to test the claim that nausea is independent of whether the subject took a placebo or Chantix.

	Placebo	Chantix
Nausea	10	30
No nausea	795	791

1. **Two-Way Tables.** What is it about the given table that makes it a *two-way table?*

2. **Expected Frequency.** What notation is used for an expected frequency? Find the value of the expected frequency in the cell with an entry of 10. Describe what this expected value represents.

3. **Interpreting Results.** The results from an analysis of the table entries cause us to reject the claim that nausea is independent of whether the subject took a placebo or Chantix. Can we conclude that getting nausea is caused by the treatment?

4. **Sampling Method.** If the subjects in the clinical trial were volunteers who responded to a newspaper advertisement seeking volunteers who would be paid $500 for their participation, would that affect the results?

Does It Make Sense? For Exercises 5–8, decide whether the statement makes sense (or is clearly true) or does not make sense (or is clearly false). Explain clearly. Not all of these statements have definitive answers, so your explanation is more important than your chosen answer.

5. **Survey.** In a health exam survey, subjects are weighed. The weights are recorded in two separate lists according to gender. The results are summarized in a two-way table.

6. χ^2 **Test Statistic.** A two-way table is used to calculate the test statistic, and the value $\chi^2 = -2.500$ is obtained.

7. χ^2 **Test Statistic.** In a two-way table, all of the observed frequencies are very close to the expected frequencies, so the χ^2 test statistic is very small, and we fail to reject the null hypothesis of independence between the row and column variables.

8. **Null Hypothesis.** Assume that a two-way table is configured so that *gender* (male/female) represents the row variable and *response* (yes/no) to a survey question represents the column variable. The null hypothesis is the statement that gender and response are independent.

Concepts and Applications

Survey Results. In Exercises 9–12, assume that a yes/no survey question is presented to a simple random sample of male and female subjects and the results are summarized in a two-way table with the format of the table below. Use the given value of the χ^2 test statistic and the given significance level to test for independence between gender and response.

	Yes	No
Female		
Male		

9. **Test statistic:** $\chi^2 = 3.499$; significance level: 0.05

10. **Test statistic:** $\chi^2 = 3.957$; significance level: 0.05

11. **Test statistic:** $\chi^2 = 12.336$; significance level: 0.01

12. **Test statistic:** $\chi^2 = 3.849$ significance level: 0.01

Complete Hypothesis Test. In Exercises 13–20, carry out the following steps.

a. State the null and alternative hypotheses.

b. Assuming independence between the two variables, find the expected frequency for each cell of the table.

c. Find the value of the χ^2 test statistic.

d. Use the given significance level to find the χ^2 critical value.

e. Using the given significance level, complete the test of the claim that the two variables are independent. State the conclusion that addresses the original claim.

13. **Testing a Lie Detector.** The table below includes results from polygraph (lie detector) experiments conducted by researchers Charles R. Honts (Boise State University) and Gordon H. Barland (Department of Defense Polygraph Institute). In each case, it was known if the subject lied or did not lie, so the table indicates when the polygraph test was correct. Use a 0.05 significance level to test the claim that whether a subjects lies is independent of the polygraph test indication. Do the results suggest that polygraphs are effective in distinguishing between truths and lies?

	Did the Subject Actually Lie?	
	No (Did Not Lie)	Yes (Lied)
Polygraph test indicated that the subject *lied*	15	42
Polygraph test indicated that the subject did *not lie*	32	9

14. Voter Turnout. The following table shows the number of citizens in a sample who voted in the last presidential election, according to gender (consistent with national population data). Use a 0.05 significance level to test the claim that gender is independent of voter turnout.

	Voted	Did not vote
Women	140	120
Men	130	110

15. E-mail and Privacy. Workers and senior-level bosses were asked if it was seriously unethical to monitor employee e-mail; the results are summarized in the table below (based on data from a Gallup poll). Use a 0.05 significance level to test the claim that the response is independent of whether the subject is a worker or a senior-level boss.

	Yes	No
Workers	192	244
Bosses	40	81

16. Is the Vaccine Effective? In a *USA Today* article about an experimental vaccine for children, the following statement was presented: "In a trial involving 1,602 children, only 14 (1%) of the 1,070 who received the vaccine developed the flu, compared with 95 (18%) of the 532 who got a placebo." The data are shown in the table below. Use a 0.05 significance level to test for independence between the variable of treatment (vaccine or placebo) and the variable representing flu (developed flu, did not develop flu). Does the vaccine appear to be effective?

	Developed Flu?	
	Yes	No
Vaccine treatment	14	1,056
Placebo	95	437

17. Arthritis Treatment. Of the 98 participants in a drug trial who were given a new experimental treatment for arthritis, 56 showed improvement. Of the 92 participants given a placebo, 49 showed improvement. Construct a two-way table for these data, and then use a 0.05 significance level to test the claim that improvement is independent of whether the participant was given the drug or a placebo.

18. Drinking and Pregnancy. A simple random sample of 1,252 pregnant women under the age of 25 includes 13 who were drinking alcohol during their pregnancy. A simple random sample of 2,029 pregnant women of age 25 and over includes 37 who were drinking alcohol during their pregnancy. (The data are based on results from the U.S. National Center for Health Statistics.) Use a 0.05 significance level to test the claim that the age category (under 25 and 25 or over) is independent of whether the pregnant woman was drinking during pregnancy.

19. Crime and Strangers. The table below lists survey results obtained from a random sample of different crime victims (based on data from the U.S. Department of Justice). Use a 0.01 significance level to test the claim that the type of crime is independent of whether the criminal was a stranger.

	Homicide	Robbery	Assault
Criminal was a stranger	12	379	727
Criminal was acquaintance or relative	39	106	642

20. Smoking in China. The table below summarizes results from a survey of males ages 15 or older living in the Minhang District of China (based on data from "Cigarette Smoking in China" by Gong, Koplan, Feng, et al., *Journal of the American Medical Association*, Vol. 274, No. 15). The males are categorized by their current educational status and whether they smoke. Using a 0.05 significance level, test the claim that smoking is independent of education level.

	Primary school	Middle school	College
Smoker	606	1234	100
Never smoked	205	505	137

PROJECTS FOR THE INTERNET & BEYOND

21. Constructing Two-Way Tables. Choose two variables that appear to have a relationship that is worth investigating. One variable should have at least two categories of individuals—for example, two or more age categories, racial categories, or geographical locations. The other variable should have at least two categories for some social, economic, or health factor—for example, two or more income categories, drinking categories, or educational attainment categories. Find the required population or sample data needed to fill in a two-way table for the two variables. Discuss whether there appears to be a relationship between the variables. A good place to start is the Web site for the *Statistical Abstract of the United States* of the U.S. Census Bureau.

22. Analyzing Two-Way Tables. Choose two variables that appear to have a relationship that is worth investigating. One variable should have at least two categories of individuals—for example, two or more age categories, racial categories, or geographical locations. The other variable should have at least two categories for some social, economic, or health factor—for example, two or more income categories, drinking categories, or educational attainment categories. Find the frequency data needed to fill in a two-way table for the two variables. Carry out a hypothesis test to determine whether there is a relationship between the variables. A good data source is the Web site for the *Statistical Abstract of the United States* of the U.S. Census Bureau.

IN THE NEWS

23. **Two-Way Tables in the News.** It's unusual (but not impossible) to see a two-way table in a news article. But often a news story provides information that could be expressed in a two-way table. Find an article that discusses a relationship between two variables that could be expressed in a two-way table. Create the table.

24. **Hypothesis Testing in the News.** News reports often describe results of statistical studies in which the conclusions came from a hypothesis test involving two-way tables. However, the reports rarely give the actual table or describe the details of the hypothesis test. Find a recent news report in which you think the conclusions *probably* were based on a hypothesis test with a two-way table. Assuming you are correct, describe in words how the hypothesis test probably worked. That is, describe the null and alternative hypotheses and the procedure by which the researchers probably carried out their test.

25. **Your Own Hypothesis Test.** Think of an example of something you'd like to know that could be tested with a hypothesis test on a two-way table. Without actually collecting data or doing any calculations, describe how you would go about conducting your study. That is, describe how you would collect the data, explain how you would organize them into a two-way table, state the null and alternative hypotheses that would apply, and describe how you would conduct the hypothesis test and reach a conclusion.

10.3 ANALYSIS OF VARIANCE (ONE-WAY ANOVA)

So far we have examined hypothesis testing with three different types of claims for the null hypothesis: claims that a population mean equals some value (H_0: μ = claimed value), which we examined with a normal distribution in Section 9.2 and with the t distribution in Section 10.1; claims that a population proportion equals some value (H_0: p = claimed value), which we examined in Section 9.3; and claims that two variables are independent of each other (H_0: no relationship), which we discussed in Section 10.2. Statisticians have developed techniques for considering many other types of null hypotheses, making it possible to apply statistics to an incredible range of applications. To give you a taste of what is possible with statistics—and perhaps to encourage you to study statistics further—we briefly consider one more type of hypothesis testing in this final section of the text.

Hypothesis Testing for Equal Means

A simple random sample of 12 pages was obtained from each of three different books: Tom Clancy's *The Bear and the Dragon*, J. K. Rowling's *Harry Potter and the Sorcerer's Stone*, and Leo Tolstoy's *War and Peace*. The Flesch Reading Ease score was obtained for each of those pages, and the results are listed in Table 10.14. The Flesch Reading Ease scoring system results in *higher* scores for text that is *easier* to read. Low scores are associated with works that are difficult to read. Our goal in this section is to use these sample data from just 12 pages of each book to make inferences about the readability of the *population* of all pages in each book.

We can informally explore the sample data by investigating center, variation, distribution, and outliers. Table 10.15 shows the important sample statistics for our case. If you compare the original data in Table 10.14 and the sample means in Table 10.15, you'll notice that although a few scores are farther from the mean than most others (such as the lowest Clancy score of 43.9 and the lowest Rowling score of 70.9), no values seem so extreme that we would consider them outliers. Moreover, detailed study of the data suggests that the samples come from populations with distributions that are close to normal.

Even before studying the data, we might expect Rowling's book to be the easiest to read of the three books because it is the only one written for children. Similarly, we might

TABLE 10.14	Flesch Reading Ease Scores	
Clancy	**Rowling**	**Tolstoy**
58.2	85.3	69.4
73.4	84.3	64.2
73.1	79.5	71.4
64.4	82.5	71.6
72.7	80.2	68.5
89.2	84.6	51.9
43.9	79.2	72.2
76.3	70.9	74.4
76.4	78.6	52.8
78.9	86.2	58.4
69.4	74.0	65.4
72.9	83.7	73.6

TABLE 10.15	Statistics for Readability Scores		
	Flesch Reading Ease Score		
	Clancy	**Rowling**	**Tolstoy**
Sample size *n*	12	12	12
Sample mean $\bar{X}$	70.73	80.75	66.15
Sample standard deviation *s*	11.33	4.68	7.86

expect Tolstoy's book to be the most difficult because it is a translation of a Russian classic. Now look at the mean readability scores in Table 10.15. Recalling that a higher Flesch score indicates an easier reading level, the data appear to support our expectations: Rowling has the highest readability score and Tolstoy has the lowest. Still, the three sample means are not wildly different, ranging only from 66.15 for Tolstoy to 80.75 for Rowling, and the sample size is a relatively small $n = 12$ for each case. We therefore arrive at our key statistical question for this section: Do these sample data provide sufficient evidence for us to conclude that the books by Clancy, Rowling, and Tolstoy really do have different mean Flesch scores?

To answer this question, we follow the same general principles laid out for hypothesis testing in Section 9.1. We begin by identifying the null hypothesis. Because we want to know whether the three books really do have different mean Flesch scores, we start with the assumption that they do *not* have different means. In other words, our null hypothesis is that the mean Flesch scores for all three books are equal. The alternative hypothesis, then, is that at least one of the three population means is different from the others. The hypothesis test must tell us whether to reject or not reject the null hypothesis. Rejecting the null hypothesis would allow us to conclude that the books have mean Flesch scores that are not all the same, as we expect. Not rejecting the null hypothesis would tell us that the data do not provide sufficient evidence for concluding that the mean Flesch scores are not all the same.

Remember that for this example each population mean (μ) represents the mean Flesch score we would obtain if we measured the score for *all* the pages in each book. Using more formal notation, we can therefore write the null hypothesis as

$$H_0: \mu_{\text{Clancy}} = \mu_{\text{Rowling}} = \mu_{\text{Tolstoy}}$$

As you can see, we need a hypothesis test that will allow us to determine whether three different populations have the same mean. The method we use is called **analysis of variance**, commonly abbreviated **ANOVA**. The name comes from the formal statistic known as the *variance* of a set of sample values; as we noted briefly in Section 4.3, variance is defined as the square of the sample standard deviation, or s^2. (For example, if a sample of heights has a standard deviation $s = 3.0$ cm, its variance is $s^2 = 9.0$ cm^2.)

TECHNICAL NOTE

The method of analysis of variance can be used with two means, but it is equivalent to a *t* test that pools the two sample variances. There is another *t* test that can be used with two independent samples, and it generally performs better. Neither of these *t* tests is included in this text.

Definition

Analysis of variance (ANOVA) is a method of testing the equality of three or more population means by analyzing sample variances.

More specifically, the method used to analyze data like those from Table 10.14 is called *one-way* analysis of variance (one-way ANOVA), because the sample data are separated into groups according to just *one* characteristic or factor. In this example, the characteristic is the author (Clancy, Rowling, or Tolstoy). There is also a method referred to as *two-way analysis of variance* that allows comparisons among populations separated into categories by two characteristics. For example, we might separate heights of people using the following two characteristics: (1) gender (male or female) and (2) right- or left-handedness. We do not consider two-way analysis of variance in this text.

Conducting the Test

Analysis of variance is based on this fundamental concept: We *assume* that the populations all have the same variance, and we then compare the variance *between* the samples to the variance *within* the samples. More specifically, the test statistic (usually called F) for one-way analysis of variance is the ratio of those two variances:

$$\text{test statistic } F \text{ (for one-way ANOVA)} = \frac{\text{variance between samples}}{\text{variance within samples}}$$

The actual calculation of this test statistic is tedious, and these days it is almost always done with statistical software (see Using Technology at the end of this section). However, we can interpret the statistic as follows, using our example of the readability of the three books:

- The variance *between* samples is a measure of how much the three sample means (from Table 10.15) differ from one another.
- The variance *within* samples is a measure of how much the Flesch Reading Ease scores for the 12 pages in each individual sample (from Table 10.14) differ from one another.
- If the three population means were really all equal—as the null hypothesis claims—then we would expect the sample mean from any one individual sample to fall well within the range of variation for any other individual sample. The test statistic (F = variance between samples/variance within samples) tells us whether that is the case:

  A large test statistic F tells us that the sample means differ *more* than the data within the individual samples, which would be *unlikely* if the populations means really were equal (as the null hypothesis claims). That is, a large F test statistic provides evidence for rejecting the null hypothesis that the population means are equal.

  A small test statistic F tells us that the sample means differ *less* than the data within the individual samples, suggesting that the difference among the sample means could easily have arisen by chance. Therefore, a small test statistic F does not provide evidence for rejecting the null hypothesis that the population means are equal.

Notice that the test statistic F for analysis of variance plays a role similar to that of the standard score z or the t test statistic in hypothesis tests we considered earlier. Therefore, just as we did in those earlier cases, we quantify the interpretation of the F test statistic by finding its P-value, which tells us the probability of getting sample results at least as extreme as those obtained, assuming that the null hypothesis is true (the population means are all equal). A small P-value shows that it is unlikely that we would get the sample results by chance with equal population means. A large P-value shows that we could easily get the sample results by chance with equal population means. Here is a memory device to help with interpretation of the P-value: "If the P (value) is low, the null must go." This means that if the P-value is very small, such as less than or equal to 0.05, then the null hypothesis of equal means should be rejected.

The following box summarizes the requirements for one-way analysis of variance and the software-aided procedure outlined in this section.

One-Way ANOVA for Testing H_0: $\mu_1 = \mu_2 = \mu_3 = \ldots$

Step 1. Enter sample data into a statistical software package, and use the software to determine the test statistic (F = variance between samples/variance within samples) and the P-value of the test statistic.

Step 2. Make a decision to reject or not reject the null hypothesis based on the P-value of the test statistic:

- If the P-value is less than or equal to the significance level, reject the null hypothesis of equal means and conclude that at least one of the means is different from the others.

- If the P-value is greater than the significance level, do not reject the null hypothesis of equal means.

This method is valid as long as the following requirements are met: The populations have distributions that are approximately normal with the same variance, and the samples from each population are simple random samples that are independent of each other.

USING TECHNOLOGY—**ANALYSIS OF VARIANCE**

EXCEL You can use either XLSTAT or Excel's Data Analysis add-in. An advantage of the Data Analysis add-in is that you are not required to stack all of the data in one column with corresponding category names in another column.

XLSTAT: First stack all of the sample data in column B with the corresponding variable names listed in column A. Click on **XLSTAT,** then select **Modeling Data,** then select **ANOVA.** In the "Quantitative" box, enter the range of cells containing the sample data, such as B1:B50. In the Qualitative box, enter the range of cells containing the variable names, such as A1:A50. Put a check next to the "Variable labels" box only if the first row consists of labels. Click **OK.** In the results, look for the "Analysis of Variance" table that includes the *F* test statistic and the *P*-value.

Data Analysis add-in: Enter the data in columns A, B, C, . . . In Excel 2013, 2010, and 2007, click on **Data;** in Excel 2003, click on **Tools.** Now click on **Data Analysis** and select **Anova: Single Factor.** In the dialog box, enter the range containing the sample data. (For example, enter A1:C30 if the first value is in row 1 of column A, and the longest column has 30 data values.)

STATDISK: Enter the data in columns of the data window. Select **Analysis** from the main menu bar, then select **One-Way Analysis of Variance,** and proceed to select the columns of sample data. Click on **Evaluate** when you are done. The results will include the *P*-value for the analysis of variance test.

TI-83/84 PLUS First enter the data as lists in L1, L2, L3 . . . then press **STAT**, select **TESTS,** and choose the option **ANOVA.** Enter the column labels. For example, if the data are in columns L1, L2, and L3, enter those columns to get **ANOVA (L1, L2, L3),** and press **ENTER**.

EXAMPLE ❶ Readability of Clancy, Rowling, Tolstoy

Given the readability scores listed in Table 10.14 and a significance level of 0.05, test the null hypothesis that the three samples come from populations with means that are all the same.

SOLUTION   We begin by checking the requirements for using one-way analysis of variance. As noted earlier, close examination of the data suggests that each sample comes from a distribution that is approximately normal. The sample standard deviations are not dramatically different, so it is reasonable to assume that the three populations have the same variance. The samples are simple random samples and they are all independent. The requirements are therefore satisfied.

We now test the null hypothesis that the population means are all equal (H_0: $\mu_1 = \mu_2 = \mu_3$). The Using Technology feature at the end of this section describes how to obtain the test statistic and *P*-value with various software packages. The table below shows the resulting display from Excel; other software packages will give similar displays.

Source of Variation	*SS*	*df*	*MS*	*F*	*P-value*	*F crit*
Between Groups	1338.002222	2	669.0011111	9.469487401	0.000562133	3.284924333
Within Groups	2331.386667	33	70.64808081			
Total	3669.388889	35				

Notice that the display includes columns for *F* and for the *P*-value. These are the two items of interest to us here, which we interpret as follows:

- *F* is the test statistic for the one-way analysis of variance (*F* = variance between samples/variance within samples). Notice that it is much greater than 1, indicating that the sample means differ more than we would expect if all the population means were equal.
- The *P*-value tells us the probability of having obtained such an extreme result by chance if the null hypothesis is true. Notice that the *P*-value is extremely small—much less than the value of 0.05 necessary to reject the null hypothesis at the 0.05 level of significance (and also much less than the 0.01 necessary to reject at the 0.01 level of significance).

We conclude that there is sufficient evidence to reject the null hypothesis, which means the sample data support the claim that the three population means are not all the same. Based on randomly selected pages from Clancy's *The Bear and the Dragon,* Rowling's *Harry Potter and the Sorcerer's Stone*, and Tolstoy's *War and Peace*, we conclude that those books have readability levels that are not all the same. Note that we have *not* concluded that the three books have the readability order that we expect—Rowling as easiest and Tolstoy as hardest—because the hypothesis test shows only that the readabilities are unequal. Nevertheless, our expectation seems reasonable since the sample means in Table 10.15 go in the expected order.

$\cdot\cdot\bullet$

Section 10.3 Exercises

Statistical Literacy and Critical Thinking

1. **ANOVA.** What is the objective in using the method of ANOVA?

2. **Variance.** For a sample of 40 randomly selected adult females, the mean height is 63.2 in. and the standard deviation is 2.7 in. What is the variance of those heights? (Include the appropriate units in your answer.)

3. **Comparing Majors.** A researcher obtains random samples of SAT scores from students at three different colleges. She plans to use the three sets of sample data with the method of analysis of variance. Identify the null and alternative hypotheses.

4. **One-Way ANOVA.** Why is the method of this section referred to as *one-way* analysis of variance? That is, what is "one-way" about the method?

Does It Make Sense? For Exercises 5–8, decide whether the statement makes sense (or is clearly true) or does not make sense (or is clearly false). Explain clearly. Not all of these statements have definitive answers, so your explanation is more important than your chosen answer.

5. **P-Value.** In a test for equality of mean skull breadths of Egyptian males from three different epochs, the *P*-value of 0.0031 is obtained. Because the *P*-value is small, we reject the null hypothesis of equal means and conclude that at least one of the three population means is different from the others.

6. **P-Value.** In a test for equality of mean commuting times for students from three different colleges, the *P*-value of 0.679 is obtained. Based on the *P*-value, we cannot support the alternative hypothesis that at least one of the three means is different from the others.

7. **ANOVA.** The Atkins, Zone, Weight Watchers, and Ornish diets were tested in clinical trials. For each subject, the amount of weight loss was recorded and results from analysis of variance show that the four diets are effective in causing weight loss.

8. **ANOVA.** An economist wants to compare the mean annual income of government employees in four different states. Analysis of variance cannot be used because there are more than three populations.

Concepts and Applications

9. **Readability of Authors.** The example in this section used the Flesch Reading Ease scores for randomly selected pages from books by Tom Clancy, J. K. Rowling, and Leo Tolstoy. When the Flesch–Kincaid Grade Level scores are used instead, the analysis of variance results from STATDISK are as shown in Figure 10.2. Assume that we want to use a 0.05 significance level in testing the null hypothesis that the three authors have Flesch–Kincaid Grade Level scores with the same mean.

   a. What is the null hypothesis?

   b. What is the alternative hypothesis?

   c. Identify the *P*-value.

   d. Based on the preceding results, what do you conclude about equality of the population means?

Source:	DF:	SS:	MS:	Test Stat, F:	Critical F:	P-Value:
Treatment	2	68.187222	34.093611	8.978506	3.284914	0.00077
Error	33	125.309167	3.797247			
Total	35	193.496389	5.528468			

Reject the Null Hypothesis
Reject equality of means

Figure 10.2

10. Fabric Flammability Tests in Different Laboratories. The Vertical Semirestrained Test was used to conduct flammability tests on children's sleepwear. Pieces of fabric were burned under controlled conditions. After the burning stopped, the length of the charred portion was measured and recorded. The same fabric samples were tested at five different laboratories. The analysis of variance results from Excel are shown below.

a. What is the null hypothesis?

b. What is the alternative hypothesis?

c. Identify the *P*-value.

d. Is there sufficient evidence to support the claim that the means for the different laboratories are not all the same? Assume that a 0.05 significance level is used.

Source of Variation	SS	df	MS	F	P-value	F crit
Between Groups	2.087194264	4	0.521798566	2.949333035	0.030665893	2.588834036
Within Groups	7.607597403	43	0.17692087			
Total	9.694791667	47				

11. M&M Weights. A random sample of M&Ms is partitioned into six categories according to their colors. The weights (grams) are obtained, and the analysis of variance results are as shown in the Minitab display below.

a. What is the null hypothesis?

b. What is the alternative hypothesis?

c. Identify the *P*-value.

d. Is there sufficient evidence to support the claim that M&Ms with different colors have different mean weights?

Source	DF	SS	MS	F	P
Factor	5	0.00611	0.00122	0.44	0.817
Error	94	0.25947	0.00276		
Total	99	0.26558			

12. Systolic Blood Pressure in Different Age Groups. A random sample of 40 women is partitioned into three categories with ages of below 20, 20 through 40, and over 40. The analysis of variance results obtained from SPSS are shown below.

a. What is the null hypothesis?

b. What is the alternative hypothesis?

c. Identify the *P*-value.

d. Is there sufficient evidence to support the claim that women in the different age categories have different mean blood pressure levels?

	Sum of Squares	df	Mean Square	F	Sig.
Between Groups	937.930	2	468.965	1.655	0.205
Within Groups	10484.470	37	283.364		
Total	11422.400	39			

In Exercises 13–16, use software to conduct the analysis of variance test.

13. Head Injury in a Car Crash. In car crash experiments conducted by the National Transportation Safety Administration, new cars were purchased and crashed into a fixed barrier at 35 miles per hour. The subcompact cars were the Ford Escort, Honda Civic, Hyundai Accent, Nissan Sentra, and Saturn SL4. The compact cars were the Chevrolet Cavalier, Dodge Neon, Mazda 626 DX, Pontiac Sunfire, and Subaru Legacy. The midsize cars were the Chevrolet Camaro, Dodge Intrepid, Ford Mustang, Honda Accord, and Volvo S70. The full-size cars were the Audi A8, Cadillac Deville, Ford Crown

Victoria, Oldsmobile Aurora, and Pontiac Bonneville. Head injury data (in hic) for the dummies in the driver's seat are listed below. Use a 0.05 significance level to test the null hypothesis that the different weight categories have the same mean. Do the sample data suggest that larger cars are safer?

Subcompact:	681	428	917	898	420
Compact:	643	655	442	514	525
Midsize:	469	727	525	454	259
Full-size:	384	656	602	687	360

14. **Chest Deceleration in a Car Crash.** The chest deceleration data (in g's) from the tests described in Exercise 13 are given below. Use a 0.05 significance level to test the null hypothesis that the different weight categories have the same mean. Do the data suggest that larger cars are safer?

Subcompact:	55	47	59	49	42
Compact:	57	57	46	54	51
Midsize:	45	53	49	51	46
Full-size:	44	45	39	58	44

15. **Archeology: Skull Breadths from Different Epochs.** The values in the table below are measured maximum breadths (in millimeters) of male Egyptian skulls from different epochs (based on data from *Ancient Races of the Thebaid*, by Thomson and Randall-Maciver). Changes in head shape over time suggest that interbreeding occurred with immigrant populations. Use a 0.05 significance level to test the claim that the different epochs all have the same mean.

4000 B.C.	1850 B.C.	150 A.D.
131	129	128
138	134	138
125	136	136
129	137	139
132	137	141
135	129	142
132	136	137
134	138	145
138	134	137

16. **Car Emissions.** Listed below are measured amounts of greenhouse gas emissions from cars in three different categories. The measurements are in tons per year, expressed as CO_2 equivalents. Use a 0.05 significance level to test the claim that the different car categories have the same mean amount of greenhouse gas emissions. Based on the results, does the number of cylinders appear to affect the amount of greenhouse gas emissions?

Table for Exercise 16

Four cylinder	7.2	7.9	6.8	7.4	6.5	6.6	6.7	6.5	6.5	7.1	6.7	5.5	7.3
Six cylinder	8.7	7.7	7.7	8.7	8.2	9.0	9.3	7.4	7.0	7.2	7.2	8.2	
Eight cylinder	9.3	9.3	9.3	8.6	8.7	9.3	9.3						

PROJECTS FOR THE INTERNET & BEYOND

17. **Noted Personalities.** The *World Almanac and Book of Facts* includes a section called "Noted Personalities," with subsections for architects, artists, business leaders, cartoonists, and several other categories. Design and conduct a study that begins with selection of samples from select groups, followed by a comparison of mean life spans of people from the different categories. Do any particular groups appear to have life spans that are different from those of the other groups?

18. **ANOVA.** Find a journal article that refers to use of analysis of variance. Identify the test being used and describe the conclusion. Did the test result in rejection of equal means? What was the *P*-value? What was the role of the method of analysis of variance?

IN THE NEWS

19. **Sports.** Find a recent news article discussing salaries of players on different professional sports teams, such as baseball teams. Find the salaries and use analysis of variance to test the null hypothesis of equal means. Summarize your findings and write a brief report that includes your conclusions.

CHAPTER REVIEW EXERCISES

1. A study sponsored by AT&T and the Automobile Association of America included the sample data in the following table.

	Had accident in last year	Had no accident in last year
Cell phone user	23	282
Not a cell phone user	46	407

a. Compare the percentage of cell phone users who had an accident to the percentage of those who did not use a cell phone and had an accident. On the basis of these results, do cell phones appear to be dangerous?

b. Identify the null and alternative hypotheses for a test of the claim that having an accident is independent of cell phone use.

c. Find the expected value for each cell of the table by assuming that having an accident is independent of cell phone use.

d. Find the value of the χ^2 statistic for a hypothesis test of the claim that having an accident is independent of cell phone use.

e. Based on the result from part d and the size of the table, refer to Table 10.7 (on page 358) and determine what is known about the *P*-value.

f. Based on the preceding results, what can you conclude from the hypothesis test about whether the two variables (*cell phone use* and *having an accident*) are independent?

g. Is the conclusion from part f consistent with what is now known about cell phone use and driving?

2. Different Disney animated children's movies are randomly selected and, for each movie, the amount of time (seconds) showing tobacco use is recorded (based on data from "Tobacco and Alcohol Use in G-Rated Children's Animated Films," by Goldstein, Sobel, and Newman, *Journal of the American Medical Association*, Vol. 281, No. 12). The results are listed below. Use a 0.05 significance level to test the claim that the sample is from a population with a mean that is greater than 60 seconds.

0	223	158	37	11	165	2	23

3. Using the sample data from Exercise 2, construct a 95% confidence interval estimate of the mean time that Disney animated children's movies show tobacco use.

4. Listed in the table below are body temperatures (°F) of randomly selected subjects from three different age groups. The STATDISK display in Figure 10.3 results from these sample values. Assume that we want to use a 0.05 significance level to test the claim that the three age groups have the same mean body temperature.

18–20	21–29	30 and older
98.0	99.6	98.6
98.4	98.2	98.6
97.7	99.0	97.0
98.5	98.2	97.5
97.1	97.9	97.3

a. What is the null hypothesis?

b. What is the alternative hypothesis?

c. Identify the *P*-value.

d. Is there sufficient evidence to reject the claim that the three age groups have the same mean body temperature?

Source:	DF:	SS:	MS:	Test Stat, F:	Critical F:	P-Value:
Treatment:	2	1.729333	0.864667	1.87971	3.88529	0.194915
Error:	12	5.52	0.46			
Total:	14	7.249333	0.51781			

Figure 10.3

1. A simple random sample of 15 values is obtained from a normally distributed population with an unknown standard deviation. Which of the following distributions is most appropriate for a hypothesis test involving a claim about a population mean?

   a. normal distribution   **b.** t distribution

   c. chi-square distribution   **d.** uniform distribution

2. A simple random sample of 45 values is obtained from a normally distributed population with an unknown standard deviation. Which of the following distributions is most appropriate for a hypothesis test involving a claim about a population mean?

   a. normal distribution   **b.** t distribution

   c. chi-square distribution   **d.** uniform distribution

3. A simple random sample of 45 values is obtained from a normally distributed population with a known standard deviation. Which of the following distributions is most appropriate for a hypothesis test involving a claim about a population mean?

   a. normal distribution   **b.** t distribution

   c. chi-square distribution   **d.** uniform distribution

4. What is the null hypothesis for a claim that the mean weight of male airline passengers is greater than 170 lb?

5. What is the alternative hypothesis for a claim that the mean weight of male airline passengers is less than 170 lb?

6. Determine whether the following statement is true or false: A hypothesis test can be used to prove that for the population of adults in the United States, gender is a cause of height.

7. Assume that you want to test the claim that adults in California, New York, Colorado, and Texas have the same mean tax payments in a year. What method would you use to test that claim?

8. If the hypothesis test of the claim described in Exercise 7 results in a P-value of 0.0099, what do you conclude about the null hypothesis?

9. A two-way table, constructed from survey results, consists of two rows representing sex (male/female) and two columns representing the response to a question (yes/no). What is the null hypothesis for a test to determine whether there is some relationship between sex and response?

10. If the hypothesis test described in Exercise 9 results in a P-value of 0.001, what do you conclude about the null hypothesis?

FOCUS ON
CRIMINOLOGY

Can You Tell a Fraud When You See One?

Suppose your professor gives you a homework assignment in which you are supposed to toss a coin 200 times and record the results in order. The two data sets below represent the results turned in by two students. Now suppose you learn that one of the students really did the assignment, while the other faked the data. Can you tell which one is fake?

Data Set 1 (H = heads; T = tails)

```
H  T  H  T  H  H  T  T  T  T  T  T  H  T  H  T  T  T  T  H
H  H  T  T  T  T  H  T  T  H  T  T  H  H  H  T  T  H  H  T
H  T  H  T  H  H  H  H  T  T  H  T  H  T  H  H  H  H  T  H
T  T  H  H  T  T  H  H  H  T  T  T  T  T  H  H  T  H  T  T
T  H  T  H  T  T  H  H  T  T  H  T  H  T  H  H  T  T  H  T
T  T  H  T  H  T  H  H  T  T  T  H  H  T  T  H  H  H  H  T
H  T  H  T  H  T  T  T  T  T  H  T  T  T  H  T  H  T  H  H
T  H  T  H  H  H  T  H  T  H  H  H  H  T  T  T  H  T  T  H
T  T  T  T  T  H  T  H  T  H  H  H  H  T  T  T  T  H  H  T
T  H  T  H  H  T  T  H  T  H  H  H  T  H  H  T  H  T  T  H
```

Data Set 2 (H = heads; T = tails)

```
T  H  H  T  T  H  H  T  T  H  T  H  H  T  T  H  H  T  H  T
H  T  T  H  T  T  H  H  T  T  H  T  H  T  T  H  H  T  H  T  H
H  H  T  T  H  T  H  H  T  H  T  T  H  T  T  H  H  T  T  H
H  T  H  T  H  T  H  T  H  H  T  H  T  T  H  H  T  H  H  T
H  T  H  T  H  T  H  T  H  T  T  H  H  T  H  H  T  H  H  T
T  H  T  H  H  T  T  H  T  T  H  T  H  T  T  H  T  H  H  T
H  T  H  T  T  H  T  H  H  T  H  T  H  H  T  H  T  H  H  T  H
T  T  H  T  H  H  T  T  H  T  T  H  T  T  H  T  H  H  T  H  H
H  T  H  H  T  T  H  T  H  T  H  T  H  T  H  H  T  H  T  T
T  H  H  T  H  T  H  H  T  H  H  T  T  H  H  T  H  T  H  T
```

To make the job a little easier, the following table summarizes a few characteristics of the two data sets that might help you decide which one is fake.

Characteristics of Data Set 1	Characteristics of Data Set 2
Total of 97 H, 103 T	Total of 101 H, 99 T
Two cases of 6 T in a row	No case of more than 3 H or 3 T in a row
Five cases of 4 H in a row	
Three cases of 4 T in a row	

If you are like most people, you will probably guess that Data Set 1 is the fake one. After all, its total numbers of heads and tails are farther from the 100 of each that many people expect, and it has two cases in which there were 6 tails in a row, plus several more cases in which there were 4 heads or tails in a row.

But consider this: The probability of getting 6 heads in a row is $(1/2)^6$, or 1 in 64. The chance of 6 consecutive tails is also 1 in 64. So with 200 tosses, the chance of getting at least one case of 6 heads or tails in a row is quite good, so the strings of consecutive heads and tails in Data Set 1 really are not surprising. In contrast, Data Set 2 has no string as long as 4 heads or tails in a row, even though the probability of such a string is only $(1/2)^4$, or 1 in 16. We therefore conclude that Data Set 2 is almost certainly a fake.

This simple example reveals an important application of statistics in criminology: It is often possible to catch people who have faked data of any kind. For example, statistics helps auditors to catch fraudulent financial statements or fraudulent tax returns, it helps scientists to catch other scientists who have falsified data, and in recent years it has been used to catch educators who falsify results on standardized tests taken by their students.

One of the most powerful tools for detecting fraud was identified by physicist Frank Benford. In the 1930s, Benford noticed that tables of logarithms (which scientists and engineers used regularly in the days before calculators) tended to be more worn on the early pages, where the numbers started with the digit 1, than on later pages. Following up on this observation, he soon discovered that many sets of numbers from everyday life, such as stock market values, baseball statistics, and the areas of lakes, include more numbers starting with the digit 1 than with the digit 2, and more starting with 2 than with 3, and so on. He eventually published a formula describing how often numbers begin with different digits, and this formula is now called *Benford's law*. Figure 10.4 shows what his law predicts for the first digits of numbers, along with actual results from several real sets of numbers. Notice how well Benford's law describes the results. (Interestingly, Benford's law was first discovered more than 50 years earlier, and published by astronomer and mathematician Simon Newcomb in 1881. However, Newcomb's article had been forgotten by the time Benford did his work.)

Benford's law is surprising because most people guess that every digit (1 through 9) would be equally likely as a starting digit. Indeed, this is the case for sets of random numbers such as lottery numbers, which are no more likely to start with 1 than with any other digit. (So Benford's law should not be used for picking lottery numbers.) However, because Benford's law does apply to many *real data* sets, it can be used to detect fraud. As shown in Figure 10.5, the data from real tax forms (dark red bars) follow Benford's law (dark blue bars) closely. In contrast, the data from a set of financial statements (light blue bars) examined in a 1995 study do not follow Benford's law. Based on this fact, the District Attorney suspected fraud, which he eventually was able to prove. The "random guess data" (green bars) came from students of Professor Theodore P. Hill at Georgia Institute of Technology (the source of much of the information for this Focus). The guesses do not follow Benford's law at all, which is why people who try to fake data can often be caught.

Benford's law mystified scientists and mathematicians for decades. Today it seems to be fairly well understood, though still difficult to explain. Here is one explanation of why Benford's law applies to the Dow Jones Industrial Average, thanks to Dr. Mark J. Nigrini of Southern Methodist University (as reported in the *New York Times*): Imagine that the Dow is at 1,000, so the first digit is a 1, and rises at a rate of about 20% per year. The doubling time at this rate of increase is a little less than four years, so the Dow would remain in the 1,000s, still with a first digit of 1, for almost four years, until it hit 2,000. It would then have a first digit of 2 until it hit 3,000. However, moving from 2,000 to 3,000 requires only a 50% increase, which takes only a little over two years. As a result, the first digit of 2 would occur for only a little more than half the number of days that a first digit of 1 occurred. Subsequent changes of first digit take even shorter times. By the time the Dow hits 9,000, it takes only an 11% increase and just seven months to reach the 10,000 mark, so a first digit of 9 occurs for only seven months. At the 10,000 mark, however, the Dow is back to a first digit of 1 again, and this would not change until the Dow doubled again to 20,000, which means another almost four years at the rate of increase of 20% per year. Thus, if you

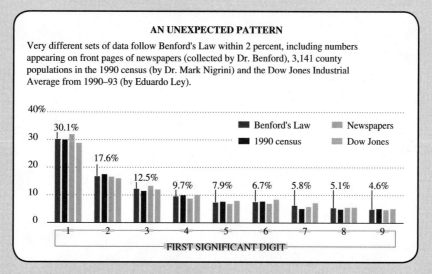

AN UNEXPECTED PATTERN

Very different sets of data follow Benford's Law within 2 percent, including numbers appearing on front pages of newspapers (collected by Dr. Benford), 3,141 county populations in the 1990 census (by Dr. Mark Nigrini) and the Dow Jones Industrial Average from 1990–93 (by Eduardo Ley).

FIRST SIGNIFICANT DIGIT

Legend: Benford's Law, 1990 census, Newspapers, Dow Jones

Figure 10.4 *Source:* T.P. Hill, "The First Digit Phenomenon," *American Scientist,* 86:4, 1998.

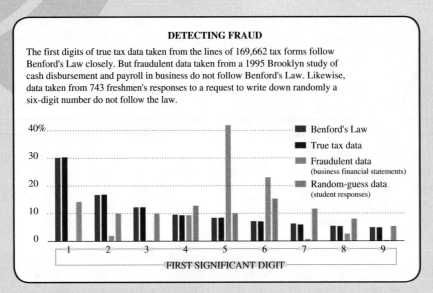

DETECTING FRAUD

The first digits of true tax data taken from the lines of 169,662 tax forms follow Benford's Law closely. But fraudulent data taken from a 1995 Brooklyn study of cash disbursement and payroll in business do not follow Benford's Law. Likewise, data taken from 743 freshmen's responses to a request to write down randomly a six-digit number do not follow the law.

- Benford's Law
- True tax data
- Fraudulent data (business financial statements)
- Random-guess data (student responses)

FIRST SIGNIFICANT DIGIT

Figure 10.5 *Source:* T.P. Hill, "The First Digit Phenomenon," *American Scientist,* 86:4, 1998.

graphed the number of days the Dow had each starting digit 1 through 9, you'd find that the number 1 would be the starting digit for longer periods of time than the number 2, and so on down the line.

In summary, Benford's law shows that numbers don't always arise with the frequencies that most people would guess. As a result, it not only helps explain a lot of mysteries about numbers (such as those in the Dow), but also has become a valuable tool for the detection of criminal fraud.

QUESTIONS FOR DISCUSSION

1. Try this experiment with friends: Ask one friend to record 200 actual coin tosses and another friend to try to fake data for 200 coin tosses. Have them give you their results anonymously, so that you don't know which sheet came from which friend. Using the ideas in this Focus, try to determine which set is real and which one is fake. After you make your guess, check with your friends to see whether you were right. Discuss your ability to detect the fakes.

2. Briefly discuss how Benford's law might be used to detect fraudulent tax returns.

3. Can Benford's law alone prove that data are fraudulent? Or can it only point to data that should be investigated further? Explain.

4. Find a set of data similar to the sets shown in Figure 10.4 and make a bar chart showing the frequencies of the first digits 1 through 9. Do the data follow Benford's law? Explain.

• • • • • • • • • • • • • • • •

FOCUS ON
EDUCATION

What Can a Fourth-Grader Do with Statistics?

Nine-year-old Emily Rosa was in fourth grade, trying to decide what to do for her school science fair project. She was thinking of doing a project on the colors of M&M candies, when she noticed her mother, a nurse, watching a videotape about a practice called "non-contact therapeutic touch" or "TT." TT is a popular alternative medical treatment, practiced in many places throughout the world. But no statistically valid test had ever clearly demonstrated whether it actually works. Emily told her mother that she had an idea for testing TT and wanted to make it her science fair project.

Despite the name, TT therapists do *not* actually touch their patients. Instead, they move their hands a few inches above a patient's body. Therapeutic touch supporters claim that these hand movements allow trained therapists to feel and manipulate what they call a "human energy field." By doing these manipulations properly, the therapists can supposedly cure many different ailments and diseases. Emily Rosa's science fair project sought to find out whether trained TT therapists could really feel a human energy field.

To do her project, Emily recruited 21 TT therapists to participate in a simple experiment. Each therapist sat across a table from Emily, laying his or her arms out flat, palms up. Emily then put up a cardboard partition with cutouts for the therapist's arms. This prevented Emily and the therapist from seeing each other's face, but allowed Emily to see the therapist's hands.

Emily then placed one of her hands a few inches above *one* of the therapist's two hands, asking the therapist to identify which hand it was. If the therapist could truly feel Emily's "human energy field," then the therapist should have been able to tell whether his or her right or left hand was closest to Emily's hand. Each trial of the experiment ended with Emily's recording whether the therapist was right or wrong in identifying the hand.

Emily took several precautions to make sure her experiment would be statistically valid. For example, to ensure that her choices between the two hands were random, Emily used the outcome of a coin toss to determine whether she placed her hand over the therapist's left or right hand in each case. And to make sure she had enough data to evaluate statistical significance, 14 of the 21 therapists got 10 tries each, while 7 got 20 tries each.

The results were a miserable failure for the TT therapists. Because there were only two possible answers in each trial—left hand or right hand—by pure chance the therapists ought to have been able to guess the correct hand about 50% of the time. In fact, the overall results showed that they got the correct answer only 44% of the time. Moreover, none of the therapists performed better than expected by chance in a statistically significant way. Emily also checked to see whether therapists with more experience did better than those with less experience. They did not. Emily's conclusion: If there is such a thing as a "human energy field" (which she doubts), the TT therapists can't feel it. And even if the "human energy field" exists, it's difficult to imagine how TT therapists could use it for healing if they can't even detect its presence.

One of the most interesting aspects of this study was that Emily was able to do it at all. Other skeptics of TT had hoped to conduct similar studies in the past, but TT therapists had refused to participate. One famous skeptic, magician James Randi, had even offered a $1 million prize to any TT therapist who could pass a test similar to Emily's. Only one person accepted Randi's challenge, and she succeeded in only 11 of 20 trials, about the same as would be expected by chance. So why was Emily able to recruit participants who had avoided more experienced researchers? Apparently, the therapists agreed to participate in Emily's experiment because they did not feel threatened by a fourth-grader.

The novelty of Emily's science fair project drew media attention, and it was not long before word reached retired Pennsylvania psychiatrist Stephen Barrett. Dr. Barrett specialized in debunking "quack" therapies, and he convinced Emily and her mother to report her results in a medical research paper. The paper was published in the *Journal of the American Medical Association* (April 1, 1998) when Emily was 11, making her the youngest-ever author of a paper in that prestigious journal.

QUESTIONS FOR DISCUSSION

1. After Emily's results were published, many TT supporters claimed that her experiment was invalid because she and her mother were biased against TT. Based on the way her experiment was designed, do you think that her personal bias could have affected her results? Why or why not?

2. Another objection to Emily's experiment was that it was only single-blind rather than double-blind. That is, the therapist could not see what Emily was doing, but Emily could see what the therapist was doing. Do you think this objection is valid in this case? Can you think of a way that Emily's experiment might be repeated but be made double-blind?

3. Emily's experiment was not a direct test of whether TT treatment works, because it did not check to see whether patients actually improved when treated by TT. Suggest a statistically valid way to test whether TT is more effective than a placebo.

4. Based on the results of Emily's study, skeptics now say that TT is so clearly invalid that it should no longer be used or funded. Do you agree? Why or why not?

• • • • • • • • • • • • • • • • •

EPILOGUE: A PERSPECTIVE ON STATISTICS

A single introductory statistics course cannot transform you into an expert statistician. After studying statistics in this book, you may feel that you have not yet mastered the material to the extent necessary to use statistics confidently in real applications. Nevertheless, by now you should understand enough about statistics to interpret critically the reports of statistical research that you see in the news and to converse with experts in statistics when you need more information. And, if you go on to take further course work in statistics, you should be well prepared to understand important topics that are beyond the scope of this introductory book.

Most importantly, while this book is not designed to make you an expert statistician, it is designed to make you a better-educated person with improved job marketability. You should know and understand the basic concepts of probability and chance. You should know that in attempting to gain insight into a data set, it's important to investigate measures of center (such as mean and median), measures of variation (such as range and standard deviation), the nature of the distribution (via a frequency table or graph), and the presence of outliers. You should know and understand the importance of estimating population parameters (such as a population mean or proportion), as well as testing hypotheses about population parameters. You should understand that a correlation between two variables does not necessarily imply that there is also some cause-and-effect relationship. You should know the importance of good sampling. You should recognize that many surveys and polls obtain very good results, even though the sample sizes might seem to be relatively small. Although many people refuse to believe it, a nationwide survey of only 1,700 voters can provide good results if the sampling is carefully planned and executed.

There once was a time when a person was considered educated if he or she could read, but we are in a new millennium that is much more demanding. Today, an educated person must be able to read, write, understand the significance of the Renaissance, operate a computer, and apply statistical reasoning. The study of statistics helps us see truths that are sometimes distorted by a failure to approach a problem carefully or concealed by data that are disorganized. Understanding statistics is now essential for both employers and employees—for all citizens. H. G. Wells once said, "Statistical thinking will one day be as necessary for efficient citizenship as the ability to read and write." That day is now.

Table A-1 is a more detailed version of Table 5.1; note that the areas under the curve shown here correspond to *percentiles*. To read this table, find the first two digits of the z-score in the left column, then read across the rows for third digit. Negative z-scores are on the left page and positive z-scores are on the right page.

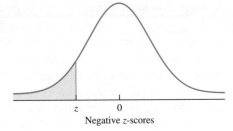

Negative z-scores

TABLE A-1	Standard Normal (z) Distribution: Cumulative Area from the LEFT									
z	.00	.01	.02	.03	.04	.05	.06	.07	.08	.09
−3.50 and lower	.0001									
−3.4	.0003	.0003	.0003	.0003	.0003	.0003	.0003	.0003	.0003	.0002
−3.3	.0005	.0005	.0005	.0004	.0004	.0004	.0004	.0004	.0004	.0003
−3.2	.0007	.0007	.0006	.0006	.0006	.0006	.0006	.0005	.0005	.0005
−3.1	.0010	.0009	.0009	.0009	.0008	.0008	.0008	.0008	.0007	.0007
−3.0	.0013	.0013	.0013	.0012	.0012	.0011	.0011	.0011	.0010	.0010
−2.9	.0019	.0018	.0018	.0017	.0016	.0016	.0015	.0015	.0014	.0014
−2.8	.0026	.0025	.0024	.0023	.0023	.0022	.0021	.0021	.0020	.0019
−2.7	.0035	.0034	.0033	.0032	.0031	.0030	.0029	.0028	.0027	.0026
−2.6	.0047	.0045	.0044	.0043	.0041	.0040	.0039	.0038	.0037	.0036
−2.5	.0062	.0060	.0059	.0057	.0055	.0054	.0052	.0051	*.0049	.0048
−2.4	.0082	.0080	.0078	.0075	.0073	.0071	.0069	.0068	.0066	.0064
−2.3	.0107	.0104	.0102	.0099	.0096	.0094	.0091	.0089	.0087	.0084
−2.2	.0139	.0136	.0132	.0129	.0125	.0122	.0119	.0116	.0113	.0110
−2.1	.0179	.0174	.0170	.0166	.0162	.0158	.0154	.0150	.0146	.0143
−2.0	.0228	.0222	.0217	.0212	.0207	.0202	.0197	.0192	.0188	.0183
−1.9	.0287	.0281	.0274	.0268	.0262	.0256	.0250	.0244	.0239	.0233
−1.8	.0359	.0351	.0344	.0336	.0329	.0322	.0314	.0307	.0301	.0294
−1.7	.0446	.0436	.0427	.0418	.0409	.0401	.0392	.0384	.0375	.0367
−1.6	.0548	.0537	.0526	.0516	.0505	*.0495	.0485	.0475	.0465	.0455
−1.5	.0668	.0655	.0643	.0630	.0618	.0606	.0594	.0582	.0571	.0559
−1.4	.0808	.0793	.0778	.0764	.0749	.0735	.0721	.0708	.0694	.0681
−1.3	.0968	.0951	.0934	.0918	.0901	.0885	.0869	.0853	.0838	.0823
−1.2	.1151	.1131	.1112	.1093	.1075	.1056	.1038	.1020	.1003	.0985
−1.1	.1357	.1335	.1314	.1292	.1271	.1251	.1230	.1210	.1190	.1170
−1.0	.1587	.1562	.1539	.1515	.1492	.1469	.1446	.1423	.1401	.1379
−0.9	.1841	.1814	.1788	.1762	.1736	.1711	.1685	.1660	.1635	.1611
−0.8	.2119	.2090	.2061	.2033	.2005	.1977	.1949	.1922	.1894	.1867
−0.7	.2420	.2389	.2358	.2327	.2296	.2266	.2236	.2206	.2177	.2148
−0.6	.2743	.2709	.2676	.2643	.2611	.2578	.2546	.2514	.2483	.2451
−0.5	.3085	.3050	.3015	.2981	.2946	.2912	.2877	.2843	.2810	.2776
−0.4	.3446	.3409	.3372	.3336	.3300	.3264	.3228	.3192	.3156	.3121
−0.3	.3821	.3783	.3745	.3707	.3669	.3632	.3594	.3557	.3520	.3483
−0.2	.4207	.4168	.4129	.4090	.4052	.4013	.3974	.3936	.3897	.3859
−0.1	.4602	.4562	.4522	.4483	.4443	.4404	.4364	.4325	.4286	.4247
−0.0	.5000	.4960	.4920	.4880	.4840	.4801	.4761	.4721	.4681	.4641

Note: For values of z below −3.49, use 0.0001 for the area.

*Use these common values that result from interpolation:

z-score	Area
−1.645	0.0500
−2.575	0.0050

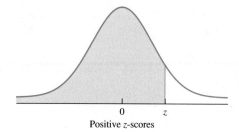

0   z
Positive z-scores

TABLE A-1	(continued) Cumulative Area from the LEFT								
z	.00	.01	.02	.03	.04	.05	.06	.07	.08	.09
0.0	.5000	.5040	.5080	.5120	.5160	.5199	.5239	.5279	.5319	.5359
0.1	.5398	.5438	.5478	.5517	.5557	.5596	.5636	.5675	.5714	.5753
0.2	.5793	.5832	.5871	.5910	.5948	.5987	.6026	.6064	.6103	.6141
0.3	.6179	.6217	.6255	.6293	.6331	.6368	.6406	.6443	.6480	.6517
0.4	.6554	.6591	.6628	.6664	.6700	.6736	.6772	.6808	.6844	.6879
0.5	.6915	.6950	.6985	.7019	.7054	.7088	.7123	.7157	.7190	.7224
0.6	.7257	.7291	.7324	.7357	.7389	.7422	.7454	.7486	.7517	.7549
0.7	.7580	.7611	.7642	.7673	.7704	.7734	.7764	.7794	.7823	.7852
0.8	.7881	.7910	.7939	.7967	.7995	.8023	.8051	.8078	.8106	.8133
0.9	.8159	.8186	.8212	.8238	.8264	.8289	.8315	.8340	.8365	.8389
1.0	.8413	.8438	.8461	.8485	.8508	.8531	.8554	.8577	.8599	.8621
1.1	.8643	.8665	.8686	.8708	.8729	.8749	.8770	.8790	.8810	.8830
1.2	.8849	.8869	.8888	.8907	.8925	.8944	.8962	.8980	.8997	.9015
1.3	.9032	.9049	.9066	.9082	.9099	.9115	.9131	.9147	.9162	.9177
1.4	.9192	.9207	.9222	.9236	.9251	.9265	.9279	.9292	.9306	.9319
1.5	.9332	.9345	.9357	.9370	.9382	.9394	.9406	.9418	.9429	.9441
1.6	.9452	.9463	.9474	.9484	.9495 *	.9505	.9515	.9525	.9535	.9545
1.7	.9554	.9564	.9573	.9582	.9591	.9599	.9608	.9616	.9625	.9633
1.8	.9641	.9649	.9656	.9664	.9671	.9678	.9686	.9693	.9699	.9706
1.9	.9713	.9719	.9726	.9732	.9738	.9744	.9750	.9756	.9761	.9767
2.0	.9772	.9778	.9783	.9788	.9793	.9798	.9803	.9808	.9812	.9817
2.1	.9821	.9826	.9830	.9834	.9838	.9842	.9846	.9850	.9854	.9857
2.2	.9861	.9864	.9868	.9871	.9875	.9878	.9881	.9884	.9887	.9890
2.3	.9893	.9896	.9898	.9901	.9904	.9906	.9909	.9911	.9913	.9916
2.4	.9918	.9920	.9922	.9925	.9927	.9929	.9931	.9932	.9934	.9936
2.5	.9938	.9940	.9941	.9943	.9945	.9946	.9948	.9949 *	.9951	.9952
2.6	.9953	.9955	.9956	.9957	.9959	.9960	.9961	.9962	.9963	.9964
2.7	.9965	.9966	.9967	.9968	.9969	.9970	.9971	.9972	.9973	.9974
2.8	.9974	.9975	.9976	.9977	.9977	.9978	.9979	.9979	.9980	.9981
2.9	.9981	.9982	.9982	.9983	.9984	.9984	.9985	.9985	.9986	.9986
3.0	.9987	.9987	.9987	.9988	.9988	.9989	.9989	.9989	.9990	.9990
3.1	.9990	.9991	.9991	.9991	.9992	.9992	.9992	.9992	.9993	.9993
3.2	.9993	.9993	.9994	.9994	.9994	.9994	.9994	.9995	.9995	.9995
3.3	.9995	.9995	.9995	.9996	.9996	.9996	.9996	.9996	.9996	.9997
3.4	.9997	.9997	.9997	.9997	.9997	.9997	.9997	.9997	.9997	.9998
3.50 and up	.9999									

Note: For values of z above 3.49, use 0.9999 for the area.

*Use these common values that result from interpolation:

z-score	Area
1.645	0.9500 ◄
2.575	0.9950 ◄

TABLE A-2	Select Critical Values of z		
	Left-tailed test	**Right-tailed test**	**Two-tailed test**
0.05 significance level	-1.645	1.645	-1.96 and 1.96
0.01 significance level	-2.33	2.33	-2.576 and 2.576

APPENDIX B: TABLE OF RANDOM NUMBERS

For many statistical applications it is useful to generate a set of randomly chosen numbers. You can generate such numbers with most calculators and computers, but sometimes it is easier to use a table such as the one given below. This table was generated by randomly selecting one of the digits 0, 1, 2, 3, 4, 5, 6, 7, 8, or 9 for each position in the table; that is, each of these digits is equally likely to appear in any position. Thus, you can generate a sequence of random digits simply by starting at any point in the table and taking the digits in the order in which they appear. A larger set of random numbers is available on the text Web site (www.aw.com/bbt).

Example 1: Generate a random list of yes/no responses.
Solution: Start at an arbitrary point in the table. If the digit is 0, 1, 2, 3, or 4, call it a *yes* response. If the digit is 5, 6, 7, 8, or 9, call it a *no* response. Continue through the table from your starting point, using each digit shown to determine either a *yes* or a *no* response for your list.

Example 2: Generate a random list of letter grades A, B, C, D, or F.
Solution: Let 0 or 1 be an A, 2 or 3 be a B, 4 or 5 be a C, 6 or 7 be a D, and 8 or 9 be an F. Start at an arbitrary point in the table and use each digit shown to determine a grade for your list.

```
9 9 3 2 7    5 6 0 8 1    6 0 2 3 2    8 3 3 1 2    4 7 6 3 4
9 7 1 8 1    6 6 7 6 6    5 4 4 7 7    6 8 1 7 1    0 8 4 9 9
8 1 7 5 0    7 8 5 2 0    9 4 3 9 0    7 6 1 9 1    0 8 7 3 4
0 5 1 0 2    8 7 4 0 3    9 2 6 2 5    8 4 2 2 5    1 9 8 3 4
2 7 8 0 8    1 8 5 6 6    4 4 5 5 4    9 3 5 2 8    6 5 5 4 3

4 8 8 3 3    8 4 6 9 1    8 2 5 7 6    9 7 1 2 3    6 5 1 8 2
5 4 5 8 7    3 8 4 5 7    4 5 2 2 2    7 7 0 2 3    0 2 4 8 6
4 1 9 4 3    0 7 1 9 0    7 3 1 4 0    8 3 2 8 0    5 0 1 0 1
2 8 7 4 6    5 7 7 6 0    0 8 9 5 5    4 0 7 3 9    1 6 3 3 2
5 8 6 8 6    9 6 7 7 5    1 3 5 2 9    7 6 6 3 5    9 4 6 0 5

8 0 9 4 8    5 0 5 6 9    1 0 6 9 5    9 0 7 8 9    9 4 8 3 7
6 1 0 4 1    7 4 0 0 3    5 6 4 2 1    5 1 1 9 0    0 2 5 0 7
3 4 0 9 1    9 3 9 5 1    0 7 4 8 1    1 9 7 0 7    1 4 5 2 6
9 9 6 5 0    7 8 6 1 0    4 9 8 7 7    4 4 7 4 0    7 8 6 4 9
1 5 0 7 8    1 6 3 6 9    5 4 9 5 4    2 4 6 0 4    3 2 6 8 4

5 7 2 8 5    8 3 1 6 4    4 2 2 3 7    0 6 6 3 2    3 5 0 4 6
1 2 3 5 7    1 9 7 7 6    5 5 3 1 9    6 0 5 1 6    3 8 0 3 4
3 4 9 3 5    0 3 7 4 6    2 1 8 5 1    4 1 7 0 2    1 4 9 4 9
2 4 2 6 6    9 9 7 2 1    7 7 3 2 0    6 7 0 7 3    9 3 3 7 5
9 9 1 5 2    8 3 9 9 4    8 9 6 0 6    7 4 6 3 1    3 6 9 8 3

2 6 3 3 9    3 4 1 9 3    0 6 9 1 2    4 1 9 9 8    7 3 2 6 9
1 3 7 8 0    1 0 6 1 6    5 4 9 3 0    7 0 7 2 3    1 7 8 9 2
8 8 4 8 4    0 2 8 5 5    7 3 7 1 2    1 8 3 5 2    0 1 5 3 2
4 2 9 2 8    8 8 9 1 2    4 6 7 1 7    5 1 5 1 9    3 2 2 8 0
1 3 7 2 1    0 6 4 7 6    1 0 8 4 8    9 7 6 3 5    5 1 2 2 9
```

SUGGESTED READINGS

Against the Gods: The Remarkable Story of Risk, P. Bernstein, Wiley, 1996.

The Arithmetic of Life, G. Shaffner, Ballantine Books, 1999.

The Arithmetic of Life and Death, G. Shaffner, Ballantine Books, 2001.

The Bell Curve Debate, R. Jacoby and N. Glauberman (eds.), Times Books, 1995.

Beyond Numeracy: Ruminations of a Number Man, J. A. Paulos, Vintage, 1992.

Billions and Billions, C. Sagan, Random House, 1997.

The Broken Dice and Other Mathematical Tales of Chance, I. Ekeland, University of Chicago Press, 1993.

Can You Win?, M. Orkin, Freeman, 1991.

The Complete How to Figure It, D. Huff, Norton, 1996.

Damned Lies and Statistics, J. Best, University of California Press, 2001.

Ecological Numeracy: Quantitative Analysis of Environmental Issues, R. Herendeen, Wiley, 1998.

Elementary Statistics, 12th ed., M. Triola, Pearson, 2014.

Emblems of Mind: The Inner Life of Music and Mathematics, E. Rothstein, Random House, 1995.

Envisioning Information, E. Tufte, Connecticut Graphics Press, Cheshire, 1983.

Games, Gods, and Gambling, F. N. David, Dover, 1998.

Go Figure! The Numbers You Need for Everyday Life, N. Hopkins and J. Mayne, Visible Ink Press, 1992.

The Honest Truth About Lying with Statistics, C. Holmes, Charles C Thomas, 1990.

How Many People Can the Earth Support, J. E. Cohen, Norton, 1995.

How to Lie with Statistics, D. Huff, Norton, 1993.

How to Tell the Liars from the Statisticians, R. Hooke, Dekker, 1983.

How to Use (and Misuse) Statistics, G. Kimble, Prentice-Hall, 1978.

Innumeracy, J. A. Paulos, Hill and Wang, 1988.

The Jungles of Randomness, I. Peterson, Wiley, 1998.

The Lady Tasting Tea: How Statistics Revolutionized Science in the 20th Century, D. Salsburg, Freeman, 2001.

Life by the Numbers, K. Devlin, Wiley, 1998.

Math for Life: Crucial Ideas You Didn't Learn in School, J. Bennett, Roberts & Co Publishers, 2012.

The Mathematical Tourist, I. Peterson, Freeman, 1988.

A Mathematician Reads the Newspaper, J. A. Paulos, Basic Books, 1995.

Mathematics, the Science of Patterns: The Search for Order in Life, Mind, and the Universe, K. Devlin, Scientific American Library, 1994.

The Mismeasure of Man, S. J. Gould, Norton, 1981.

The Mismeasure of Woman, C. Tavris, Touchstone Books, 1993.

Misused Statistics: Straight Talk for Twisted Numbers, A. Jaffe and H. Spirer, Dekker, 1998.

Nature's Numbers, I. Stewart, Basic Books, 1995.

Once Upon a Number, J. A. Paulos, Basic Books, 1998.

Overcoming Math Anxiety, S. Tobias, Houghton Mifflin, 1978. Revised edition, Norton, 1993.

Pi in the Sky: Counting, Thinking, and Being, J. D. Barrow, Little, Brown, 1992.

The Population Explosion, P. R. Ehrlich and A. H. Ehrlich, Simon and Schuster, 1990.

Probabilities in Everyday Life, J. McGervey, Ivy Books, 1989.

The Psychology of Judgment and Decision Making, S. Plous, McGraw-Hill, 1993.

Randomness, D. Bennett, Harvard University Press, 1998.

The Statistical Exorcist: Dispelling Statistics Anxiety, M. Hollander and F. Proschan, Dekker, 1984.

Statistics, 4th ed., D. Freedman, R. Pisani, R. Purves, and A. Adhikari, Norton, 2007.

Statistics: Concepts and Controversies, 7th ed., D. Moore, W. Notz, Freeman, 2009.

Statistics with a Sense of Humor, F. Pyrczak, Fred Pyrczak Publisher, 1989.

Tainted Truth: The Manipulation of Fact in America, C. Crossen, Simon and Schuster, 1994.

The Tipping Point: How Little Things Can Make a Big Difference, M. Gladwell, Little, Brown, 2000.

200% of Nothing, A. K. Dewdney, Wiley, 1993.

The Universe and the Teacup: The Mathematics of Truth and Beauty, K. C. Cole, Harcourt Brace, 1998.

The Visual Display of Quantitative Information, E. Tufte, Connecticut Graphics Press, Cheshire, 1983.

Vital Signs, compiled by the Worldwatch Institute, Norton, 1998 (updated annually).

What Are the Odds? Chance in Everyday Life, L. Krantz, Harper Perennial, 1992.

What Is a P-Value, Anyway?, A. Vickers, Pearson, 2009.

CREDITS

Photo Credits

Chapter 1 p. 1, Bill Pugliano/Getty Images; p. 3, Anthony J. Causi/ Icon SMI 942/Newscom; p. 7, Olivier Asselin/Alamy; p. 11, Golden Pixels LLC/Alamy; p. 13, Jeffrey Bennett; p. 14, Photodisc/Getty Images; p. 22, Corbis; p. 25, Andersen Ross/Digital Vision/Getty Images; p. 27, Photodisc/Getty Images; p. 31, Photodisc/Getty Images; p. 34, Picture Press/Alamy; p. 35, Doug Martin/Photo Researchers, Inc.; p. 42, Gazmandhu/Shutterstock; p. 44, Warren Goldswain/Shutterstock

Chapter 2 p. 45, Science Source/Photo Researchers, Inc.; p. 52, Egd/Shutterstock; p. 54, NASA; p. 55, Exactostock/SuperStock; p. 60, View Stock/Alamy; p. 61, Arctic Images/Alamy; p. 66, Gene Blevins/Corbis; p. 69, Bettmann/Corbis; p. 72, Lynnette Peizer/ Alamy; p. 76, Igor Stepovik/Shutterstock

Chapter 3 p. 78, Sergej Khakimullin/Shutterstock; p. 91, Everett Collection; p. 109, Charles & Josette Lenars/Corbis; p. 115, Corbis; p. 117, British Antarctic Survey/Photo Researchers, Inc.

Chapter 4 p. 119, United States Geological Survey; p. 123, Timothy A. Clary/AFP/Newscom; p. 133, Laurent Rebours/AP Images; p. 140, Beth Anderson/Pearson Education, Inc.; p. 145, Toru Hanai/Reuters; p. 152, Stock Connection Blue/Alamy; p. 157, Prisma Bildagentur AG/Alamy; p. 159, Chuck Wagner/Shutterstock

Chapter 5 p. 161, Liming/EPA/Newscom; p. 164, Beth Anderson/ Pearson Education, Inc.; p. 171, Craig Wactor/Shutterstock; p. 176, The United States Army; p. 189, VIPDesignUSA/Shutterstock; p. 191, Auremar/Shutterstock

Chapter 6 p. 193, James "BO" Insogna/Shutterstock; p. 200, Morgan Lane Photography/Shutterstock; p. 201, Corbis Bridge/ Alamy; p. 204, Morgan Hill/Alamy; p. 205, Anastasia E Kozlova/ Shutterstock; p. 209, Varley/Sipa/Newscom; p. 216, Brt Photo/ Alamy; p. 223, Jocelyn William KRT/Newscom; p. 233, Stephen VanHorn/Shutterstock; p. 235, SIU Biomed Comm/Custom Medical Stock Photo/Newscom

Chapter 7 p. 237, Frances Roberts/Alamy; p. 239, AptTone/ Shutterstock; p. 249, Gorilla/Shutterstock; p. 256, Stock Connection Distribution/Alamy; p. 261, Dmitriy Shironosov/Shutterstock; p. 265, Image100/AGE Fotostock; p. 266, AP Images; p. 271, Wave Break Media, Ltd./Shutterstock; p. 273, Jan Martin Will/Shutterstock

Chapter 8 p. 275, Jim West/Alamy; p. 287, Wave Break Media, Ltd./Shutterstock; p. 304, Photodisc/Getty Images; p. 305, Wellcome Library, London; p. 306, GL Archive/Alamy

Chapter 9 p. 308, Moodboard/Alamy; p. 309, S. Borisov/ Shutterstock; p. 316, Stock Connection Blue/Alamy; p. 328, Layne Kennedy/Corbis; p. 340, Blend Images/Shutterstock; p. 342, Nick Gregory/Alamy

Chapter 10 p. 344, Nagelestock.com/Alamy; p. 372, Alex Skopje/ Shutterstock; p. 375, Beth Anderson/Pearson Education, Inc.

Technology Credits

Portions of information contained in this publication appear with permission of Minitab Inc. All such material remains the exclusive property and copyright of Minitab Inc. All rights reserved.

All images showing usage of TI83/84 Plus calculator are copyright © Texas Instruments Inc. Reproduced by permission.

MICROSOFT® WINDOWS®, AND MICROSOFT OFFICE® ARE REGISTERED TRADEMARKS OF THE MICROSOFT CORPORATION IN THE U.S.A. AND OTHER COUNTRIES. THIS BOOK IS NOT SPONSORED OR ENDORSED BY OR AFFILIATED WITH THE MICROSOFT CORPORATION.

MICROSOFT AND/OR ITS RESPECTIVE SUPPLIERS MAKE NO REPRESENTATIONS ABOUT THE SUITABILITY OF THE INFORMATION CONTAINED IN THE DOCUMENTS AND RELATED GRAPHICS PUBLISHED AS PART OF THE SERVICES FOR ANY PURPOSE. ALL SUCH DOCUMENTS AND RELATED GRAPHICS ARE PROVIDED "AS IS" WITHOUT WARRANTY OF ANY KIND. MICROSOFT AND/OR ITS RESPECTIVE SUPPLIERS HEREBY DISCLAIM ALL WARRANTIES AND CONDITIONS WITH REGARD TO THIS INFORMATION, INCLUDING ALL WARRANTIES AND CONDITIONS OF MERCHANTABILITY, WHETHER EXPRESS, IMPLIED OR STATUTORY, FITNESS FOR A PARTICULAR PURPOSE, TITLE AND NON-INFRINGEMENT. IN NO EVENT SHALL MICROSOFT AND/OR ITS RESPECTIVE SUPPLIERS BE LIABLE FOR ANY SPECIAL, INDIRECT OR CONSEQUENTIAL DAMAGES OR ANY DAMAGES WHATSOEVER RESULTING FROM LOSS OF USE, DATA OR PROFITS, WHETHER IN AN ACTION OF CONTRACT, NEGLIGENCE OR OTHER TORTIOUS ACTION, ARISING OUT OF OR IN CONNECTION WITH THE USE OR PERFORMANCE OF INFORMATION AVAILABLE FROM THE SERVICES. THE DOCUMENTS AND RELATED GRAPHICS CONTAINED HEREIN COULD INCLUDE TECHNICAL INACCURACIES OR TYPOGRAPHICAL ERRORS. CHANGES ARE PERIODICALLY ADDED TO THE INFORMATION HEREIN. MICROSOFT AND/OR ITS RESPECTIVE SUPPLIERS MAY MAKE IMPROVEMENTS AND/OR CHANGES IN THE PRODUCT(S) AND/OR THE PROGRAM(S) DESCRIBED HEREIN AT ANY TIME. PARTIAL SCREEN SHOTS MAY BE VIEWED IN FULL WITHIN THE SOFTWARE VERSION SPECIFIED.

absolute change The actual increase or decrease from a reference value to a new value:

$$\text{absolute change} = \text{new value} - \text{reference value}$$

absolute difference The actual difference between the compared value and the reference value:

$$\text{absolute difference} = \text{compared value} - \text{reference value}$$

absolute error The actual amount by which a measured value differs from the true value:

$$\text{absolute error} = \text{measured value} - \text{true value}$$

accident rate The number of accidents due to some particular cause, expressed as a fraction of all people at risk for the same cause. For example, an accident rate of "5 per 1,000 people" means that an average of 5 in 1,000 people suffer an accident from this particular cause.

accuracy How closely a measurement approximates a true value. An accurate measurement is very close to the true value.

alternative hypothesis (H_a) A statement that can be supported only if the null hypothesis is rejected.

analysis of variance (ANOVA) A method of testing the equality of three or more population means by analyzing sample variances.

and **probability** The probability that event A *and* event B will both occur. How it is calculated depends on whether the events are independent or dependent. Also called *joint probability*.

ANOVA See *analysis of variance*.

a priori **method** See *theoretical method*.

bar graph A diagram consisting of bars representing the frequencies (or relative frequencies) for particular categories. The bar lengths are proportional to the frequencies.

best-fit line The line on a scatter diagram that lies closer to the data points than all other possible lines (according to a standard statistical measure of closeness). Also called *regression line*.

bias In a statistical study, any problem in the design or conduct of the study that tends to favor certain results. See also *participation bias*; *selection bias*.

bimodal distribution A distribution with two peaks, or modes.

binning Grouping data into categories (bins), each of which covers a range of possible data values.

blinding The practice of keeping experimental subjects and/or experimenters in the dark about who is in the treatment group and who is in the control group. See also *double-blind experiment*; *single-blind experiment*.

boxplot A graphical display of a five-number summary. A number line is used for reference, the values from the lower to the upper quartiles are enclosed in a box, a line is drawn through the box for the median, and two "whiskers" are extended to the low and high data values. Also called *box-and-whisker plot*.

case-control study An observational study that resembles an experiment because the sample naturally divides into two (or more) groups. The participants who engage in the behavior under study form the cases, like the treatment group in an experiment. The participants who do not engage in the behavior are the *controls*, like the control group in an experiment.

causality A relationship present when one variable is a cause of another.

census The collection of data from every member of a population.

Central Limit Theorem Theorem stating that, for random samples (all of the same size) of a variable with any distribution (not necessarily a normal distribution), the distribution of the means of the samples will, as the sample size increases, tend to be approximately a normal distribution.

chi-square statistic (χ^2) A number used to determine the statistical significance of a hypothesis test in a contingency table (or two-way table). If it is less than a critical value (which depends on the table size and the desired significance level), the differences between the observed frequencies and the expected frequencies are not significant.

cluster sampling Dividing the population into groups, or clusters; selecting some of these clusters at random; and then obtaining the sample by choosing all the members within each cluster.

coefficient of determination (R^2) A number that describes how well data fit a best-fit equation found through multiple regression.

compared value A number that is compared to a reference value in computing a relative difference.

complement For an event A, all outcomes in which A does *not* occur, expressed as $\overline{A}$. Its probability is $P(\overline{A}) = 1 - P(A)$.

conditional probability The probability of one event given the occurrence of another event, written $P(B \text{ given } A)$ or $P(B|A)$.

confidence interval A range of values associated with a confidence level, such as 95%, that is likely to contain the true value of a population parameter.

confounding Confusion in the interpretation of statistical results that occurs when the effects of different factors are mixed such that the effects of the individual factors being studied cannot be determined.

confounding factors Any factors or variables in a statistical study that can lead to confounding. Also called *confounding variables*.

Consumer Price Index (CPI) An index number designed to measure the rate of inflation. It is computed and reported monthly, based on a sample of more than 60,000 goods, services, and housing costs.

contingency table See *two-way table*.

continuous data Quantitative data that can take on any value in a given interval.

contour map A map that uses curves (contours) to connect geographical regions with the same data values.

control group The group of subjects in an experiment who do not receive the treatment being tested.

convenience sampling Selecting a sample that is readily available.

correlation A statistical relationship between two variables. See also *negative correlation; no correlation; positive correlation*.

correlation coefficient (*r*) A measure of the strength of the relationship between two variables. Its value is always between -1 and 1 (that is, $-1 \le r \le 1$).

cumulative frequency For any data category, the number of data values in that category and all preceding categories.

death rate The number of deaths due to some particular cause, expressed as a fraction of all people at risk for the same cause. For example, a death rate of "5 per 1,000 people" means that an average of 5 in 1,000 people die from this particular cause.

degrees of freedom (for a t distribution) The sample size minus one $(n - 1)$.

dependent events Two events for which the outcome of one affects the probability of the other.

deviation How far a particular data value lies from the mean of a data set, used to compute standard deviation.

discrete data Quantitative data that can take on only particular values and not other values in between (for example, the whole numbers 0, 1, 2, 3, 4, 5).

distribution The way the values of a variable are spread over all possible values. It can be displayed with a table or with a graph.

distribution of sample means The distribution that results when the means ($\bar{x}$) of all possible samples of a given size are found.

distribution of sample proportions The distribution that results when the proportions ($\hat{p}$) in all possible samples of a given size are found.

dotplot A diagram similar to a bar graph except that each individual data value is represented with a dot.

double-blind experiment An experiment in which neither the participants nor the experimenters know who belongs to the treatment group and who belongs to the control group.

either/or probability The probability that *either* event A *or* event B will occur. How it is calculated depends on whether the events are overlapping or non-overlapping.

empirical method See *relative frequency method.*

event In probability, a collection of one or more outcomes that share a property of interest. See also *outcome.*

expected frequency In a two-way table, the frequency one would expect in a given cell of the table if the row and column variables were independent of each other.

expected value The mean value of the outcomes for some random variable.

experiment A study in which researchers apply a treatment and then observe its effects on the subjects.

experimenter effect An effect that occurs when a researcher or experimenter somehow influences subjects through such factors as facial expression, tone of voice, or attitude.

five-number summary A description of the variation of a data distribution in terms of the minimum value, lower quartile, median, upper quartile, and maximum value.

frequency For a data category, the number of times data values fall within that category.

frequency table A table that lists all the categories of data in one column and the frequency for each category in another column.

gambler's fallacy The mistaken belief that a streak of bad luck makes a person "due" for a streak of good luck.

geographical data Data that can be assigned to different geographical locations.

histogram A bar graph showing a distribution for quantitative data (at the interval or ratio level of measurement). The bars have a natural order, and the bar widths have specific meaning.

hypothesis In statistics, a claim about a population parameter, such as a population proportion, p, or population mean, μ. See also *alternative hypothesis; null hypothesis.*

hypothesis test A standard procedure for testing a claim about the value of a population parameter.

independent events Two events for which the outcome of one does not affect the probability of the other.

index number A number that provides a simple way to compare measurements made at different times or in different places. The value at one particular time (or place) must be chosen to be the reference value (or base value). The index number for any other time (or place) is

$$\text{index number} = \frac{\text{value}}{\text{reference value}} \times 100$$

inflation The increase over time in prices and wages. Its overall rate is measured by the CPI.

interval level of measurement A level of measurement for quantitative data in which differences, or intervals, are meaningful but ratios are not. Data at this level have an arbitrary starting point.

joint probability See *and probability.*

law of large numbers An important result in probability that applies to a process for which the probability of an event A is $P(A)$ and the results of repeated trials are independent. It states: If the process is repeated through many trials, the larger the number of trials, the closer the proportion should be to $P(A)$. Also called *law of averages.*

left-skewed distribution A distribution in which the values are more spread out on the left side.

left-tailed test A hypothesis test that involves testing whether a population parameter lies to the left (lower values) of a claimed value.

level of measurement See *nominal level of measurement; ordinal level of measurement; interval level of measurement; ratio level of measurement.*

life expectancy The number of years a person of a given age today can be expected to live, on average. It is based on current health and medical statistics and does not take into account future changes in medical science or public health.

line chart A graph showing the distribution of quantitative data as a series of dots connected by lines. The horizontal position of each dot corresponds to the center of the bin it represents and the vertical position corresponds to the frequency value for the bin.

lower quartile See *quartile, lower.*

margin of error The maximum likely difference between an observed sample statistic and the true value of a population parameter. Its size depends on the desired level of confidence.

mean The sum of all values divided by the total number of values. It's what is most commonly called the average value.

median The middle value in a sorted data set (or halfway between the two middle values if the number of values is even).

median class For binned data, the bin into which the median data value falls.

meta-analysis A study in which researchers analyze many individual studies (on a particular topic) as a combined group, with the aim of finding trends that were not evident in the individual studies.

middle quartile See *quartile, middle.*

mode The most common value (or group of values) in a distribution.

multiple bar graph A simple extension of a regular bar graph, in which two or more sets of bars allow comparison of two or more data sets.

multiple line chart A simple extension of a regular line chart, in which two or more lines allow comparison of two or more data sets.

multiple regression A technique that allows the calculation of a best-fit equation that represents the best fit between one variable (such as price) and a *combination* of two or more other variables (such as weight and volume).

negative correlation A correlation in which the two variables tend to change in opposite directions, with one increasing while the other decreases.

no correlation Absence of any apparent relationship between two variables.

nominal level of measurement A level of measurement for qualitative data that consist of names, labels, or categories only and cannot be ranked or ordered.

nonlinear relationship A relationship between two variables that cannot be expressed with a linear (straight-line) equation.

non-overlapping events Two events for which the occurrence of one precludes the occurrence of the other.

normal distribution A special type of symmetric, bell-shaped distribution with a single peak that corresponds to the mean, median, and mode of the distribution. Its variation can be characterized by the standard deviation. See also *68-95-99.7 rule*.

null hypothesis (H_0) A specific claim (such as a specific value for a population parameter) against which an alternative hypothesis is tested.

observational study A study in which researchers observe or measure characteristics of the sample members, but do not attempt to influence or modify these characteristics.

of **versus** *more than (less than)* **rule** A rule for comparisons. It states: If the compared value is $P\%$ *more than* the reference value, then it is $(100 + P)\%$ *of* the reference value. If the compared value is $P\%$ *less than* the reference value, then it is $(100 - P)\%$ *of* the reference value.

one-tailed test See *left-tailed test*; *right-tailed test*.

ordinal level of measurement A level of measurement for qualitative data that can be arranged in some order. It generally does not make sense to do computations with the data.

outcome In probability, the most basic possible result of an observation or experiment. See also *event*.

outlier A value in a data set that is much higher or much lower than almost all other values.

overlapping events Two events that could possibly both occur.

Pareto chart A bar graph of data at the nominal level of measurement, with the bars arranged in descending order.

participants People (as opposed to objects) who are the subjects of a study.

participation bias Bias that occurs any time participation in a study is voluntary.

peer review A process by which several experts in a field evaluate a research report before it is published.

percentiles Values that divide a data distribution into 100 segments, each representing about 1% of the data values.

pictograph A graph embellished with artwork.

pie chart A circle divided so that each wedge represents the relative frequency of a particular category. The wedge size is proportional to the relative frequency, and the entire pie represents the total relative frequency of 100%.

placebo Something that lacks the active ingredients of a treatment but is identical in appearance to the treatment. Thus, participants in a study cannot distinguish the placebo from the real treatment.

placebo effect An effect in which patients improve simply because they believe they are receiving a useful treatment, when in fact they may be receiving only a placebo.

population The complete set of people or things being studied.

population mean The true mean of a population, denoted by the Greek letter μ (pronounced "mew").

population parameters Specific characteristics of the population that a statistical study is designed to estimate.

population proportion The true proportion of some characteristic in a population, denoted by p.

positive correlation A type of correlation in which two variables tend to increase (or decrease) together.

practical significance In a statistical study, significance in the sense that the result is associated with some meaningful course of action.

precision The amount of detail in a measurement.

probability For an event, the likelihood that the event will occur. The probability of an event, written as $P(\text{event})$, is always between 0 and 1 inclusive. A probability of 0 means the event is impossible and a probability of 1 means the event is certain. See also *relative frequency method*; *subjective method*; *theoretical method*.

probability distribution The complete distribution of the probabilities of all possible events associated with a particular variable. It may be shown as a table, graph, or formula.

*P***-value** In a hypothesis test, the probability of selecting a sample at least as extreme as the observed sample, assuming that the null hypothesis is true.

qualitative data Data consisting of values that describe qualities or nonnumerical categories.

quantitative data Data consisting of values representing counts or measurements.

quartile, lower The median of the data values in the lower half of a data set. Also called *first quartile*.

quartile, middle The overall median of a data set. Also called *second quartile*.

quartile, upper The median of the data values in the upper half of a data set. Also called *third quartile*.

quartiles Values that divide a data distribution into four parts with approximately 25% of the data values in each part.

random errors Errors that occur because of random and inherently unpredictable events in the measurement process.

randomization The process of ensuring that the subjects of an experiment are assigned to the treatment or control group at random and in such a way that each subject has an equal chance of being assigned to either group.

range For a distribution, the difference between the lowest and highest data values.

range rule of thumb A guideline stipulating that, for a data set with no outliers, the standard deviation is approximately equal to range/4.

rare event rule Rule stating that it is appropriate to conclude that a given assumption (such as the null hypothesis) is probably not correct if the probability of a particular event at least as extreme as the observed event is very small.

ratio level of measurement A level of measurement for quantitative data in which both intervals and ratios are meaningful. Data at this level have a true zero point.

raw data The actual measurements or observations collected from a sample.

reference value The number that is used as the basis for a comparison.

regression line See *best-fit line*.

relative change The size of an absolute change in comparison to the reference value, expressed as a percentage:

$$\text{relative change} = \frac{\text{new value} - \text{reference value}}{\text{reference value}} \times 100\%$$

relative difference The size of an absolute difference in comparison to the reference value, expressed as a percentage:

$$\text{relative difference} = \frac{\text{compared value} - \text{reference value}}{\text{reference value}} \times 100\%$$

relative error The relative amount by which a measured value differs from the true value, expressed as a percentage:

$$\text{relative error} = \frac{\text{measured value} - \text{true value}}{\text{true value}} \times 100\%$$

relative frequency For any data category, the fraction or percentage of the total frequency that falls in that category:

$$\text{relative frequency} = \frac{\text{frequency in category}}{\text{total frequency}}$$

relative frequency method A method of estimating a probability based on observations or experiments by using the observed or measured relative frequency of the event of interest. Also called *empirical method*.

representative sample A sample in which the relevant characteristics of the members are generally the same as the characteristics of the population.

right-skewed distribution A distribution in which the values are more spread out on the right side.

right-tailed test A hypothesis test that involves testing whether a population parameter lies to the right (higher values) of a claimed value.

rounding rule For statistical calculations, the practice of stating answers with one more decimal place of precision than is found in the raw data. For example, the mean of 2, 3, and 5 is 3.3333 . . . , which would be rounded to 3.3.

sample A subset of the population from which data are actually obtained.

sample mean The mean of a sample, denoted $\bar{x}$ ("*x*-bar").

sample proportion The proportion of some characteristic in a sample, denoted $\hat{p}$ ("*p*-hat").

sample statistics Characteristics of the sample that are found by consolidating or summarizing the raw data.

sampling The process of choosing a sample from a population.

sampling distribution The distribution of a sample statistic, such as a mean or proportion, taken from all possible samples of a particular size.

sampling error Error introduced when a random sample is used to estimate a population parameter; the difference between a sample result and a population parameter.

sampling methods See *cluster sampling*; *convenience sampling*; *simple random sampling*; *stratified sampling*; *systematic sampling*.

scatterplot A graph, often used to investigate correlations, in which each point corresponds to the values of two variables. Also called *scatter diagram*.

selection bias Bias that occurs whenever researchers select their sample in a biased way. Also called *selection effect*.

self-selected survey A survey in which people decide for themselves whether to be included. Also called *voluntary response survey*.

simple random sampling A sample of items chosen in such a way that every possible sample of the same size has an equal chance of being selected.

Simpson's paradox A statistical paradox that arises when the results for a whole group seem inconsistent with those for its subgroups; it can occur whenever the subgroups are unequal in size.

single-blind experiment An experiment in which the participants do not know whether they are members of the treatment group or the control group but the experimenters do know—or, conversely, the participants do know but the experimenters do not.

single-peaked distribution A distribution with a single mode. Also called *unimodal distribution*.

68-95-99.7 rule Guideline stating that, for a normal distribution, about 68% (actually, 68.3%) of the data values fall within 1 standard deviation of the mean, about 95% (actually, 95.4%) of the data values fall within 2 standard deviations of the mean, and about 99.7% of the data values fall within 3 standard deviations of the mean.

skewed See *left-skewed distribution*; *right-skewed distribution*.

stack plot A type of bar graph or line chart in which two or more different data sets are stacked vertically.

standard deviation A single number commonly used to describe the variation in a data distribution, calculated as

$$\text{standard deviation} = \sqrt{\frac{\text{sum of all (deviations from the mean)}^2}{\text{total number of data values} - 1}}$$

standard score For a particular data value, the number of standard deviations (usually denoted by z) between it and the mean of the distribution:

$$z = \text{standard score} = \frac{\text{data value} - \text{mean}}{\text{standard deviation}}$$

Also called *z-score*.

statistical significance A measure of the likelihood that a result is meaningful.

statistically significant result A result in a statistical study that is unlikely to have occurred by chance. The most commonly quoted levels of statistical significance are the 0.05 level (the probability of the result's having occurred by chance is 5% or less, or less than 1 in 20) and the 0.01 level (the probability of the result's having occurred by chance is 1% or less, or less than 1 in 100).

statistics (plural) The data that describe or summarize something.

statistics (singular) The science of collecting, organizing, and interpreting data.

stemplot A graph that looks much like a histogram turned sideways, with lists of the individual data values in place of bars. Also called *stem-and-leaf plot*.

stratified sampling A sampling method that addresses differences among subgroups, or strata, within a population. First the strata are identified, and then a random sample is drawn within each stratum. The total sample consists of all the samples from the individual strata.

subjective method A method of estimating a probability based on experience or intuition.

subjects In a statistical study, the people or objects chosen for the sample. See also *participants*.

symmetric distribution A distribution in which the left half is a mirror image of the right half.

systematic errors Errors that occur when there is a problem in the measurement system that affects all measurements in the same way.

systematic sampling Using a simple system to choose the sample, such as selecting every 10th or every 50th member of the population.

t distribution A distribution that is very similar in shape and symmetry to the normal distribution but that accounts for the greater variability expected with small samples. It approaches the normal distribution for large sample sizes.

theoretical method A method of estimating a probability based on a theory, or set of assumptions, about the process in question. Assuming that all outcomes are equally likely, the theoretical probability of a particular event is found by dividing the number of ways the event can occur by the total number of possible outcomes. Also called *a priori method*.

time-series diagram A histogram or line chart in which the horizontal axis represents time.

treatment Something given or applied to the members of the treatment group in an experiment.

treatment group The group of subjects in an experiment that receive the treatment being tested.

two-tailed test A hypothesis test that involves testing whether a population parameter lies to either side of a claimed value.

two-way table A table showing the relationship between two variables by listing the values of one variable in its rows and the values of the other variable in its columns. Also called *contingency table*.

type I error In a hypothesis test, the mistake of rejecting the null hypothesis, H_0, when it is true.

type II error In a hypothesis test, the mistake of failing to reject the null hypothesis, H_0, when it is false.

uniform distribution A distribution in which all data values have the same frequency.

unimodal distribution See *single-peaked distribution*.

unusual values In a data distribution, values that are not likely to occur by chance, such as those values that are more than 2 standard deviations away from the mean.

upper quartile See *quartile, upper*.

variable Any item or quantity that can vary or take on different values.

variables of interest In a statistical study, the items or quantities that the study seeks to measure.

variation How widely data are spread out about the center of a distribution. See also *five-number summary*; *range*; *standard deviation*.

vital statistics Data concerning births and deaths of people.

voluntary response survey See *self-selected survey*.

weighted mean A mean that accounts for differences in the relative importance of data values. Each data value is assigned a weight, and then

$$\text{weighted mean} = \frac{\text{sum of (each data value} \times \text{its weight)}}{\text{sum of all weights}}$$

z-score See *standard score*.

ANSWERS

Section 1.1

1. A population is the complete set of people or things being studied, while a sample is a subset of the population. The difference is that the sample is only a part of the complete population.

2. They do not have the same meaning. The term "baseball statistics" refers to measurements summarizing past results. The second use of the term "statistics" refers to the science of using statistical methods for analyzing the effectiveness of the drug.

3. A sample statistic is a numerical measure of a characteristic of a *sample* found by consolidating or summarizing raw data. A population parameter is a numerical measure of a characteristic of a *population*. Because it is usually impractical to directly measure population parameters for large populations, we usually infer likely values of the population parameters from the measured sample statistics.

4. The margin of error is important because it describes the range of values likely to contain the population parameter. That range is found by adding and subtracting the margin of error from the sample statistic obtained in the study.

5. Does not make sense. Population parameters are typically inferred from sample statistics. When population parameters are known, sample statistics are not needed and the concept of "margin of error" does not apply.

6. Makes sense. Although the poll makes it seem likely that Smith would end up winning, there is always some chance that the poll results (which are sample statistics) do not reflect the actual population parameters, especially if the poll was not conducted with enough care. Moreover, because the poll was conducted two weeks before the election, it is possible that opinions changed between the time the poll was conducted and the actual vote.

7. Does not make sense; in fact, the statement is clearly false. The confidence interval in this case runs from $54\% - 3\% = 51\%$ to $54\% + 3\% = 57\%$. Assuming the margin of error is defined as it is in most cases, this interval is a 95% confidence interval, meaning that we can be 95% confident that Johnson received between 51% and 57% of the vote. While this makes it quite likely that Johnson really did receive a majority of the vote (assuming the poll was conducted well), it is not a certainty, as there is still a 5% chance that her true percentage of the vote was above 57% or below 51% — and it is possible that it could be enough below so that she did not win a majority of the votes.

8. Does not make sense. A larger margin of error means we can infer less about the population than a smaller margin of error. Clearly, no one would decide to pay the same amount for lower-quality data.

9. Does not make sense. The sample should be drawn from the population of all people who have suffered through a family tragedy, which is not the same as the population of patients in support groups for loss of a spouse. There are many types of family tragedies besides loss of a spouse, and not all of those people join support groups.

10. Makes sense. The purpose of statistics is to help with decision making, and if the survey was conducted well, it is possible to draw conclusions with high confidence from a survey of a 1,000-person sample. If the survey results make the advertising campaign look like a good idea, then it makes sense to follow its guidance. Of course, there is still no *guarantee* that the advertising campaign will be successful.

11. *Sample:* The 1,018 adults selected. *Population:* All adults in the United States. *Sample statistic:* 22%. The value of the population parameter is not known, but it is the percentage of all adults in the United States who smoked cigarettes in the past week.

12. *Sample:* The 186 babies selected. *Population:* The complete set of all babies. *Sample statistic:* 3,103 grams. The value of the population parameter is not known, but it is the average (mean) birth weight of all babies.

13. *Sample:* The 47 subjects treated with Garlicin. *Population:* The complete set of all adults. *Sample statistic:* 3.2. The value of the population parameter is not known, but it is the average (mean) change in LDL cholesterol.

14. *Sample:* The 150 senior executives who were surveyed. *Population:* The complete set of all senior executives. *Sample statistic:* 47%. The value of the population parameter is not known, but it is the percentage of all senior executives who say that the most common job interview mistake is to have little or no knowledge of the company.

15. 57% to 63%

16. 84% to 86%

17. 93% to 99%

18. 98.1°F to 98.3°F

19. Yes. Although there is no guarantee, the results suggest that the majority of adults believe that immediate government action is required, because the percentage is most likely between 53% and 57%.

20. Yes, because the true percentage of those who prefer commercials most is likely to be between 50.4% and 51.6%, so it is likely to be a majority.

21. Yes. Based on the survey, the actual percentage of voters is expected to be between 67% and 73%, which does not include the 61% value based on actual voter results. If the survey was conducted well, then it is unlikely that it would be so different from the actual results, implying either that respondents intentionally lied to appear favorable to the pollsters or that their memories may have been inaccurate.

22. It appears that the surveyed men may have been influenced by the gender of the interviewer. When interviewed by women, they may have been more inclined to agree with a statement that they perceived to be more favorable to the interviewers who were women.

23. **a.** *Goal:* Determine the percentage of all adults in favor of the death penalty for people convicted of murder. *Population:* The complete set of all adults. *Population parameter:* The percentage of all adults who are in favor of the death penalty for people convicted of murder. **b.** *Sample:* The 511 selected subjects. *Raw data:* Individual responses to the question. *Sample statistic:* 64%. **c.** 60% to 68%

24. a. *Goal:* Determine the percentage of older adults (57–85) who use at least one prescription drug. *Population:* The complete set of all older adults (57–85). *Population parameter:* The percentage of all older adults who use at least one prescription drug. **b.** *Sample:* The 3,005 older adults selected for the survey. *Raw data:* Individual responses to the question. *Sample statistic:* 82%. **c.** 80% to 84%

25. a. *Goal:* Determine the percentage of households with a TV tuned to the Super Bowl. *Population:* The complete set of all U.S. households. *Population parameter:* The percentage of all U.S. households tuned to the Super Bowl. **b.** *Sample:* The sample of 9,000 households selected for the survey. *Raw data:* Individual indications of whether the household has a TV tuned to the Super Bowl. *Sample statistic:* 45%. **c.** 44% to 46%

26. a. *Goal:* Determine the percentage of human resource professionals who say that piercings or tattoos make job applicants less likely to be hired. *Population:* The complete set of all human resource professionals. *Population parameter:* The percentage of all human resource professionals who say that piercings or tattoos make job applicants less likely to be hired. **b.** *Sample:* The sample of 514 human resource professionals selected for the survey. *Raw data:* Individual responses to the question. *Sample statistic:* 46%. **c.** 42% to 50%

27. *Step 1.* Goal: Identify the percentage of all drivers who use cell phones while they are driving. *Step 2.* Choose a sample of drivers. *Step 3.* Collect data from the sample of drivers to determine the percentage of them who use a cell phone while driving. *Step 4.* Use the techniques of the science of statistics to make an inference about the percentage of all people who use cell phones while they are driving. *Step 5.* Based on the likely value of the population parameter, form a conclusion about the percentage of people who use cell phones while they are driving.

28. *Step 1.* Goal: Identify the mean FICO score of all adults in the United States. *Step 2.* Choose a sample of consumers. *Step 3.* Obtain the FICO scores of the selected consumers. For this sample, find the mean FICO score. *Step 4.* Use the techniques of the science of statistics to make an inference about the mean FICO score of all consumers in the United States. *Step 5.* Based on the likely value of the population mean, form a conclusion about the mean FICO score of all consumers in the United States.

29. *Step 1.* Goal: Identify the mean weight of all commercial airline passengers. *Step 2.* Choose a sample of airline passengers. *Step 3.* Weigh each selected airline passenger, then find the mean of these weights. *Step 4.* Use the techniques of the science of statistics to make an inference about the mean weight of all airline passengers. *Step 5.* Based on the likely value of the population mean, form a conclusion about the mean weight of all airline passengers.

30. *Step 1.* Goal: Identify the mean length of time that pacemaker batteries last before failure. *Step 2.* Choose a sample of pacemaker batteries. *Step 3.* Record the length of time that each pacemaker battery in the sample lasts before failure. *Step 4.* Use the techniques of the science of statistics to make an inference about the mean length of time that all pacemaker batteries last before failure.

Step 5. Based on the likely value of the population parameter, form a conclusion about the mean length of time that all pacemaker batteries last before failure.

Section 1.2

1. A census is the collection of data from every member of the population, but a sample is a collection of data from only part of the population.

2. Yes. Because it drew only from registered Democrats, while all voters are entitled to participate in a general election, the sample is surely biased and highly unlikely to be useful in predicting election results.

3. With cluster sampling, we select all members of randomly selected subgroups (or clusters), but with stratified sampling, we select samples from each of the different subgroups (or strata).

4. The sample is a convenience sample. It is not likely to be biased because there is nothing about right-handedness that would cause students in a college class to be different from the proportion in the entire population.

5. Does not make sense because it is not possible or practical to survey every student, as would be required for a census.

6. Makes sense. In some cases, the convenience sample may give meaningful results. (For example, see Exercise 4.)

7. Makes sense. Because it is obvious that the percentage of Americans more than 6 feet tall is well under 75%, the study must have somehow used a biased sample.

8. Makes sense. The procedure described does result in a simple random sample, and it is a procedure that is commonly used.

9. A census is practical because the population consists of the small number of players on the LA Lakers team, and it is easy to obtain their heights.

10. A census is not practical because the number of basketball players in the United States is much too large.

11. A census is not practical. The number of statistics instructors is large, and it would be extremely difficult to get them all to take an IQ test.

12. A census is practical. The number of statistics instructors at the University of Colorado is not very large and it would probably be easy to get their ages through a survey that promised anonymity.

13. The sample is the service times of the 4 selected senators. The population consists of the service times of all 100 senators. This is an example of random sampling (or simple random sampling). However, because the sample is so small, it is not likely to be representative of the population.

14. The sample is the 5,108 selected households. The population is the complete set of all households. This is an example of simple random sampling. Because the sample is fairly large and it was obtained by a reputable firm, the sample is likely to be representative of the population.

15. The sample is the 1,059 selected adults, and the population is the complete set of all adults. This is an example of simple random sampling. Because the sample is fairly large and it was obtained by a reputable firm, the sample is likely to be representative of the population.

16. The sample is the 65 responses. The population is the complete set of all adult Americans. This is an example of convenience sampling, because she sent the survey to people she already knew. Moreover, because people chose whether or not to respond—and because the survey concerned mailing and mailing was required to respond—the sample is not likely to be representative of the population.

17. Most representative: Sample 3, because the list is likely to represent the most people, and there is no reason to think that people with the first 1,000 numbers would differ in any particular way from other people. Sample 1 is biased because it involves owners of expensive vehicles. Sample 2 is biased because it involves people from only one geographic region. Sample 4 is biased because it involves a self-selected sample consisting of people who are more likely to have strong feelings about the issue of credit card debt.

18. Best: Sample 4, which is a good use of systematic sampling. Sample 1 is biased because it consists of people from one geographic region located at the extreme southern part of the state. Sample 2 is biased because it consists of people from one specific geographic region. Sample 3 is likely to be biased because it is a self-selected sample.

19. Yes. Because the film critic indirectly works for Disney, he or she may be more inclined to submit a more favorable review.

20. No. Because the magazine does not accept any advertising and it does not accept free products, it is not influenced by the manufacturers of the cars that it reviews.

21. Yes. The university scientists receive the funding from Monsanto, so they might be inclined to please the company in the hopes of getting further funding in the future. Thus, there may be an inclination to provide favorable results. To determine whether this bias is a problem, you would need to explore the methods and conclusions very carefully.

22. Yes. Because the physicians receive funding from the pharmaceutical company, they might be more inclined to provide more favorable results so that they can get additional funding in the future. The magazine now requires that all such physician authors disclose any funding, and those disclosures are included in the articles.

23. The sample is a simple random sample that is likely to be representative because there is no bias in the selection process.

24. The sample is systematic, and it is likely to be representative because there is no inherent bias in the way that the sample is selected.

25. The sample is a cluster sample. It is likely to be representative, although the exact method of selecting the polling stations could affect whether the sample is biased.

26. The sample is stratified. It is likely to be biased because the collection of people who participate in sports does not consist of equal numbers of golfers, tennis players, and swimmers. Many sports are not included in the sample.

27. The sample is a convenience sample. It is likely to be biased, because the sample consists of family members likely to have the same socio-economic backgrounds and interests.

28. The sample is a cluster sample. It is likely to be biased primarily because any waiters and waitresses that have cheated may not provide honest responses to the IRS. Also, the relatively small number of restaurants could easily result in a sample that is not representative.

29. The sample is a stratified sample. It is likely to be biased because people from those age groups are not evenly distributed throughout the population. However, the results could be weighted to reflect the age distribution of the population.

30. The sample is a convenience sample and is likely to be biased, because the students are attending the same college, which is not likely to be representative of all college students.

31. The sample is a systematic sample. The sample is likely to be representative, because there is nothing about the alphabetical ordering that is likely to result in a biased sample.

32. The sample is a simple random sample. Because it is a simple random sample and the sample size is fairly large, it is likely to be representative.

33. The sample is a stratified sample. It is likely to be biased because the population does not have equal numbers of people in each of the three categories. However, the results could be weighted to reflect the actual distribution of the population.

34. The sample is a cluster sample. The study could easily be biased, because sampling across a small number of randomly chosen classes may not yield a representative sample of students. For example, many colleges have a much larger selection of classes for seniors than for freshmen, so the selection of classes would tend to be biased toward including more seniors.

35. The sample is a convenience sample. It is likely to be biased because it is a self-selected sample and consists of those with strong feelings about the issue.

36. The sample is a simple random sample. Because it is a simple random sample, it is likely to be representative, although it would probably be better to use a larger sample size.

37. The sampling plan results in a simple random sample, so it is likely to be representative.

38. The sample is a systematic sample. The sample is likely to be representative, unless there are special factors, such as a manufacturing process that systematically results in defective items.

39. Simple random sampling of the student body. Be sure you use a large enough sample size to get meaningful results.

40. Simple random sampling may be adequate. Better, however, would be stratified sampling with different ethnic groups as the strata, because there may be differences in blood type distribution among these groups.

41. Cluster sampling of death records should give good data for this case.

42. You will need stratified sampling in which you measure the mercury content of tuna in different markets that represent different sources for the tuna.

Section 1.3

1. A placebo is physically similar to a treatment, but it lacks any active ingredients, so it should not by itself produce any effects. A placebo is important so that results from subjects given a real treatment can be compared with the results from subjects given a placebo.

2. Blinding is the characteristic of an experiment whereby the participants or the experimenters do not know who belongs to

the treatment group and who belongs to the control group. It is important to use blinding for participants so that they are not affected by the knowledge that they are receiving the real treatment, and it is important for experimenters so that they can evaluate results objectively instead of being persuaded by knowledge about who is getting the real treatment.

3. Confounding is the mixing of effects from different factors so that we cannot determine the effects from the specific factors being studied. If males are given the treatment and females are given placebos, we would not know whether effects are due to the treatment or the sex of the participant.

4. No. In such a situation, the clinical trial should be stopped and subjects receiving a placebo should be given the effective treatment.

5. It makes sense to use a double-blind experiment, but the experimental design is tricky because the subjects can clearly see the clothing that they are wearing, and the evaluators can also see those colors. Blinding might be used by not telling the subjects about the experiment, so their knowledge of their clothes colors does not affect their results. Because it is difficult to use blinding for the evaluators, it is important to obtain results based on objective measures not subject to the judgments of the evaluators.

6. In this case, the lawn itself is not subject to changing perceptions or effects based on the knowledge that it receives a treatment, so blinding of the participants (the lawn) is automatic in this case. It is important to use blinding for those who evaluate the results so that any judgment is not affected by the knowledge of whether sections of lawn are given the new fertilizer treatment.

7. The experimenter effect occurs when a researcher or experimenter somehow influences subjects through such factors as facial expression, tone of voice, or attitude. The experimenter effect can be avoided by using blinding so that those who evaluate the results do not know which subjects are given an actual treatment and which subjects are given no treatment or a placebo.

8. Because the IQ scores are measured objectively from subject responses, there is no opportunity to change results, so it is not necessary to take precautions against an experimenter effect.

9. Observational because the batteries are measured, but they are not treated.

10. Experiment because the batteries are treated.

11. Experiment. The treatment group consists of the subjects given the magnetic bracelets, and the control group consists of the subjects given the bracelets that have no magnetism. It is probably not possible to use blinding with this study, because the passengers could easily test their bracelets to see if they are magnetic (by holding them near metal).

12. Observational because the subjects were tested, but they were not given any treatment.

13. This is an observational, retrospective study examining how a characteristic determined before birth (identical or fraternal twins) affected mental skills later.

14. This is an observational, retrospective study comparing those who were texting and those who were not.

15. Experiment because the subjects are given a treatment. The treatment group consists of the 152 couples given the YSORT treatment. The control group consists of others not given any treatment.

16. This is an observational study.

17. This is an experiment. The treatment group consists of the genetically modified corn, and the control group consists of corn not genetically modified.

18. Observational because the subjects were surveyed, but they were not given any treatment.

19. This is an experiment. The treatment group consists of those treated with magnets. The control group consists of those given the non-magnetic devices.

20. This is a meta-analysis.

21. Confounding is likely to occur. If there are differences in effects from the two groups, there is no way to know if those differences are attributable to the treatment (fertilizer or irrigation) or the type of region (moist or dry). This confounding can be avoided by using blocks of fertilized trees in both the moist region and dry region and using blocks of irrigated trees in both the moist region and dry region.

22. Confounding is very possible. Those more comfortable with computers and Internet usage are more likely to choose participation in the Internet user group. It would be much better to randomly assign participants to the two groups. Also, it should be noted that the subjects are volunteers, so the whole study is subject to a self-selection sample bias.

23. Confounding is likely. If there are differences in the amounts of gasoline consumed, there would be no way to know whether those differences are due to the octane rating of the gasoline or the type of vehicle. Confounding can be avoided by using 87 octane gasoline in half of the vans and half of the sport utility vehicles and by using 91 octane gasoline in the other vehicles. Even better, conduct an experiment in which identical vehicles are driven under the same conditions (speed, distance, etc.) with the different octane gasolines.

24. The primary problem with this experiment is that the sample sizes are too small to draw meaningful statistical results about potential benefits of aspirin in preventing heart attacks.

25. Subjects clearly know whether they are treated with running, so confounding is possible from a placebo effect. Moreover, there is no objective way to measure back pain, so different subjects may report changes in back pain differently. Also, there could be an experimenter effect that can be avoided with blinding of those who evaluate results.

26. Confounding is possible due to experimenter effects, because the physician's knowledge of who received the treatment could affect their judgments of how well the skin is responding. It would be better to use blinding so that the physicians do not know who is given the treatment and who is given the placebo.

27. In this case, the tennis balls play the role of placebos. Confounding can occur because of a placebo effect and/or an experimenter effect, because it will be obvious to both subjects and experimenters whether they are lifting heavy weights. It would be better to use the heavy weights and the tennis balls with the same subjects at different times, to see if the different regimens affect blood pressure.

28. Confounding is possible because the researchers may have a bias toward one or the other brand. The experiment should use blinding so the researchers evaluate the results without knowing which brand they are evaluating.

29. The control group consists of those who do not listen to Beethoven and the treatment group consists of those who do listen to Beethoven. By using subjects who are coded, blinding could be used so that those who measure intelligence are not influenced by their knowledge about the participants.

30. This should be a double-blind experiment with a control group consisting of subjects given placebos and a treatment group consisting of those treated with Lipitor. Participants should be randomly assigned to the two groups.

31. The control group consists of cars using gasoline without the ethanol additive, and the treatment group consists of cars using gasoline with the ethanol additive. Blinding is not necessary for the cars, and it is probably unnecessary for the researchers because the mileage will likely be measured with objective tools.

32. The control group consists of houses with wood siding, and the treatment group consists of houses with aluminum siding. Blinding is not necessary for the houses, and it is unnecessary for the researchers if the longevity is measured with objective tools. Blinding would be difficult to implement because anyone would know whether a home has wood siding or aluminum siding.

Section 1.4

1. Peer review is a process in which experts in a field evaluate a research report before the report is published. It is useful for lending credibility to the research because it implies that other experts agree that it was carried out properly.

2. Selection bias occurs when researchers select their sample in a biased way, and participation bias occurs when the participants themselves decide to be included in the study.

3. When participants select themselves for a survey, those with strong opinions about the topic being surveyed are more likely to participate, and this group is typically not representative of the general population.

4. Confounding variables are those that result in effects that are mixed in such a way that we cannot determine the effects of the specific variables being studied.

5. Does not make sense. A survey involving a large sample could be very poor if it involves a poor sampling method (such as a self-selected sample), while a smaller sample could produce much better results if it involves a sound sampling method (such as a simple random sample).

6. Makes sense. By surveying only people on college campuses, the sample is not likely to be representative of the population of adult Americans.

7. Does not make sense. We often don't even know if confounding variables are present, so we cannot be certain they've all been taken into account.

8. Does not make sense. A mean weight loss of only 1.7 pounds is so small that it has little practical significance.

9. The survey was funded by a source that can benefit through increased sales fostered by the survey results, so there is a potential for bias in the survey. Guideline 2 is most relevant.

10. Because the treatment group involved only college students, the results do not necessarily apply to the general population of smokers of all ages. Guideline 1 is most relevant.

11. "Good" is not well defined and is difficult to measure. Guideline 4 is most relevant.

12. The weather and soil conditions in California are different from those in Arizona. It is impossible to determine whether differences are due to the irrigation systems or the weather and soil conditions. Guideline 5 is most relevant.

13. The sample is self-selected, so participation bias is a serious issue. Guideline 3 is most relevant.

14. The conclusion of a "landslide" given in the headline is not consistent with the results of the poll, so Guideline 7 is most relevant.

15. The wording of the question is biased and tends to elicit negative responses. Guideline 6 is most relevant.

16. It is extremely difficult to measure the value of counterfeit goods. Guideline 4 is most relevant.

17. Because much of the funding was provided by Mars and the Chocolate Manufacturers Association, the researchers may have been more inclined to provide favorable results. The bias could have been avoided if the researchers were not paid by the chocolate manufacturers. If that was the only way the research could be done, then the researchers should institute procedures to ensure that they publish all results, including negative ones.

18. The sample is self-selected and involves a small proportion of replies, so the responses were more likely to come from those with strong feelings about the issues. A better sampling procedure, such as interviews with 4,500 randomly chosen women, would have helped.

19. The wording of the question was biased to strengthen opposition against a particular candidate, and is likely to be a "push poll" financed by supporters of another candidate, rather than a legitimate poll. A better sampling method would involve questions devoid of such bias.

20. The list of property owners is clearly biased toward those who can afford to own property. Also, the mail survey will result in a self-selected sample. A better sampling method, such as the simple random sampling used by most polling companies, was needed.

21. The word "wrong" in the first question could be misleading. Some people might believe that abortion is wrong, but still favor choice. The second question could also be confusing, as some people might think that "advice of her doctor" means that the woman's life is in danger, which could alter their opinion about abortion in this situation. Groups opposed to abortion would be likely to cite the results of the first question, while groups favoring abortion would be more likely to cite the results of the second question.

22. The first question refers only to "governmental programs," which many people consider to be generally wasteful. The second question lists specific programs that are very popular. Groups favoring tax cuts would be likely to cite the results of the first question, and groups opposed to tax cuts would be more likely to cite the results of the second question.

23. The first question requires a study of Internet dates. The second question involves a study of married people to determine whether their first date was an Internet date. The two questions are likely to yield very different types of information. For example, the second question would tell you what percentage of current marriages began with Internet dates, but the first would only tell you how often Internet dating leads to marriage, which might not be any different from how often other forms of dating lead to marriage. The goals of the study need to be better defined and the questions framed to meet the goals.

24. The first question involves a study of those who teach introductory courses, but the second question involves a study of the full-time faculty. The first question requires examination of introductory courses to determine which are taught by full-time faculty. The second question requires a survey of full-time faculty.

25. The first question involves a study of college students in general, and the second question involves a study of those who do binge drinking. The first question might be addressed by surveying college students. The second question would be addressed by surveying binge drinkers, and it would be much more difficult to survey this group.

26. The first question involves a study of college students in general, and the second question involves a study of people who take statistics courses (including high school students, employees who take a statistics course at their place of work, and so on). The first question involves a group that is much easier to identify, locate, and survey.

27. The headline refers to drugs whereas the story refers to "drug use, drinking, or smoking." Because "drugs" is generally considered to consist of drugs other than cigarettes or alcohol, the headline is very misleading.

28. The story does not include the margin of error. With a sample size of 500, the margin of error is around 4 percentage points, so the likely range for a satisfying sex life is 78% to 86%, and the likely range for job satisfaction is 75% to 83%. Because these ranges overlap, it is quite possible that the headline is incorrect.

29. No information is given about the meaning of "confidence." The sample size and margin of error are not provided.

30. The report appears to be making a statement about the quality of restaurants in New York City (the "Big Apple"), but much information is missing. What about restaurants with ratings of 30 or 28? What criteria were used for the ratings? Who actually did the rating?

31. No information is given to justify the statement that "more" companies try to bet on weather forecasting. If only the four cited companies are new, the increase is relatively insignificant.

32. The article suggests that China is thrown off balance, so that some change is having a dramatic effect, but no information is given about any such change.

Chapter 1 Review Exercises

1. a. 12% to 16%   **b.** All adults in the United States   **c.** It is an observational study because the subjects were not treated or modified in any way. The variable of interest is whether the subject has a tattoo. For this survey, that variable has two values: yes or no.   **d.** The value is a sample statistic because it is based on the sample of 2,320 adults, not the population of all adults.   **e.** No, because it would be a self-selected sample with a likely participation bias.   **f.** Use a computer to randomly generate Social Security numbers, identify the people with those Social Security numbers, then select those people.   **g.** Select a sample of subjects in each state.   **h.** Select all of the adults

in several randomly selected streets.   **i.** Select every 10th adult, by address, on each street in a city.   **j.** Select your classmates.

2. a. It is a sample chosen in such a way that every sample of the same size has the same chance of being selected from the population.   **b.** No, because not every sample of 2,007 people has the same chance of being selected. For example, it is impossible to select a sample with 2,007 people who are all in the same primary sampling unit.   **c.** Repeat the process of randomly selecting a primary sampling unit, then randomly selecting one of its members. If anyone is selected more than once, ignore the subsequent selections.

3. a. No, because there is no information about the occurrence of headaches among people who do not use Bystolic. Based on the given information, it is possible that 7% of the population experiences headaches whether or not Bystolic is used.   **b.** It appears that Bystolic users have about the same rate of headaches as those given a placebo, so headaches do not appear to be an adverse reaction to Bystolic use.   **c.** This is an experiment because subjects are given a treatment.   **d.** With blinding, the trial participants do not know whether they are getting Bystolic or a placebo, and those who evaluate the results also do not know.   **e.** An experimenter effect occurs if the experimenter somehow influences subjects through such factors as facial expression, tone of voice, or attitude. It can be avoided through the use of blinding.

4. a. Because the word "welfare" has negative connotations, the second question should be used.   **b.** The first question, because it is more likely to elicit negative responses.   **c.** This is largely a subjective judgment. Some professional pollsters are opposed to all such questions that are deliberately biased, but others believe that such questions can be used. An important consideration is that survey questions can modify how people think, and such modification should not occur without their awareness or agreement.

Chapter 1 Quiz

1. c   **2.** a   **3.** a   **4.** b   **5.** c   **6.** a   **7.** c
8. c   **9.** b   **10.** c   **11.** b   **12.** b   **13.** b
14. c   **15.** b

Section 2.1

1. Qualitative data consist of values that can be placed into different nonnumerical categories, whereas quantitative data consist of values representing counts or measurements.

2. No. The numbers are a different way to represent names, but they don't represent counts or measurements of anything, so they are qualitative data.

3. Yes. Data consist of either qualities or quantities (numbers), so all data are either qualitative or quantitative.

4. No. The ZIP codes do not consistently measure distance from the east coast or from any other reference point, so they are qualitative data.

5. Blood groups are qualitative because they don't measure or count anything.

6. The measures are quantitative because they consist of counts of the number of white blood cells.

7. Braking reaction times are quantitative because they consist of measurements.

8. The specialties of physicians are qualitative because they don't measure or count anything.

9. The answers to multiple choice test questions are qualitative because they don't measure or count anything.

10. The responses are qualitative because they don't measure or count anything.

11. The television shows are qualitative because they don't measure or count anything.

12. The number is quantitative because it consists of a count.

13. Head circumferences are quantitative because they consist of measurements.

14. Shoe sizes are quantitative because they consist of measurements.

15. The grade point averages are quantitative because they are measures of course grades.

16. The area codes are qualitative because they don't measure or count anything.

17. The numbers are discrete because only the counting numbers are used, and no values between counting numbers are possible.

18. The weights are continuous data because they can be any value within some range of values.

19. The numbers are discrete because they are counts. Only the counting numbers are used, and no values between counting numbers are possible.

20. The lengths of times are continuous because they can be any value within a range of values.

21. The times are continuous data because they can be any value within some range of values.

22. The times are continuous data because they can be any value within some range of values.

23. The numerical test scores are discrete data because they can be counting numbers only.

24. The numbers are discrete data because they can be counting numbers only.

25. The speeds are continuous data because they can be any value within some range of values.

26. The numbers are discrete data because they can be only one of several values, with no values in between.

27. The numbers are discrete data because they can be counting numbers only.

28. The amounts are continuous because they can be any value within a range of values.

29. Ratio

30. Ordinal

31. Nominal

32. Interval

33. Ordinal

34. Nominal

35. Ordinal

36. Ratio

37. Nominal

38. Ratio

39. Ratio

40. Ordinal

41. The ratio level does not apply. The ratio is not meaningful because the stars don't measure or count anything. Differences between star values are not meaningful.

42. The ratio level applies. The 40 mi/h speed is four times as fast as the speed of 10 mi/h.

43. The ratio level does not apply. IQ tests do not measure intelligence on the type of scale required for the ratio level. Someone with an IQ score of 140 is not necessarily twice as intelligent as someone with an IQ score of 70.

44. The ratio level does not apply. 0° F is an arbitrary setting and does not represent "no heat," so the ratio of "twice" is not meaningful.

45. The ratio level applies. The age of 1,000 years is twice as long as the age of 500 years.

46. The ratio level applies. The 200 year age is twice as long as the 100 year age, so the ratio of "twice" is meaningful.

47. The ratio level applies. The $150,000 salary amount is twice the salary of $75,000, so the ratio of "twice" is meaningful.

48. The ratio level does not apply. The SAT scores do not measure the qualification for college in a way that results in values at the ratio level of measurement, so the ratio of "twice" is not meaningful.

49. The data are quantitative and are at the ratio level. The data are continuous. The times have a natural zero starting point and the times can be any values within a particular range.

50. The data are qualitative and are at the nominal level of measurement. The nations don't measure or count anything.

51. The data are qualitative and are at the nominal level of measurement. The numbers are different ways to express the names, and they don't measure or count anything.

52. The data are quantitative and are at the ratio level of measurement. The data are continuous (although they appear to be discrete after rounding to some unit, such as days). The times have a natural zero starting point and they can be any values within some range.

53. The data are quantitative and are at the interval level of measurement. The data are discrete because they consist of whole numbers only. The years are measured from an arbitrary reference (the year 0), not a natural zero starting point. Differences between the years are meaningful values, but ratios are not meaningful.

54. The data are qualitative and are at the nominal level of measurement. Even though numbers are used, the actual data do not represent counts or measurements.

55. The data are qualitative at the ordinal level of measurement. The ratings consist of an ordering, but they do not represent counts or measurements.

56. The data are qualitative at the nominal level of measurement. The labels consist of categories only, and they do not measure or count anything.

Section 2.2

1. It is a random error, because there is no way to predict when a clerk will write a number incorrectly or whether the number will be too high or too low.

2. The absolute error is 0.002 kg and the relative error is 0.2%.

3. Because the recorded weight has so many decimal places, it is very precise. Because the recorded weight is not very close to the true weight, it is not very accurate.

4. The given population size is very precise (in fact, it is exact), but it is not likely to be very accurate because it is impossible to know the exact population at any specific point in time.

5. Does not make sense. The number is too precise. Some species have not yet been identified, and new species are being found.

6. Does not make sense. It would be sensible if "absolute error" is used instead of "relative error." The astronomer could have a 1% relative error measuring something in light years, while the biologist could easily have a 3% relative error measuring something microscopic.

7. Makes sense. With random error, we expect that about half of the errors will be in favor of the supermarket and half will be in favor of the consumer.

8. Does not make sense. The fact that the number is very precise does not necessarily mean that it is accurate.

9. Mistakes tend to result from random errors, but dishonesty tends to result in systematic errors that benefit the taxpayer.

10. This is a systematic error resulting in altimeter readings that are too low throughout the entire flight.

11. Because about half of the batteries have more than 3.7 volts and half have more than 3.7 volts, it appears that we have a random error.

12. This is a systematic error that results in an underestimation of the true rate of fatalities that involve alcohol.

13. Random errors could occur when people make estimates because they don't recall the exact amounts. Systematic errors could occur when people tend to report that they donate more money than they actually donate so that they appear to be more benevolent.

14. Examples of random errors would include taxpayers who make honest mistakes or instances when the income amounts are recorded incorrectly. Examples of systematic errors would include situations where all or most of the tax returns were prepared using a computer program that calculated incomes incorrectly (but the same way for everyone) or where incomes were systematically recorded from, say, taxable income (which means deductions have already been subtracted) rather than total income.

15. Random errors could occur when people don't know their actual weight and they report an amount that is wrong. Systematic errors could occur when people intentionally lie about their weight by reporting a value that is considerably lower than the true weight.

16. Random errors could occur with reading the scale or with a scale that is inaccurate. A systematic error occurs by incorrectly including the weight of the cup with the weight of each M&M.

17. Random errors could occur with an inaccurate radar gun or with honest mistakes made when the officer records the speeds. Systematic errors could occur with a radar gun that is incorrectly calibrated so that it consistently reads too high or too low.

18. Random errors occur with mistakes made in the calculations or honest mistakes made in assigning values to the products. Systematic errors occur with counterfeit products that are bought and sold without the knowledge of the police or the tendency of the police to exaggerate the value of the counterfeit goods sold.

19. Random errors occur with errors in the calculations. Systematic errors may undercount the actual number if some cigarettes are sold illegally without the tax being collected.

20. Random errors occur when the groundskeeper positions the ruler too far or too short from the previous measurement. A systematic error occurs if the groundskeeper consistently locates the ruler too far (or too close) from its previous location or by measuring along a path that is longer than the straight-line distance.

21. Absolute error: $1,750. Relative error: 141% (using the absolute error of $1,750 and the correct bill amount of $1,245).

22. Absolute error: 2 ounces. Relative error: 11.1%.

23. Absolute error: −$1. Relative error: −36.4%.

24. Absolute error: −1 doughnut. Relative error: −7.7%.

25. a. These errors are random. If they were systematic, there would be a tendency for the measurements to be all too high or all too low.   **b.** Use the average because it is likely to be in error by less than most of the individual measurements and it is more reliable than any single measurement.   **c.** Systematic errors might result if the students have all been taught to use the tape measure in the same incorrect way.   **d.** No. If there is a systematic error in the measurements, that same error will be present in the average.

26. a. These errors are random. If they were systematic, there would be a tendency for the measurements to be all too high or all too low.   **b.** Use the average because it is likely to be in error by less than most of the individual measurements and it is more reliable than any single measurement.   **c.** Systematic errors might result from a scale that consistently reads too high (or too low) or from consistently recording the weights incorrectly.   **d.** No. If there is a systematic error in the measurements, that same error will be present in the average.

27. The Department of Transportation scale is more precise because its weight of 3,298.2 lb has more decimal places than the other weight of 3,250 lb. The manufacturer's scale is more accurate because its weight of 3,250 lb is closer to the true weight. (3,250 lb is in error by 23 lb, but the weight of 3,298.2 lb is in error by 25.2 lb.)

28. The measurement of 175.5 cm is more precise (because it has one more decimal place). The measure of 175 cm is more accurate because it is closer to the true height.

29. The measurement of 52.88 kilograms is both more precise (because it has more decimal places) and more accurate (because it is closer to your weight).

30. The measurement of 52.88 kilograms is more precise (because it has more decimal places), but the measurement of $52\frac{1}{2}$ kilograms is more accurate (because it is closer to your weight).

31. The given number is very precise, but it is not likely to be accurate. We cannot even measure the population this precisely today, and the uncertainties were certainly greater in 1860.

32. The given number is very precise, but it is not likely to be accurate. Some crash victims died a considerable time after the crash, so they are not included in the total. Some reporting errors are likely. Some deaths might have questionable causes, such as death due to a heart attack during a crash that resulted in no obvious personal injuries.

33. The given number is very precise, but it is not likely to be accurate. There are many people in China who are not counted. The census of any nation is likely to be in error by considerable amounts due to the inherent difficulties in conducting a national census. The population of China likely changed during the course of the year.

34. It should be easy to accurately measure the height of a building with precision to the nearest foot, so the claim is believable, though it would be good to verify the source before quoting this number.

35. It is easy to accurately measure the height of a structure with a reasonable degree of precision, such as the nearest 1/10 foot, but the given number has far too much precision (it implies knowing the height to a size smaller than an atom!), so the claim is not believable.

36. The number of cell phone users is constantly changing, and it probably cannot be determined with great accuracy, but the reported number is not very precise, so it is believable, though we would need more information to know if it really is accurate.

37. The number of college students is constantly changing with new enrollments and dropouts, so the given number must be an estimate. The precision of the given number is probably unjustified. The given number seems to be accurate, but we would need more information to know if it really is accurate.

38. The government classifies species as endangered or threatened through a detailed evaluation scheme, meaning it actually keeps a list of the species identified as such. So the number is believable, though of course it should be verified before you use it.

Section 2.3

1. a. $414 million   **b.** 8.4%   **c.** 2.8% decrease
d. $5066.35 million

2. If all of the plaque is removed, 100% of it is removed. It is not possible to reduce plaque by more than 100%, so the 300% figure is incorrect.

3. The statement that the margin of error is 3% incorrectly implies that the error can be up to 3% of 25%, which is 0.75%, but the actual error can be up to 3 percentage points away from 25% (from 22% to 28%).

4. 472; 1126

5. Does not make sense. The number of people with cell phones may have increased by 1.2 million, but 1.2 million people is not a percentage, so the statement is not sensible.

6. Does not make sense. The two percentages involve different base amounts, so the 5% cut will be more money than the 5% raise. For example, if an employee earns $100,000, the 5% cut will reduce that salary to $95,000, then the 5% raise will increase the salary from $95,000 to $99,750. The 5% cut followed by the 5% raise results in a salary that is less than the current salary.

7. Makes sense. The 100% increase indicates that the loan rate doubled. For example, if the loan rate doubled from 5% to 10%, it increased the loan rate by 100%.

8. Does not make sense. For example, if the annual loan rate was 5%, an increase of 100 percentage points would make the loan rate 105%—far higher than any reputable bank would charge (though some "payday" loan operations have interest rates in that range).

9. a. 75/100 or 3/4, 0.75, 75%   **b.** 3/8, 0.375, 37.5%   **c.** 4/10 or 2/5, 0.4, 40%   **d.** 80/100 or 4/5, 0.8, 80%

10. a. 350/100 or 7/2, 3.50, 350%   **b.** 25/10 or 5/2, 2.5, 250%
c. −44/100 or −11/25, −0.44, −44%   **d.** −2/1, −2.00, −200%

11. a. 71%   **b.** 33%   **c.** 41%   **d.** 81%

12. a. 59%   **b.** 41%   **c.** 51%   **d.** 45%

13. −38% (38% decrease from 1990)

14. 18% (18% increase from 1980)

15. 18% (18% increase from January 1996)

16. 20% (20% increase from 2000)

17. 14%. The *Wall Street Journal* has 14% more circulation than *USA Today*.

18. 3%. The Toyota Camry had 3% more sales than the Honda Civic.

19. −25%. O'Hare handled 25% fewer passengers than Atlanta's Hartsfield Airport.

20. 34%. France had 34% more tourists than the United States.

21. 66

22. 71

23. 834

24. 8127

25. 140%. The truck weighs 100% of the car's weight plus another 40%.

26. 124%. The area of Norway is the area of Colorado plus another 24%.

27. 80%. The population of Montana is the population of New Hampshire minus 20%.

28. 58%

29. Yes. Three percentage points corresponds to a range from 86% to 92%, which was intended. A margin of error of 3% corresponds to 3% of 89%, which is $0.03 \times 0.89 = 0.0267$, but this is not what was meant.

30. The percentage can also be expressed as 16% $\pm$ 2%, so the margin of error is 2 percentage points.

31. −15.5 percentage points; −22.7%

32. −7.6 percentage points; −28.0%

33. 22 percentage points; 56.4%

34. 21 percentage points; 77.8%

Section 2.4

1. An index number is a ratio without any units, such as dollars. The number appears to be the actual cost of gasoline in 2011, not an index number.

2. The computer costs in 2012 are 15% of those in the year 2000.

3. Yes. The Consumer Price Index is based on the prices of goods, services, and housing, so increases in those prices will result in an increase in the Consumer Price Index.

4. No. The Consumer Price Index is based on prices of goods, services, and housing, so wages can rise or fall independent of the Consumer Price Index.

5. 409.8   6. 214.8

7. $1.10   8. $2.31

9. 19.9, 23.1, 78.2, 78.8, 100, 182.1

10. 86.1, 100, 338.9, 341.7, 433.3, 788.9

11. $45.44   12. $42.60

13. The tuition in 2010 is about 660% of the amount in 1980, which means a rise of about 550%. The CPI in 2010 is 264.7% of the CPI in 1980, for a rise of 164.7%. So the cost of tuition rose much more than the cost of typical goods, services, and housing.

14. The tuition in 2010 is about 630% of the amount in 1980, which means a rise of about 530%. The CPI in 2010 is 264.7% of the CPI in 1980, for a rise of 164.7%. So the cost of tuition rose much more than the cost of typical goods, services, and housing.

15. The home prices in 2010 are about 162% of the amount in 1990, which means a rise of about 62%. The CPI in 2010 is 167% of the CPI in 1980, for a rise of 67%. So the cost of homes rose at a slightly lower rate than the CPI.

16. The home prices in 2010 are about 182% of the amount in 1990, which means a rise of about 82%. The CPI in 2010 is 167% of the CPI in 1980, for a rise of 67%. So the cost of homes rose at a slightly higher rate than the CPI.

17. $582,000; $180,000

18. $209,497; $120,112

19. $1,591,667; $1,491,667

20. $1,067,039; $167,598

Chapter 2 Review Exercises

1. a. 599   b. Discrete, because only the counting numbers are used, and no values between counting numbers are possible.
c. 60%   d. 732   e. Ratio   f. Nominal

2. a. 242   b. 6%   c. Nominal.   d. Because the poll uses respondents who themselves chose to participate, the sample is self-selected and is not likely to accurately reflect the opinion of the population.

3. The health care spending in 2010 is 3125% of the amount in 1973, but the CPI in 2010 is 491% of the 1973 CPI, so health care spending grew at a much greater rate than the general rate for goods, services, and housing.

4. a. $2.78   b. $5.77   c. The minimum wage increases did not keep up with inflation, so workers earning the minimum wage had relatively lower wages in 2009 than in 1996.

Chapter 2 Quiz

1. Continuous   2. Ratio   3. −2.4 cm   4. 2.7%

5. Nominal   6. 5%   7. 52   8. 178 cm; 179.18 cm

9. 123.7   10. $52,972

Section 3.1

1. A frequency table has two columns, one for categories and one for frequencies. Categories are the different values that a variable may have (for example, different eye colors or different ranges of income). Frequencies are the counts of the numbers of data values in each category.

2. 0.10 or 10%, 0.30 or 30%, 0.40 or 40%, 0.15 or 15%, 0.05 or 5%

3. 4, 16, 32, 38, 40

4. No. We know that there are 24 values that fall between $0 and $999, but there is no way to identify the exact 24 original values.

5. The statement does not make sense, because a column in the frequency table must consist of frequency counts, but neither of the columns of *State* and *Median Income* consist of frequency counts.

6. The statement makes sense, because the frequency for a category can be 0.27 or 27% of the total of all frequencies.

7. The statement does not make sense, because each individual frequency is a whole number, and any sum of frequencies must also be a whole number, so 25.5 is not a possible value of a cumulative frequency.

8. The statement makes sense. If the width of the bins increases, fewer bins are needed to accommodate all of the data.

9.

Grade	Frequency	Relative frequency	Cumulative frequency
A	3	10.0%	3
B	10	33.3%	13
C	11	36.7%	24
D	2	6.7%	26
F	4	13.3%	30
Total	**30**	**100%**	**30**

10.

Rating	Frequency	Relative frequency	Cumulative frequency
1 Star	2	5.0%	2
2 Stars	12	30.0%	14
3 Stars	18	45.0%	32
4 Stars	6	15.0%	38
5 Stars	2	5.0%	40
Total	**40**	**100%**	**40**

11.

Weight (pounds)	Frequency	Relative frequency	Cumulative frequency
0.7900–0.7949	1	1/36	1
0.7950–0.7999	0	0	1
0.8000–0.8049	1	1/36	2
0.8050–0.8099	3	3/36	5
0.8100–0.8149	4	4/36	9
0.8150–0.8199	17	17/36	26
0.8200–0.8249	6	6/36	32
0.8250–0.8299	4	4/36	36
Total	**36**	**100%**	**36**

12.

Weight (pounds)	Frequency	Relative frequency	Cumulative frequency
0.7750–0.7799	4	4/36	4
0.7800–0.7849	13	13/36	17
0.7850–0.7899	15	15/36	32
0.7900–0.7949	4	4/36	36
Total	**36**	**100%**	**36**

13. The age category of 40 to 49 includes the most actors.

Age	Number of Actors
20–29	1
30–39	27
40–49	35
50–59	14
60–69	6
70–79	1

14.

Temperature	Frequency
96.9–97.2	3
97.3–97.6	6
97.7–98.0	9
98.1–98.4	10
98.5–98.8	17
98.9–99.2	2
99.3–99.7	3

15.

Category	Frequency	Relative frequency
A	12	24%
B	9	18%
C	12	24%
D	11	22%
F	6	12%
Total	**50**	**100%**

16.

Category	Frequency	Cumulative frequency
A	1	1
B	6	7
C	7	14
D	9	23
F	2	25
Total	**25**	**25**

17. a. 200   **b.** 142   **c.** 16%   **d.** 0.135, 0.155, 0.210, 0.200, 0.140, 0.160   **e.** 27, 58, 100, 140, 168, 200

18. a. 70.24 million families   **b.** 63.14 million families   **c.** 50.6%   **d.** 10.1%

19. a.

Rating	Frequency	Relative frequency
0–2	20	38.5%
3–5	14	26.9%
6–8	15	28.8%
9–11	2	3.8%
12–14	1	1.9%
Total	**52**	**100%**

b.

Rating	Frequency	Relative frequency
0–2	33	63.5%
3–5	19	36.5%
6–8	0	0%
9–11	0	0%
12–14	0	0%
Total	**52**	**100%**

c. The Dvorak keyboard appears more efficient because it has more lower ratings and fewer high ones.

20. a. The variable of age is quantitative and is at the interval level of measurement. The variable of transportation is qualitative and is at the nominal level of measurement.

b.

		Transportation				
		1	2	3	4	5
Age	1	2	1	1	1	0
	2	1	2	0	0	0
	3	1	0	1	1	2
	4	0	0	1	1	0
	5	1	1	0	0	3

Section 3.2

1. The distribution of data is the way data values are spread over all possible values.

2. A histogram or line chart provides a graph that has a shape, and it is much easier to understand the shape of the distribution from a graph than from a list of data values.

3. By arranging bars from highest to lowest, the Pareto chart draws attention to the most important categories. The pie chart does not do that.

4. The original data values cannot be recreated from a histogram, but they can be recreated from a stemplot. An advantage of a stemplot over a histogram is that it allows you to recreate the original list of sample values.

5. Does not make sense. Histograms require quantitative data, and names of political parties are qualitative.

6. Does not make sense. A pie chart cannot be used to depict time-series data.

7. Makes sense.

8. Does not make sense. Because the data set is the same for both graphs, both must have the same peak.

9. A histogram would work well to show the frequencies of the different categories of incomes.

10. A Pareto chart or pie chart would work well, but the Pareto chart would do a better job of showing the most frequent political parties.

11. A time series graph would be effective in showing any trend in the number of movie theaters.

12. A histogram would work well to show the frequencies of the different age categories.

13. a.

Reading Category	Relative Frequency
Popular fiction	50.4%
Cooking/Crafts	10.2%
General nonfiction	8.9%
Religious	8.6%
Psychology/Recovery	6.3%
Technical/Science/Education	5.6%
Art/Literature/Poetry	3.7%
Reference	2.6%
All other categories	2.5%
Travel/Regional	1.3%

b.

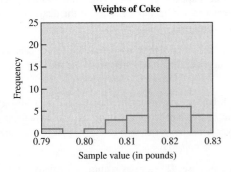

c. The Pareto chart makes it easier to see which categories are more popular.

14. a. 40   **b.** 38   **c.** 22   **d.** 550

15.

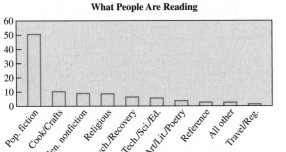

16.

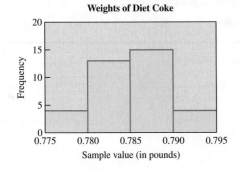

17.

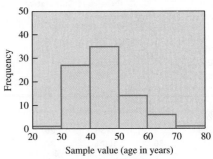

18.

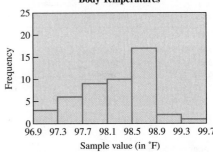

19.

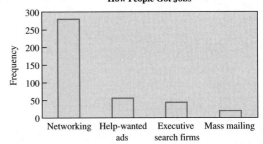

20.

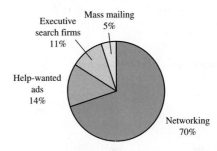

By using a vertical scale and arranging the bars in descending order, the Pareto chart is more effective in showing the relative importance of job sources.

21.

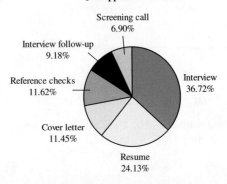

22.

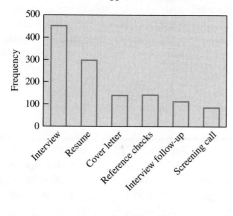

By using a vertical scale and by arranging the bars in descending order, the Pareto chart is more effective in showing the relative importance of the mistakes.

23.

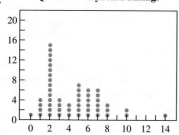

24.

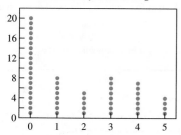

The Dvorak configuration appears to be better because the keyboard ratings are lower with the Dvorak configuration than the QWERTY configuration.

25.

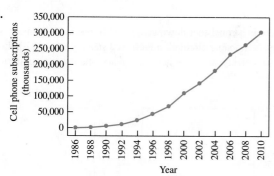

The graph does not appear to show linear growth.

26.

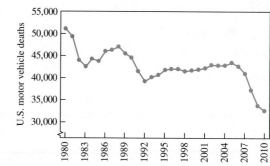

The decreasing trend in deaths results from vehicles that are built to be stronger in crashes, the use of seat belt and air bags, and stricter enforcement of laws against drinking and driving.

27.

Stem	Leaves
6	7
7	25
8	5899
9	09
10	0

The lengths of the rows are similar to the height of bars in a histogram, so that longer rows of data correspond to higher frequencies.

28.

Stem (tens)	Leaves (units)
6	449
7	01112334444555555666778899
8	0011122233346899
9	0024
10	
11	
12	0

The stemplot shows the distribution of the data. The lengths of the rows of data show which values occur more often. By turning the stemplot on its side, we get a display of the distribution that is somewhat similar to the heights of bars in a histogram.

Section 3.3

1. Yes, any histogram can be depicted as a three-dimensional histogram (as in Figure 3.18). The three-dimensional version of a histogram does provide more visual appeal, but it does not provide any additional information.

2. A multiple bar graph has two or more sets of bars, and it is helpful because it provides contrast between two or more sets of data.

3. Geographical data are raw data corresponding to different geographic locations. Two examples of displays of geographical data are color-coded maps and contour maps.

4. A contour connects locations with the same value of some variable, such as temperature. Contours close together indicated that the variable changes substantially over relatively short distances. Contours far apart indicated that the variable changes by small amounts over relatively short distances.

5. Does not make sense.

6. Does not make sense.

7. Makes sense.

8. Makes sense.

9. a. The numbers of females consistently outnumber the numbers of males. The numbers of both genders are increasing gradually over time.

b.

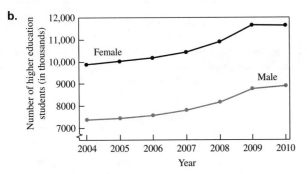

10. a. Home prices have generally risen for all regions until around 2006 when all regions began a steady decline. **b.** Prices are higher in the West and Northeast than in the Midwest and South. The recent decline in prices is most dramatic in the West. Prices in the South and Midwest are not very far apart.

11. a. Males consistently have much higher median salaries than females, and both males and females have steadily increasing salaries over time. **b.** The line graph makes it easier to examine the trend over time. Also, the line graph is less cluttered, so that the information is easier to understand and interpret.

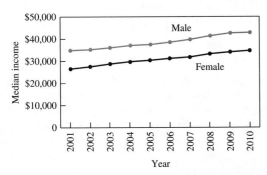

12. a. Total numbers in both categories have gone up due to increasing population, so the rates give better information about trends. The marriage rate has remained somewhat steady, but the divorce rate grew more during the period from 1900 to 1980, and it has declined slightly since then. **b.** By placing the adjacent bars side by side, the multiple bar graph makes the comparisons easier. The total heights of the bars in the stacked plot aren't meaningful because they combine two very different rates.

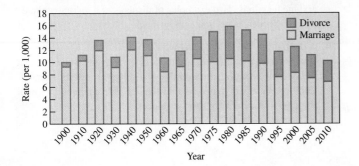

13. The stack plot shows that entitlement spending (Social Security, Medicare, Medicaid, and insurance subsidies) is projected to grow dramatically in coming decades. The graph shows that, without an increase in tax revenues, this entitlement spending would consume all government revenue by mid-century, which would leave no room for any other government programs, including the military. The clear message is that without changes, entitlement spending is on a track that would cause severe budgetary problems.

14. a. 1970: 475,000 males and 365,000 females; 2010: 730,000 males and 965,000 females. **b.** Around 1980. **c.** The numbers of bachelor's degrees awarded to males and females both increased over the years, and the proportion of degrees earned by females has also increased. **d.** The stack plot makes it easy to see the changing total, but a double bar chart (one bar for men and one for women) would make the male/female comparison easier.

15. There appears to be a general trend of higher melanoma mortality in southern states and western states. This might be the result of people in these states spending more time outdoors, exposed to sunlight. As a researcher, you might be particularly interested in regions that deviate from general trends. For example, a county in eastern Washington state stands out with a very high melanoma mortality. You might first want to verify that the data point is accurate, and not an error of some type. If it is accurate, you may want to find out why this one county has a higher melanoma mortality than surrounding counties.

16. There are significant regional differences. For example, the probability that a black student has white classmates is generally much higher in the north than the south and in rural areas than in urban areas.

17.

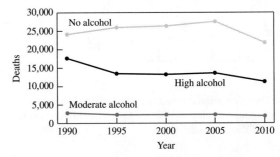

18. There is a consistent downward trend in the number of newspapers. Newspaper circulation increased steadily until 1970, then it was relatively stable for about 20 years, and it has been decreasing since 1990.

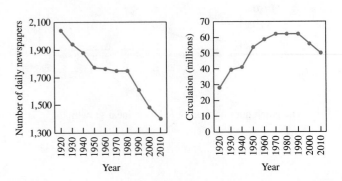

19.

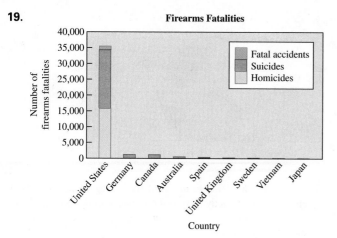

Firearms Fatalities

20.

Mothers in the Labor Force

Section 3.4

1. The difference of 1 foot is not meaningful. By drawing a bar graph with a vertical scale starting at a value such as 130 feet, the difference will be greatly exaggerated.

2. The population numbers are one-dimensional, but it is likely that the drawings of people will appear to have two or three dimensions, so the data will be distorted. The general name for such drawings of people or objects is a pictograph.

3. The scale is called an exponential scale. The advantage of such a scale is that it allows us to include values that vary over a very large range.

4. The first sugar cube has a volume of 1 cm^3 and the second sugar cube has a volume of 8 cm^3. The illustration is misleading because the production doubled, but the graph makes it appear that the production increased by a factor of 8.

5. By starting the vertical scale at 30 instead of 0, the difference between the two amounts of fuel consumption is greatly exaggerated. The graph would not be misleading if it is changed so that the vertical scale begins at 0.

6. The scale used for the population values does not begin at 0, so the differences are exaggerated. By starting the population scale at 0, the graph will not be misleading.

7. The amount of oil used by each country appears to be related to the volume of the barrels in the pictograph, when it is really related to the height of the barrels. The U.S. consumption is about four times that of Japan, not 64 times as suggested by the volumes of the barrels.

8.

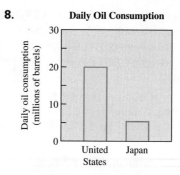

Daily Oil Consumption

9. a, b. Because of the three-dimensional appearance of the pie charts, the sizes of the wedges on the page do not match the percentages. Instead they show how the wedges would look if the entire pie was tilted at an angle. This distortion makes it difficult to see the true relationships among the categories.

c.

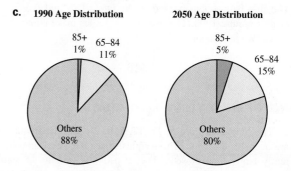

1990 Age Distribution   2050 Age Distribution

d. In 2050, there will be relatively more older people and fewer younger people in the U.S. population than in 1990. Note, however, that because of population growth, all age groups are expected to be larger in number in 2050 than in 1990.

10. a.

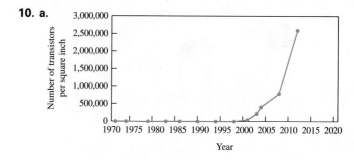

b.

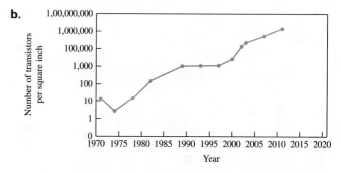

c. The graph in part b makes it easier to see the changes in the first eight years. By using an exponential scale, the graph in part b makes it easier to fit in all of the data. The graph in part a gives us a better picture of the true nature of the overall rate of change.

11. The percent change in the CPI is greatest in 1990. In 2009, the percent change in CPI is negative, so prices decreased from those in 2008. Prices have increased in every year (except for 2009), but the increases near the end of the time period are lower than those near the beginning.

12. a. Using a scale that starts at zero would make the variations appear much smaller.   **b.** A horizontal line (no variation) is consistent with the error bars, so the claim of a seasonal variation is not supported by the data.

13. The actual minimum wage in unadjusted dollars has either remained constant or has risen steadily since 1955. The purchasing power increased from 1955 to 1968, then decreased from 1968 to 1989, and it has been relatively stable in recent years. The purchasing power in 2011 is just slightly higher than it was in 1955.

14. The first baby boom peaked in the late 1950s and the second peaked around 1990. The overlapping scales make comparisons of the two baby booms easier.

Chapter 3 Review Exercises

1. a. The frequencies are 5, 12, 12, 5, 0, 2.   **b.** The frequencies are 5, 5, 16, 8, 2.   **c.** The weights of regular Pepsi are consistently larger than those of Diet Pepsi. The weights of regular Pepsi are larger because of the sugar content that is not included in the cans of Diet Pepsi.

2. a. The relative frequencies are 0.139, 0.333, 0.333, 0.139, 0, 0.056.   **b.** The cumulative frequencies are 5, 17, 29, 34, 34, 36.

3. a.

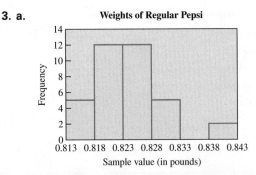

b.

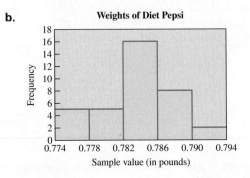

c. The shapes of the histograms are not dramatically different, indicating that the distributions of the weights are similar. However, the range of values is very different, indicating that the weights of regular Pepsi are considerably higher than those of Diet Pepsi.

4.

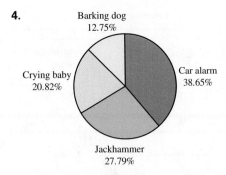

5. By drawing attention to the most frustrating sounds, the Pareto chart is more effective. By using a meaningful scale, the Pareto chart also does a better job of showing the relative importance of the different frustrating sounds.

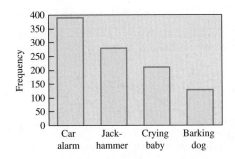

6. The graph has a vertical scale that does not begin with zero, so the difference between the two frequencies is exaggerated. The graph makes it appear that adoptions decreased to a level that is roughly 10% of the number in 2005, but that is not the case. Instead of decreasing to roughly 10% of the 2005 number, the actual decrease is to about 45% (or a little under half) of the 2005 number.

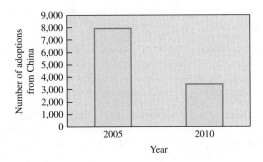

Chapter 3 Quiz

1. Histogram

2. Pareto chart

3. 61 min, 62 min, 62 min, 62 min, 62 min, 67 min, 69 min

4. There are 15 values between 20 and 29 inclusive.

5. 25% of the values are between 20 and 29 inclusive.

6. There are 80 values equal to or less than 49.

7. Using a vertical scale that does not start at zero causes the difference between the two frequencies to be exaggerated.

8.

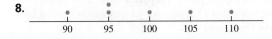

9. 0, 1, 1, 10, 11, 12, 49, 49

10. The second snowplow will appear to have an area that is nine times as large as the area of the first snowplow, but the amounts increased by a factor of three instead of a factor of nine.

Section 4.1

1. No. The mean refers specifically to the sum of all data values divided by the number of data values. The term "average" could refer to any one of several different statistics, including the mean, median, or mode.

2. In this case, the median would provide a value that is more typical. The mean would be distorted by one high value.

3. Yes. The executive secretary's income is much higher than the incomes of the other 23 students, so it is an outlier. In general, the definition of an outlier is somewhat vague, so an outlier cannot be clearly and objectively identified. (However, statisticians have created rules that can be applied to identify outliers in some cases.)

4. No. Each of the 50 values is given the same weight, but some states have very small populations while others have very large populations. The calculation of the mean should therefore account for the population of each state.

5. Does not make sense.

6. Makes sense.

7. Makes sense.

8. Makes sense.

9. Mean. For the population of basketball players, there are only a few heights that are extreme, and they are not extreme by very large amounts. In this case, the mean would be better because it takes every height into account.

10. Median. A few of the players receive substantially higher salaries. These outlier salaries would have a strong effect on the mean, but the median would be unaffected by them.

11. Median. The exact values of "over" ages are not known, so the mean cannot be computed.

12. Mean. It is unlikely that any of the pulse rates will be very far from almost all of the others. In this case, the mean would be better because it takes every value into account.

13. Mean: 53.3; median: 52.0; mode: none

14. Mean: 217.3 hours; median: 235.0 hours; mode: 235.0 hours

15. Mean: 58.3 sec; median: 55.5 sec; mode: 49 sec

16. Mean: 98.44; median: 98.40; mode: 98.4

17. Mean: 0.188; median: 0.165; mode: 0.16

18. Mean: 91.3 min; median: 93.5 min; mode: 92 min, 95 min, 98 min

19. Mean: 0.9194 g; median: 0.9200 g; there is no mode

20. Mean: 5.619 g; median: 5.595 g; mode: 5.58 g

21. a. Mean: 157,586; median: 104,100   **b.** Alaska is an outlier on the high end. Without Alaska the mean is 81,317 and the median is 78,650.   **c.** Connecticut is an outlier on the low end. Without Connecticut (but with Alaska) the mean is 182,933 and the median is 109,050.

22. a. Mean: 0.8124; median: 0.8161   **b.** 0.7901 is an outlier.
c. Mean: 0.8161; median: 0.8163

23. a. 85.0   **b.** 100   **c.** No; a mean of 90 for five quizzes requires a total of 450 points, which you cannot reach even with 100 on the fifth quiz.

24. a. 73.0   **b.** 85   **c.** 77.5

25. Mean score of your students: 74.7; median score: 70.0. So it depends on the meaning of "average." If it is mean, then your students are above average in their scores. If it is median, then your students are below average.

26. Mean height on your team: 6' 8.6"; median height on your team: 6'6". So it depends on the meaning of "average." If it is mean, then your team is above average height. If it is median, then your team has below average height.

27. 0.39 pound

28. No. The classes have different numbers of students.

29. Each student is taking three classes with enrollments of 20 each and one class with an enrollment of 100, so the mean class size for each student is (20 + 20 + 20 + 100)/4 = 40. There are three classes with 100 students each and 45 classes with 20 students each, so the 48 courses have a mean class size of (300 + 900)/48 = 25. The two means are very different.

30. 81.25, which is rounded to 81.3.

31. 0.417; it is the average number of hits per at-bat.

32. No, unless the number of at-bats in all previous games is the same (only 4!) as in the most recent game.

33. Probably not. The overall rate of defects would be 3% only if both sites produce the same number of batteries.

34. a. 0.500   **b.** 0.929   **c.** Yes, because it is possible (though not easy!) to average more than 1 point per at-bat.

35. The outcome is no, by 600 votes to 400 votes.

36. 3.27

37. Mean: 1.9; median: 2; mode: 1. The mode of 1 correctly indicates that the smooth yellow peas occur more than any other phenotype, but the mean and median don't make sense with these data at the nominal level of measurement.

38. The population center has been moving westward with time, reflecting increased population in western states relative to eastern states.

Section 4.2

1. The distribution is not symmetric.

2. Uniform

3. Because the professors have successfully completed a rigorous program of education, their IQ scores are likely to all be higher than average and closer together, so they probably have less variation than IQ scores of randomly selected adults. The lower variation of IQ scores of the professors results in a graph that is narrower with less spread than the graph of IQ scores from the randomly selected adults.

4. Skewness refers to a lack of symmetry with the graph more spread out on one side than on the other side.

5. Does not make sense.

6. Makes sense.

7. Makes sense.

8. Does not make sense.

9. Two modes, left-skewed, wide variation.

10. One mode, right-skewed, moderate variation.

11. Single-peaked, nearly symmetric, moderate variation.

12. The distribution is bimodal and is roughly symmetric. The gap between the left portion of the distribution and the right portion reflects the fact that this graph actually includes two different populations: the pennies made before 1983 and the pennies made after 1983.

13. a. The distribution is right-skewed. It is not symmetric. **b.** 409 (half of 817)

14. a. At least 183 days had no rain, so the data set has at least 183 values of 0 in. (There were actually 283 days with no rain, but there is no way to determine this from the given information.) **b.** Because the data set has so many days with no rain (at least 183 days), the graph is right-skewed because there is a mode at 0 in., and the frequency for this mode is substantially higher than any other modes that might be present.   **c.** It rained 82 days, but there is no way to determine that from the given information. Based on the given information, we know only that there are at least 183 values of 0 in., so the number of days that it rained must be 182 or less.

15. a. One mode of $0   **b.** Right-skewed, because there will be many students making little or nothing.

16. a. One mode   **b.** Symmetric

17. a. One mode   **b.** Symmetric

18. a. Two modes   **b.** Symmetric

19. a. One mode   **b.** Symmetric

20. a. Two modes   **b.** Symmetric

21. a. One mode   **b.** Left-skewed

22. a. One mode   **b.** Right-skewed

23. a. One mode   **b.** Right-skewed

24. a. One mode   **b.** Right-skewed

25. a. One mode   **b.** Right-skewed

26. a. One mode   **b.** Right-skewed

Section 4.3

1. The range is found by subtracting the lowest sample value from the highest. It is a measure of variation. A major disadvantage of the range is that its value depends on only the lowest and highest sample values and it does not take every value into account.

2. The standard deviation of 10 mg is better because it indicates less variation, so the tablets have a more consistent amount of aspirin.

3. The statement is incorrect because it defines the standard deviation as a value that depends on the minimum and maximum values, but the standard deviation uses every data value.

4. About 25% of the salaries (or 204 salaries) are $4,355,101 or less.

5. Does not make sense.

6. Makes sense.

7. Makes sense.

8. Does not make sense.

9. Range: 45.0 words; standard deviation: 15.7 words

10. Range: 381.0 hours; standard deviation: 105.9 hours

11. Range: 26.0 sec; standard deviation: 9.5 sec

12. Range: 1.00; standard deviation: 0.30

13. Range: 0.170; standard deviation: 0.057

14. Range: 33.0 min; standard deviation: 8.9 min

15. Range: 0.1160 g; standard deviation: 0.0336 g

16. Range: 0.320 g; standard deviation: 0.075 g

17. *Cat on a Hot Tin Roof:* range = 10.0 and standard deviation = 2.6.   *The Cat in the Hat:* range = 3.0 and standard deviation = 0.9. There is much less variation among the word lengths in *The Cat in the Hat*.

18. From the 1920s and 1930s, the range is 3.90 and the standard deviation is 1.48. From recent winners: the range is 3.50 and the standard deviation is 1.19. This represents a significant decline in variation. Coupled with the fact that the mean BMI also dropped significantly, it supports the charge that contest winners today are expected to conform to a thinner body type.

19. One day: range = 11.0 degrees and standard deviation = 2.6. degrees. Five day: range = 15.0 degrees and standard deviation = 4.5 degrees. As expected, there appears to be greater variation in errors from the five-day forecast.

20. No treatment: range = 0.350 kg and standard deviation = 0.127 kg. Fertilizer and irrigation: range = 2.110 kg and standard deviation = 0.859 kg. The trees with no treatment appear to have weights that vary much less than the weights of the treated trees.

21. a. 5th percentile   **b.** 69th percentile   **c.** 48th percentile

22. a. 74th percentile   **b.** 89th percentile   **c.** 91st percentile

23. Answers for data set (1) only: **a.** Histogram has frequency of 7 for data value of 9; no other values.   **b.** low value = 9; lower quartile = 9; median = 9; upper quartile = 9; high value = 9   **c.** 0

24. Answers for data set (1) only: **a.** Histogram has frequency of 7 for data value of 6; no other values.   **b.** low value = 6; lower quartile = 6; median = 6; upper quartile = 6; high value = 6.   **c.** 0

25. a. Faculty: mean = 2, median = 2, range = 4; student: mean = 6.2, median = 6, range = 9.   **b.** Faculty: low = 0, lower quartile = 1, median = 2, upper quartile = 3, high = 4; student: low = 1, lower quartile = 4, median = 6, upper quartile = 9, high = 10.   **c.** Faculty: st. dev. = 1.2; student: st. dev. = 3.0.   **d.** Faculty: range/4 = 1.0; student: range/4 = 2.3.

26. a. School: mean = 20.6, median = 20, range = 8; Downtown: mean = 30.1, median = 31, range = 12.   **b.** School: low = 17, lower quartile = 18, median = 20, upper quartile = 23.5, high = 25; Downtown: low = 24, lower quartile = 27, median = 31, upper quartile = 33, high = 36.   **c.** School: st. dev. = 2.9; Downtown: st. dev. 3.9.   **d.** School: range/4 = 2; Downtown: range/4 = 3.

27. a. First 7: mean = 58.3, median = 57, range = 4; Last 7: mean = 56.1, median = 54, range = 23.   **b.** First 7: low = 57, lower quartile = 57, median = 57, upper quartile = 61, high = 61; Last 7: low = 46, lower quartile = 47, median = 54, upper quartile = 64, high = 69.   **c.** First 7: st. dev. = 1.9; Last 7: st. dev. = 8.7.   **d.** First 7: range/4 = 1; Last 7: range/4 = 5.75.

28. a. Beethoven: mean = 38.8, median = 36, range = 42; Mahler: mean = 75, median = 80, range = 44.   **b.** Beethoven: low = 26, lower quartile = 29, median = 36, upper quartile = 45, high = 68; Mahler: low = 50, lower quartile = 62, median = 80, upper quartile = 87.5, high = 94.   **c.** Beethoven: st. dev. = 13.13; Mahler: st. dev. = 15.44   **d.** Beethoven: range/4 = 10.5; Mahler: range/4 = 11.

29. The second site has better quality. Although the second site has a mean that is a little farther from the target of 12 volts, the standard deviation is much less, so the voltages will be more consistent.

30. Jerry, with the larger standard deviation, serves up more very small servings and is more likely to generate complaints.

31. Having more stocks in the portfolio reduces the variation in the returns, because some stocks are likely to go up and others go down. The reduced variation means reduced risk, though also a lower likelihood of large gains.

32. Today, players generally have batting averages closer to the mean of 0.260 than in the past. High batting averages (such as 0.350) are therefore less common today.

Section 4.4

1. A false positive occurs when the test indicates use of banned substances for someone who does not actually use them. A false negative occurs when the test indicates that banned substances are not used by someone who actually does use them. A true positive occurs when the player tests positive and he actually does use banned substances. A true negative occurs when the player tests negative and he does not use banned substances.

2. The test indicated that Jennifer is pregnant. She might actually be pregnant or she might not be pregnant if the result is a false positive.

3. True negative

4. Yes, it is possible that one of the quarterbacks can be better in each half while being worse overall.

5. Makes sense.

6. Does not make sense.

7. Does not make sense.

8. Does not make sense.

9. Josh; Josh; Jude

10. Allan; Allan; Abner

11. a. New Jersey; Nebraska

12. a. The SAT scores declined in each of the five grade categories between 1988 and 1998.   **b.** The overall average SAT score increased by 10 points between 1988 and 1998.   **c.** Even though the average SAT score decreased in each grade category, the overall average increased. This is an illustration of Simpson's Paradox, resulting from the changing percentages of students who were in each grade category, with higher percentages of students in the three "A" categories and lower percentages in the "B" and "C" categories. The total number of students is also probably different in the two years, and this also contributes to the paradox.

13. a. Whites: 0.18%; nonwhites: 0.54%; total 0.19%.   **b.** Whites: 0.16%; nonwhites: 0.34%; total 0.23%.   **c.** The rate for both whites and nonwhites was higher in New York than in Richmond, yet the overall rate was higher in Richmond than in New York. The percentage of nonwhites was significantly lower in New York than in Richmond.

14. This is a case of Simpson's paradox because the Gazelles had a greater improvement in both the Weight Training group and the No Weight Training group than did the Cheetahs, but less improvement overall. The apparent paradox arises because the two teams had different percentages of runners involved in weight training. And yet overall the Cheetahs had a greater average improvement than did the Gazelles. 65% of the Cheetahs and 50% of the Gazelles were in the Weight Training group.

15. a. Spelman has a better record for home games (34.5% vs. 32.1%) and away games (75.0% vs. 73.3%), individually.   **b.** Morehouse has a better overall average.   **c.** Morehouse has a better team, as teams are generally rated on overall records.

16. a. Drug A cured 5/100 = 0.050 of the women. Drug A cured 400/800 = 0.500 of the men. Drug B cured 101/900 = 0.112 of the women. Drug B cured 196/200 = 0.980 of the men. Thus Drug B cured a higher percentage of the men and of the women.   **b.** Drug A cured 405/900 = 0.450 of all patients. Drug B cured 297/1100 = 0.270 of patients.   **c.** In this case, because of the great differences in cure rates for men and women, it does not make sense to combine the results. Drug B is better for men and women.

17. a. 57 appeared to be lying. 15 of them were telling the truth, and 42 were actually lying. 26% of those who appear to be lying were not actually lying.   **b.** 41 appeared to be telling the truth. 32 of them were actually telling the truth, and 9 were lying. 78% of those who appeared to be telling the truth were actually truthful.

18. a. Of those with the disease, 48 out of 60, or 80%, test positive.   **b.** Of the 48 + 788 = 836 who tested positive, only 48 or 48/836 = 0.057 (= 5.7%) actually had the disease. This percentage and the one in part a are different; the first (part a) is the percentage of diseased people who tested positive and the latter is the percentage of those who tested positive who have the disease.   **c.** A patient who tests positive has only about a 6% chance of actually having the disease. This is, however, about 4 times the actual incidence rate of the disease (1.5%). Thus the test is 4 times more likely to identify the disease than would occur by just picking a person at random.

19. A higher percentage of women than men were hired in both the white-collar and blue-collar positions, suggesting a hiring preference for women. Overall, 20% of the 200 females (40) who applied for white-collar positions were hired, and 85% of the 100 females (85) who applied for blue-collar positions were hired. Thus $40 + 85 = 125$ of the $200 + 100 = 300$ females who applied were hired, a percentage of 41.7. Overall, 15% of the 200 males (30) who applied for white-collar positions were hired, and 75% of the 400 males (300) who applied for blue-collar positions were hired. Thus $30 + 300 = 330$ of the $200 + 400 = 600$ males who applied were hired, a percentage of 55.0.

20. Treatment A had a higher cure rate than Treatment B in the first trial and again in the second trial. Overall, Treatment A cured $40 + 85 = 125$ out of $200 + 100 = 300$, a cure rate of $125/300 = 0.417$; Treatment B cured $30 + 300 = 330$ out of $200 + 400 = 600$, a cure rate of $330/600 = 0.550$. Thus, Treatment B had the overall higher cure rate.

21. **a.** In the general population, $57 + 3 = 60$ of the 20,000 in the sample are infected. This is an incidence rate of $60/20000 = 0.003$ or 0.3%. In the "at-risk" population, $475 + 25 = 500$ of the 5,000 in the sample are infected. This is an incidence rate of $500/5000 = 10.0\%$. **b.** In the "at-risk" category, 475 out 500 infected with HIV test positive, or 95%. Of those who test positive, 475 out of $475 + 225 = 700$ have HIV, a percentage of $475/700 = 0.679$ or 67.9%. These two figures are different because they measure different things. The 700 who test positive include 225 who were false positives. While the test correctly identifies 95% of those who have HIV, it also incorrectly identifies some who do not have HIV. Thus only 67.9 % of those who test positive actually have HIV. **c.** In the "at-risk" population, a patient who tests positive for the disease has about a 68% chance of actually having the disease. This is nearly 7 times as great as the incidence rate (10%) of the disease in the at-risk category. Thus the test is very valuable in identifying those with HIV. **d.** In the general population, patients with HIV test positive 57 times out of 60, or 95% of the time. Of those who test positive, 57 out of $57 + 997 = 1054$ actually have HIV, a percentage of $57/1054 = 0.054$ or 5.4%. These two figures are different because they measure different things. The 1054 who test positive include 997 who were false positives. While the test correctly identifies 95% of those who have HIV, it also incorrectly identifies some who do not have HIV. Thus only 5.4 % of those who test positive actually have HIV. **e.** In the general population, a patient who tests positive for the disease has about a 5.4% chance of actually having the disease. This is 18 times as great as the incidence rate (0.3%) of the disease in the general population. Thus the test is very valuable in identifying those with HIV.

Chapter 4 Review Exercises

1. **a.** Non-filtered: Mean is 1.28 mg and median is 1.15 mg. Filtered: Mean is 0.93 mg and median is 1.00 mg. **b.** Non-filtered: Range is 0.70 mg and standard deviation is 0.26 mg. Filtered: Range is 0.80 mg and standard deviation is 0.24 mg.

c.

Non-filtered

1 1.15   1.7
1.1   1.4

Filtered

0.4   1 1.2
0.8   1.1

d. Non-filtered: Standard deviation is estimated to be 0.18 mg and it is actually 0.26 mg. Filtered: Standard deviation is estimated to be 0.20 and it is actually 0.24 mg. Both estimates are in the general ballpark of the actual vales. **e.** There does appear to be a difference. The nicotine in filtered cigarettes appears to be generally lower, so the filters appear to be at least somewhat effective.

2. **a.** 80%   **b.** 1.1 mg

3. **a.** 0   **b.** Although both batteries have the same mean lifetime, the batteries with the smaller standard deviation are better because their lifetimes will be closer to the mean, and fewer of them will strand drivers by failing sooner than expected. **c.** The outlier pulls the mean either up or down depending on whether it is above or below the mean, respectively. **d.** The outlier has no effect on the median. **e.** The outlier increases the range. **f.** The outlier increases the standard deviation.

Chapter 4 Quiz

1. Mean   **2.** 77.0   **3.** 27.0

4. The amount of variation among the pulse rates.

5. The mean and median are equal.

6. Only the second, third, and fourth statements could be correct.

7. 1.50 sec and 2.50 sec

8. The result of 6.8 is very roughly in the general ballpark of the correct answer of 10.9.

9. Zero

10. Minimum value, first quartile, median (or second quartile), third quartile, maximum value.

Section 5.1

1. The word *normal* has a special meaning in statistics. It refers to a specific category of distributions, all of them being bell-shaped.

2.

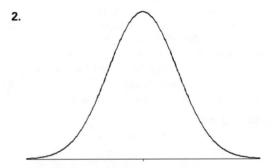

3. No. The ten different possible digits are all equally likely, so the graph of the distribution will tend to be flat, not bell-shaped.

4. 0.88   **5.** Makes sense.   **6.** Makes sense.

7. Does not make sense.   **8.** Does not make sense.

9. Distribution b is not normal. Distribution c has the larger standard deviation.

10. Distribution a is not normal. Distribution b has the larger standard deviation.

11. Normal. It is common for a manufactured product, such as quarters, to have a distribution that is normal. The weights typically vary

above and below the mean weight by about the same amounts, so the distribution has one peak and is symmetric.

12. Not normal. Many people have no income or a relatively small income, and there are a few people with extremely large incomes, so the distribution is not symmetric.

13. Not normal. The outcomes of the numbers between 1 and 49 are all equally likely, so the distribution tends to be uniform, not normal. A graph of the distribution will tend to be very flat, not bell-shaped.

14. Normal. There is a small number of scores that are very high and there is a small number of scores that are very low, and the distribution is likely to peak at the value of the mean, which is around 1518.

15. Normal. Such physical measurements generally tend to be normally distributed. A small number of females will have extremely low counts, a small number will have extremely high counts, and the distribution will tend to peak around the value of the mean.

16. Not normal. Many flights are not delayed at all, so there will be a very high number of zero times. With such a large peak at the low end, the distribution will not be bell-shaped.

17. Not normal. The waiting times will tend to be uniformly distributed.

18. Not normal. Many cars will not have been ticketed, so there will be a very high number of zero amounts. With such a large peak at the low end, the distribution will not be bell-shaped.

19. Not normal. Movie lengths have a minimum length (zero) and no maximum length.

20. Nearly normal. Data values are distributed symmetrically about the mean. Many physiological variables have normal distributions.

21. Nearly normal. The deviations from the mean are evenly distributed around the mean.

22. While the data values are distributed fairly symmetrically about the mean, the distribution does not have right and left tails for the extreme weights. In other samples, this variable could be normally distributed, but in this particular sample, it is not.

23. a. 1   **b.** 0.20   **c.** 0.80   **d.** 0.35   **e.** 0.45

24. a. 1   **b.** 0.50   **c.** 0.30   **d.** 0.70   **e.** 0.20

25. a. 115   **b.** 15%   **c.** 45%   **d.** 15%

26. a. 155   **b.** 15%   **c.** 15%   **d.** 45%

Section 5.2

1. 0

2. Below the mean by 2 standard deviations.

3. No. The rule applies to normal distributions, but the outcomes from a die roll have a uniform distribution, not a normal distribution.

4. A z-score can be positive or negative. A negative z-score corresponds to a value that is below the mean. Percentiles cannot be negative. They represent a cumulative percentage of scores, so the smallest theoretical percentile is 0%.

5. Does not make sense.   **6.** Does not make sense.

7. Does not make sense.   **8.** Makes sense.

9. a. 50%   **b.** 84%   **c.** 2.5%   **d.** 84%   **e.** 81.5%

10. a. 50%   **b.** 84%   **c.** 34%   **d.** 81.5%   **e.** 47.5%

11. a. 68%   **b.** 95%   **c.** 99.7%   **d.** 47.5%

12. a. 95%   **b.** 99.85%   **c.** 34%

13. 50%   **14.** 84.13%   **15.** 2.28%   **16.** 30.85%

17. 97.72%   **18.** 30.85%   **19.** 93.32%   **20.** 0.13%

21. 68.26%   **22.** 95.44%   **23.** 84.00%   **24.** 62.47%

25. 50.00%   **26.** 84.13%   **27.** 84.13%   **28.** 6.68%

29. 99.38%   **30.** 0.62%   **31.** 0.13%   **32.** 0.26%

33. 68.26%   **34.** 38.30%   **35.** 89.25%   **36.** 15.52%

37. In all cases, 5% of coins are rejected. Cents: 2.44 gm to 2.56 gm; Nickels: 4.88 gm to 5.12 gm; Dimes: 2.208 gm to 2.328 gm; Quarters: 5.530 gm to 5.810 gm; Half dollars: 11.060 gm to 11.620 gm.

38. a. 11.51%   **b.** 1.79%   **c.** 2.28%

39. a. 6.68%   **b.** 48.01%   **c.** 36.54%

40. a. Approximately 90.5 percentile
   b. A score of approximately 687.

41. a. 0.47%   **b.** Approximately 4%   **c.** 30.055 to 30.745 inches   **d.** The best estimate would be the mean of the readings, or 30.4 inches.

42. 88%.

43. Approximately 96.33%

44. a. Approximately 35%   **b.** Approximately 18%
   c. Approximately 0.96

Section 5.3

1. Because the digits are equally likely, they have a uniform distribution. Because the sample means are based on samples of size 3 drawn from a population that does not have a normal distribution, we should not treat the sample means as having a normal distribution.

2. σ represents the standard deviation of the population, and n represents the size of the sample.

3. Yes. If the original population is normally distributed, the sample means will be normally distributed for any sample size, not just large sample sizes.

4. The sample means have a distribution that can be approximated by a normal distribution.

5. a. The mean is 100 and the standard deviation is 2.
   b. The mean is 100 and the standard deviation is 1.6.
   c. With larger sample sizes (as in part b), the means tend to be closer together, so they have less variation, which results in a smaller standard deviation.

6. a. The mean is 1518 and the standard deviation is 32.5.
   b. The mean is 1518 and the standard deviation is 6.5.
   c. With larger sample sizes (as in part b), the means tend to be closer together, so they have less variation, which results in a smaller standard deviation.

7. a. The mean is 6.5 and the standard deviation is 0.384.
b. The mean is 6.5 and the standard deviation is 0.345.
c. With larger sample sizes (as in part b), the means tend to be closer together, so they have less variation, which results in a smaller standard deviation.

8. a. The mean is 5.5 and the standard deviation is 0.410.
b. The mean is 5.5 and the standard deviation is 0.144.
c. With larger sample sizes (as in part b), the means tend to be closer together, so they have less variation, which results in a smaller standard deviation.

9. 69.2%; 99.9%   **10.** 54.0%; 84.1%   **11.** 15.9%; 95.4%

12. 11.2%; 15.3%

13. a. 0.35%   **b.** No, it appears that the mean is greater than 12.00 oz, but consumers are not being cheated because the cans are being overfilled, not under filled.

14. a. 0.62%   **b.** The z score of 19.4 is beyond the largest value in Table 5.1, so the likelihood is less than 0.02%. It is extremely unlikely that the mean for 60 times will be greater than 4.00 s.   **c.** Part a. The behavior of the individual times is more important than the behavior of batches of 60 times.

15. a. 57.93% chance   **b.** 97.72% chance   **c.** Although the mean head breadth of 100 men is very likely to be less than 6.2 in., there could be many individual men that could not use the helmets because they have head breadths greater than 6.2 in. Based on the result from part a, these helmets would not fit about 42% of men.

16. a. 42.07%   **b.** 2.28% chance. The system appears to be effective because it will be rarely overloaded.

17. a. 52.91%   **b.** Approximately a 73% chance.   **c.** Part a, because the seats will be occupied by individual women, not groups of women.

18. a. 51.99% chance   **b.** 88.49% chance   **c.** Instead of filling each bag with exactly 465 M&Ms, the company probably fills the bags so that the weight is as stated on the label. In any event, the company appears to be doing a good job of filling the bags.

19. a. 16 (5.74% of 280)   **b.** The standard z scores are beyond the range of those included in Table 5.1, so we conclude that it is very likely (more than a 99.96% chance) that the mean falls between the limits of 5.550 g and 5.790 g.   **c.** Part a, because the individual rejected quarters could result in lost sales and lower profits.

20. a. 46.02% chance   **b.** 6.68% chance. Yes, because an overweight condition could occur on 6.68% of such flights, and that risk is too high.

21. a. 27.4%   **b.** 0.2%   **c.** Part a more relevant, because the individual desks will be occupied by individual girls, not groups of girls.

22. a. 0.22   **b.** Essentially zero   **c.** 0.14

Chapter 5 Review Exercises

1. a. Because the 38 outcomes are all equally likely, the distribution is uniform, not normal.   **b.** Weights of members in a homogeneous population, such as twelve-year old girls, typically

have a normal distribution.   **c.** Approximately normal. Such test scores tend to be normally distributed.

2. a. 68%   **b.** 95%   **c.** Yes, because such an extreme score occurs only about 0.3% of the time.

3. a. 90 percentile   **b.** 1.29   **c.** No, the data value lies less than 2 standard deviations from the mean.   **d.** 0.0065   **e.** The temperature is unusual; it lies more than 2 standard deviations above the mean.   **f.** 99.22   **g.** 97.18   **h.** Fewer than 0.01%; yes, this temperature seems appropriate for claiming that a patient has a fever.   **i.** The sample mean is 6.6 standard deviations below the mean; the chance of selecting such a sample is extremely small. The assumed mean (98.60) may be incorrect.

Chapter 5 Quiz

1. b, c, and d are correct.   **2.** 95%   **3.** 95.44%   **4.** 850

5. 4.9   **6.** −1   **7.** 84%   **8.** 97.72%   **9.** 81.85%

10. Part c

Section 6.1

1. No, the argument is not valid. With 1000 births, any specific number of girls will have a very low probability, but 501 girls is very close to the 500 girls that is expected, so the result is not evidence that the method is effective.

2. No. Statistical significance refers specifically to a set of measurements or observations that are unlikely to occur by chance. The importance of the results is not relevant. It is possible that results might be statistically significant and not practically significant.

3. No. Statistical significance at the 0.05 level means that there is less than a 0.05 probability that the result occurred by chance, but a probability less than 0.05 does not necessarily mean that there is less than a 0.01 chance.

4. Yes. Statistical significance at the 0.01 level means that there is less than a 0.01 probability that the result occurred by chance, so it follows that there must be less than a 0.05 probability that the results occurred by chance.

5. Does not make sense.   **6.** Makes sense.

7. Makes sense.   **8.** Does not make sense.

9. Statistically significant. The results are very unlikely to occur by chance.

10. Statistically significant. The results are very unlikely to occur by chance.

11. Not statistically significant. We expect to get about one 6 in every six rolls, so not getting any 6s could easily occur by chance.

12. Not statistically significant. The result could easily occur by chance.

13. Statistically significant. It is very unlikely that when 20 adults are randomly selected, they are all women.

14. Statistically significant. The result is very unlikely to occur by chance.

15. Not statistically significant. The result could easily occur by chance.

16. Not statistically significant. The rate of headaches is about 6% in the treatment group and 5% in the control group, and the difference between those two rates could easily be explained by chance.

17. Even though the sample size is relatively small, such a large (21%) improvement in mileage is significant.

18. Based on the sample sizes and the large difference between the two success rates, it appears that the results are statistically significant.

19. With 945 babies, the number of girls would usually be around 472.5, so the result of 879 girls is a substantial departure from the results expected by chance. The results appear to be statistically significant.

20. The results do have statistical significance at the 0.05 level, but not the 0.01 level. The bed nets do appear to be effective, although they do not guarantee protection from malaria.

21. a. If 100 samples were selected, the mean temperature would be 98.20 or less in 5 or fewer of the samples. **b.** Selecting a sample with a mean this small is extremely unlikely and would not be expected by chance.

22. The likelihood of such a difference occurring simply by chance is less than 1 in 10,000.

23. This result is not significant at the 0.05 level because the probability of it occurring by chance when there is no real improvement is greater than 0.05.

24. The likelihood of such a difference occurring simply by chance is less than 1 in 100.

Section 6.2

1. $P(A)$ represents the probability that you answer the question correctly, $P(\text{not } A)$ is the probability that you do not answer the question correctly, and $P(\overline{A})$ also represents the probability that you do not answer the question correctly. Each of the three values is 0.5.

2. The reasoning is correct in that there are two events, but it is wrong to conclude that the probability of life is 1/2, because that assumes that the two possible outcomes of life and no life are equally likely, but they are not equally likely.

3. There is one chance out of 524,288 that all 20 babies are of the same gender. Such an event is unusual because its probability is so small.

4. The answer varies, but an answer in the neighborhood of 0.001 is reasonable.

5. Makes sense.   **6.** Makes sense.

7. Makes sense.   **8.** Does not make sense.

9. Does not make sense.   **10.** Makes sense.

11. 2/3, assuming that the die is fair and the outcomes are equally likely.

12. 1/5, assuming that the guess is random in the sense that all five possible answers have the same chance of being selected.

13. 18/38 or 9/19, assuming that the 38 slots all have the same chance of being selected.

14. 1/7

15. 1/365, assuming that births on the 365 days are equally likely.

16. 1/2 or 0.5, assuming that boys and girls are equally likely.

17. 1/2 or 0.5

18. 1/36, assuming that both dice are fair in the sense that all possible outcomes on each die are equally likely.

19. 0   **20.** 1   **21.** 6/7   **22.** 6/7

23. 0.45   **24.** 4/5 or 0.8   **25.** 0.720

26. 0.99   **27.** 0.22; 0.33; 0.44; 0.56   **28.** 20%

29. a. 1/8 = 0.125 (GGG)   **b.** 3/8 = 0.375 (BBG, BGB, GBB)   **c.** 1/8 = 0.125 (GBB)   **d.** 7/8 = 0.875 (GGG, BGG, GBG, GGB, BBG, BGB, GBB)   **e.** 4/8 = 0.5 (BBG, BGB, GBB, BBB)

30. a. 16: BBBB, BBBG, BBGB, BGBB, GBBB, BBGG, BGBG, BGGB, GBGB, GBBG, GGBB, BGGG, GBGG, GGBG, GGGB, GGGG.   **b.** 1/16 = 0.0625, (BBBB); 1/16 = 0.0625, (GGGG)   **c.** 1/16 = 0.0625 (BGBG)   **d.** 6/16 = 0.375 (BBGG, BGBG, BGGB, GBGB, GGBG, GGBB)

31. 0.60   **32.** 1/100 or 0.01   **33.** $P(\text{success}) = 0.86$

34. $67/73 \approx 0.92$

35. The probability that a person you meet at random is over 65 will be 34.7/281 = 0.123 in 2000 and will be 78.9/394 = 0.200 in 2050. Thus your chances will be greater in 2050.

36. a. 0.045   **b.** 0.066

37. a.

Outcomes for Rolling Four Fair Coins					
Coin 1	Coin 2	Coin 3	Coin 4	Outcome	Probability
H	H	H	H	HHHH	1/16
H	H	H	T	HHHT	1/16
H	H	T	H	HHTH	1/16
H	H	T	T	HHTT	1/16
H	T	H	H	HTHH	1/16
H	T	H	T	HTHT	1/16
H	T	T	H	HTTH	1/16
H	T	T	T	HTTT	1/16
T	H	H	H	THHH	1/16
T	H	H	T	THHT	1/16
T	H	T	H	THTH	1/16
T	H	T	T	THTT	1/16
T	T	H	H	TTHH	1/16
T	T	H	T	TTHT	1/16
T	T	T	H	TTTH	1/16
T	T	T	T	TTTT	1/16

b.

Probability Distribution for the Number of Heads in Rolling Four Fair Coins	
Result	Probability
4 Heads (0 Tails)	1/16
3 Heads (1 Tail)	4/16
2 Heads (2 Tails)	6/16
1 Head (3 Tails)	4/16
0 Heads (4 Tails)	1/16
Total	1

c. 6/16 = 0.375   **d.** 15/16 = 0.9375   **e.** 2 heads (Probability is 0.375)

38. a. $1/42 = 0.024$ **b.** $163/5964 = 0.027$ for a 34
c. $116/5964 = 0.019$ for a 6   **d.** The deviations are what might be expected by chance.

Section 6.3

1. If some process is repeated over many trials, the proportion of trials in which some event occurs will be close to the probability of the event. As the number of trials increases, the proportion of trials in which the event occurs gets closer to the probability of the event.

2. No. The expected value is 2.5 girls, not 3 girls. The expected value is a type of average over many trials, not the likely result for a single trial, so a result such as 2.5 girls is acceptable.

3. His betting strategy is unwise because the casino continues to have an advantage. His past performance does not affect future events. His reasoning is not correct. Although his proportion of winning bets is likely to increase, it could increase while he continues to lose more money. This flawed reasoning has caused many gamblers to lose large amounts of money.

4. The gambler's fallacy is the mistaken belief that a streak of bad luck makes a person due for a streak of good luck.

5. Makes sense.   **6.** Makes sense.

7. Does not make sense.   **8.** Makes sense.

9. No. You should not expect to get exactly 250 boys and 250 girls, because the probability of that particular outcome is very small. As the number of births increases, the proportion of girls should get closer to 0.5.

10. It means that the driver is "due" for an accident or a traffic citation. If accidents happen randomly, then the statement is not true. If accidents depend on driving habits, it may be a true statement.

11. Expected value is $-\$3$. Because the expected value is negative, the game will produce a loss in the long run, so you should not play.

12. Because you have one chance of winning and there are 10,000 different possible four-digit numbers, the probability of winning is 1/10,000. The expected value is $-50¢$.

13. 0.94; 0.74

14. a. $350   **b.** $35,000,000.

15. 15 minutes   **16.** $-\$0.43$; $-\$157$ **17.** $-\$0.78$; $-\$285$

18. a. $-26¢$ **b.** $-26¢$ **c.** Don't bet, because the expected value with no bet is 0, which is better than $-26¢$.

19. $-7.07¢$; $-1.4¢$

20. The expected value is $-22¢$. In the long run, you can expect to lose 22¢ for each 50¢ bet.

21. a. Decision 1—Option A: Expected value $= \$1,000,000$; Option B: Expected value $= \$1,140,000$. **Decision 2**—Option A: Expected value $= \$110,000$; Option B: Expected value $= \$250,000$ **b.** Responses are not consistent with expected values in Decision 1, but they are consistent in Decision 2.   **c.** It appears that people choose the certain outcome ($1,000,000) in Decision 1.

22. a. 1.2¢   **b.** 1.2¢ minus the cost of a stamp.

23. a. 0.5; 0.5   **b.** Loss of $10   **c.** Loss of $16   **d.** Loss of $20   **e.** 45%; 46%; 48%. The percentage of even numbers approaches 50%, but the difference between the numbers of even and odd numbers increases.   **f.** 60 even numbers

24. a. If you toss a head, with probability 0.5, the difference decreases to 23. If you toss a tail, with probability 0.5, the difference increases to 25.   **b.** On each of 1000 additional tosses, the probability is 0.5 that the difference will increase and 0.5 that it will decrease. So, overall, the difference is equally as likely to be greater than 24 as it is to be less than 24.   **c.** Once you have 24 more tails than heads, the difference is as likely to increase as to decrease; thus, the number of tails is likely to remain greater than the number of heads.   **d.** By part c, the number of tails is likely to exceed the number of heads at any time. The gambler's fallacy is that the difference between heads and tails will eventually be corrected.

Section 6.4

1. Because they are expressed as the numbers of births for each 1000 people in the population, the birth rates can be compared directly. The actual numbers of births should be considered in the context of the population sizes of the countries, so a comparison would require the numbers of births along with population sizes, and even then the comparison would not be easy because of the numbers involved.

2. Vital statistics are data related to births and deaths.

3. Life expectancy is the number of additional years a person with a given age can expect to live on average. A 30-year-old person will have a shorter life expectancy than a 20-year-old person because the 30-year-old person is not expected to live as many additional years as the 20-year-old.

4. All of the 20-year-old people in the United States will have times that they live beyond the age of 20, and those times have a mean of 58.8 years.

5. Does not make sense.   **6.** Does not make sense.

7. Makes sense.   **8.** Makes sense.

9. 2000: 0.0102; 2004: 0.0013; 2008: 0.0003. The year 2008 was the safest because it had the lowest number of fatalities per 1000 departures.

10. 2000: 0.1328; 2004: 0.0191; 2008: 0.0042. The year 2008 was the safest because it had the lowest number of fatalities per billion passenger miles.

11. 2000: 0.1381; 2004: 0.0201; 2008: 0.0043. The year 2008 was the safest because it had the lowest number of fatalities per million passengers.

12. 0.000000000004 (or 4×10^{-12}) deaths per passenger mile. The very small number is too inconvenient and is not very easy to comprehend.

13. 58.8 years   **14.** 60.7 years

15. 8.3 deaths per 10,000 people

16. 6.1 deaths per 10,000 people

17. Utah: 20.3; Maine: 10.3

18. a. California: 6.6; Alaska: 4.6   **b.** No. A death rate is computed from the number of deaths and the population size, so a state with the highest (or lowest) number of deaths does not necessarily have the highest (or lowest) death rate.

19. a. 4,319,400   **b.** 2,629,200   **c.** 1,690,200
d. 0.0054; 0.54%

20. a. 16,441,631   **b.** 9,357,026   **c.** 7,084,605
d. 0.0053; 0.53%

Section 6.5

1. The occurrence of event A does not affect the probability of event B, so the two events are independent.

2. Two events are non-overlapping if they cannot occur at the same time.

3. Sampling with replacement. The second outcome is independent of the first.

4. Yes. There is no overlap between an event occurring and the event not occurring.

5. Does not make sense.   **6.** Does not make sense.

7. Makes sense.   **8.** Makes sense.

9. 1/2 or 0.5   **10.** 1/16 or 0.0625   **11.** 1/260,000

12. a. 0.00238   **b.** 0.00200   **c.** If cases are to be selected for a follow-up study, it doesn't make much sense to select the same item twice, so select without replacement.

13. a. 0.0039   **b.** 0.00098   **c.** 0.125   **d.** 0.0625
e. 1/60 = 0.0167. There are 60 equally likely songs available for each selection. No matter which song is played first, the probability that the next one is the same is 1/60 = 0.0167.

14. a. Dependent if names are removed from the list after each call is made.   **b.** 0.303   **c.** 0.309   **d.** The results are nearly the same.

15. P(guilty plea or sent to prison) = 1014/1028 = 0.986

16. P(not guilty plea or not sent to prison) = 636/1028 = 0.619

17. 0.865   **18.** 0.191   **19.** 0.381

20. 0.549   **21.** 0.410   **22.** 0.940

23. 0.920   **24.** 0.730   **25.** 0.0195   **26.** 0.109

27. a. 0.733   **b.** 1   **c.** 0.643   **d.** 0.22

28. a. 156/1205 = 0.129. Yes. A high refusal rate results in a sample that is not necessarily representative of the population, because those who refuse may well constitute a particular group with opinions different from others.
b. 202/1205 = 0.168   **c.** 1060/1205 = 0.880
d. 358/1205 = 0.297

29. a. 0.5625   **b.** 0.375   **c.** 0.0625
d.
Event	Probability
AA	0.5625
Aa	0.1875
aA	0.1875
aa	0.0625

e. 0.9375

30. a. 0.20   **b.** 0.038   **c.** 0.400   **d.** 0.0083   **e.** 0.316
31. a. 0.518   **b.** 0.491
32. a. 0.17   **b.** 0.31   **c.** at least 23 partners

Chapter 6 Review Exercises

1. 576/3562 = 0.162   **2.** 1232/3562 = 0.346

3. 3466/3562 = 0.973   **4.** 96/3562 = 0.0270

5. 2330/3562 = 0.654   **6.** 0.0445 (not 0.0446)

7. 480/576 = 0.833

8. Based on data from J.D. Power and Associates, 16.7% of car colors are black, so any estimate between 0.05 and 0.25 is reasonable.

9. a. 0.73   **b.** 0.073   **c.** 1.35   **d.** 0.0014; you should doubt the stated yield.

10. a. 0.000152   **b.** 0.0000000231   **c.** 0.999696

Chapter 6 Quiz

1. 0.30   **2.** 0.49

3. Answer varies, but an answer such as 0.01 or lower is reasonable.

4. No, because the results could have easily occurred by chance.

5. 0.6   **6.** 427/586 = 0.729

7. 572/586 = 0.976   **8.** 0.00161

9. 10/586 = 0.0171   **10.** 10/24 = 0.417

Section 7.1

1. r is called the (linear) correlation coefficient, and it measures how well the paired sample data fit the pattern of a straight line.

2. A scatterplot is a graph in which the paired sample data are plotted as points. The scatterplot allows us to visualize the pattern of the points, and that is helpful in trying to determine whether there is a correlation between two variables.

3. No. It is possible that there is some correlation that corresponds to a scatterplot with a pattern that is not a straight-line pattern.

4. Both scatterplots will show that the points fit a straight-line pattern perfectly, but the straight line for the first set of paired data will be rising from left to right, whereas the straight line for the second set of paired data will be falling from left to right.

5. Does not make sense.   **6.** Does not make sense.

7. Does not make sense.   **8.** Makes sense.

9. Positive correlation. As the weight goes up, the cost will also go up.

10. The variables are not correlated.

11. Negative correlation. Airliners that weigh more tend to use more fuel, so as weight increases, the fuel consumption in miles per gallon will decrease.

12. Positive correlation because airliners that weigh more tend to use more fuel, so the fuel consumed increases as the weight increases.

13. The variables are not correlated.

14. Negative correlation because the temperature decreases as the altitude increases.

15. The variables are not correlated.

16. Negative correlation, because the prize money increases for golfers who shoot lower scores.

17. There is a strong positive correlation, with the correlation coefficient approximately 0.8 or 0.9.

18. An estimate of the correlation coefficient is $r = 0.7$ to $r = 0.8$. Forecasts are reasonably accurate for two days in the future. Results should be similar for other two-week periods.

19. a.

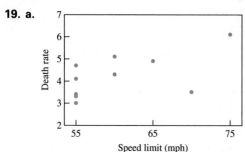

b. There is a moderate positive correlation; $r = 0.59$ exactly.
c. With the exception of Britain, the higher speed limits are generally associated with higher death rates. Death rates are also influenced by other factors beside speed limits.

20. a.

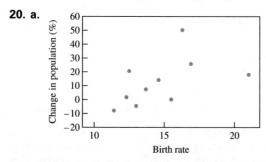

b. There is a moderate positive correlation ($r = 0.50$ exactly).
c. The birth rate does give some indication of population growth rate, but other factors, such as immigration and death rates, also affect population growth.

21. a.

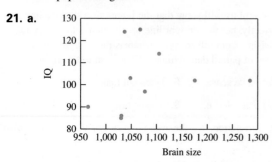

b. Because the points are scattered with no obvious pattern, there does not appear to be a correlation between brain size and IQ. The value of the correlation coefficient is approximately 0.2. **c.** The data suggest that there is not a correlation between brain size and IQ.

22. a.

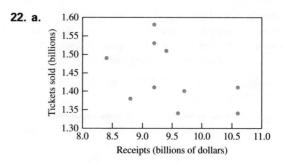

b. There is a weak negative correlation showing that high ticket sales correlated with lower revenues, suggesting that more tickets were sold when ticket prices were lower.

23. a.

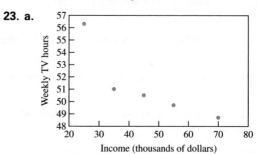

b. There is a strong negative correlation between income and the number of TV hours per week ($r = -0.86$ exactly). **c.** Families with more income have more opportunities to do other things. No.

24. a.

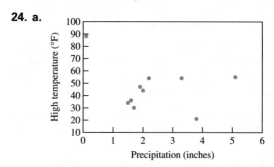

b. There is a weak negative correlation ($r = -0.30$ exactly). **c.** Mean high temperature and precipitation are fairly uncorrelated. Note that the outlier at (0.1, 88) affects the correlation coefficient significantly, making it more negative than it would be if that point were removed from the data.

25. a.

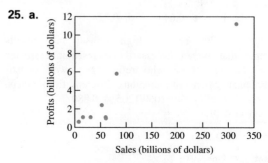

b. There is a strong correlation between sales and earnings ($r = 0.92$ exactly). The strong correlation in this case is highly affected by the Wal-Mart data. **c.** Higher sales do not necessarily translate into higher earnings. Some companies have larger expenses, driving earnings down.

26. a.

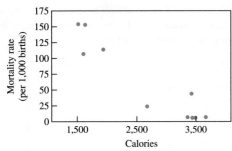

b. There is a strong negative correlation between the mean daily calories and the infant mortality rate ($r = -0.94$).   **c.** The mean daily calories is an indicator of both the health and wealth of a nation, so it is not surprising that lower infant mortality is associated with higher caloric intake.

27. The variables x and y appear symmetrically in the formula (interchanging x and y does not change the formula).

28. If x (or y) is scaled (multiplied by a constant), it occurs in both the numerator and denominator; so there is no overall effect.

Section 7.2

1. The correlation is consistent with the possibility that Lisinopril lowers blood pressure, but without further study we cannot be sure this is the case, since the correlation could be due to other factors or coincidence.

2. The statement attributes a cause to a correlation, but the statement should say that SIDS deaths decreased at a time that doctors advised putting infants in the supine position. In response to that article, Moorestown, NJ resident Jean Mercer wrote a letter to the editor that was subsequently published by the *New York Times*. That letter included this statement: "The article 'Clues on Sudden Infant Death Syndrome' (Nov. 24) stated that the practice of putting infants to sleep in the supine position has decreased deaths from SIDS. It would be more accurate to say that pediatricians advised the supine sleeping position during a time when the SIDS rate fell."

3. Outliers are values that are very far away from almost all of the other values in a data set, so the salary of $1 is an outlier. Outliers can have a significant effect on the correlation coefficient and the conclusions we form. Outliers might make it appear that there is a significant correlation that is not real, or they might mask a real correlation.

4. A scatter diagram is a graph of paired data. Each point represents one pair of data values. One axis is used for one variable and the other axis is used for the second variable. A scatter diagram helps us to visualize relationships between the two variables.

5. Does not make sense.   **6.** Does not make sense.

7. Makes sense.   **8.** Makes sense.

9. There is a positive correlation that is probably due to a common underlying cause. Many crimes are committed with handguns that are not registered.

10. There is a negative correlation that is probably due to a direct cause. As people run, they burn calories, and as they run farther, they burn more calories.

11. There is a positive correlation that is due to a direct cause. As students study more, they gain a better understanding of the subject and their test scores are likely to be higher.

12. There is a positive correlation that is due to a direct cause. With more and more vehicles sharing the same roads, congestion is more likely to occur, and drivers must wait longer times.

13. There is a positive correlation that is probably due to a common cause, such as the general increase in the number of cars and traffic.

14. There is a positive correlation between the distance and the speed. Astronomers can explain the correlation with a direct cause.

15. There is a negative correlation that is probably due to a direct cause. As gas prices increase by large amounts, people can't afford to drive as much, so they cut costs by driving less.

16. There is a positive correlation between the incidence of melanoma and latitude. Either a direct cause (fairer skin in higher latitudes) or a common underlying cause explains the correlation.

17. a. The outlier is the upper left point (0.4, 1.0). Without the outlier, the correlation coefficient is 0.0.   **b.** With the outlier, the correlation coefficient is -0.58.

18. a. The outlier is the point (0.5, 1.0). Without the outlier, correlation coefficient is 0.99.   **b.** The outlier will have the effect of reducing the correlation coefficient. In fact, with the outlier, the correlation coefficient is 0.85.

19. a. The actual correlation coefficient is $r = 0.92$, which is significant at the 0.01 level, so there is a very strong correlation between weight and shoe size.

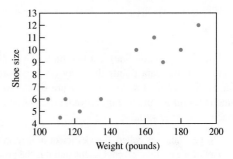

b. Within each of the two groups the correlation is less strong than the overall correlation.

20. a. It would appear that there is a strong negative correlation (actual $r = -0.87$).

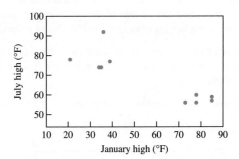

b. There does not appear to be any correlation within each of the two groups of five cities. The correlation in part a is due to the fact that summer and winter are reversed in the Northern and Southern Hemispheres.

21. a. The actual correlation coefficient is $r = 0.77$. This is significant at the 0.01 level, indicating a strong correlation.

b. The 16 points to the right correspond to relatively poor countries, such as Uganda. The remaining points on the left correspond to relative affluent countries, such as Sweden.   **c.** There appears to be a negative correlation between the variables for the poorer countries and a positive correlation for the wealthier countries.

c. Wealthier countries have a negative correlation; higher birth rates go with lower death rates. Poorer countries have a positive correlation; higher birth rates go with higher death rates.

22. a. The actual correlation coefficient is $r = -0.09$, indicating virtually no correlation.

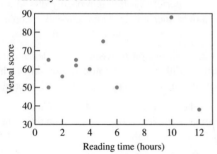

b. The figure shows a group of points with an upward trend and another group with a downward trend. It's plausible that the first group corresponds to the book readers and the second to the comic readers.   **c.** If the conjecture is correct, then book reading time is correlated strongly and positively with test scores ($r = 0.97$) and comic reading time is correlated strongly and negatively with test scores ($r = -0.98$).

Section 7.3

1. A best fit line (or regression line) is a line on a scatter diagram that lies closer to the data points than any other possible line (according to the standard statistical measure of closeness). A best fit line is useful for predicting the value of a variable given some value of the other variable.

2. $r^2 = 0.16$. In general, r^2 is called the coefficient of determination. It is the square of the correlation coefficient, and it is the proportion of variation in a variable that can be explained by the best-fit line.

3. The investigator should use multiple regression. Multiple regression involves determination of the best-fit equation that represents the best fit between one variable and a combination of two or more other variables and, in this case, she wants the best fit between the variable representing the heights of daughters and the two variables representing heights of mothers and heights of fathers.

4. $R^2 = 0.68$ tells us that 68% of the variation in heights of daughters can be explained by the variation in heights of mothers and heights of fathers.

5. Makes sense.   **6.** Does not make sense.

7. Does not make sense.   **8.** Makes sense.

9. a.

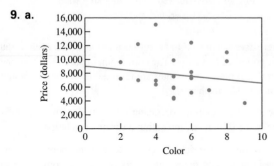

b. Actual $r = -0.16$; $r^2 = 0.026$. So about 3% of the variation in price can be explained by the best-fit line.   **c.** The best-fit line should not be used to make predictions.

10. a.

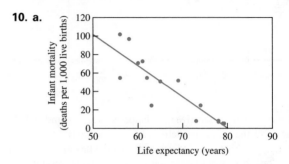

b. Actual $r = -0.90$; $r^2 = 0.81$. About 81% of the variation in infant mortality can be explained by the best-fit line.   **c.** The best-fit line could be used to make predictions.

11. a.

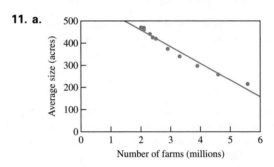

b. Actual $r = -0.99$; $r^2 = 0.97$. About 97% of the variation in farm size can be explained by the best-fit line.   **c.** The best-fit line could be used to make predictions within the range of the number of farms included in the data. Because there does appear to be a slight curvature to the points, predicting outside that range should not be done.

12. a.

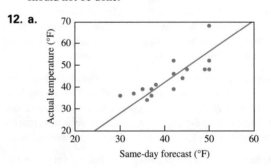

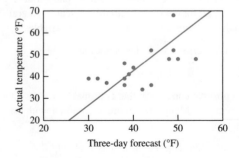

b. For the same day forecast, the actual correlation coefficient is $r = 0.79$; $r^2 = 0.62$; about 62% of the variation in actual temperatures is explained by the best-fit line. For the three-day forecast, the actual correlation coefficient is $r = 0.61$;

$r^2 = 0.37$; about 37% of the variation in actual temperature can be explained by the best-fit line.   **c.** The best-fit line is more reliable for the same-day prediction than for the three-day predictions.

13. a.

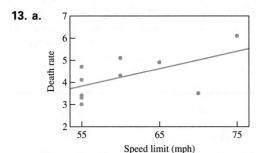

b. There is a moderate positive correlation; $r = 0.59$ exactly; $r^2 = 0.35$; 35% of the variation can be accounted for by the best-fit line.   **c.** (75, 6.1) and (70, 3.5) are both possible outliers, the first because it is away from most of the data points and the latter because the death rate is lower than might be expected considering the rest of the data. Because one point is above the best-fit line and one is below, the net effect of the two points is probably to cancel each other out since both points will "pull" the line toward themselves.   **d.** The value of r is too small to consider predictions based on the best-fit line to be reliable.

14. a.

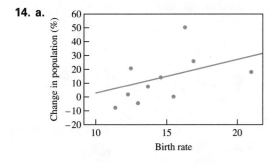

b. The actual $r = 0.50$; $r^2 = 0.25$; 25% of the variation can be accounted for by the best-fit line.   **c.** (21.0, 17.9) is an outlier. The best-fit line would have a steeper slope if that point were removed.   **d.** Predictions based on the best-fit line are not reliable.

15. a.

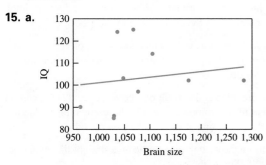

b. The actual $r = 0.179$; $r^2 = 0.032$; only 3% of the variation can be accounted for by the best-fit line.   **c.** (1285, 102) could be considered an outlier because it is far from the other data points. Although this outlier is far to the right, it seems to fit the general pattern, so it does not have much of an effect on the strength of the correlation and the best-fit line.   **d.** Predictions based on the best-fit line are not reliable.

16. a.

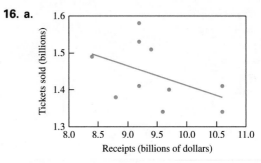

b. The actual $r = -0.460$; $r^2 = 0.2116$; about 21% of the variation can be accounted for by the best-fit line.   **c.** No outliers.   **d.** Predictions based on the best-fit line are not reliable.

17. a.

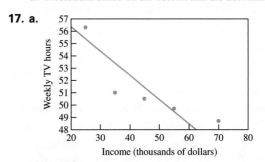

b. The actual $r = -0.86$; $r^2 = 0.74$; 74% of the variation can be accounted for by the best-fit line.   **c.** (25000, 56.3) is an outlier. The best-fit line would have a less steep downward slope if that point were removed.   **d.** Predictions based on the best-fit line could be reliable, but the presence of the outlier and its effect on the best-fit line make the reliability questionable.

18. a.

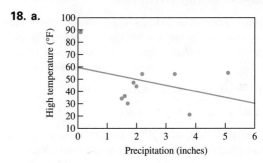

b. The actual $r = -0.30$; $r^2 = 0.09$; 9% of the variation can be accounted for by the best-fit line.   **c.** (0.1, 88) and (5.1, 55) are outliers.   **d.** Predictions based on the best-fit line are not reliable.

19. a.

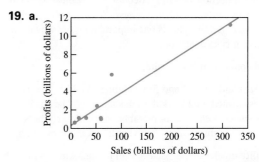

b. The actual $r = 0.941$; and $r^2 = 0.886$; 88.5% of the variation can be accounted for by the best-fit line.   **c.** (315.6, 11.2) is an outlier.   **d.** Predictions based on the best-fit line could be reliable, but the strong effect of the outlier in determining the best-fit line makes the reliability questionable. More data is needed.

20. a.

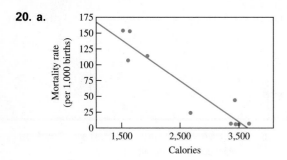

b. The actual $r = -0.94$; $r^2 = 0.88$; 88% of the variation can be accounted for by the best-fit line.   **c.** No outliers. **d.** Predictions outside the data range may not be valid, since the best-fit line predicts negative infant mortality above about 4,000 calories per day.

Section 7.4

1. (1) The correlation is the result of a coincidence; (2) the correlation is due to a common cause; (3) the correlation is due to a direct influence of one of the variables on the other.

2. We do not want to rule out the explanation of a direct influence of one variable on the other variable, because that is the explanation that we would like to establish.

3. A confounding variable is a variable that is not included in the analysis, but it affects the variables that are included in the analysis. Failure to include or account for a confounding variable might cause a researcher to miss an underlying causality by considering only data showing no correlation.

4. Finding a correlation involves finding a statistical association or relationship, and there may or may not be a causality between the two variables. Establishing causality between two variables is finding that one of the variables has a direct influence on the other variable.

5. Does not make sense.   **6.** Does not make sense.

7. Does not make sense.   **8.** Makes sense.

9. The causal connection is valid. When students spend more time and effort studying for a test, the test grades tend to be higher.

10. The causal connection is not valid, although people spend large amounts of money believing incorrectly that wearing a magnet can cure a wide variety of health problems.

11. The causal connection is valid. Alcohol is a depressant to the central nervous system, and it has several effects that include decreased reaction time. This is one important reason why drinking and driving is so dangerous.

12. The causal connection is not valid.

13. Guideline 1; Guidelines 2 and 5; Guidelines 3 and 5. The headaches are associated with work days in some way. The headaches are not associated with Coke or possibly with caffeine. The headaches are possibly the result of bad ventilation in the building.

14. Susceptibility involves other factors than just smoking and varies among individuals. Also, some smokers die of other causes first.

15. Smoking can only increase the risk already present.

16. The study compared the life expectancy of conductors with that of *all* American males including those who die as children. Conductors don't usually become conductors until they are middle

aged, say at least 30 years old. So they are not a representative sample of the population (they are older on the average) and should be expected to have a higher average life span than that of all males.

17. This was an observational study. Later child bearing reflects an underlying cause. While it's possible that the conclusions are correct, there are other possible explanations for the findings. For example, it's also possible that the younger women lived during a time when having babies after age forty was less likely (by choice). It is still possible for them to live to be 100.

18. The people who live near the high voltage lines may all be exposed to some other common cause of cancer in the same area; for example, radium in the soil or pollutants in the water sources or air. Any experiment to isolate the cause (for example, removing the high voltage power lines) will require many years to be conclusive.

19. Availability is not itself a cause. Social, economic, or personal conditions cause individuals to use the available weapons.

20. The vasectomies do not cause prostate cancer; it's the visits to the doctor that increase the chance of detecting cancer.

Chapter 7 Review Exercises

1. The scatterplot suggest that there is a very strong positive correlation between the old ratings and the new ratings.

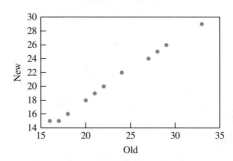

2. The actual value is $r = 0.998$. Because the value is so close to 1, it suggests that there is a strong positive correlation between the old ratings and the new ratings.

3. No. The presence of a strong positive correlation does not imply that either variable is the direct cause of the other.

4. $r^2 = 0.198$. That value tells us that about 20% of the variation in one of the variables can be explained as a linear relationship with the other variable.

5. (1) The correlation is the result of a coincidence; (2) the correlation is due to a common cause; (3) the correlation is due to a direct influence of one of the variables on the other.

6. The points on the scatter diagram lie on a straight line with negative slope (falling to the right).

7. Correlation alone never implies causation, and, in this case, certainly more trips to the dentist do not cause higher incomes. Households with more disposable income can afford more trips to the dentist or can afford dental insurance that covers the costs of the trips.

8. Variables affecting the value of a home might include its location, size, age, condition, and lot size. Location is often cited as the most important factor with considerations including nearness to schools, shopping, churches, etc. The age of the previous owner would be unrelated to the value.

9. The data values that were collected were uncorrelated. It's still possible that the variables represented by the data values are related in some non-linear way, i.e., the scatter plot forms a curve instead of a straight line.

10. The correlation coefficient is -0.056, but any positive or negative value near 0 is reasonable.

Chapter 7 Quiz

1. -1 and 1   2. c, d, and e   3. a

4. The actual value of r is -0.934, but any value between -0.5 and -0.99 is a reasonable answer.

5. Yes   6. False   7. True   8. False

9. False   10. False

Section 8.1

1. Uniform; approximately normal.

2. $\hat{p} = 0.20$. The sampling method is seriously flawed because it uses a voluntary response sample.

3. $\bar{x}$ denotes the mean of a sample and μ denotes the mean of a population. $\bar{x}$ is the mean of a sample, but μ is the mean of a population.

4. $\hat{p}$ denotes the proportion of a sample and p denotes the proportion of a population. $\hat{p}$ is the proportion of a sample, but p is the proportion of a population.

5. Does not make sense.   6. Makes sense.

7. Does not make sense.   8. Does not make sense.

9. The given sample proportion is not likely to be a good estimate of the proportion for the population of children in the United States. There are regional factors that could have a strong influence on that proportion in different parts of the country.

10. Yes, there would be more confidence in the estimate with the larger sample size of 500. As the sample size increases, the sample mean becomes a more reliable estimate of the population mean.

11. a. 9.5   b. The probability is very small, such as 0.01%.   c. No. It appears that the cans are being filled with an amount that is greater than 12.00 ounces.

12. a. -2.1 (or 2.1 standard deviations below the mean)   b. 1.79%

13. a. 0.528   b. 0.49   c. The sample proportion is a little too low.

14. a. 0.110   b. 0.160   c. Not necessarily. The discrepancy could be attributable to sampling error and the discrepancy could occur with a good sampling method.

15. 783 people. We would be more confident of the estimate if we sampled 300 people.

16. 59,867 people. The larger sample is more likely to provide a more reliable estimate of the population proportion.

17. a. -0.67 standard deviations below the mean ($z = -0.67$)

b. 0.253

18. a. 0.14 standard deviations above the mean ($z = 0.14$)   b. 0.444

19. a. 1.0, 1.5, 3.0, 1.5, 2.0, 3.5, 3.0, 3.5, 5.0   b. 2.7   c. Yes; yes

20. a. 46.0, 51.5, 51.5, 54.0, 54.0, 55.0, 55.0, 57.0, 57.0, 57.0, 59.5, 59.5, 60.5, 60.5, 62.0, 62.5, 62.5, 63.0, 63.0, 64.0, 65.0, 65.0, 66.0, 66.0, 68.0   b. 59.4   c. Yes; yes

21. a. 5.0, 28.5, 28.5, 32.5, 32.5, 52.0, 56.0, 56.0, 60.0   b. 39.0   c. Yes; yes

22. a. 46.0, 47.5, 47.5, 49.0, 51.0, 51.0, 52.0, 52.0, 52.5, 52.5, 53.5, 53.5, 56.0, 57.0, 57.0, 58.0   b. 52.25 years   c. Yes; yes

Section 8.2

1. We have 95% confidence that the limits of 8.0518 g and 8.0902 g actually do contain the true population mean weight of all dollar coins. We expect that 95% of such samples will result in confidence interval limits that do contain the population mean.

2. $\$8738 < \mu < \$9270$

3. The media often omit reference to the confidence level, which is typically 95%. The word "mean" should be used instead of the word "average."

4. No. A sample can provide a good estimate of a population mean if its size is large enough based on the standard deviation and desired margin of error, but the size of the population is generally not a factor.

5. Does not make sense.   6. Does not make sense.

7. Makes sense.   8. Does not make sense.

9. Margin of error: 0.6. Confidence interval: $24.6\,\text{cm} < \mu < 25.8\,\text{cm}$

10. Margin of error: 0.7 km. Confidence interval: $3.8\,\text{km} < \mu < 5.2\,\text{km}$

11. Margin of error: 0.4 ft. Confidence interval: $7.6\,\text{ft} < \mu < 8.4\,\text{ft}$

12. Margin of error: $15. Confidence interval: $\$535 < \mu < \$565$

13. 64   14. 110   15. 7866

16. 1212   17. 185   18. 172

19. 100   20. 278

21. 5.639 g; $5.619\,\text{g} < \mu < 5.659\,\text{g}$

22. 3103 g; $3001\,\text{g} < \mu < 3205\,\text{g}$

23. 5.15; $5.10 < \mu < 5.20$ 24. 1.91 lb; $1.64\,\text{lb} < \mu < 2.18\,\text{lb}$

25. $146\,\text{lb} < \mu < 221\,\text{lb}$ 26. $135 < \mu < 210$

27. a. 3.1   b. 1.4   c. 3.1   d. $2.6 < \mu < 3.6$ e. The sample is small and was concentrated in one neighborhood, so it is unlikely to be very representative of all American families.

28. a. 2.0   b. 1.3   c. 2.0   d. $1.5 < \mu < 2.5$ e. The sample is quite small, so it will not be very reliable.

Section 8.3

1. We have 95% confidence that the limits of 0.393 and 0.553 actually do contain the true population proportion. We expect that 95% of such samples will result in confidence interval limits that do contain the population proportion.

2. $0.671 < p < 0.723$

3. The media often omit reference to the confidence level, which is typically 95%.

4. No. The quality of results is based largely on the sample size, and the size of the population is usually not relevant.

5. Does not make sense.   **6.** Makes sense.

7. Makes sense.   **8.** Does not make sense.

9. Margin of error: 0.0310. $0.369 < p < 0.431$

10. Margin of error: 0.0306. $0.219 < p < 0.281$

11. Margin of error: 0.0256. $0.074 < p < 0.126$

12. Margin of error: 0.0263. $0.134 < p < 0.186$

13. 625   **14.** 400

15. 4445   **16.** 817

17. $0.140 < p < 0.160$

18. $0.143 < p < 0.157$. Doubling the sample size did not produce too much of a change in the confidence interval, so it does not appear to be worth the extra cost and effort.

19. $E = 0.021$; $0.779 < p < 0.821$

20. Abortion: $E = 0.00190$; $0.508 < p < 0.512$

Casual sex: $E = 0.0187$; $0.398 < p < 0.402$

21. Margin of error: 0.060. $0.775 < p < 0.896$

22. a. 0.550   **b.** $0.526 < p < 0.574$ **c.** Yes. Because the range of values in the confidence interval includes only values greater than 0.5, it does appear that the majority (more than 50%) is in favor of immediate government action.

23. a. 0.98   **b.** $0.966 < p < 0.994$ **c.** To be a random sample, all films must have an equal chance of being chosen.

24. $0.409 < p < 0.471$

25. Poll 1: $0.494 < p < 0.546$; Poll 2: $0.494 < p < 0.534$; Poll 3: $0.498 < p < 0.532$. Because all of these confidence intervals include values less than 0.5, Martinez cannot be confident of winning a majority.

26. a. $E = 0.00148$. This level of precision enables the Bureau to detect very small changes in the unemployment rate and is reasonable for that purpose.   **b.** E is reduced by a factor of 2 (by one-half).   **c.** E is increased by a factor of 2.

27. a. $0.50 < p < 0.58$ **b.** $n = 625$

28. a. The actual margin of error is 0.038 which is consistent with the stated margin of error of 4 percentage points.   **b.** 2,500

Chapter 8 Review Exercises

1. a. 0.791   **b.** $0.761 < p < 0.820$ **c.** 0.030   **d.** There is 95% confidence that the limits of 0.754 and 0.827 contain the true value of the population proportion. If such samples of size 745 were randomly selected many times, the resulting confidence interval limits would contain the true population proportion in 95% of those samples.   **e.** The distribution will be approximately normal, or bell-shaped.   **f.** 2500

2. a. 114   **b.** If the sample size is larger than necessary, the confidence interval will be better in the sense that it will be narrower than necessary. If the sample size is smaller than necessary, the confidence interval will be worse in the sense that it will be wider than it should be.   **c.** Because college students are a more homogeneous group than the population as a whole, the standard deviation for college students is likely to be smaller than 16. If we use the actual value that is smaller than 16, the sample size will be smaller than 114.

3. a. $258.9 < \mu < 300.1$ **b.** 20.6   **c.** There is 95% confidence that the limits of 258.9 and 300.1 contain the true value of the population mean. If such samples of size 40 were randomly selected many times, the resulting confidence interval limits would contain the true population mean in 95% of those samples.   **d.** 171

4. a. 2500   **b.** No. Your sample size will be large enough, but it has a high potential for being biased. The people who refuse to answer could well constitute a segment of the population with a different perspective, and that perspective will be incorrectly excluded from the sample.

Chapter 8 Quiz

1. Approximately normal or bell-shaped

2. Approximately normal or bell-shaped

3. The sample proportion   **4.** $0.52 < p < 0.68$

5. b: mean of the sample   **6.** Margin of error is 0.03

7. Margin of error is 0.3   **8.** $68.4 < \mu < 70.8$

9. $0.695 < p < 0.745$ **10.** The confidence interval limits should be proportions between 0 and 1, but the upper limit is not.

Section 9.1

1. A hypothesis test is a standard procedure for testing a claim about the value of a population parameter.

2. A null hypothesis is a starting assumption for a hypothesis test, and it gives a specific value for a population parameter. A null hypothesis is denoted by H_0. An alternative hypothesis is a statement that the population parameter has a value that somehow differs from the value assumed in the null hypothesis. An alternative hypothesis is denoted by H_a.

3. $\mu \neq 195$ lb, $\mu < 195$ lb, $\mu > 195$ lb

4. A P-value is the probability of selecting a sample at least as extreme as the observed sample, assuming that the null hypothesis is true.

5. Does not make sense.   **6.** Does not make sense.

7. Makes sense.   **8.** Does not make sense.

9. Makes sense.   **10.** Does not make sense.

11. Does not make sense.   **12.** Makes sense.

13. a. 22 girls in 40 births does not appear to be statistically significant.   **b.** 35 girls in 40 births does appear to be statistically significant.   **c.** Part (b), because 35 girls in 40 births is statistically significant, so the P-value must be relatively small number.

14. a. The result of 135 girls in 200 births appears to be significant. **b.** The result of 105 girls in 200 births does not appear to be significant. **c.** Part (b), because 105 girls in 200 births is not statistically significant, so the *P*-value is not very small.

15. H_0: $\mu = 250$ calories. H_a: $\mu \neq 250$ calories. There is sufficient sample evidence to warrant rejection of the claim that the mean caloric content is equal to 250 calories. There is not sufficient sample evidence to warrant rejection of the claim that the mean caloric content is equal to 250 calories.

16. H_0: $\mu = 1518$. H_a: $\mu < 1518$. There is not sufficient sample evidence to support the claim that the mean SAT score is less than 1518. There is sufficient sample evidence to support the claim that the mean SAT score is less than 1518.

17. H_0: $p = 0.7$. H_a: $p > 0.7$. There is not sufficient sample evidence to support the claim that the proportion of girls is greater than 0.7. There is sufficient sample evidence to support the claim that the proportion of girls is greater than 0.7.

18. H_0: $p = 0.04$. H_a: $p < 0.04$. There is not sufficient sample evidence to support the claim that the proportion of defective carbon monoxide detectors is less than 0.04. There is sufficient sample evidence to support the claim that the proportion of defective carbon monoxide detectors is less than 0.04.

19. H_0: $\mu = 10$ oz. H_a: $\mu \neq 10$ oz. There is not sufficient sample evidence to warrant rejection of the claim that the mean is equal to 10 oz. There is sufficient sample evidence to warrant rejection of the claim that the mean amount supplied is equal to 10 oz.

20. H_0: $\mu = 350$ mg. H_a: $\mu < 350$ mg. There is not sufficient sample evidence to support the claim that the mean is less than 350 mg. There is sufficient sample evidence to support the claim that the mean is less than 350 mg.

21. H_0: $p = 0.5$. H_a: $p > 0.5$. There is not sufficient sample evidence to support the claim that the proportion of students is greater than 0.5. There is sufficient sample evidence to support the claim that the proportion of students is greater than 0.5.

22. H_0: $p = 0.20$. H_a: $p < 0.20$. There is not sufficient sample evidence to support the claim that the proportion of college graduate smokers is less than 0.20. There is sufficient sample evidence to support the claim that the proportion of college graduate smokers is less than 0.20.

23. No. The *P*-value of 0.382 is greater than the significance level of 0.05, and this high value suggests that there is a high likelihood of getting 48 or fewer males by chance.

24. No. The *P*-value of 0.184 is greater than the significance level of 0.01, and this high value suggests that there is a high likelihood of getting 45 or fewer males by chance.

25. No. The *P*-value of 0.028 is greater than the significance level of 0.01, and this high value suggests that there is a high likelihood of getting 40 or fewer males by chance.

26. Yes. The *P*-value of 0.028 is less than the significance level of 0.05, and this low value suggests that there is a small likelihood of getting 40 or fewer males by chance.

27. Yes. The *P*-value of 0.002 is less than the significance level of 0.01, and this low value suggests that there is a small likelihood of getting 35 or fewer males by chance.

28. Yes. Because the occurrence of 32 males is more extreme than the lowest table value of 35 males, the *P*-value is less than 0.002, so it is less than the significance level of 0.01, and this low value suggests that there is a small likelihood of getting 32 or fewer males by chance.

Section 9.2

1. n represents the sample size, $\bar{x}$ represents the mean of the sample values, s represents the standard deviation of the sample values, σ represents the standard deviation of the population values, and μ represents the mean of the population.

2. The sample is a convenience sample, not a simple random sample. It is possible that the McDonald's restaurants that are within 30 miles of her home are not representative of the population of all McDonald's restaurants.

3. A type I error is the mistake of rejecting the null hypothesis when that null hypothesis is actually true. A type II error is the mistake of failing to reject the null hypothesis when that null hypothesis is actually false.

4. Because the *P*-value of 0.009 is so low, the mean fuel consumption reduction does have statistical significance. However, a mean reduction of only 0.08 miles per gallon is not a meaningful difference, so there isn't practical significance.

5. Does not make sense.   **6.** Does not make sense.

7. Makes sense.   **8.** Makes sense.   **9.** Makes sense.

10. Makes sense.   **11.** Does not make sense.   **12.** Makes sense.

13. $z = -3.33$. The alternative hypothesis is supported.

14. $z = -1.20$. The alternative hypothesis is not supported.

15. $z = 8.00$. The alternative hypothesis is supported.

16. $z = 6.00$. The alternative hypothesis is supported.

17. $z = 1.03$. The alternative hypothesis is not supported.

18. $z = -0.36$ The alternative hypothesis is not supported.

19. $z = -5.11$. The alternative hypothesis is supported.

20. $z = 8.03$. The alternative hypothesis is supported.

21. 0.3446. H_a is not supported.   **22.** 0.0082. H_a is supported.

23. 0.0287. H_a is supported.   **24.** 0.0668. H_a is not supported.

25. 0.1096. H_a is not supported.   **26.** 0.0456. H_a is supported.

27. 0.0892. H_a is not supported.   **28.** 0.0574. H_a is not supported.

29. 0.0035. H_a is supported.   **30.** 0.0002. H_a is supported.

31. 0.0179. H_a is supported.   **32.** 0.1357. H_a is not supported.

33. 0.8808. H_a is not supported.   **34.** 0.0004. H_a is supported.

35. The test is right-tailed, but the standard score is to the left of the center of the distribution, so the *P*-value must be greater than 0.5. The alternative hypothesis is not supported. The formal hypothesis test is not necessary because there is no way that we could support a claim that the mean is greater than some value, when the sample mean is less than that value.

36. The test is left-tailed, but the standard score is to the right of the center of the distribution, so the P-value must be greater than 0.5. The alternative hypothesis is not supported. The formal hypothesis test is not necessary because there is no way that we could support a claim that the mean is less than some value, when the sample mean is greater than that value.

37. H_0: $\mu = 7.5$ years, H_a: $\mu < 7.5$ years; $n = 100$, $\bar{x} = 7.01$ years, $s = 3.74$ years, $\sigma = 0.374$ years, $z = -1.3$; not significant at the 0.05 level, P-value $= 0.0968$; there is not sufficient evidence to support the claim that the mean time of ownership is less than 7.5 years.

38. H_0: $\mu = 2.0$ days, H_a: $\mu > 2.0$ days; $n = 81$, $\bar{x} = 2.2$ days, $s = 1.2$ days, $\sigma = 0.133$ days, $z = 1.5$; not significant at the 0.05 level, P-value $= 0.0668$; there is not sufficient evidence to support the claim that the mean hospital stay is greater than 2.1 days.

39. H_0: $\mu = 12$ oz, H_a: $\mu \neq 12$ oz; $n = 36$, $\bar{x} = 12.19$ oz, $s = 0.11$ oz, $\sigma \approx s$, $z = 10.36$; significant at the 0.05 level, P-value is less than 0.0001; there is sufficient evidence to support the claim that the mean weight of the packages is different from 12.00 oz.

40. H_0: $\mu = 8.50$ cm, H_a: $\mu \neq 8.50$ cm; $n = 64$, $\bar{x} = 8.56$ cm, $s = 0.24$ cm, $\sigma \approx s$, $z = 2.0$; significant at the 0.05 level, P-value $= 0.0456$; there is sufficient evidence to support the claim that the mean axle diameter is different from the required 2.50 cm.

41. H_0: $\mu = 600$ mg, H_a: $\mu \neq 600$ mg; $n = 65$, $\bar{x} = 589$ mg, $s = 21$ mg, $\sigma \approx s$, $z = -4.2$; significant at the 0.05 level, P-value $= 0.0004$; there is sufficient evidence to support the claim that the mean amount of acetaminophen is different from the listed 600 mg.

42. H_0: $\mu = 5.4$, H_a: $\mu < 5.4$; $n = 50$, $\bar{x} = 5.23$, $\sigma = 0.54$, $z = -2.23$; significant at the 0.05 level, P-value $= 0.0130$; there is sufficient evidence to support the claim that the sample is from a population with a mean less than 5.4.

43. H_0: $\mu = 21.4$ mpg, H_a: $\mu < 21.4$ mpg; $n = 40$, $\bar{x} = 19.8$ mpg, $s = 3.5$ mpg, $\sigma \approx s$, $z = -2.9$; significant at the 0.05 level, P-value $= 0.0019$; there is sufficient evidence to support the claim that the mean gas mileage for SUVs is less than 21.4 miles per gallon.

44. H_0: $\mu = 92.84$ in, H_a: $\mu < 92.84$ in; $n = 40$, $\bar{x} = 92.67$ in, $s = 1.79$ in, $\sigma \approx s$, $z = -0.60$; significant at the 0.05 level, P-value $= 0.2743$; there is not sufficient evidence to support the claim that the mean bounce height of the new balls is less than 92.84 in.

45. H_0: $\mu = 0.21$, H_a: $\mu > 0.21$; $n = 32$, $\bar{x} = 0.83$, $s = 0.24$, $\sigma \approx s$, $z = 14.6$; significant at the 0.05 level, P-value < 0.0002; there is sufficient evidence to support the claim that the mean is greater than 0.21.

46. H_0: $\mu = 5.670$ grams, H_a: $\mu \neq 5.670$ grams; $n = 50$, $\bar{x} = 5.622$ grams, $s = 0.068$ grams, $\sigma \approx s$, $z = -5.0$; significant at the 0.05 level, P-value < 0.0004; there is sufficient evidence to support the claim that the mean weight of quarters in circulation is different from 5.67 grams.

47. H_0: $\mu = 3.39$ kg, H_a: $\mu \neq 3.39$ kg; $n = 121$, $\bar{x} = 3.67$ kg, $s = 0.66$ kg, $\sigma \approx s$, $z = 4.7$; significant at the 0.05 level, P-value < 0.0004; there is sufficient evidence to support the claim that the mean birth weight of male babies born to mothers on a vitamin supplement is different from the national average for all male babies.

48. H_0: 150 lb, H_a: $\mu > 150$ lb; $n = 54$, $\bar{x} = 182.9$ lb, $\sigma = 121.8$ lb, $z = 1.98$; significant at the 0.05 level, P-value $= 0.0236$; there is sufficient evidence to support the claim that the mean weight is greater than 150 lb.

49. H_0: $\mu = 24$ mo, H_a: $\mu < 24$ mo; $n = 70$, $\bar{x} = 22.1$ mo, $s = 8.6$ mo, $\sigma \approx s$, $z = -1.85$; significant at the 0.05 level, P-value $= 0.0322$; there is sufficient evidence to support the claim that the mean is less than 24 months.

50. H_0: $\mu = \$500{,}000$, H_a: $\mu < \$500{,}000$; $n = 40$, $\bar{x} = \$415{,}953$, $\sigma = \$463{,}364$, $z = -1.15$; not significant at the 0.05 level, P-value $= 0.1257$; there is not sufficient evidence to support the claim that the mean salary of a football coach in the NCAA is less than \$500,000.

51. Type I error: Reject the claim that the patient is free of a disease when the patient is actually free of the disease. Type II error: Fail to reject the claim that the patient is free of a disease when the patient is not free of the disease.

52. Type I error: Reject the claim that the defendant is not guilty when the defendant is not guilty. Type II error: Fail to reject the claim that the defendant is not guilty when the defendant is guilty.

53. Type I error: Reject the claim that the lottery is fair when the lottery is fair. Type II error: Fail to reject the claim that the lottery is fair when the lottery is biased.

54. Type I error: Reject the claim that the mean length is 3.456 cm when the mean length is 3.456 cm. Type II error: Fail to reject the claim that the mean length is 3.456 cm when the mean length is not 3.456 cm.

Section 9.3

1. The symbol p represents the proportion in a population, $\hat{p}$ represents the proportion in a sample, and a P-value is the probability of getting a sample proportion that is at least as extreme as the sample being considered.

2. Normal distribution

3. No. The sample is a voluntary response sample, not a simple random sample. It is likely that those who responded are not representative of the general population.

4. Assuming that the proportion of car crashes occurring within 5 miles of home is 0.5, there is a probability of 0.00001 of getting a sample at least as extreme as the sample that was obtained. Because that probability is so small, it indicates that the sample results are very unlikely to have occurred by chance, so there is sufficient evidence to support the claim that the majority of car crashes occur within 5 miles of home.

5. Makes sense.   **6.** Makes sense.

7. Does not make sense.   **8.** Makes sense.

9. H_0: $p = 0.5$, H_a: $p > 0.5$; $n = 400$, $\hat{p} = 0.5125$, standard deviation of distribution of sample proportions is 0.025, $z = 0.5$; not significant at the 0.05 level, P-value $= 0.3085$; there is not sufficient evidence to support the claim that a majority of the voters support the candidate.

10. H_0: $p = 0.27$, H_a: $p < 0.27$; $n = 785$, $\hat{p} = 0.18$, standard deviation of distribution of sample proportions is 0.014, $z = -5.46$; significant at the 0.05 level, P-value < 0.0001; there is sufficient evidence to support the claim that the rate of smoking among those with four years of college is less than the 27% rate for the general population.

11. H_0: $p = 0.50$, H_a: $p < 0.50$; $n = 1015$, $\hat{p} = 0.44$, standard deviation of distribution of sample proportions is 0.0157, $z = -3.8$; significant at the 0.05 level, P-value < 0.0002; there is sufficient evidence to support the claim that fewer than half of all teenagers in the population feel that grades are the greatest source of pressure.

12. H_0: $p = 0.586$, H_a: $p < 0.586$; $n = 1445$, $\hat{p} = 0.56$, standard deviation of distribution of sample proportions is 0.0131, $z = -2.0$; significant at the 0.05 level, P-value $= 0.0228$; there is sufficient evidence to support the claim that the sample is from a population with a married percentage less than 58.6%.

13. H_0: $p = 0.099$, H_a: $p < 0.099$; $n = 1050$, $\hat{p} = 0.0933$, standard deviation of distribution of sample proportions is 0.0092, $z = -0.6$; not significant at the 0.05 level, P-value $= 0.2743$; there is not sufficient evidence to support the claim that illegal drug use is less than the national average.

14. H_0: $p = 0.133$, H_a: $p < 0.133$; $n = 4342$, $\hat{p} = 0.121$, standard deviation of distribution of sample proportions is 0.0052, $z = -2.3$; significant at the 0.05 level, P-value $= 0.0107$; there is sufficient evidence to support the claim that the poverty rate in Idaho is less than the national rate of 13.3%.

15. H_0: $p = 0.5$, H_a: $p > 0.5$; $n = 276{,}000$, $\hat{p} = 0.51$, standard deviation of distribution of sample proportions is 0.000952, $z = 10.5$; significant at the 0.05 level, P-value < 0.0002; there is sufficient evidence to support the claim that over half of all freshmen believe that abortion is should be legal.

16. H_0: $p = 3/4$, H_a: $p \neq 3/4$; $n = 3011$, $\hat{p} = 0.73$, standard deviation of distribution of sample proportions is 0.00810, $z = -2.54$; significant at the 0.05 level, P-value $= 0.0112$; there is sufficient evidence to reject the claim that ¾ of all adults use the Internet.

17. H_0: $p = 0.53$, H_a: $p \neq 0.53$; $n = 3600$, $\hat{p} = 0.54$, standard deviation of distribution of sample proportions is 0.0083, $z = 1.2$; not significant at the 0.05 level, P-value $= 0.2302$; there is not sufficient evidence to support the claim that the percentage of households using natural gas has changed.

18. H_0: $p = 0.019$, H_a: $p > 0.019$; $n = 863$, $\hat{p} = 0.022$, standard deviation of distribution of sample proportions is 0.0050, $z = 0.6$; not significant at the 0.05 level, P-value $= 0.2743$; there is not sufficient evidence to support the claim that the percentage of treated patients with flu symptoms is greater than the 1.9% rate for patients not given treatments

Chapter 9 Review Exercises

1. a. H_0: $\mu = 12$ ounces   **b.** H_a: $\mu > 12$ ounces   **c.** $z = 10.4$ **d.** $z = 1.645$ **e.** Less than 0.0002   **f.** Reject the null hypothesis and claim that cans of Coke have a mean amount of cola greater than 12 ounces.   **g.** A Type I error results if we concluded that the population mean was greater than 12 ounces when, in fact, it was not.   **h.** A type II error results if we did not conclude that the population mean was greater than 12 ounces when, in fact, it is greater than 12 ounces.   **i.** The P-value

would be the probability that z is greater than 10.4 or less than -10.4. Since the probability that z is greater than 10.4 is less than 0.0002, the P-value is less than $2 \times 0.0002 = 0.0004$.

2. a. H_0: $p = 0.5$ **b.** H_a: $p > 0.5$ **c.** $z = 0.8$ **d.** $z = 1.645$ **e.** 0.2119   **f.** There is not sufficient evidence to support the claim that the majority of the subjects are smoking a year after the nicotine patch treatment.   **g.** A type I error is the mistake of rejecting the null hypothesis when it is true. In this case, it is the mistake of supporting the claim that the majority of the subjects are smoking a year later when the proportion of smokers a year later is 0.5 (or less).   **h.** A type II error is the mistake of failing to reject the null hypothesis when it is false. In this case, it would be the mistake of not supporting the claim that the majority are smoking a year later, when a majority of the subjects actually are smoking a year later.   **i.** The P-value doubles to become 0.4238.

3. a. H_0: $p = 0.5$, H_a: $p > 0.5$; $n = 703$, $\hat{p} = 0.610$, standard deviation of distribution of sample proportions is 0.0184, $z = 5.85$; significant at the 0.05 level, P-value < 0.0001; there is sufficient evidence to support the claim that most workers get their jobs through networking.   **b.** With $\hat{p} = 0.610$, there is no way that the sample data could support a claim that $p < 0.5$. We would need a value of $\hat{p}$ less than 0.5.

4. The sample is not random. It is very possible that the medical students are not representative of the population, so the results are biased.

Chapter 9 Quiz

1. $\mu > 1126$ cm^3 **2.** $p = 0.5$ **3.** $p \neq 0.6$

4. Two-tailed   **5.** $\mu = 110$ **6.** $\mu > 110$

7. Support the claim that the mean is greater than 110.

8. Reject the null hypothesis. Do not reject the null hypothesis.

9. Support the claim that the mean is greater than 110. Do not support the claim that the mean is greater than 110.

10. Type I error

Section 10.1

1. No. The sample is a convenience sample and is not a simple random sample. The sample is not likely to be representative of the population.

2. The normal distribution requires that the value of the population standard deviation is known, but that value is rarely known in real applications.

3. There is no difference. The term "t distribution" is commonly used instead of Student t distribution, but the two terms are equivalent.

4. Yes. Although the sample size of 25 seems very small relative to the population size, a good estimate can be obtained. The size of such a large population is irrelevant.

5. Does not make sense.   **6.** Does not make sense.

7. Makes sense.   **8.** Makes sense.

9. $124.7 < \mu < 135.3$ **10.** 76.0 in. $< \mu < 79.8$ in.

11. 14.3 in. $< \mu < 14.7$ in.   **12.** $1392 < \mu < 1614$

13. a. $\$6370 < \mu < \$11{,}638$ **b.** $\$11{,}638$

14. a. $-1.777° < \mu < 0.939°$ **b.** Yes. Because a mean difference of $0°$ represents no difference between the forecast and actual temperatures, we cannot conclude that the three-day forecast temperatures are inaccurate.

15. a. $0.085 < \mu < 0.157$ **b.** The claim does not appear to be valid. Because 0.165 grams/mile is not included in the confidence interval, that value does not appear to be the mean.

16. a. 108.8 min $< \mu < 134.7$ min   **b.** No. The data and the confidence interval suggest that some movies will run longer than 130 minutes.

17. $H_0: \mu = 0.3$. $H_a: \mu < 0.3$. The test statistic $t = -0.119$ is not less than the value of $t = -1.753$ found in Table 10.1, so do not reject the null hypothesis. (P-value $= 0.4534$.) There is not sufficient evidence to support the claim that the mean amount of sugar in all cereals is less than 0.3 gm.

18. $H_0: \mu = 0$. $H_a: \mu \neq 0$. The test statistic $t = 3.473$ is greater than the value of $t = 2.045$ found in Table 10.1, so reject the null hypothesis. (P-value $= 0.0016$.) There is sufficient evidence to warrant rejection of the claim that the population mean is equal to 0 sec. The watches do not appear to be accurate.

19. $H_0: \mu = 2.5$ g. $H_a: \mu \neq 2.5$ g. The test statistic $t = -0.299$ is not less than the value of $t = -2.045$ found in Table 10.1, so do not reject the null hypothesis. (P-value $= 0.7670$.) There is not sufficient evidence to reject the claim that the population has a mean of 2.5 g. Pennies appear to conform to the specification of the U. S. Mint.

20. $H_0: \mu = 420$. $H_a: \mu > 420$. The test statistic $t = 1.936$ is greater than the value of $t = 1.761$ found in Table 10.1, so reject the null hypothesis. (P-value $= 0.0366$.) There is sufficient evidence to support the claim that the mean time between failures is greater than 420 h. The modifications appear to have improved reliability.

21. $H_0: \mu = 3.39$. $H_a: \mu \neq 3.39$. The test statistic $t = 1.735$ is not greater than the value of $t = 2.131$ found in Table 10.1, so do not reject the null hypothesis. (P-value $= 0.1032$.) There is not sufficient evidence to support the claim that the mean birth weight for all male babies of mothers given vitamins is different from 3.39 kg. Based on the given results, the vitamin supplement does not appear to have an effect on birth weight.

22. $H_0: \mu = 60$. $H_a: \mu > 60$. The test statistic $t = 6.485$ is greater than the value of $t = 1.729$ found in Table 10.1, so reject the null hypothesis. (P-value $= 0.000002$.) There is sufficient evidence to support the claim that the mean pulse rate of statistics students is greater than 60 beats per minute.

23. $H_0: \mu = 1000$ hic. $H_1: \mu < 1000$ hic. Test statistic: $t = -2.661$. Critical value: $t = -2.015$. (P-value: 0.0224). Reject H_0. There is sufficient evidence to support the claim that the population mean is less than 1000 hic. These results suggest that the requirement is being satisfied, but one of the six sample items violates the requirement of being less than 1000 hic, so the population of booster seats appears to have a mean less than 1000 hic while some individual booster seats violate that requirement.

24. $H_0: \mu = 40$. $H_a: \mu > 40$. The test statistic $t = 2.746$ is greater than the value of $t = 1.833$ found in Table 10.1, so reject the null hypothesis. (P-value $= 0.0113$.) There is sufficient evidence to support the claim that the mean nicotine content is greater than 40 mg.

Section 10.2

1. The table includes entries that are frequency counts corresponding to two different variables. One variable is whether the subject experienced nausea and the other variable is whether the subject was given Chantix or a placebo.

2. The expected value is denoted by E. $E = 19.8$, and it indicates that if experiencing nausea is independent of whether the subject is treated with Chantix or a placebo, we expect that among the 1626 subjects, 19.8 of them should be given a placebo and experience nausea.

3. No. Rejection of independence does not necessarily mean that there is a causal relationship between the two variables. The hypothesis test can allow us to conclude that there is some relationship, but it does not allow us to conclude that there is a *causal* relationship.

4. Yes. The sample group would be self-selected and would not likely be representative of the population.

5. Does not make sense.   **6.** Does not make sense.

7. Makes sense.   **8.** Makes sense.

9. Critical value: $\chi^2 = 3.841$. Do not reject the null hypothesis of independence. Gender and response do not appear to be related.

10. Critical value: $\chi^2 = 3.841$. Reject the null hypothesis of independence. Gender and response appear to be related.

11. Critical value: $\chi^2 = 6.635$. Reject the null hypothesis of independence. Gender and response appear to be related.

12. Critical value: $\chi^2 = 6.635$. Do not reject the null hypothesis of independence. Gender and response appear to be related.

13. a. H_0: Whether a subject lies is independent of the polygraph test indication. H_a: Whether a subject lies and the polygraph test indication are related.   **b.** 27.34, 29.66, 19.66, 21.34   **c.** 25.571   **d.** 3.841   **e.** Reject the null hypothesis of independence. There does appear to be a relationship between whether a subject lies and the polygraph test indication.

14. a. H_0: Gender and voter turnout are independent. H_a: Gender and voter turnout are related.   **b.** 140.40, 119.60, 129.60, 110.40   **c.** 0.005   **d.** 3.841   **e.** Do not reject the null hypothesis of independence. There does not appear to be a relationship between gender and voter turnout.

15. a. H_0: Response is independent of whether the subject is a worker or senior-level boss. H_a: There is some relationship between whether the subject is a worker or senior-level boss.   **b.** 181.60, 254.40, 50.40, 70.60   **c.** 4.698   **d.** 3.841   **e.** Reject the null hypothesis of independence. There does appear to be a relationship between whether the subject is a worker or senior-level boss.

16. a. H_0: Whether a flu develops is independent of treatment. H_a: There is some relationship between whether a flu develops and

treatment.   **b.** 72.80, 997.20, 36.20, 495.80   **c.** 153.462
d. 3.841   **e.** Reject the null hypothesis of independence.
There does appear to be a relationship between whether a flu
develops and treatment.

17. a. H_0: Improvement is independent of whether the participant was
given the drug or a placebo. H_a: Improvement and treatment (drug
or placebo) are somehow related.   **b.** 54.16, 43.84, 50.84, 41.16
c. 0.289   **d.** 3.841   **e.** Do not reject the null hypothesis of
independence. There does not appear to be a relationship between
improvement and treatment (drug or placebo).

18. a. H_0: The age category (under 25 and 25 or over) is independent of whether women drink during pregnancy. H_a: The
age category is independent of whether women drink during
pregnancy.   **b.** 19.08, 1232.92, 30.92, 1998.08   **c.** 3.181
d. 3.841   **e.** Do not reject the null hypothesis of independence. There does not appear to be a relationship between age
category and whether women drink during pregnancy.

19. a. H_0: The type of crime is independent of whether the
criminal is a stranger. H_a: The type of crime is related
to whether the criminal is a stranger.   **b.** 29.93, 284.64,
803.43, 21.07, 200.36, 565.57   **c.** 119.330   **d.** 9.210
e. Reject the null hypothesis of independence. The type of
crime appears to be related to whether the criminal was a
stranger.

20. a. H_0: In China, smoking is independent of education level.
H_a: In China, smoking and education level are related.
b. 564.53, 1210.50, 164.97, 246.47, 528.50, 72.03   **c.** 95.725
d. 5.991   **e.** Reject the null hypothesis of independence. In
China, smoking and education level appear to be related.

Section 10.3

1. ANOVA is a method for testing the equality of three or more
population means.

2. 7.29 in.^{2}

3. H_0: $\mu_1 = \mu_2 = \mu_3$ (The three populations of students have equal
means.) H_a: At least one of the three population means is different from the others.

4. The sample data are separated into groups according to *one* characteristic, or factor.

5. Makes sense.   **6.** Makes sense.

7. Does not make sense.   **8.** Does not make sense.

9. a. H_0: $\mu_1 = \mu_2 = \mu_3$ (The three books have pages with the same
mean Flesch-Kincaid Grade Level score.)   **b.** H_a: The three
books have mean Flesch-Kincaid Grade Level scores that are
not all equal.   **c.** 0.00077   **d.** Reject the null hypothesis
of equal means. The pages from the three books do not have the
same mean Flesch-Kincaid Grade Level scores.

10. a. H_0: $\mu_1 = \mu_2 = \mu_3 = \mu_4 = \mu_5$ (The five laboratories produce
the same mean length of the charred portion of fabric.)   **b.** The
five different laboratories produce means that are not all
the same.   **c.** 0.030665893   **d.** Yes. The small P-value
suggests that we should reject the null hypothesis of equal means.
The five different laboratories are producing measurements that
do not have the same mean.

11. a. H_0: $\mu_1 = \mu_2 = \mu_3 = \mu_4 = \mu_5 = \mu_6$ (M&Ms in the six different color categories have the same mean weight.)   **b.** M&Ms
from the six different color categories have means that are not all
equal.   **c.** 0.817   **d.** No. There is not sufficient evidence
to support the claim that M&Ms with different colors have different mean weights.

12. a. H_0: $\mu_1 = \mu_2 = \mu_3$ (Women in the three age groups have the
same mean systolic blood pressure.)   **b.** Women in the three
age groups have systolic blood pressures with means that are
not all the same.   **c.** 0.205   **d.** No. The large P-value
suggests that we should not reject the null hypothesis of equal
means. The three population means do not appear to be different.

13. With a P-value of 0.4216, do not reject the null hypothesis of
equal means. There is not sufficient evidence to reject the claim
that the different weight categories have the same mean. Based on
these sample data, we cannot conclude that larger cars are safer.

14. With a P-value of 0.2956, do not reject the null hypothesis of
equal means. There is not sufficient evidence to reject the claim
that the different weight categories have the same mean. Based on
these sample data, we cannot conclude that larger cars are safer.

15. With a P-value of 0.0305, reject the null hypothesis of equal
means. There is sufficient evidence to reject the claim that the
different epochs have the same mean.

16. With a P-value of 0.0000, reject the null hypothesis of equal
means. There is sufficient evidence to support the claim of different population means. Given that the sample means are 6.82
(4 cylinder), 8.03 (6 cylinder), and 9.11 (8 cylinder), it appears
that cars with more cylinders produce larger amounts of greenhouse gases.

Chapter 10 Review Exercises

1. a. 7.5% of those who use a cell phone had an accident,
compared to 10.2% of those who did not use a cell phone.
Based on these results, cell phone use does not appear
to be dangerous.   **b.** H_0: Having an accident in the
last year is independent of whether the person uses a cell
phone. H_a: Having an accident in the last year and using
a cell phone are dependent.   **c.** 27.76, 277.24, 41.24,
411.76   **d.** 1.505   **e.** Because the test statistic of 1.505
is less than 3.841, the P-value is greater than 0.05.   **f.** There
is not sufficient evidence to warrant rejection of the claim that
having an accident in the last year is independent of whether the
person uses a cell phone. Based on these results, cell phones do
not appear to be a factor in having accidents.   **g.** More data
suggest that cell phone usage and driving are somehow related,
with cell phones causing a distraction that becomes dangerous.
Some states have banned cell phone use by drivers.

2. H_0: $\mu = 60$ seconds. H_a: $\mu > 60$ seconds. The test statistic
$t = 0.549$ is not greater than the value of $t = 1.895$ found in
Table 10.1, so do not reject the null hypothesis. (P-value $= 0.3000$.)
There is not sufficient evidence to support the claim that the mean
is greater than 60 seconds.

3. 2.6 seconds $< \mu < 152.2$ seconds

4. a. H_0: $\mu_1 = \mu_2 = \mu_3$ (People in the three different age groups
have the same mean body temperature.)   **b.** People in the
three different age groups have mean body temperatures that are

not all the same.   **c.** 0.194915   **d.** No. The *P*-value is greater than 0.05, which suggests that we should not reject the null hypothesis of equal means.

Chapter 10 Quiz

1. (b) *t* distribution   **2.** (b) *t* distribution

3. (a) normal distribution   **4.** $H_0: \mu = 170$ lb

5. $H_a: \mu < 170$ lb   **6.** False   **7.** Analysis of variance

8. Reject the null hypothesis that adults in the four states have the same mean taxes in a year.

9. H_0: Sex and response are independent.

10. Reject the null hypothesis.